*Bailey's Industrial
Oil and Fat Products*

EDITORIAL ADVISORY BOARD

Company or Association	Represented by
Archer Daniels Midland Company Decatur, Illinois	Dr. Thomas H. Smouse Manager, Oil Processing Research
Bunge Foods Corporation Bradley, Illinois	Mr. Tom Crosby Director, Research and Development
Cargill, Inc. Minneapolis, Minnesota	Mr. Stephan C. Anderson Vice President, Oil Seeds Processing
Central Soya Company Fort Wayne, Indiana	Dr. Bernard F. Szuhaj Director, Research and Development
General Foods White Plains, New York	Dr. Charles J. Cante Vice President, Research and Development, Engineering, Test Kitchens
Kraft Glenview, Illinois	Dr. Ronald D. Harris Vice President
Land O'Lakes, Inc. Minneapolis, Minnesota	Dr. David Hettinga Vice President, Research Technology Engineering
Lockwood Greene Engineers, Inc. Spartanburg, South Carolina	Mr. Bob Hart Director of Design
National Renderers Association Alexandria, Virginia	Dr. Don A. Franco Director, Technical Services
POS Pilot Plant Corporation Saskatchewan, Canada	Mr. R.A. Carr President
Wurster & Sanger, Inc. Minneapolis, Minnesota	Mr. Barry Smith General Manager

Individual

Dr. Chi Tang Ho
Department of Food Science
Cook College
Rutgers University
New Brunswick, New Jersey

BAILEY'S INDUSTRIAL OIL AND FAT PRODUCTS

Fifth Edition
Volume 3
Edible Oil and Fat Products: Products and Application Technology

Edited by
Y. H. HUI
Technology and Commerce, International

A Wiley-Interscience Publication
JOHN WILEY & SONS, INC.
New York · Chichester · Brisbane · Toronto · Singapore

Cover drawing courtesy of De Smet Process & Technology, Inc. The De Smet Multistock deodorizer incorporates a commercially proven single-vessel design to deodorize different feedstocks at capacities of over 720 TPD.

This text is printed on acid-free paper.

Copyright © 1996 by John Wiley & Sons, Inc.

All rights reserved. Published simultaneously in Canada.

Reproduction or translation of any part of this work beyond that permitted by Section 107 or 108 of the 1976 United States Copyright Act without the permission of the copyright owner is unlawful. Requests for permission or further information should be addressed to the Permissions Department, John Wiley & Sons, Inc., 605 Third Avenue, New York, NY 10158-0012.

Library of Congress Cataloging in Publication Data
Bailey, Alton Edward, 1907–1953.
   [Industrial oil and fat products]
    Bailey's industrial oil and fat products. —5th ed./edited by Y. H. Hui.
      p.   cm.
   ISBN 0-471-59427-X (alk. paper)
    1. Oils and fats. I. Hui, Y. H. (Yiu H.) II. Title.
TP670.B28   1996
665—dc20   95-9528

Printed in the United States of America

10 9 8 7 6 5 4 3 2

Contributors

MICHAEL M. BLUMENTHAL: Libra Technologies, Inc., Metuchen, New Jersey, Frying technology.

MICHAEL M. CHRYSAM: Nabisco Foods, Inc., East Hanover, New Jersey, Margarines and spreads.

BOB GEBHARDT: Kraft Food Ingredients, Memphis, Tennessee, Oils and fats in snack foods.

DAVID HETTINGA: Land O'Lakes, Inc., Minneapolis, Minnesota, Butter.

R.G. KRISHNAMURTHY: Kraft Foods, Inc., Glenview, Illinois, Cooking oils, salad oils, and oil-based dressings.

EDMUND W. LUSAS: Texas A&M University, College Station, Fats in feedstuffs and pet foods.

DOUGLAS J. METZROTH: Cherry-Burrell Process Equipment, Louisville, Kentucky, Shortening: Science and technology.

RICHARD D. O'BRIEN: Consultant, Plano, Texas, Shortening: types of formulations.

GEORGE M. PIGOTT: University of Washington, Seattle, Marine oils.

DANIEL E. PRATT: Purdue University, West Lafayette, Indiana, Antioxidants: technical and regulatory considerations.

MIAN N. RIAZ: Texas A&M University, College Station, Fats in feedstuffs and pet foods.

CLYDE E. STAUFFER: Consultant, Cincinnati, Ohio, Oils and fats in bakery products; Emulsifiers for the food industry.

ROBERT E. WAINWRIGHT: ABITEC Corporation, Columbus, Ohio, Oils and fats in confections.

VERNON C. WITTE: Kraft Foods, Inc., Glenview, Illinois, Cooking oils, salad oils, and oil-based dressings.

Reviewers

To assure accuracy of the information, each Chapter in this five-volume text has been reviewed by experts in industry, government, and academia. Although most Chapters have one or two reviewers, some have as many as five or six. A list of the reviewers is presented. However, in studying this list, please note:

1. The authors of the Chapters have also served as reviewers for Chapters other than their own. They are not included in this list.
2. This list is incomplete for a variety of reasons. The review was spread over three years and some names are misplaced, some reviewers wish to remain anonymous, and so on. A note of appreciation and/or apology is extended to those reviewers not included in the list.

INDIVIDUALS

VERLIN ALLBRITTON: Pure Flo Product Group, Chicago, Illinois
ROBERT ALLEN: Consultant, Scroggins, Texas
DAVID J. ANNEKEN: Henkel Corporation/Emery Group, Cincinnati, Ohio
RAHAMAN ANSARI: Quest International Fragrances, Mount Olins, New Jersey
RALPH ASTARITA: Consultant, Chester, New York
ANTHONY ATHANASSIADIS: Consultant, Belgium
IMRE BALAZS: Central Soya Company, Fort Wayne, Indiana
THOMAS J. BARTUS: Van Leer Chocolate Corporation, Jersey City, New Jersey
CHRISTOPHER BEHARRY: Procter & Gamble, Cincinnati, Ohio
ELLIOT BERLIN: U.S. Department of Agriculture, Beltsville, Maryland
KAREN BETT: U.S. Department of Agriculture, New Orleans, Louisiana
BRUCE H. BOOTH: Zapata Protein USA, Reedville, Virginia
CHARLES BRAH: Patrick Cudahy Company, Cudahy, Wisconsin

THOMAS L. CAIN: Van Der Bergh Foods, Joliet, Illinois
GEORGE CAVANAGH: Consultant, Fresno, California
MAROLYN R. CHAMBERS: Riceland Foods Inc., Little Rock, Arkansas
CHUCK CHITTICK: Geka Thermal Systems Inc., Atlanta, Georgia
WILLIAM E. COCHRAN: Liberty Vegetable Oil Company, Santa Fe Spring, California
RICHARD COPELAND: Central Soya Company, Fort Wayne, Indiana
IAN COTTRELL: Rhône-Poulenc Food Ingredients, Cranbury, New Jersey
C. DAYTON: Central Soya Company, Fort Wayne, Indiana
ETIENNE DEFFENSE: Tirtiaux, Belgium
ROBERT DELASHUNT: Consultant, Pickwick Dam, Tennessee
GIORGIO DELL'ACQUA: Fratelli Gianazza, Italy
JOSEPH F. DEMPSEY: Karlshamns USA Inc., Harrison, New Jersey
ALBERT J. DIJKSTRA: n.v. Vandemoortele, Belgium
MARK DREHER: Nabisco Foods, Parsippany, New Jersey
WALTER FARR: Owensboro Grain Company, Owensboro, Kentucky
FIELDEN FRALEY: Pure Flo Product Group, Chicago, Illinois
HERBERT GEHRING: Koerting Hannover AG, Germany
JAMES GEYER: Wisconsin Dairies, Baraboo, Wisconsin
LEWRENE GLASER: U.S. Department of Agriculture, Washington, D.C.
PAUL HALBERSTADT: Swift Eckrich Inc., Downers Grove, Illinois
GEORGE W. HALEK: Rutgers University, New Brunswick, New Jersey
EARL G. HAMMOND: Iowa State University, Ames, Iowa
ANTHONY J. HARPER: De Smet Process & Technology, Belgium
RICHARD HEINZE: Griffith Laboratories, Alsip, Illinois
VIRGINIA H. HOLSINGER: U.S. Department of Agriculture, Philadelphia, Pennsylvania
ARTHUR HOUSE: Lou Ana Foods, Opelousas, Louisiana
TOM HURLEY: Lou Ana Foods, Opelousas, Louisiana
LEWIS G. JACOBS: Distillation Product Industry, Rochester, New York
LAWRENCE JOHNSON: Iowa State University, Ames, Iowa
DONALD V. KINSMAN: Henkel Corporation/Emery Group, Cincinnati, Ohio
LINDA KITSON: Archer Daniels Midland Company, Decatur, Illinois
D. LAMPERT: Cargill Inc., Minneapolis, Minnesota
DAVID LAWRENCE: Safeway, Deninson, Texas
ZALMAN LEIBOVITZ: H.L.S., Israel
STACEY LEVINE: Bunge Foods, Bradley, Illinois
MARY LOCNISKAR: University of Texas, Austin, Texas
ROBERT LOEWE: Keebler Inc., Elmhurst, Illinois
EARNIE LOUIS: Perdue Farms Inc., Salisbury, Maryland

TED MAG: Consultant, Canada
EUGENE MATERN: Ed Miniat Inc., Chicago, Illinois
WILLIAM MCPHERSON: EMI Corporation, Des Plaines, Illinois
KEN MCVAY: Henkel-Emery Group, Cincinnati, Ohio
JEANNE COCHRANE MILEWSKI: RTM Winners, Atlanta, Georgia
RON MOELLER: Cargill Inc., Minneapolis, Minnesota
DON MORTON: Premier Edible Oils Corporation, Portland, Oregon
JERRY MURPHY: Rangen Inc., Buhl, Idaho
BRIAN F. OSBORNE: Hutrel Engineering, Vancouver Canada
BOB PIERCE: Fats and Oils Consultant, Tucson, Arizona
ANTHONY SACCONE: Karlshamns, USA, Harrison, New Jersey
FOUAD Z. SALEEB: General Foods, White Plains, New York
TIMOTHY SAUNDERS: North Carolina State University, Raleigh, North Carolina
MARY SCHMIDL: Sandox Nutrition, Minneapolis, Minnesota
RICHARD W. SCHOENFELD: SVO Enterprises, Eastlake, Ohio
JAMES SCHWARTZ: Procter & Gamble, Cincinnati, Ohio
JACQUES SEGERS: OPAU (Unilever), Holland
PETER SJOBERG: Deodorization, Tetra-Laval, Sweden
MARTIN SILGE: Mariani Packing Inc., San Jose, California
PETER SLEGGS: Campro Agra Ltd., Canada
BARRY SMITH: Crown Iron Works, Minneapolis, Minnesota
FRANK E. SULLIVAN: Consultant, San Diego, California
BEVERLY TEPPER: Rutgers University, New Brunswick, New Jersey
DAVID TRECKER: Pfizer Food Science Group, Groton, Connecticut
STEPHEN W. TURNER: Henkel Corporation, Cincinnati, Ohio
PHILLIPPE VAN DOOSSELEARE: De Smet Process & Technology, Belgium
PETER WAN: U.S. Department of Agriculture, New Orleans, Louisiana
KLAUS WEBER: Krupp Machinentechnik, Germany
ANTHONY E. WINSTON: Church & Dwight Company, Inc., Princeton, New Jersey
VERNON C. WITTE: Kraft General Foods, Glenview, Illinois
JOHN S. WYATT: Grindsted Products Inc., Industrial Airport, Kansas
JIM YEATES: Ag Processing Inc., Omaha, Nebraska

COMPANIES

De Smet Process & Technology, Belgium
Hixson, Cincinnati, Ohio

A Tribute to Alton E. Bailey

ALTON E. BAILEY

Alton Edward Bailey was born in Midland, Texas, in 1907 and died 46 years later in Memphis, Tennessee. During his relatively short professional career, A.E. Bailey made an imprint on the science and technology of fats and oils unequaled by any other person either before or since. Of his accomplishments, it is agreed that the most important was the legacy of his book, *Bailey's Industrial Oil and Fat Products*, first published in 1945. It immediately became the Bible of the fats and oils industry and continues to be so regarded, even today. While updated and expanded several times since his death in order to include more recent scientific findings and engineering developments, the 1945 first edition can still be perused with the reader hardly being aware that A.E. Bailey wrote the words 50 years previously.

In our present age, when having a Ph.D. from a prestigious university is almost mandatory in order to be taken seriously as a research investigator or authoritative author, it is interesting to note that "Ed" Bailey's university education ended upon receiving a B.S. in Chemical Engineering from the University of New Mexico in 1927.

Following graduation, A.E. Bailey was employed in the laboratories of the Cudahy Packing Company, first in Omaha, Nebraska, and later in Memphis, Tennessee. He left Cudahy in 1941 to accept a position as head of the oil processing research section of the United States Department of Agriculture at their Southern Regional Research Laboratory in New Orleans. It was in the pilot plant there that Ed Bailey tested out and quantified many of his ideas. His published laboratory work relied heavily on the use of the then new tool of dilatometry to explain and understand the functional characteristics of fats.

Five years later, Alton Bailey resigned from the USDA to join the Votator Division of the Girdler Corporation in Louisville, Kentucky, as Chief Process Engineer of their Oil and Fat Section. In that position, he was instrumental in the development of the semicontinuous deodorizer which, in addition to producing products of very high quality, permitted rapid changeover from one product to another. In 1950, Mr. Bailey returned to Memphis as Vice President and Director of Research for The Humko Company, a position he occupied until his death.

Alton E. Bailey joined the American Oil Chemists' Society in 1935 and subsequently made many contributions to the technical publications, education programs, and organization management of AOCS. He was elected to the Governing Board in 1949 and served the Society as its president in 1951–1952.

According to his contemporaries, Ed Bailey possessed an exceptional intellect, intense curiosity, and extremely high motivation. Those attributes were coupled with an almost photographic memory, along with significant ability to organize material and put it into writing in a straightforward and understandable manner.

When Fred Astaire died, Mikhail Baryshnikov commented, "Mr. Astaire was an *artiste*, the rest of us are dancers." We, who are authors of individual Chapters of this fifth edition, feel the same relationship with Mr. Alton E. Bailey. We hope you will find our efforts worthy of being published in a book bearing his name.

ROBERT C. HASTERT

October 1995

A Tribute to Alton E. Bailey   **xiii**

BAILEY'S LEGACY IS DECADES OF FATS AND OILS RESEARCH

Following in the innovative footsteps of Alton E. Bailey, agricultural research scientists continue to have a significant impact on the fats and oils industry of today and tomorrow. Working in the scientific environment that cultured Bailey's creative genius, researchers at the regional research centers of the USDA's Agricultural Research Service (ARS) have contributed to decades of advancements in the processes and applications of fats and oils, both for food and industrial uses.

An extended program of research enabled scientists to help transform the soybean into the major source of high-protein feeds and food oil products of today. In the early 1940s the flavor of soybean oil was variously described by consumers as "grassy" or "beany" or "fishy," and it tasted even worse after it had been stored for awhile. In consultation with colleagues in the fats and oils industry, scientists decided that establishment of some uniformity of judgment about how various soybean oils tasted was an essential first step in the research. The technique of flavor evaluation, which they developed involving selected, trained taste panels and numerical rating of the flavors, was the first significant milestone in improving soybean oil

Guided by judgments of taste panels, researchers identified the source of many of the off-flavors in soybean oil as trace metals, particularly iron and copper. Even extremely small amounts of these contaminants sped oxidation of the oil, shortening its storage life and promoting undesirable flavors. Responding to these findings, industry removed brass valves in refineries and substituted stainless steel for the cold, rolled steel in equipment that came in contact with soybean oil. These actions alone improved the flavor of the oil. It was further discovered that the addition of citric acid to the deodorized oil would deactivate the trace metals in soybean oil. Today, practically all soybean oil is protected by adding citric acid during processing.

But many questions remained unanswered. Metal contaminants could speed the development of off-flavors, but chemists wondered what caused the flavors to develop in the first place. One of the principal causes turned out to be linolenic acid, a fatty acid that makes up from 7 to 9% of soybean oil. Of all the major edible vegetable oils on the market only soybean and rapeseed oils contained linolenic acid at these levels. It is this constituent of soybean oil that was a major contributor to deterioration on the shelf or when the oil was heated repeatedly in deep fryers. The industry turned to the nickel-catalyzed, partial hydrogenation process to lower the linolenic acid content of soybean oil to about 3.0%. In the 1960s, it was this lightly hydrogenated product that enabled soybean oil to displace cottonseed oil as the major edible oil in the world.

At the same time, the discovery that linolenic acid was a major reason soybean oil went bad spurred plant breeders in research to find soybean and rapeseed lines with lower linolenic acid content, research that continues to this day.

Today, while hydrogenation of soybean oil is still required for use as a high temperature cooking oil, in shortenings and margarines, and for long-term storage, evaluations by taste panels have made it clear that the industry's implementation of improved processing techniques and protection of the oil during processing produces an oil that is stable at ambient temperatures and has good shelf-life stability.

Other researchers modified cottonseed oil, giving it properties similar to that of cocoa butter. The confectionery fats derived from cottonseed and soybean oil are now being used for many applications in the industry. Further, the fatty acids from vegetable oils were combined with sucrose or table sugar to form sucrose esters, used today as emulsifiers, stabilizers, and texturizers in baked goods, baking mixes, biscuit mixes, frozen dairy desserts, and whipped milk products.

Over the years, ARS has devoted much of its resources to discovering new industrial uses for fats and oils. A new market for fats and oils was created by the discovery of a process called epoxidation, in which hydrogen peroxide was used to insert an atom of oxygen into the hydrocarbon chain of fatty acids. Epoxidized oils, when used as plasticizers, blend well with commonly used resins. They also eliminate the need for poisonous salts of lead, barium, or cadmium in vinyl plastics, which turn the plastics cloudy or opaque. Seventy-five percent of the 50,000 tons of epoxidized oils now used are based on soybean oil. The discovery also helped to create a billion-dollar plastics industry. Today, about 75% of the plasticizers for flexible vinyl plastics are made from soybean oil.

Modification of vegetable oil fatty acids has been a significant alternative to petrochemicals for chemical feedstocks. Oleic acid is converted to emulsifiers, cosmetic ingredients, and other specialty chemicals and is used in textile mills in lubricants and antistatic agents. Acetoglycerides, derived from fatty acids, can be formed into thin, stretchable films for a variety of uses in the food and cosmetics industries. Another development was a group of multipurpose chemicals called isopropenyl esters from fatty acids. These can be used to make paper and cotton repel water, to coat glass to reduce breakage in bottling lines, and in other applications where they have proven superiority to chemicals now in use.

A significant market for soybean oil was created in 1988 when scientists formulated a 100% soybean-oil-based printing ink that not only has a lower cost than petroleum-based inks, but also gives superior penetration of pigment into newsprint. These inks are adjustable to a wide range of viscosity and tackiness for news offset printing and have rub-off characteristics equal to those formulated and marketed as low-rub inks.

In 1959, the North Central Section, American Oil Chemists' Society, established the Alton E. Bailey award to annually recognize research and/or service in the field of oils, fats, and related disciplines. The list of 35 scientist recipients from academia, industry, and government is testament to the lasting impact of Bailey's legacy of research in the fats and oils field.

TIMOTHY L. MOUNTS

October 1995

Preface

This fifth edition of *Bailey's Industrial Oil and Fat Products* differs from the fourth edition in many ways:

1. There are five volumes instead of three.
2. All five volumes are published at one time instead of over several years.
3. In the fourth edition, Chapters on edible and nonedible products were distributed over the three volumes. However, the fifth edition is clearly divided into the two groups: edible (Volumes 1–4) and nonedible (Volume 5).
4. Volume 1 serves as an introduction covering several subjects, some of which are basic to the discipline while others, because of their unique themes, cannot be placed in other volumes. As a result, Volume 1 covers such topics as chemistry, nutrition, toxicology, vegetable oil, animal fats, flavors, analysis, and sensory evaluations. Each of Volumes 2 to 4 covers one specific topic on edible oils or fats. Volume 2 covers individual oil seeds; Volume 3 discusses the application of oils and fats; and Volume 4 concentrates on the technology and engineering of processing vegetable oils.
5. Volume 5 covers the application of oils and fats in nonedible products including consumer goods such as soap, paints, leather, textiles, pharmaceuticals, and cosmetics. Other topics include rendering, fatty acids, and glycerine.

As observed, this new edition does not exactly update the fourth edition. Rather, it is more comprehensive and covers more applied information. Many professionals have expressed to me the following:

1. Much of the information in the older editions will always be useful.
2. Each new edition does not always update information in the last edition.

3. Each new edition always covers more information than the last edition.
4. Each new edition provides more applied information than the last edition.

Volume 3 contains the following information: Six Chapters on commercial products (butter, margarine, shortening, edible oils, and marine oils), each of which concentrates on chemistry, biology, and production technology and engineering. Four Chapters concentrate, respectively, on the application of fats and oils in pet foods, bakery products, confections, and snack foods; each provides insight into the role of fats and oils in product manufacturing and applications. Two Chapters focus on important processing chemicals (emulsifiers and antioxidants) for the fats and oils industry. One Chapter is devoted entirely to the latest research, development, and applications in frying oil technology.

In any book with multiple authors, the editor faces the same difficulties: length, content, format, delay, and updates. In spite of the difficulties encountered in an undertaking of this magnitude, I feel that through the excellent work of the authors I have achieved the major goal of this edition. This five-volume text will provide professionals in the oil and fat industry an excellent reference source on the subject matter with a special emphasis on edible products. Most of the authors are from industry, with a limited number from academia and government. They have worked hard to make this edition a success and I will forever be grateful for their participation. But, of course, you are the final judge of the usefulness of this work.

Y. H. HUI

Acknowledgment

Most of us are aware that to prepare and publish a professional book of this magnitude is invariably the result of teamwork.

To overcome the technical difficulties, I have been advised by two groups of experts: members of the editorial advisory board and my professional colleagues. It is not an exaggeration to state that this work could not have been completed without their counsel during the last four years. Additionally, through the years of professional consultation for this project, several individuals have actually become my indispensable advisors. They have helped me above and beyond the normal call of professional courtesy. They are well-known individuals in the oils and fats industry. As a matter of fact, two of them were past presidents of the American Oil Chemists' Society. I will always be grateful to Ken Carlson, Bob Hastert, Tom Smouse, and Peter Wan; I value their friendship.

Every time we read a book we evaluate two things in our minds: the information and the quality of the book. I have worked with the production staff at Wiley in 1978, 1986, 1987, and 1991. They have consistently produced books of excellent quality. This one is no exception.

Let me express my most sincere appreciation to everyone who has participated in the completion of this work. And, lastly, I am grateful to the support of my family.

Contents

1 BUTTER 1

David Hettinga

Chemical Composition, 2
Modification of Milk Fat, 13
Quality Control, 22
Butter Manufacture, 28
Butterfat Products, 49
Marketing, 57

2 MARGARINES AND SPREADS 65

Michael M. Chrysam

Historical Development of Margarine, 65
Recent U.S. Trends, 67
Regulatory Status in the United States, 69
Product Characteristics, 74
Oils Used in Vegetable Oil Margarines and Spreads, 77
Other Common Ingredients, 90
Processing, 96
Low-Calorie Spreads, 102
Deterioration and Shelf-Life, 104

xxii Contents

3 SHORTENING: SCIENCE AND TECHNOLOGY   115

Douglas J. Metzroth

Introduction, 115
Plastic Theory, 120
Formulation, 122
Manufacturing Processes and Equipment, 127
Shortening Production Systems, 140
Analytical Evaluation and Quality Control, 148
Packaging and Storage, 151
Recent Innovations, 154

4 SHORTENING: TYPES AND FORMULATIONS   161

Richard D. O'Brien

Introduction, 161
Shortening Attributes, 165
Base Stock System, 171
Shortening Formulation, 174
Shortening Crystallization, 181
Plasticized Shortening Consistency, 184
Liquid Opaque Shortenings, 188
Shortening Chips and Flakes, 190

5 COOKING OILS, SALAD OILS, AND OIL-BASED DRESSINGS   193

R.G. Krishnamurthy, Vernon C. Witte

Natural and Processed Cooking and Salad Oils, 194
Stability of Salad and Cooking Oils, 196
Quality Evaluation of Salad and Cooking Oils, 198
Additives for Salad and Cooking Oils, 201
Nutrition-Oriented Salad and Cooking Oils, 202
New Salad and Cooking Oils, 202
Oil-Based Dressings, 202
Viscous or Spoonable Dressings, 203
Pourable Dressings, 210

Reduced-Calorie Dressings, 215
Fat-Free Dressings, 216
Refrigerated Dressings, 219
Heat-Stable Dressings, 219

6   MARINE OILS   225

    George M. Pigott

    Introduction, 225
    The Unique Properties of Fish Oils, 226
    Extraction and Refining of Marine Oils, 229
    Marine Oils in Industrial Products, 233
    Fish Oil in Animal Diets, 238
    Fish Oil in Human Diets, 242

7   FATS IN FEEDSTUFFS AND PET FOODS   255

    Edmund E. Lusas, Mian N. Riaz

    History, 257
    Information Sources, Authorities, and Obligations, 258
    Availability, Characteristics, and Composition, 262
    Digestion Metabolism and Fats Feeding Requirements, 279
    Fat Utilization Practices, 305

8   OILS AND FATS IN BAKERY PRODUCTS   331

    Clyde E. Stauffer

    Introduction, 331
    Bread and Rolls, 332
    Layered Doughs, 334
    Cakes, 336
    Cake Donuts, 339
    Cookies, 340
    Pie Crust, Biscuits, 344
    Specifications for Bakery Shortenings, 344

xxiv   Contents

9   OILS AND FATS IN CONFECTIONS   353
     Robert E. Wainwright

     Chocolate, 354
     Cocoa Butter Alternatives, 378
     Emulsifiers, 394
     Tropicalization of Chocolate, 398
     Other Applications, 399
     Oil Migration, 403
     Fat Reduction, 404

10   OILS AND FATS IN SNACK FOODS   409
     Bob Gebhardt

     Maintaining Oil Quality Prior to Usage, 409
     The Frying Operation, 415
     Snack Food Products and Processing Systems, 417

11   FRYING TECHNOLOGY   429
     Michael M. Blumenthal

     Introduction, 429
     Frying Process and Systems, 433
     New Paradigm, 449
     Chemical Degradation of Frying Fats and Oils, 459
     Regulation, 465

12   EMULSIFIERS FOR THE FOOD INDUSTRY   483
     Clyde E. Stauffer

     Emulsifiers as Amphiphiles, 483
     Surfaces and Interfaces in Foods, 484
     Surface Activity, 485
     Emulsions, 490
     Foams, 495
     Wetting, 498

Physical State of Emulsifier plus Water, 499
Emulsifiers for Food Applications, 502
Interactions with Other Food Components, 511
Some Food Applications, 516

13 ANTIOXIDANTS: TECHNICAL AND REGULATORY
    CONSIDERATIONS   523

Daniel E. Pratt

Classification, 524
Mechanism of Antioxidation, 526
Major Food Antioxidants, 528
Antioxidant Formulations, 534
Regulatory Status and Safety Issues, 535
Evaluation of Antioxidant Potency, 538
Thermal Degradation of Phenolic Antioxidants, 539

INDEX   547

Bailey's Industrial Oil and Fat Products

1
Butter

Buttermaking is one of the oldest forms of preserving the fat component of milk. Its manufacture dates back to some of the earliest historical records, and reference has been made to the use of butter in sacrificial worship, for medicinal and cosmetic purposes, and as a human food long before the Christian era. Documents indicate that, at least in the Old World, the taming and domestication of animals constituted the earliest beginnings of human civilization and culture. There is good reason to believe, therefore, that the milking of animals and the origin of buttermaking predate the beginning of organized and permanent recording of human activities.

The evolution of the art of buttermaking has been intimately associated with the development and use of equipment. With the close of the eighteenth century, the construction and use of creaming and buttermaking equipment (other than that made of wood) began to receive consideration, and the barrel churn made its appearance.

By the middle of the nineteenth century, attention was given to improvement in methods of creaming. These efforts gave birth to the deep-setting system. Up to that time, creaming was done by a method called shallow pan. The deep-setting system shortened the time for creaming and produced a better quality cream. An inventive Bavarian brewer in 1864 conceived the idea of adapting the principle of the laboratory centrifuge. In 1877 a German engineer succeeded in designing a machine that, although primitive, was usable as a batch-type apparatus. In 1879 engineers in Sweden, Denmark, and Germany succeeded in the construction of cream separators for fully continuous operation (1).

In 1870, the year before the introduction of factory buttermaking, butter production in the United States totaled 514 million lb, practically all farm made. Authentic records concerning the beginning of factory buttermaking are meager. It appears that the first butter factory was built in Iowa in 1871. This also introduced the pooling system of milk for creamery operation (1).

Other inventions that assisted in the development of the butter industry included the Babcock test (1890), which accurately determines the percentage of fat in milk and cream; the use of pasteurization to maintain milk and cream quality; the use of pure cultures of lactic acid bacteria; and refrigeration to help preserve cream quality.

Multiple butterfat products, including butter oils, anhydrous butterfat, butterfat–vegetable oil blends, and fractionated butterfats, are manufactured around the world today. In the past, butterfat in the form of butter was the primary preservation technique. Today, the preferred preservation method involves the processing of butterfat to the anhydrous butter oil state, then hermetically packaging under nitrogen to substantially increase the shelf life and reduce the incidence of degradation.

Historically, milk fat has been held in the highest esteem, whether in liquid milk, as cream, or as butter. Its consumption was associated with a higher standard of living. In recent times, with the prosperity of the Western world, per capita consumption has been decreasing. Ironically, this phenomenon contradicts all historical patterns for butterfat consumption and use. Several reasons exist for this decline. This chapter explores the chemical composition, marketing, technology, processing, quality, legal restrictions, and new uses for butter and butterfat.

CHEMICAL COMPOSITION

Some of the information in this chapter comes directly from the fourth edition of Bailey's (2). Jensen and Clark (3) have provided a complete review of the lipid composition, and data have been selected for inclusion in this review.

The composition of milk fat is somewhat complex. Although dominated by triglycerides, which constitute some 98% of milk fat (with small amounts of diglycerides, monoglycerides, and free fatty acids), various other lipid classes are also present in measurable amounts. It is estimated that about 500 separate fatty acids have been detected in milk lipids; it is probable that additional fatty acids remain to be identified. Of these, about 20 are major components; the remainder are minor and occur in small or trace quantities (4,5). The other components include phospholipids, cerebrosides, and sterols (cholesterol and cholesterol esters). Small amounts of fat-soluble vitamins (mainly A, D, and E), antioxidants (tocopherol), pigments (carotene), and flavor components (lactones, aldehydes, and ketones) are also present.

The composition of the lipids of whole bovine milk is given in Table 1.1 (4,5). The structure and composition of the typical milk fat globule is exceedingly complex. The globule is probably 2–3 μ in diameter with a 90-Å-thick membrane surrounding a 98–99% triglyceride core. The composition of the milk fat membrane is quite different from milk fat itself in that approximately 60% triglycerides are present, much less than in the parent milk fat (Table 1.2) (6,7).

It has been generally recognized that butterfat consists of about 15 major fatty acids with perhaps 12 or so minor (trace quantity) acids. Triglycerides are normally defined with respect to their carbon number (CN), i.e., the

Table 1.1 Composition of lipids in whole bovine milk (4,5)

Lipid	Weight Percent
Hydrocarbons	Trace
Sterol esters	Trace
Triglycerides	97–98
Diglycerides	0.28–0.59
Monoglycerides	0.016–0.038
Free fatty acids	0.10–0.44
Free sterols	0.22–0.41
Phospholipids	0.2–1.0

number of fatty acid carbon atoms present in the molecule; the three carbon atoms of the glycerol moiety are ignored. Because the fatty acid spectrum of milk fat is dominated by acids containing an even number of carbon atoms, so is the triglyceride spectrum. However, the proportion of triglycerides with an odd carbon number is about three times greater than the proportion of odd-numbered fatty acids.

Although obvious correlations exist between fatty acid composition and triglyceride distribution, detailed information is lacking that would enable the triglyceride distribution to be predicted from the fatty acid composition. Much more needs to be understood of the strategy used in the bovine mammary gland in assembling a complex array of fatty acids into triglycerides. This is not an arcane study; it is necessary if processes such as fractionation are to yield products with consistent qualities throughout the year. In effect, the detailed structure of milk fat is not yet understood. Perhaps this is not surpris-

Table 1.2 Composition of lipids from milk fat globule membrane (6,7)

Lipid Component	Percent of Membrane Lipids
Carotenoids (pigment)	0.45
Squalene	0.61
Cholesterol esters	0.79
Triglycerides	53.4
Free fatty acids	6.3[a]
Cholesterol	5.2
Diglycerides	8.1
Monoglycerides	4.7
Phospholipids	20.4

[a] Contained some triglycerides.

4   Butter

Table 1.3   Characteristics and composition of butterfats

Characteristic	Value[a]	Range of Values[b]	GLC[c]
Iodine number	32.9	—	—
Saponification equivalent	236.3	—	—
Reichert-Meissle value	32.5	—	—
Polenske value	—	—	—
Kirschner value	—	—	—
Fatty acid, wt. %			
Butyric	3.5	2.8–4.0	3
Caproic	1.4	1.4–3.0	1
Caprylic	1.7	0.5–1.7	1
Capric	2.6	1.7–3.2	3
Lauric	4.5	2.2–4.5	4
Myristic	14.6	5.4–14.6	12
Palmitic	30.2	26–41	29
Stearic	10.5	6.1–11.2	11
Above C18	1.6	—	2
Total saturated	70.6	—	66
Decenoic	0.3	0.1–0.3	—
Dodecenoic	0.2	0.1–0.6	—
Tetradecenoic	1.5	0.6–1.6	2
Hexadecenoic	5.7	2.8–5.7	4
Octadecenoic (oleic, etc.)	18.7	18.7–33.4	25
Octadecadienoic	2.1	0.9–3.7	2
C20 and C22 unsaturated	0.9	—	1
Total unsaturated	29.4	—	34

[a] From Refs. 8 and 9.
[b] From Refs. 10 and 11.
[c] From Ref. 12.

ing if we consider only the 15 major fatty acids; there are 15^3 (3375) possible triglyceride structures using a purely random model.

The data in Table 1.3 represent general characteristics and composition of butterfat as reported by several sources (8–12). Note the range in values. Precise and repeatable values are not highly correlated due to such variables as stage of lactation, feed source, cattle breed, etc. Although 16 categories of fatty acids are outlined, it was generally appreciated that many other fatty acids are present in small or trace quantities. For nutritional and dairy science purposes, these data are of value, but from a detailed scientific point of view, they afford only a vague, broad generalization of the actual state of fatty acid composition of butterfat. A more complete view of composition is provided in Table 1.4 (13,14).

From 1956 to 1983, a great volume of information became available on the occurrence of many minor constituents in butterfat. Somewhat less intensity

Table 1.4 Fatty acid composition of milk and butterfat[a]

Fatty Acid[b]	June[c]	December[d]	Average[e]	Moore and Co-workers[f]
4:0	4.22	3.51	3.57	3.98
6:0	2.53	2.24	2.22	2.36
8:0	2.34	1.07	1.17	1.36
9:0	0.05	0.05	0.03	—
10:0	2.24	2.57	2.54	2.76
10:1	0.32	—	—	—
11:0	0.34	0.29	0.33	—
12:0	2.40	2.77	2.81	3.14
13:0 (12:1)	0.29	0.29	0.33	0.14
14 (br)[g]	0.23	0.14	0.17	0.12
14:0	9.01	10.58	10.06	8.39
14:1 (15 br)	1.54	1.61	1.63	1.84
15:0	1.29	1.11	1.09	1.34
16 (br)	0.42	0.39	0.38	0.35
16:0	22.05	25.98	24.97	30.05
16:1 (17 br)	2.29	2.98	2.55	2.80
17:0	0.69	1.08	0.91	1.00
17:1 (18 br)	—	—	—	0.37
18:0 (br)	0.31	0.40	0.38	—
18:0	14.27	11.58	12.07	11.74
18:1	30.41	24.75	27.09	24.93
18:8[h]	0.24	1.56	1.26	—
18:2	1.23	2.75	2.39	1.78
18:3 (20:0)	2.61	2.30	2.06	1.23

[a] In weight percent.
[b] Structural assignments are not necessarily authentic, but represent, in almost all instances, the most likely structure for the fraction.
[c] Data from the Department of Animal Industries, Storrs (Conn.) Agricultural Experiment Station; 408 samples of milk plant production from June 1960 to June 1961.
[d] Data from Storrs Agriculture Experiment Station; 4–8 samples.
[e] For 108 samples.
[f] Ref. 14.
[g] Branched chain.
[h] Carbon number obtained by semilog plots retention time/chain length.

of interest has prevailed since then, but further information continues to appear, and we can expect more data on butterfat as a consequence of research on the relationship between dairy cow feeding studies and resulting butterfat fatty acid composition.

The great variety of fatty acids in butterfat cannot be treated in detail here; reference will be made to only a few of the many available reports. Octadecadienoic acids are present in significant amounts; there are traces of

6 Butter

Table 1.5 Positional and geometric isomers of bovine milk lipid fatty acids (wt. %) (16)

Position of Double Bond	Cis Isomers				Trans Isomers	
	14:1	16:1	17:1	18:1	16:1	18:1
5	1.0	Trace	—	—	2.2	—
6	0.8	1.3	3.4	—	7.8	1.0
7	0.9	5.6	2.1	—	6.7	0.8
8	0.6	Trace	20.1	1.7	5.0	3.2
9	96.6	88.7	71.3	95.8	32.8	10.2
10	—	Trace	Trace	Trace	1.7	10.5
11	—	2.6	2.9	2.5	10.6	35.7
12	—	Trace	Trace	—	12.9	4.1
13	—	—	—	—	10.6	10.5
14	—	—	—	—	—	9.0
15	—	—	—	—	—	6.8
16	—	—	—	—	—	7.5

hexadecadienoic acid, octadecatrienoic acids, and highly unsaturated C20 and C22 acids. Traces of dihydroxystearic acid and hydroxypalmitic acid have been detected (8,9). A small proportion of the octadecenoic acid consists, not of oleic acid, but of *trans*-11,12 isomer, vaccenic acid (8,9). One report states that about 66% of one octadecienoic acid content is normal linoleic acid, and the remainder consists of the *cis*-9, *trans*-12 or the *trans*-9, *cis*-12 isomers (15); but other positional and geometric isomers are undoubtedly also present (4). The positional and geometric isomers of bovine milk lipid fatty acids are presented in Table 1.5 (16).

Few compilations of the extensive fatty acid distributions in butterfat have been made since Iverson and co-workers (17) reported quantitative data on 82 fatty acids that were detected by means of urea fractionation and GLC (Table 1.6). Table 1.7 provides the fatty acid composition of bovine milk lipids.

The advent of new techniques of gas chromatography for monoglycerides, diglycerides, and triglycerides (18,19) should assist markedly in the identification of the specific triglycerides of butterfat. It has already been possible to identify and quantitate about 168 molecular species of bovine milk serum triglycerides, excluding enantiomers. Nutter and Privett (20) employed liquid–liquid and argentation TLC along with pancreatic lipase hydrolysis for this purpose. Because of their high degree of saturation, ruminant milk fats do not lend themselves readily to argentation TLC, and resolution by gas chromatography using polyester columns is a likely recourse.

There is a pronounced seasonal change in the fatty acid composition of butterfat. It is normally several iodine number units higher in the summer than in the winter, with corresponding variation in the relative proportions of unsaturated and saturated fatty acids. In colder climates, the difference appears to be slightly larger. The change is usually associated with the difference in the feed of the animals in different seasons, but not completely so: cows put on green pasturage produce softer butterfat even if their feed has previously consisted of hay or silage comparable in solid composition with the green feed.

There are also differences in the butterfat of different cows on identical rations, and the age of the animal and duration of lactation have some influence on butterfat composition. Much of the dairy literature provides information relating dairy animal species and the composition of the butterfat from them.

When corn and peanut oils are protected (entrapped in formaldehyde-treated casein) significant changes in the fatty acid composition of milk fat occur (Table 1.8) (21).

Protected oils are hydrolyzed in the abomasum, and the fatty acids are absorbed in the small intestine, thereby avoiding hydrogenation. The 18:2 content in the milk fat was increased about fivefold, and the 14:0, 16:0, and 18:0 were decreased accordingly. Plasma and depot fats were also increased in 18:2 content by this program (21).

Results at the USDA are similar: cow's milk can be increased in 18:2 acid from 3 to 35% by feeding protected safflower oil (22,23). However, at high 18:2 levels, milk develops an oxidized off-flavor, usually after about 24 h, and creams require a longer aging time for satisfactory churning. As expected, butter that contains more than 16% linoleic acid is soft and sticky (5).

Extensive data have been published on the Reichert-Meissl, Polenske, and Kirschner values of mixtures of butterfat, coconut, and palm kernel oils (Table 1.9) (24–26). Other average characteristics of butterfat are approximately as follows: density at 60°C, 0.887; melting point, 38°C; titer, 34°C; and unsaponifiable matter, 0.4%. The optical properties of butterfat are misleading and are in part contributed by the nonglyceride components.

A significant variation in milk fat composition can occur in colostrum milk. Ahren and co-workers (27) analyzed the content of glycerol ethers in neutral lipids and phospholipids isolated from bovine colostrum and milk (Table 1.10). Lactone content of butterfat has also been determined (Table 1.11).

Odd-numbered methyl ketones containing from 3 to 15 carbon atoms are found in small quantities in butterfat. These compounds, along with microtraces of acetone, acetaldehyde, methyl sulfide, C4–C10 free fatty acids, and the various lactones already mentioned, generally are considered to be the substances that comprise the pleasant, bland, olfactory, nonoxidative flavor and odor of milk fat. Representative concentrations of homologous methyl ketones have been well documented (30–32).

The phospholipids of milk fat are found in the fat globule membrane in association with proteins and cerebrosides. Phospholipids are amphipolar in nature and are strongly surface active. These properties enable them to stabilize both oil-in-water and water-in-oil emulsions (Table 1.12) (4–6).

The sterols found in the unsaponifiable fraction of milk lipids are mostly

8 Butter

Table 1.6 Fatty acid distributions of 82 acids in butterfat[a]

Saturated[b]		Branched[c]		Monoenes	
Acid	Weight Percent	Acid	Weight Percent	Acid	Weight Percent
—	—	12:0 i	0.01	10:1	0.48
8:0	0.69	13:0 i	Trace	12:1	0.05
10:0	1.88	14:0 i	0.03	13:1	0.003
11:0	0.12	15:0 i	0.14	14:1	0.75
12:0	2.96	15:0 2	0.23	15:1	0.02
13:0	0.10	16:0 i	0.2	16:1	1.84
14:0	11.2	17:0 i	0.36	17:1	0.2
15:0	1.52	18:0 i	0.02	18:1	30.3
16:0	27.8	19:0 br	0.01	19:1	0.14
17:0	0.71	20:0 br	0.01	—	—
18:0	12.1	21:0 br	0.01	—	—
19:0	0.05	22:0 br	0.02	—	—
20:0	0.02	23:0 br	0.01	—	—
21:0	0.06	24:0 br	0.02	—	—
22:0	0.04	25:0 br	0.0004	—	—
23:0	0.01	26:0 br	0.0004	—	—
24:0	0.02	20:1	0.52	—	—
25:0	0.02	21:1	0.01	—	—
26:0	0.02	22:1	0.02	—	—
27:0	0.00004	23:1	0.05	—	—
28:0	0.00004	24:1	0.0008	—	—
—	—	25:1	0.0008	—	—
—	—	26:1	0.0008	—	—

cholesterol esters, small quantities of lanosterol, and even smaller quantities of two new constituents: dihydrolanosterol and β-sitosterol (33).

From the standpoint of nutritional value, the vitamin A content of butter is important. Because the source of vitamin A in butter is β-carotene or other carotenoid pigments in the feed of the cows, the content of this vitamin varies considerably, being highest in the summer when the dairy herds are in pasture and lowest in winter when there are no green feedstuffs in their rations. A portion of the carotene in the feed is transferred to the butterfat without change. The amount of carotene transferred by the cow into the butterfat varies with the feeding regimen parallel to variations in the production of vitamin A, so that the intensity of the yellow color of butter to some extent serves to indicate its vitamin A content.

The vitamin A potency of butter is in part due to vitamin A as such and in part to carotene, which is partially converted to the vitamin in the human body. The vitamin A content of butter is usually within the range 6–12

Table 1.6 (Continued)

	Dienes		Polyenes		Multibranched[e]	
Acid	Weight Percent	Acid	Weight Percent	Acid	Weight Percent
14:2	0.04	18:3	1.03	16:0 br3	0.01
16:2	0.02	18:4	0.10	17:0 br3	0.01
18:2	2.22	20:3	0.05	18:0 br3	0.16
20:2	0.12	20:4	0.07	—	—
22:2	0.14	20:5	0.02	—	—
24:2	0.02	22:3	0.03	—	—
26:2	0.0004	22:4	0.04	—	—
—	—	22:5	0.02	—	—
19:0 br4[d]	0.02	—	—	—	—
20:0 br4	0.14	—	—	—	—
21:0 br4	0.02	—	—	—	—
22:0 br4	0.02	—	—	—	—
23:0 br4	0.01	—	—	—	—
24:0 br4	0.10	—	—	—	—
25:0 br4	0.10	—	—	—	—
26:0 br3	0.01	—	—	—	—
27:0 br4	0.04	—	—	—	—
28:0 br3	0.02	—	—	—	—
28:0 br4	0.12	—	—	—	—
28:0 br5	0.01	—	—	—	—

[a] Detected by urea fractionation and gas liquid chromatography in 1965 (17).
[b] Acids below 8:0 were not determined (totally or partially lost during removal of solvent); also did not measure trans isomers, conjugated dienes and trienes, and ketoacids.
[c] *i*, iso; *br*, iso and/or antiiso. Last number indicates number of methyl branches for multibranched acids.
[d] The number following *br* indicates the number of methyl branches for multibranched acids.
[e] Tentatively identified in appropriate urea fractions by semilogarithmic plots of GLC retention times.

mg/g, and the carotene content is in the range of 2–10 mg/g (33); 1 IU of vitamin A is defined as the amount possessing the biological activity of 0.6 μg of pure β-carotene.

The vitamin D content of butter is much less significant than that of vitamin A, but it is nevertheless appreciable. It varies from about 0.1 to 1.0 IU/g, being highest in the summer and lowest in the winter (33).

The composition of milk fat is the most important factor affecting the firmness of butter and, therefore, its spreadability. The composition of milk fat changes primarily according to the feed; therefore, the entire problem is connected to the animal's diet. The fatty acid composition of milk fat produced

10 Butter

Table 1.7 Fatty acid composition of bovine milk lipids, August 1983 (3)

Number	Type	Identity
27	Normal saturate	2–28
25	Monobranched saturate	24; 13, 15, 17, 18 three or more positional isomers
16	Multibranched	16–28
62	Cis monoene	10–26, except for 11:1, positional isomers of 12:1, 14:1, 16:1–18:1, and 23:1–25:1
58	Trans monoene	12–14, 16–24; positional isomers of 14:1, 16:1–18:1, and 23:1–25:1
45	Diene	14–26 evens only; cis, cis; cis, trans; or trans, cis and trans; trans, geometric isomers; unconjugated and conjugated and positional isomers
10	Tripolyene	18, 20, 22; geometric positional, conjugated and unconjugated isomers
5	Tetrapolyene	18, 20, 22; positional isomers
2	Pentapolyene	20, 22
1	Hexapolyene	22
38	Keto (oxo) saturated	10, 12, 14, 15–20, 22, 24; positional isomers
21	Keto (oxo) unsaturated	14, 16, 18; positional isomers of carbonyl and double bond
16	Hydroxy, 2-position	14:0, 16:0–26:0, 16:1, 18:1, 21:1, 24:1, 25:1
	Hydroxy, 4- and 5-position	10:0–16:0, 12:$\triangle$–6 and 12:1–$\triangle$–9
60	Other positions	
1	Cyclic, hexyl	11; terminal cyclohexyl

Table 1.8 Effect of feeding protected corn and peanut oils on fatty acid composition of bovine milk fat (4,21)

	Fatty Acid Composition of Milk Lipids (wt. %)		
Fatty Acids	Corn Oil	Peanut Oil	Control
14:0	7.9	9.7	11.9
16:0	20.5	22.1	31.1
18:0	9.8	11.0	13.5
18:1	28.8	25.3	29.5
18:2	20.1	20.5	4.2
18:3	1.8	2.9	2.7
Others	11.1	8.5	7.1

Chemical Composition

Table 1.9 Distinctive characteristics of butterfat compared with other fats (23)

Characteristic	Butterfat	Coconut Oil	Palm Kernel Oil	Soy and Corn Fats and Oils
Saponification number	210–250	245–260	240–250	~200
Refractive index, 60°C	~1.4465	~1.4410	~1.4430	>1.4465[a]
Reichert-Meissl value	22.34	6.8	5.7	<1
Polenske value	2–4	14–18	10–12	<1
Kirschner value	20–26	1–2	0.5–1	<0.5

[a] Unless the iodine number is very nearly zero.

Table 1.10 Content of glycerol ethers in neutral lipids and phospholipids isolated from bovine colostrum and milk (27)

Characteristic	Colostrum (% wt/wt)	Milk (% wt/wt)
Total lipids	5.6	3.9
Neutral lipids in total lipids	99.0	99.3
Phospholipids in total lipids	1.0	0.7
Glycerol ethers in total lipids	0.061	0.009
Glycerol ethers in natural lipids	0.06	0.007
Glycerol ethers in phospholipids	0.16	0.25
Glycerol ethers in natural lipids of total glycerol ethers	97.4	80
Glycerol ethers in phospholipids of total glycerol ethers	2.6	20

Table 1.11 Amounts of γ- and s-aliphatic lactones isolated from butterfat (ppm) (2,28,29)

Carbon Number	s-Lactones	γ-Lactones
6	2.0	Trace
7	0.2[a]	—
8	2.6	0.5
9	0.4[a]	0.2
10	15.0	1.2
11	0.7	0.5
12	35.0	1.6
13	1.5	0.5
14	34.0	1.4
15	6.4	1.3
16	23.2	1.3
18	2.3[a]	—
2,3-Dimethyl-2,4-nonadien-4-olide	—	0.5

[a] Semiquantitative.

Table 1.12 Phospholipid content of bovine milk (4–6)

Phospholipid	Mole Percent
Phosphatidylcholine	34.5
Phosphatidylethanolamine	31.8
Phosphatidylserine	3.1
Phosphatidylinositol	4.7
Sphingomyelin	25.2
Lysophosphatidylcholine	Trace
Lysophosphatidylethanolamine	Trace
Total choline phospholipids	59.7
Plasmalogens	3
Diphosphatidyl glycerol	Trace
Ceramides	Trace
Cerebrosides	Trace

in various countries has been rather accurately determined, as have the seasonal variations. In Europe the amount of saturated fatty acids is generally highest in winter and lowest in summer or fall (see Table 1.11) (34). Green fodder decreases the amount of saturated fatty acids and correspondingly increases the amount of unsaturated fatty acids. The differences between the maximum and minimum values can be fairly large. For palmitic and oleic acids, the quantitatively most important fatty acids, a difference of more than 10% between the maximum and minimum values was found in some cases. This makes it understandable that there are also significant differences in the physical characteristics of the butter. The structure of the triglycerides in the milk fat, along with the fatty acid composition, is important in determining the physical characteristics of the fat, because the softening point of fat has been found to rise as the result of interesterification (35).

Textural characteristics of butter significantly depend on milk fat composition and the method of manufacture. If the chemical composition of the milk fat is known, it is possible to select the appropriate technological parameters

Table 1.13 Compositional characteristics of summer and winter milk fat (36)[a]

| Samples | Fatty Acids | | | | Iodine Number |
	Volatile	Saturated	Monounsaturated	Polyunsaturated	
Average of total	10.98	56.50	29.81	2.50	32.2
Summer	9.49	58.82	33.53	3.14	36.8
Winter	12.45	59.15	26.15	1.86	27.7

[a] $N = 140$.

of the buttermaking to improve its texture. To obtain butter with constant rheological characteristics and to control the parameters of the buttermaking process, it is necessary to take into account the difference in the chemical composition and the properties of the milk fat in various seasons. Table 1.13 shows various compositional changes of milk fat derived from summer and winter milk (36).

MODIFICATION OF MILK FAT

2.1 Melting and Crystallization of Milk Fat Triglycerides

The complex fatty acid composition of milk fat is reflected in its melting behavior. Melting begins at $-30°C$ and is complete only at $37°C$. At any intermediate temperature, milk fat is a mixture of solid and liquid. To a large extent, the solid: liquid ratio determines the rheological properties of the fat. For example, at refrigeration temperature butter has a higher solids content than does a tub margarine. Hence the latter product is more easily spread (37).

As crystallization proceeds, the growing crystals impinge to form aggregates. A network results, in which both the solid and liquid phases may be regarded as continuous. Formation of the network greatly increases the firmness of the fat.

As a liquid fat is cooled, crystallization begins. There are two parts to the crystallization: (*1*) nucleation and (*2*) growth. In a bulk fat, nucleation occurs at the surfaces of impurities, a phenomenon described as heterogeneous nucleation. A considerable degree of supercooling is necessary to initiate nucleation. Subsequent growth of the nuclei tends to be slow in natural fats because of competitive inhibition. In materials of low molecular weight, impurities are rejected at the face of the growing crystal. In fats, however, the various triglyceride species are so closely related that the term *impurity* tends to lose its meaning (38).

2.2 Hydrogenation

Hydrogenation of various fats and oils is used extensively in industry but is not generally applied to butterfat (the high cost of the raw material argues against its use as a feedstock). The process reduces the degree of unsaturation of the fat and increases its melting point.

Given the criticism directed at milk fat because of its saturated nature, there appears to be little future in increasing the degree of saturation by means of hydrogenation. The reverse procedure, desaturation or dehydrogenation, offers more attractive prospects.

Flavor deterioration in fat-rich milk and dairy products is mainly due to autoxidative degradation of lipids. This degradation may be retarded by partial hydrogenation. The objective of partial hydrogenation or trace hydrogenation

14 Butter

is a selective saturation of the polyunsaturated fatty acids without saturation of the monounsaturated fatty acids to improve the oxidative stability. Selective hydrogenation has been studied for years in the vegetable oil industry with some success. This process has been applied by some researchers to milk fat (Figure 1.1) (39).

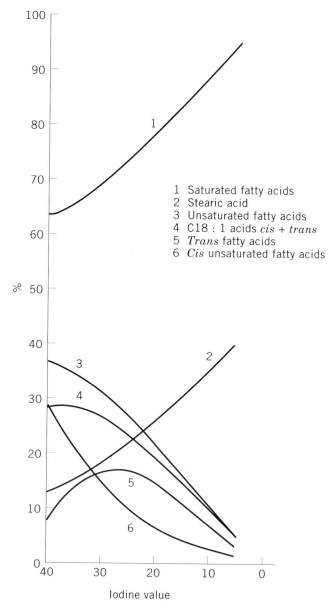

Figure 1.1 Changes in the composition of fatty acids during the hydrogenation of milk fat (38).

2.3 Interesterification

Interesterification (also called ester interchange, randomization, and transesterification) involves the exchange and redistribution of acyl groups among triglycerides. This technology was initially developed as high temperature interesterification in Germany during 1920–1930. Since about 1950–1960 the process has been developed still further in the United States and Europe (39). The resultant product exhibits the same total fatty acid composition as the starting material, but the triglyceride composition and the physical properties are changed. Interesterification catalyzed by chemical catalysts or by lipases is used in the fat industry for the manufacture of margarines, shortenings, and confectionery fat (40).

The fatty acid composition is not modified by interesterification, but there is a significant modification of the glyceride composition (Table 1.14). In the untreated milk fat the triglycerides may be divided into two groups: triglycerides with lower molecular weights (lower than C42) and the triglycerides with higher molecular weights (C44–C54) (39).

Interesterification offers opportunities for modifying the glyceride composition of milk fat and recombined butter. The technique confers some positive nutritive value to milk fat. One disadvantage of the interesterification reaction is the loss of flavor during neutralization, interesterification, and subsequent

Table 1.14 Glyceride composition of natural and interesterified milk fat (mol %) (39)

Triglyceride	Natural	Interesterified
C22	–	0.1
C24	0.3	0.8
C26	0.1	1.4
C28	0.6	1.5
C30	0.9	1.3
C32	1.9	2.0
C34	4.4	3.0
C36	9.5	5.9
C38	13.1	9.1
C40	12.1	9.9
C42	7.7	7.6
C44	6.8	8.3
C46	7.5	10.7
C48	8.8	12.8
C50	11.2	14.2
C52	10.8	10.9
C54	4.6	0.4
Ratio C38:C50	1.17	0.64

16 Butter

deodorization. Most manufacturers would support that little or no commercial interest will result from using this process.

2.4 Reduction of Cholesterol in Milk Fat

One obvious area for development is in the modification of dairy products to satisfy the changing dietary habits of consumers. The mounting health concerns are related to intake of calories, cholesterol, and saturated fats. Concern about cholesterol in the diet originates from the fact that high serum cholesterol, especially the low density lipoproteins, is one of the risk factors associated with atherosclerosis. Dietary intake of cholesterol may be one of the factors contributing to the elevation of serum cholesterol; other dietary factors are high total fat, high saturated fat, and low dietary fiber intake.

There have been many cholesterol-reduction technologies developed all over the world because of high interest by the dairy industry. However, there are only a few technologies available for technology transfer. Fractionation by thermal crystallization, steam stripping, short-path molecular distillation, supercritical fluid extraction, selective absorption, and crystallization using solvents or enzymatic modification can achieve fat alterations of significance to the dairy industry.

Vacuum Steam Distillation. There has been direct application of cholesterol removal by vacuum steam distillation, an old technology. This process is widely used in the fats and oils industry for deodorization.

Cholesterol is a low volatile compound, but it is more volatile than the major triglycerides of milk fat. Superheated steam can be bubbled through the oil, heating it indirectly, which provides for the latent heat of vaporization of the distilling compounds and prevents steam condensation. Thus the temperature and pressure can be varied independently. When the sum of the partial vapor pressures of water vapor and the distillates is equal to the total pressure, water vapor and the low volatile components such as cholesterol and free fatty acids distill over.

The process for cholesterol removal from anhydrous milk fat was patented by General Mills (40). Fractionment Tirtiaux also disclosed the development of a vacuum steam distillation system called the LAN cylinder (38). The steam distillation process (Figure 1.2) was commercialized, producing a 90–95% cholesterol reduction in anhydrous milk fat with a 95% yield that was reconstituted into 2% fat fluid milk (42). The major disadvantage to the process is that it strips or removes most all volatile flavor components from the fat. These flavor components must be captured (i.e., vacreation) before the distillation process to attempt to reproduce the delicate flavors so desired for reconstitution into a butter product.

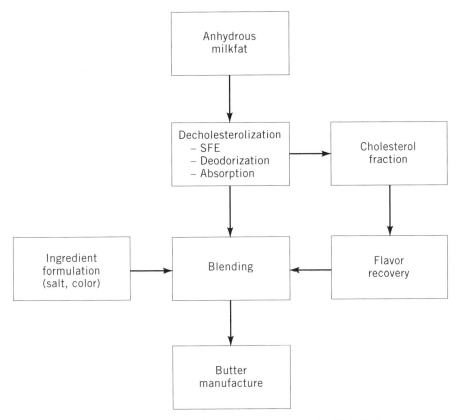

Figure 1.2 Schematic for the decholesterolization of milk fat (41).

Short-path Molecular Distillation. Short-path molecular distillation offered great promise for the selective removal of cholesterol from milk fat. The process achieves the simplest separation of desired substance from a mixture of components of high molecular weight.

The rate of distillation at any given temperature is a function of the ratio $P:M^{1/2}$, where P is the partial pressure of the compound and M its molecular weight. Owing to the temperature dependence of the rate of distillation, a fractionation of components with different molecular weights can be carried out by holding the temperature constant until the more volatile constituents are removed. Figure 1.3 illustrates potential capability of the process (43).

The process has been applied to strip oil-soluble vitamins, sterols, and fatty acids from fats and oils. Cholesterol has been successfully removed from anhydrous milk fat in the range of 70–90% (44,45). Extensive studies were performed and various temperatures and pressures were used to fractionate milk fat (46). Unfortunately, the process has not proved to be economically

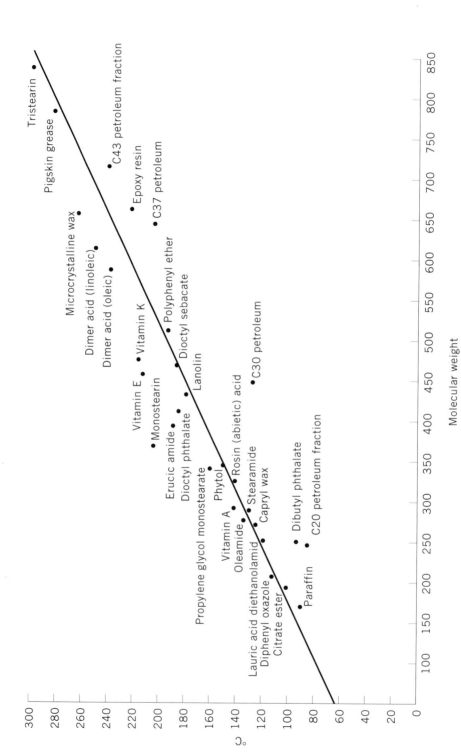

Figure 1.3 Distillation under vacuum (43).

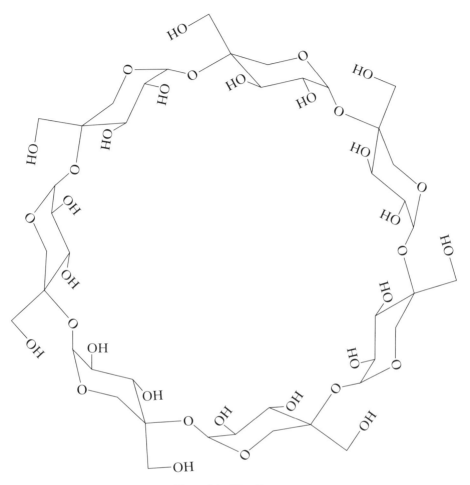

Figure 1.4 Elec. Prepress

feasible due to the low butterfat yield when significant cholesterol was removed (Land O'Lakes research).

Absorption. One of the most promising technologies has been the use of cyclodextrins to complex cholesterol from a mixture and then selectively separate the cholesterol–cyclodextrin complex. European researchers have pioneered almost all of the research in this area, and patents have been issued (47,48).

The process is based on the fact that β-cyclodextrin specifically forms an insoluble inclusion complex with cholesterol. β-cyclodextrin is a cyclic oligosaccharide of seven glucose units. It consists of 1,4-*a*-D-linked glucopyranose residues, as shown in Figure 1.4. As a consequence of the C1 conformation of the glucopyranose units, the secondary OH groups are located on the edge

20 Butter

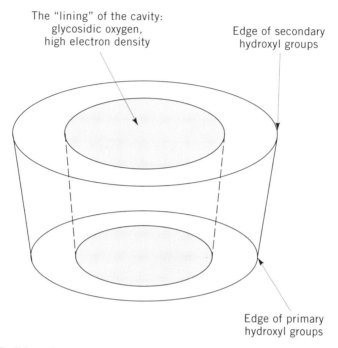

Figure 1.5 Schematic of the β-cyclodextrin molecule, showing the hydrophobic cavity (48).

of the toruslike cyclodextrin molecule, while all the primary OH groups are on the other side (Figure 1.5) (48). The central cavity is, therefore, hydrophobic, giving the molecule its affinity for nonpolar molecules such as cholesterol. The radius of the cavity can accommodate a cholesterol molecule almost exactly, explaining the highly specific nature of β-cyclodextrin's ability to form an inclusion complex with cholesterol.

This process appears to provide considerable economic and practical advantages over alternative cholesterol reduction technologies, such as steam distillation and supercritical carbon dioxide fluid extraction. For instance, there is no absorption of vitamins, it is a low temperature operation, and it has a low capital cost. The only economic concern is that the ratio of the addition of β-cyclodextrin to the cholesterol removed is high, creating the potential for a high cost process. Even so, the Europeans have commercialized the process and reduced-cholesterol butter and cheese products have been introduced into the marketplace (49).

Multiple absorbants have been researched, and they include digitonin, tomatine, sodium cholate (bile salts), and active carbon (45,47,50). Most will face severe food regulatory problems. New Zealand (48) performed extensive research on active carbon, carbon impregnated with different metal salts, and inert supports impregnated with selected organic compounds. This technique (active carbon) is not promising, because it does not retain the delicate butter flavors or the carotenoid pigments. An off-flavor develops in the milk fat. The University of California has evaluated the use of food-grade saponins as absorbents (51). This process showed promise, but most work has been discontinued due to U.S. regulatory prohibitions.

Solvents. The use of organic solvents such as acetone in the laboratory has proven to be an effective method for removal of milk fat components, including cholesterol. Unfortunately, this method creates regulatory and negative consumer perceptions due to the potential of solvent residues in the natural butter/butterfat-containing products.

Supercritical Fluid Extraction. The supercritical fluid extraction process created extensive excitement in the mid-1980s in the research community as a preferred process for cholesterol removal. Extensive research at various universities was initiated to evaluate its potential, and significant publicity was generated within the dairy industry (45,46,52–56).

Liquidlike densities of supercritical gases result in liquidlike solvent powers; this property and faster diffusion characteristics due to low gas viscosity make supercritical fluids attractive extraction agents. Solubility of substances in supercritical gases arises from van der Waals' molecular attractive forces and increases with increasing pressure at a constant temperature. The temperature influences the solution equilibria in a more complicated way than does the pressure. Compounds can be selectively dissolved by changing the density of the gas, i.e., pressure and temperature conditions.

The extraction of cholesterol into the mobile gas phase is determined by the balance of a tripartite interaction: triglyceride–CO_2, triglyceride–cholesterol, and cholesterol–CO_2. Being a minor constituent of the milk fat, cholesterol is probably associated with the triglycerides for which it has higher affinity, namely, short- and medium-chain triglycerides and, to some extent, long-chain unsaturated triglycerides. Because CO_2 affinity is low for cholesterol at 200 bar and 80°C (low gas concentration), there would be less competition between triglycerides and CO_2 for cholesterol. Those cholesterol molecules associated with short- and medium-chain triglycerides would be eluted into the gas phase along with them at low gas concentrations. However, the cholesterol molecules associated with long-chain triglycerides are not eluted at low gas concentrations because of their larger size. Cholesterol esters, being large molecules, would not be eluted at low gas concentrations.

By the late 1980s technologies for the removal of cholesterol with supercritical carbon dioxide were offered by a number of companies (38). Commercialization was never attempted by any major food company for removal of cholesterol. Successful scale-up and commercialization was achieved by the General Foods Corporation for removal of caffeine from coffee (45). The primary disadvantages for the dairy industry were the low yields, low cholesterol removal, and the very high capital and operating costs of the equipment.

Enzymatic. Biological procedures for cholesterol removal make use of microorganisms which produce enzymes to convert cholesterol into innocuous compounds. Several enzymatic systems are being investigated in different countries of the world. Most systems use a cholesterol reductase that converts the cholesterol into coprostanol and coprosterol (52,57). These converted compounds are very poorly absorbed by the digestive system and pass through intact. Several investigators have isolated and characterized *Eubacteria* able to convert cholesterol into coprostanol from rat, baboon, and human feces. Leaves of cucumber, soybeans, corn, and beans are known to contain similar enzymes (45,57,58). *Lactobacillus acidophilus* has also been reported to metabolize cholesterol (51).

Once suitable enzyme systems are identified, the next step involves transfer of the gene that codes for the enzyme into suitable microorganisms, such as *Lactobacillus* and *Streptomyces* species, for large-scale production and purification of the enzyme. Further steps may involve attaching the enzyme onto a solid support or adding purified enzyme in soluble form to the food systems. It is equally envisioned that cholesterol-degrading enzymes such as cholesterol reductase can be genetically transplanted from one group of bacteria into lactic bacteria, which are the traditional dairy starter cultures. The cholesterol-reducing cultures could then be used in cultured dairy products such as cheese.

The industrial scaleup of enzymatic technology is both highly complicated and expensive. Moreover, a primary regulatory hurdle will involve demonstrating that the end products of the cholesterol–enzyme reaction and the novel compounds formed through genetic engineering are harmless.

Regulatory–Nutritional. Many of the processes for cholesterol removal will meet with regulatory hurdles because of residues, unapproved additives, or by-product formation. But the real burden for the U.S. dairy industry was the 1993 Nutritional Labeling and Education Act (54). This act created commercial prohibitions, because it requires that cholesterol-reduction claims cannot be applied to products that contain more than 2 mg/g of saturated fat; butterfat is approximately 65% saturated. The industry has not yet developed a cost-effective means to reduce butterfat saturation.

For years, the nutritional community has claimed that the effect of dietary cholesterol intake on serum cholesterol level was much less significant than the ratio of total fat to saturated fat in the diet (59). The general public is becoming aware of this and has consequently reduced its demand for low-cholesterol foods.

QUALITY CONTROL

In the production of all butterfat-containing foods, quality control is essential to ensure shelf life, safety, and the food's appearance, flavor, and texture. The key to the final product is the quality of the raw materials used. The handling of raw milk is of particular importance. Careful attention is paid to temperature control in raw milk handling systems and, naturally, to the cleaning of all equipment used for storage and transport. Raw milk is held refrigerated to less than 5°C on the farms, and although this has eliminated the growth of a number of organisms, others (notably psychrotrophs) can still multiply under these conditions. Unfortunately, some of the lipolytic and proteolytic enzymes produced by these classes of organisms are heat stable, even surviving ultrahigh temperature (UHT) treatment. Today, a shelf life of many months is expected of a number of UHT-treated products, thus the presence of lipolytic and proteolytic enzymes can be disastrous. The only way to avoid this problem is to ensure that the numbers of organisms are kept to an absolute minimum during all stages of the collection and manufacturing process.

Some quality problems can be eliminated. For example, certain cattle feeds produce undesirable volatile flavors, which tend to concentrate in the butterfat portion of the milk. Historically, the industry adopted steam stripping of cream for buttermaking to reduce the intensity of these flavors. The equipment universally used is the vacreator (Figure 1.6), which is designed to pass as much as 0.30 kg steam to each kilogram of cream. The design of the vacreator evolved during the 1930s and 1960s, culminating in the development of the Vac 25 and 16 models (59).

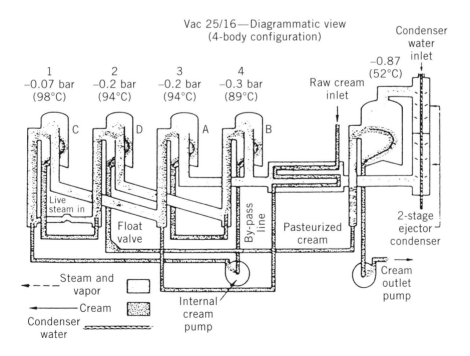

Figure 1.6 Diagram of the Vac 25/16 vacreator (59).

Vacreation accomplishes a number of tasks in one operation. Pasteurization of the cream for buttermaking is perhaps the most important of these, followed by steam stripping of off odors. Other functions include destruction of natural milk lipases and development of a slightly "nutty" or "cooked" flavor in the resulting butter.

3.1 Hazard Analysis and Critical Control Points (HACCP)

The World Health Organization (WHO) recognized HACCP as an effective rational means to ensure food safety from the farm to consumer. The Food and Drug Administration (FDA) and the U.S. Department of Agriculture (USDA) are moving toward the adoption of HACCP systems to ensure the safety of foods sold in the United States.

The HACCP program is a management tool that provides a logical and cost-effective basis for better decision making with respect to dairy product safety. One of the key advantages of the HACCP concept is that it enables a dairy food manufacturing company to move away from a philosophy of control based on testing to a preventive approach that identifies and controls potential hazards in the manufacturing environment.

3.2 Composition Control

Composition control of the butterfat content has received much attention in the ever present drive to maximize returns and create highly consistent products. As noted, multiple technologies are under development to standardize and simplify the processes.

3.3 Grading, Standards, and Definition

Standards for butter in the United States were established by an act of Congress and are supported by USDA standards for grades of butter. In the revised standards, the following definitions apply: *butter* refers to the food product usually known as butter, which is made exclusively from milk, cream, or both, with or without common salt, and with or without additional coloring matter. The milk fat content of butter is not less than 80% by weight, allowing for all tolerances. *Cream* refers to the cream separated from milk produced by healthy cows. Cream is pasteurized at a temperature of not less than 73.9°C for not less than 30 min, or it can be pasteurized at a temperature of not less than 89°C for not less than 15 s. There are other approved methods of pasteurization that give equivalent results (60).

The flavor of the cream may be enhanced by culturing, adding food-grade lactic acid bacteria, or adding natural flavors obtained by distilling a fermented milk; cream may also be added to the finished butter. In addition, color, derived from an FDA-approved source, may be used (61).

Legal requirements for butter vary considerably in different countries. For example, in Europe butter must contain 82% fat, and in France it may contain a maximum of 16% moisture (62). In some tropical parts of the world, milk fat is used in nearly anhydrous form, because it is less susceptible to bacterial spoilage. This product is known as ghee. In the Middle East and India, ghee is prepared from heated cow or buffalo milk.

The new nutrition labeling regulations promulgated under the Nutrition Labeling and Education Act of 1993 (54), mandate that only strictly defined terms be used to make nutrient content claims. For example, the term *light* may only be used on products that have been specifically formulated or altered to meet one of two conditions: (*1*) if the product derives 50% or more of its calories from fat, reduce the fat level by 50% (as compared to a reference product), and (*2*) if the product derives less than 50% of its calories from fat, reduce the calorie level by one-third (compared with a reference product). Generally, butter products derive more than 50% of their calories from fat and, therefore, must achieve a minimum 50% fat reduction to use the term *light*. The term *reduced* when used as a nutrient descriptor requires a formulation alteration that achieves a minimum 25% reduction in the nutrient from a reference product (63).

When considering products to be labeled with a cholesterol claim, the following applies:

Reduced cholesterol: the product has an allowable maximum of 2 g saturated fat and a minimum 25% cholesterol reduction per serving.

Low cholesterol: maximum of 2 g saturated fat and a maximum of 20 mg cholesterol per serving are allowed.

Cholesterol free: maximum of 2 g saturated fat and less than 2 mg cholesterol are allowed per serving.

When considering a fat free or no fat claim, the new regulations require the product to have less than 0.5 g fat per serving and no added fat unless noted (i.e., "trivial fat") (54).

There are three U.S. grades of butter: AA, A, and B. Butter is graded by first classifying its flavor organoleptically. In addition to the overall quality of the butter flavor itself, the standards list 17 flavor defects and the degree to which they may be present for each grade. This grade is then lowered by defects in the workmanship and the degree to which they are apparent. Deratings are characterized by negative body, flavor, or salt attributes, which are fully described in the standards. Butter that does not meet the requirements for U.S. Grade B is not graded. To bear the USDA seal, the finished product must fall within the following microbiological specifications:

Proteolytic count not more than 100 per gram.
Yeast and mold count not more than 20 per gram.
Coliform count not more than 10 per gram.

Table 1.15 Standards for anhydrous milk fat, anhydrous butter oil, and butter oil (65)[a]

Composition and Quality	Anhydrous Milk Fat	Anhydrous Butter Oil	Butter Oil
Milk fat, minimum	99.8% m/m	99.8% m/m	99.6% m/m
Water	0.1% m/m	0.1% m/m	0.3% m/m
Free fatty acids, as oleic acid, maximum	0.3% m/m	0.3% m/m	0.4% m/m
Peroxide value, m Eq oxygen/kg fat, maximum	0.3	0.3	0.6
Copper, mg/kg, maximum	0.05	0.05	0.05
Iron, mg/kg, maximum	0.2	0.2	0.2

[a] Taste and odor at 40–45°C should be acceptable for market requirements. Texture, depending on temperature, should be smooth and fine granules to liquid.

Butter should be stored at 4.4°C or lower or at less than −17.8°C, if it is to be held for more than 30 days (62). The International Dairy Federation (IDF) has produced specifications for milk fat (64), which include reference to the feedstock. (These specifications relate to the time of manufacture but are often used as purchase standards.) The highest grade, anhydrous milk fat (AMF), must be produced from fresh milk, cream, or butter to which no neutralizing substances have been added. It should have a clean, bland flavor when tasted at 20–25°C and a peroxide value (PV) of less than 0.2 meq oxygen/ 1 kg fat. Anhydrous butter oil may be produced from butter or cream of different ages and has no pronounced, unclean, or other objectionable taste or flavor. The term *butter oil* should be used where there is no pronounced unclean or other objectionable taste or odor. The FAO/WHO Codex standard for milk fat is shown in Table 1.15 (65).

3.4 Specialized Analytical Methods

The dairy industry produces a valuable fat that has a desirable flavor and positive consumer awareness; these attributes must be protected. Significant development activity has occurred for rapid, simple methods to detect adulteration (66,67). Because lack of spreadability was determined to be a major butter negative, a flurry of research was initiated creating the need for measurement techniques (68–70). Many procedures have been used or proposed to assess the microbiological quality of the milk or cream. Generally, microbiological tests are performed to determine the hygiene of production and storage conditions or for safety reasons. Tests include total counts and counts for specific classes of microorganisms such as yeasts and molds, coliforms, psychotrophs, and pathogens such as *Salmonella*. Rapid-screening tests based on dye reduction or direct observation using a microscope or automatic total counters are also in use.

3.5 Lipase Activity

An increasing problem is lipolysis in butterfat after manufacturing, which is caused by thermoresistant lipase enzymes that are created in the milk or cream by psychotrophic bacteria or by residual native lipases that survive pasteurization. Based on a determination of the lipase activity in cream, the keeping quality of manufactured butter in regard to lipolysis can be predicted with reasonable accuracy. A similar prediction for sweet cream butter can be based on lipase activity in the serum phase (71). The characteristic lipolytic flavors that can develop in milk products are primarily associated with the short- and medium-chain fatty acids that are relatively abundant in milk fat; they have lower flavor threshold values than the long-chain fatty acids. Because of improvements in the quality of raw milk and the standards of processing, lipolytic rancidity is seldom present in the fat source before its use in recombination (72).

3.6 Oxidation

The flavor of dairy products is largely determined by the fat component. Consequently, it is particularly important to restrict the development of oxidized off-flavors in the fat source before use. Oxidation is the chief mode of deterioration of fats and a major factor in determining the shelf life of fat-containing foods (72). Unsaturated fatty acid esters react with oxygen to form peroxides. Although flavorless themselves, peroxides are unstable and readily decompose to yield flavorful carbonyl compounds. The latter are the source of the characteristic oxidized flavors that are detectable at low concentrations. The rate of oxidation depends on the concentration of dissolved oxygen, the temperature, the presence of prooxidants such as copper and iron, the degree of unsaturation of the fat, and the presence of antioxidants that may retard the onset of oxidation. Compared with many fats, milk fat has a good oxidative stability, because it is high in total saturates, low in polyunsaturates, and contains natural antioxidants, principally α-tocopherol.

The development of oxidative rancidity in milk fat is the major determinant of the stability of the fat on storage. Dissolved air in the milk fat can give dissolved oxygen levels of up to 40 ppm at 30°C. In practice, the dissolved oxygen level in the freshly processed milk fat would be about 5 ppm at 45°C, a level sufficient to permit the development of oxidative rancidity, but if the milk fat were allowed to equilibrate with the air, then this level could increase to 33 ppm with a consequent increase in the rate of development of oxidative rancidity. The solubility curve for oxygen in milk fat is a compound of the solubility curves for the liquid and solid phases (Figure 1.7). Though the

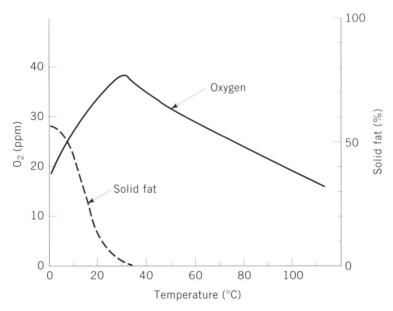

Figure 1.7 The effect of temperature on the solid fat content and solubility of oxygen in milk fat (73).

solubility decreases with increasing temperature for both phases, the solubility of oxygen in the liquid phase is much higher than for the closely packed solid phase (73).

The oxygen level in the milk fat may be limited by either active or passive actions. For passive control, processing procedures and plant design are established to minimize air exposure. Deaeration devices (74), vacreation (60), the use of antioxidants (72), effective destruction of lipases (75), and nitrogen spanning of container headspace are examples of active control of product quality.

BUTTER MANUFACTURE

4.1 Milk and Cream Separation

The most basic and oldest processing method is cream separation. Ancient people are known to have used milk freely. It is probable that they used the cream that rose to the top of milk that had been held for some time in containers, although there is little in ancient literature to suggest that such use was common. It is well established that in early times butter was produced by churning milk.

The principal questions concerning separation in a butter manufacturing facility are the choice of cream fat content and the choice of the separation technique, milk separation before or after pasteurization, temperature of separation, and regulation of the fat content. Separation of cream from milk is possible because of a difference in specific gravity between the fat and the liquid portion, or serum. Whether separation is accomplished by gravity or centrifugal methods, the result depends on this difference (48).

A modern dairy separator will separate virtually all fat globules larger than 0.8 μm, and because most of the milk fat is present in the form of such globules it is relatively easy to separate the bulk of the milk fat. Fat globules below 0.8 μm are generally referred to as nonseparable globules (76).

The percentage of fat in the cream must be known and controlled. It influences fat losses during churning. Knowledge of the fat content assists in yield estimations for operational conditions in continuous manufacture. A number of satisfactory analytical procedures are available, with the Babcock test being the most common.

The chemical composition of the triglycerides, which make up milk fat, varies throughout the year, depending on the stage of lactation and the cow's diet. The seasonal variation causes a cyclic change in the melting properties of the fat. In the control of the buttermaking process and the physical properties of the finished butter, this factor must be monitored. The term *melting property* is used rather than softness or hardness, because these more correctly refer to altogether different attributes of solids. A number of procedures have been used to follow the seasonal change in the melting properties. The iodine number, refractive index, differential scanning calorimetry, or pulsed nuclear magnetic resonance spectroscopy can be used to prepare a melting curve. However, the expense and complexity of these melting curve techniques precludes this approach in most quality-control situations (77). The traditional chemical determinations for fat, saponification value, and Polenske value are of limited value. They are scarcely relevant for quality control, and the information they provide can be more usefully quantified by the determination of the fatty acid profile using gas chromatography.

Today, the use of stainless steel has essentially eliminated the exposure of the fat to copper and iron. The presence of copper and, to a lesser extent, iron can catalyze oxidative deterioration of butter during storage, particularly in the presence of salt and a low pH.

4.2 Crystallization

The crystal structure of fat and the resulting physical properties of butter made by both conventional and alternative processes have received considerable study. When churned conventionally or by the continuous Fritz process for butter manufacture, most of the milk fat is contained within the fat globule in cream during the cooling and crystallization process. The fat globule provides a natural limit to the growth of fat crystals. Cooling and holding of cream is normally carried out overnight, and thus sufficient time exists for the crystallization process to approach equilibrium (78).

The principles of crystallization of plastic fats in the type of equipment used for margarine manufacture have been described (79). It is important for the butter to develop small fat crystals that remain substantially discrete and do not form a strong interlocking structure. Small crystals (e.g., 5 μm diameter) have a greater total surface area than large crystals and will bind water and free liquid fat by adsorption more effectively (78). Large crystals impart a gritty texture to the product. When fats are cooled rapidly in a scraped-surface heat exchanger, fat crystallization commences, but the fat is substantially supercooled on exiting. If crystallization is then permitted to continue under quiescent conditions, crystals will grow together and form a lattice structure. The product will thus be hard and brittle and may tend to leak moisture. If, however, crystallization is permitted to occur under agitated conditions (e.g., for 1–3 min in a pin worker), the formation of small independent crystals will be favored and the product will have a fine smooth texture. If crystallization under agitated conditions is permitted to continue for too long, the product will be too soft for most patting or bulk filling operations, and it is likely to be too soft and greasy at warm room temperatures.

4.3 Neutralization

When lactic acid has developed in the raw, unpasteurized cream by microbial activity to a degree considered excessive, neutralizer may be added to return the cream acidity to a desirable level. Sodium carbonates have been found suitable in practice for batch neutralization. For continuous neutralization by pH control, sodium hydroxide is more suitable. These chemicals must be food grade.

4.4 Heat Treatment

The heat treatment of cream plays a decisive role in the butter-manufacturing process and the eventual quality of the butter. It is important that milk and cream be handled in the gentlest possible way to avoid mechanical damage to the fat, a serious problem in continuous manufacture (Fritz process) of butter (80). Cream is pasteurized or heat treated for the following reasons: to destroy pathogenic microorganisms and reduce the number of bacteria, to deactivate enzymes, to liquify the fat for subsequent control of crystallization, and to provide partial elimination of undesirable volatile flavors.

4.5 Batch Butter Manufacture

Today, batch processing is not used to any extent for the production of large quantities of butter. Batch systems are still encountered in small butter plants, primarily in less industrially developed countries. Continuous systems are

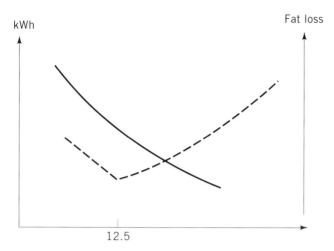

Figure 1.8 Energy consumption and fat loss in buttermilk in relation to churning temperature and heating temperature (81). ———, Energy consumption; ----, fat loss

more efficient and cost-effective for large outputs; batch systems have low capital cost.

The processing of cream by a batch churn requires filling to approximately 30–50% capacity at a cream temperature of 4.4–12.8°C. Cream temperature varies, depending on season, the butter characteristics desired, and the desired rate of fat inversion (Figure 1.8). Churning is accomplished by rotation of the churn at approximately 35 rpm until small butter granules appear. The process usually requires about 45 min for coalescence of the fat globules and clean separation of the buttermilk so that it can be drained (82). The granules may be washed with cold water to remove surface buttermilk. Salt is then added with water if standardization of the fat content is required. The butter is worked to ensure uniformity and desirable body and texture characteristics and the rate of fat inversion.

4.6 Continuous Butter Manufacture

Between 1930 and 1960 a number of continuous processes were developed. In the Alfa, Alfa-Laval, New Way, and Meleshin processes, phase inversion takes place by cooling and mechanical treatment of the concentrated cream. In the Cherry-Burrell Gold'n Flow and Creamery Package processes, phase inversion takes place during or immediately after concentration, producing a liquid identical to melted butter, before cooling and working. The Alfa, Alfa-Laval, and New Way processes were unsuccessful commercially. The Meleshin process, however, was adopted successfully in the former U.S.S.R. The Cherry-Burrell Gold'n Flow process appears to have been the more successful of the two American processes (78).

The Fritz continuous buttermaking process, which is based on the same principles as traditional batch churning, is now the predominant process for butter manufacture in most butter-producing countries. In the churning process, crystallization of milk fat is carried out in the cream, with phase inversion and milk fat concentration taking place during the churning and draining steps. However, because of the discovery that cream could be concentrated to a fat content equal to or greater than that of butter, methods have been sought for converting the concentrated or plastic cream directly into butter. Such methods would carry out the principal buttermaking steps essentially in reverse order, with concentration of cream in a centrifugal separator, followed by a phase inversion, cooling, and crystallizing of the milk fat (82).

Increased demands on the keeping qualities of butter require careful construction, operation, and cleaning of the milk- and cream-processing equipment as well as research to develop machines that will ensure butter production and packing under conditions without contamination and air admixture. It has been demonstrated that butter produced under closed conditions has a better keeping quality than butter produced in open systems (83).

There are two classes of continuous processes in use: one using 40% cream, such as the Fritz process (Figure 1.9) (81), and the other using 80% cream, such as the Cherry-Burrell Gold'n Flow (84). As much as 85% of the butter in France is made by the Fritz process. In this process, 40% fat cream is churned as it passes through a cylindrical beater, all in a matter of seconds. The butter granules are fed through an auger where the buttermilk is drained and the product is squeeze-dried to a low moisture content. It then passes through a second working stage where brine and water are injected to standardize the moisture and salt content. As a result of the efficient draining of the buttermilk, this process is suitable for the addition of lactic acid bacteria cultures at this point. The process then becomes known as the NIZO method when the lactic starter is injected (78). Advantages of the NIZO method over traditional culturing are improved flavor development, improved acid values as a result of lower pH, more flexible temperature treatment of the cream because culturing and tempering often are accomplished concurrently, and most important, production of sweet cream buttermilk.

The Cherry-Burrell Gold'n Flow process is similar to margarine manufacture (84). The process starts with 18.3°C cream that is pumped through a high speed destabilizing unit and then to a cream separator from which a 90% fat plastic cream is discharged. It is then vacuum pasteurized and held in agitated tanks to which color, flavor, salt, and milk are added. Then this 80% fat-water emulsion, which is maintained at 48.9°C, is cooled to 4.4°C by use of scraped surface heat exchangers. It then passes through a crystallizing tube and then a perforated plate that works the butter. Before chilling, 5% nitrogen gas is injected into the emulsion. Improvements of the processing continue to occur. It is now possible to manufacture butter from high fat cream (>82% milk fat) on a continuous basis (85).

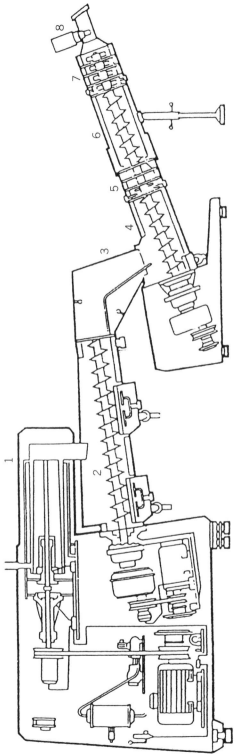

Figure 1.9 Continuous butter maker (Westfalia). *1*, Churning cylinder; *2*, separation section (first working section); *3*, squeeze-drying section; *4*, second working section; *5*, injection section; *6*, vacuum working section; *7*, final working stage; and *8*, moisture-control unit (81).

Although the Meleshin process continues to be in widespread use in the former U.S.S.R., the use of alternative continuous buttermaking processes based on high fat cream has declined in Western countries during the past 20 years (78). The principal reasons for this decline appear to be economics and butter quality, particularly when compared with the Fritz process. A Fritz manufacturing process can be installed in existing batch churn factories with almost no modification to cream-handling or butter-packing equipment. The churns could be retained in case the Fritz breaks down. However, little batch plant equipment could be reused in the alternative systems (i.e.; Gold'n Flow). When a completely new plant is being bought, the alternative systems still tend to be more expensive, and operational advantages over the Fritz system are not significant. Butter from the Fritz process is nearly identical in its physical and flavor characteristics to batch-churned butter, whereas butter produced by the alternative processes tends to be different (86). These differences may be perceived as defects by the consumer, and manufacturers have been reluctant to alter a traditional product.

There are a number of advantages that the alternative systems have over the modern Fritz line (87). The most attractive advantage is the flexibility to produce a wide range of products, with fat contents ranging from 30 to 95% butter–vegetable oil blends and the ability to incorporate fractionated fats (88). The alternative processes also present the possibility of a number of operational advantages. The use of an efficient centrifuge during the cream concentration stage can substantially reduce fat losses in the buttermilk. The composition of the butter can be more accurately controlled, either by including a batch standardization step or by the use of accurate continuous metering systems.

4.7 Cultured Butter Manufacture

There are several ways of making cultured butter from sweet cream. Pasilac-Danish Turnkey Dairies, Ltd. developed the IBC method (Figure 1.10) (81). The main principles of the IBC method are as follows. After sweet cream churning and buttermilk drainage, a starter culture mixture is worked into the butter, which produces both the required lowering of butter pH and, because of the diacetyl content of the starter culture mixture, the required aroma. The starter mixture consists of two types of starter culture: (*1*) *Lactococcus lactis* and (*2*) *L. cremoris* and *L. lactis* ssp. *diacetylactis.* With respect to production costs, the experience with this method shows that for the manufacture of mildly cultured butter the direct costs are only about one-third of the costs of other methods (81).

4.8 Reduced Fat Butter

Fat shortages during World War II first stimulated interest in low fat spreads in the United States. Oil-in-water spreads were first developed in the 1950s

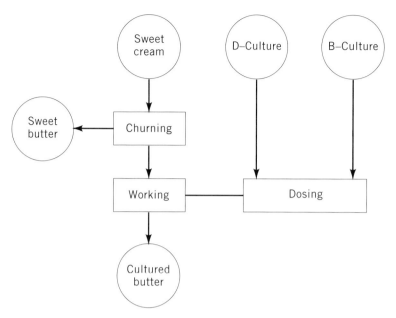

Figure 1.10 Schematic of the Pasilac-Danish IBC method (81).

and 1960s. However, they had a number of limitations, including a shelf life of only 10–14 days (conventional butter shelf life is 4 months), a spongy texture, a lack of melting properties, and an inability to withstand freezing (89). The public's interest in low calorie foods has motivated manufacturers to produce low fat butter products. Although these products can no longer be called *butter* (according to international standards), they are nonetheless often called low calorie butter, half-butter, light butter, or a similar name. Many patents have been obtained for these products, because emulsifying properties are needed to deal with the water content (nearly 50%) of butterlike spreads. In addition, the emulsion must often be stabilized with additives. In some countries, a low fat butter (40% total fat) containing vegetable oil has been designated as Minarine, but Minarine can also be prepared using only butterfat (62). Consumer interest in a reduction of additives in food products and a growing awareness of the importance of proper nutrition have created a demand for a low fat product. It is now possible to produce a butter product based exclusively on butterfat with a fat content of 40%, without using emulsifiers. However, it is precisely this wish for a lower fat content in spreads and in other products that is forcing manufacturers to invest time and money in product development to find a use for their excess butterfat.

The first reduced-fat butter (50% fat), called Light Butter, was introduced in the United States by the Lipton Co. in the mid-1980s. The product was withdrawn due to FDA objections of not meeting Standards of Identity for nomenclature. Also, the product contained stabilizers not allowed for in the

Standard of Identify for butter. In the late 1980s, Ault, Inc. introduced a reduced-fat butter (39% fat) called Pure and Simple, which contained no unusual additives (90). Unfortunately, this all-natural product had severe negatives: it had a short shelf life, experienced moisture seepage, and lacked the highly desirable butter notes. In 1990, Land O'Lakes, Inc. launched its Light Butter (52% fat), which contained emulsifiers, added vitamin A, and preservatives. The FDA was in the process of establishing standards for reduced-fat products at this time and no objection was registered. The new standards were established in 1993 (63), which automatically required Land O'Lakes to reformulate to a 40% butterfat content; it did so and relaunched. The product was a success and has established dominance in the U.S. market.

Butterlike products with reduced-fat content are manufactured in several countries. Stabilizers, milk and soy proteins, sodium albumin or caseinate, fatty acids, and other additives are used. A product is now available on a commercial scale in the former U.S.S.R. that has the following composition: 45% milk fat, 10% nonfat solids, and 45% moisture. It has a shelf life of 10 days at 5°C (91). Each country has established its own standards for butter and butterfat products. Many are still developing standards for a reduced fat butter product to meet the growing consumer demand.

Manufacturers have experienced many problems with the production of low fat butter (92). Low fat butter cannot be manufactured in conventional continuous butter makers. The technology of producing low fat butter and margarine products is similar to that of ordinary margarine production, and it has nothing in common with modern butter (Fritz process) production (Figure 1.11). The conditions are, of course, more critical for products that contain only 40% fat. These low calorie water-in-fat emulsions have such a dense package of water droplets that unwanted phase inversion during processing and/or structural weak points in the product can occur, which may, for example, severely limit the microbiological shelf life. The scraped-surface heat exchanger type of machine is preferred for production of low fat products.

Because of the close packing of the aqueous-phase droplets, the composition of the water phase is critical. Protein concentrates, caseinate, gelling agents, and special emulsifiers have been recommended to simplify the emulsification and to stabilize the end product (93–98). For manufacture, the basic material for production is a mix that is chemically identical to the end product. This mix consists of milk fat in the form of butter, butter oil, and fractionated butter oil or cream, in many cases, it also has milk solids, milk concentrates (including dissolved milk powder and caseinates), and emulsifiers (see Figure 1.10) (81). The fat mix (i.e., butter, butter oil, etc.) is melted and pasteurized.

The required amount is metered into an emulsion tank. If emulsifiers are used, they are melted and mixed with a small amount of fat before being transferred to the emulsion tank and added to the fat. The required amount of water is apportioned, nonfat milk solids are added, and the mix is pasteurized. The water–milk mix is transferred to the emulsion tank and mixed into

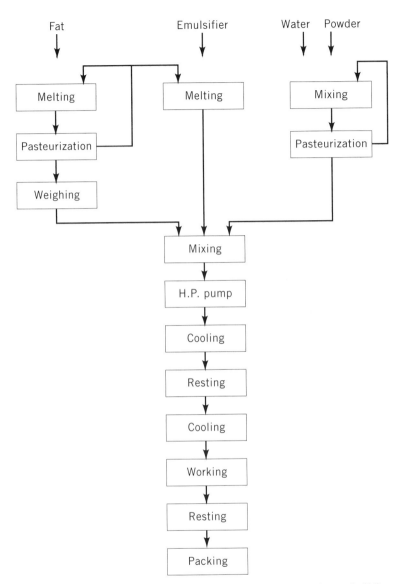

Figure 1.11 Schematic for the production of low-fat butter and spreads (81).

the fat mix. The emulsion is then pumped to a specially designed scraped-surface cooler, where the emulsion, under heavy mechanical treatment and rapid cooling, is supercooled and crystallized, forming the water-in-fat emulsion.

Alternative methods have been developed. For example, the APV Pasilac method is widely used in Europe (Figure 1.12) (99). The continuous APV

38 Butter

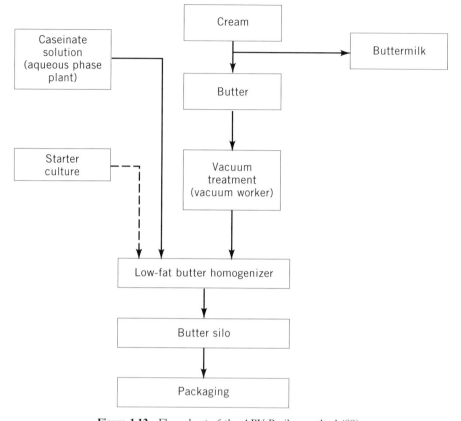

Figure 1.12 Flow sheet of the APV Pasilac method (99).

Pasilac method is quite simple. Ordinary butter with a fat content of 80–82% is mixed with an aqueous phase to the desired fat content. The mixture is then subjected to vigorous mechanical treatment in a special butter homogenizer, and the low fat butter is ready for packaging. As Figure 1.12 shows, the low-fat butter equipment includes an aqueous phase plant. The production of the aqueous phase involves the following processes:

Dissolution of one or more powders in water–milk.
Pasteurization of the solution.
Cooling of the solution to emulsification temperature.

The process makes it possible to manufacture a butter with a fat content as low as 28% (99).

Figure 1.13 shows the method used by Land O'Lakes to produce low-fat butter (40%). The method is similar to margarine manufacture.

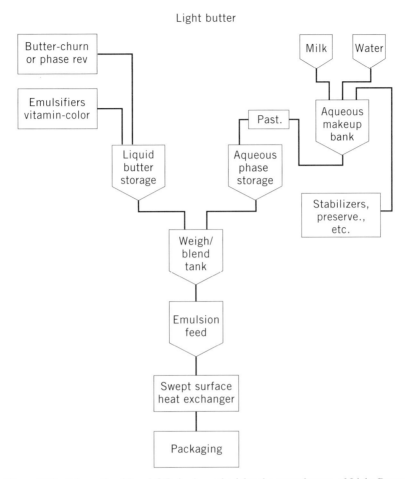

Figure 1.13 Schematic of Land O'Lakes's method for the manufacture of Light Butter.

Fat substitutes and zero-calorie fats offer the potential to reduce the total fat content of foods. Nutrition and marketing experts predict that consumers will show the same enthusiasm for fat substitutes as they exhibited for alternative sweeteners (100). The two products receiving most attention at the moment are Olestra, a sucrose polyester, and Simplesse, a dairy-based fat substitute that contains ultrafiltered egg white or whey protein, evaporated skim milk, sugar, pectin, lecithin, and citric acid. Olestra is a zero-calorie fat, because it is not absorbed across the intestine during digestion (87). It is chemically synthesized by a solvent-free reaction system involving the interesterification of sucrose and long-chain fatty acid methyl esters followed by refining and extraction.

Figure 1.14 shows a schematic of the manufacture of a no-fat spread using modified whey protein concentrate (WPC). The most widely known WPC-

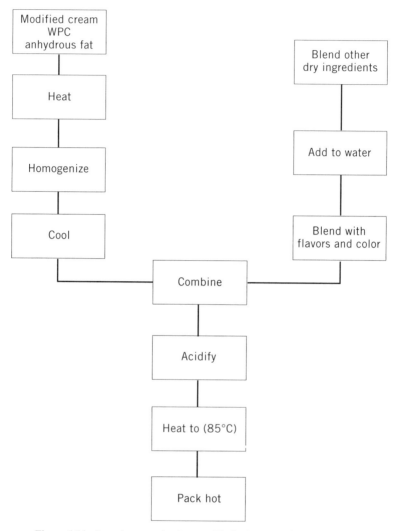

Figure 1.14 Low fat spread using modified whey protein concentrate.

based fat substitute is Simplesse, and when used in combination with other fat mimetics, it has the potential to create a no-fat system for products that is acceptable to consumers.

4.9 Physical and Organoleptic Characteristics

Consistency. Consistency has been defined as "that property of the material by which it resists permanent change of shape and is defined by the complete force flow relation" (34). This implies that the concept of consistency includes many aspects and cannot be expressed by one parameter. Today, more importance is attached to spreadability than anything else in the evaluation of the consistency of butter. The general consensus is that butter should be spreadable at refrigerator temperatures. There is no suitable method available to measure such a subjective criterion as the spreadability of butter. For this reason, the firmness of butter, which should correlate well with spreadability, was selected as the parameter to be measured. It was recommended that the use of the cone penetrometer, along with other methods, could give good results (101).

Flavor. One of the most important consumer attributes of butter is the pleasing flavor. Butter flavor is made up of many volatile and nonvolatile compounds. Researchers have identified more than 40 neutral volatiles, of which the most prominent are lactones, ethyl esters, ketones, aldehydes, and free fatty acids (102). The nonvolatiles, of which salt (sodium chloride) is the most prominent, contribute to a balanced flavor profile. Diacetyl and dimethyl sulfide also contribute, especially in cultured butter flavor (103).

Body and Texture. By means of appropriate qualifications of the terms *body* and *texture,* butter graders describe the physical properties of butter that are noted by the senses. The exact meanings of these terms have not been clearly outlined. Frequently, they are used as if they had the same meaning. Certain properties such as hardness and softness refer to the body of butter, whereas properties such as openness refer to texture. But some of the properties, such as leakiness or crumbliness, are confusing. Usually, most body and texture terms are used to describe a defect, e.g., gritty, gummy, and sticky (86). Good butter should be of fine and close texture; have a firm, waxy body; and be sufficiently plastic to be spreadable at cold temperatures.

Color. The color of butter may vary from a light creamy white to a dark creamy yellow or orange yellow. Differences in butter color are the result of variations in the color of the butterfat, which is affected by the cows' feed and season of the year; variations in the size of the fat globule; presence or absence of salt; conditions of working the butter; and the type and amount of natural coloring added.

Butter colorings are oil soluble and most often are natural annatto (an extraction of the seeds of the tropical tree *Bixa orellana*) or natural carotenes (extractions from various carotene-rich plants). Because they are oil soluble, colorings are added to the cream to obtain the most uniform dispersion.

4.10 Texturization and Spreadability

The most common method to improve the spreadability of butter is to incorporate air or nitrogen, generally increasing the volume by 33%. The product called whipped butter is sold in a tub rather than in stick form.

42 Butter

The consistency of butter is determined by the percentage of solid fat present, which is directly influenced by the fat composition, the thermal treatment given to cream before churning, the mechanical treatment given to butter after manufacture, and the temperature at which the butter is held (104). The European butter market demands that butter be softer and more spreadable in winter and harder in summer. With information on the changes in fat composition from gas liquid chromatography analysis and the use of nuclear magnetic resonance to estimate solid fat, suitable tempering procedures can be selected to modify the fat composition and to produce the most acceptable product for the consumer. A spreadable consistency of butter can be achieved by either varying the fatty acid composition or varying heat-step cream-ripening times and temperatures (105). There are a number of recognized processing options available that influence butter spreadability. Examples include mechanical treatment (texturization) (104,106–109), temperature profiling of the cream (Alnarping) (85), blending winter and summer butters (107), fractionation of the anhydrous milk fat (37,110–113), interesterification of the fatty acids (105,114), the diet of the cow (38,115–117), and cream ripening (118). When blending butters, it is important to understand milk fat melting and solidification curves (Figure 1.15).

Mechanical treatment, or working, involves physically disrupting the three-dimensional fat crystal network and breaking the bonds between the crystals. It is essential that this mechanical treatment is applied to butter in which the crystallization has been totally completed (usually 7 days after churning). The primary crystal structure is strong, and once it is destroyed by kneading it

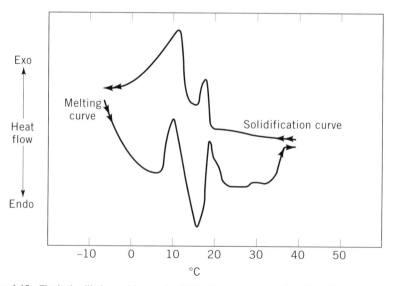

Figure 1.15 Typical milk fat melting and solidification curves obtained by differential scanning calorimetry (107).

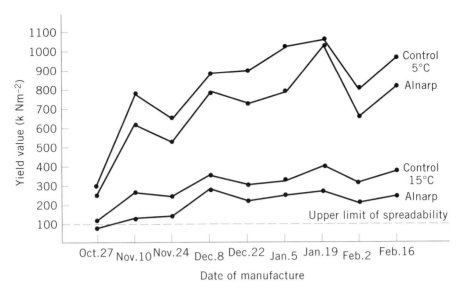

Figure 1.16 Effect of Alnarp-type treatment on firmness of winter butter (107).

does not reform easily. A new, secondary structure is then formed. This structure is weak and reforms quickly after having been destroyed by kneading.

In the Netherlands, bulk butter is often processed in a Mikrofix homogenizer to give it more plastic structure before it is repackaged for retailing (109).

The first published cream tempering process (1937) was developed by the Alnarp Dairy Institute in Sweden, which lent its name to this and similar processes (107). In the original Alnarp process, pasteurized cream was cooled to 8°C for 2 h, warmed to 19°C for 2 h, and then cooled to 16°C and held until churning. Modifications allowed for initial cooling of the cream to initiate nucleation of the crystals at a temperature well below the main solidification temperature ranges (see Figure 1.14). The cream was then warmed to a temperature several degrees above that of the lower melting peak temperature to encourage a fractionation process in which high melting glycerides crystallized and low melting glycerides melted (Figure 1.16).

Another alternative to increasing butter spreadability is providing cows with specific feeds to change the fatty acid composition. The main opportunity for varying fat composition comes in the indoor feeding period, when cows receive some type of concentrate in addition to silage. If, for example, free soy oil is added to the diet, its constituent 18:2 and 18:3 fatty acids are converted in the rumen largely to 18:0 and some 11–18:1-*trans* (vaccenic acid). The gut wall and mammary gland contain a desaturase, which converts 18:0 to 11-*cis*-18:1. The milk fat resulting from such a diet may contain up to 60% of C18 acids, with 18:1 (cis and trans) assuming the dominant position. The milk fat contains a relatively low proportion of 16:0, 14:0, 12:0, 10:0, and 8:0, because of the depression of *de novo* synthesis within the gland. The weight proportion of 4:0 remains fairly constant (which implies that the molar proportion may increase slightly) and that of 6:0 decreases only slightly, both facts indicating that 4:0 and 6:0 are synthesized by a route that is relatively much less significant (38,114,116,117). Because of the high unsaturated C18 fatty acid content, this type of milk fat has a lower melting spectrum. The outcome is a butter that has reasonable spreadability at refrigeration temperatures (114).

Feeding protected unsaturated fat usually leads to an increase in the proportion and yield of milk fat. However, a major cost is in this instance associated with the preparation of the feedstuff, involving as it does homogenization of an oil–water–protein mixture and the subsequent removal of the water by spray drying. This type of feeding practice has little commercial application at present (116). Thus, although the scientific knowledge to alter the fatty acid composition of milk fat is well established, economic considerations have prevented its exploitation.

In efforts to improve the spreading properties of butter in relation to hard butterfat, one alternative put forward is the use of soft fat fractions obtained in the fractionation of anhydrous milk fat. Although several practical methods of fractionation have been presented, the use of soft fat fractions in buttermaking has not become general practice. This is evidently because fractionation in all cases significantly raises the cost of the butter produced. In addition, a common problem has been to find suitable uses for the hard fat fractions. Furthermore, in fractionation methods that use solvents or additives, fractionation should be linked to fat refining, and in this process butter also loses its natural food classification. In studies that have used soft butterfat fractions, a substantial softening of the butter has been obtained; however, this butter, like normal butter, hardens as the temperature increases and again decreases (118). Because milk fat exhibits a wide melting range from about $-30°C$, it may be possible to use a dry fractionation process. Suitable sizes of crystals are developed by controlled cooling of the melt, and the crystals are separated from the liquid phase by filtration or centrifugation. The fractionation of milk fat by melt crystallization has been extensively studied (37). The general conclusions from these investigations were as follows: the short-chain triglycerides and the short-chain and unsaturated long-chain fatty acids were enriched in the liquid fraction; the efficiency of molecular size separation in the melt crystallization process was poor; and the flavoring compounds, pigments, vitamin A, and cholesterol were slightly concentrated in the liquid fraction.

Currently, dry fractionation of anhydrous milk fat is performed by two conventional systems—Tirtiaux and De Smet (both from Belgium)—which are bulk crystallization processes. The widely used Tirtiaux dry fractionation process enables one-step or up to five-step fractionation of anhydrous butter oil at any temperature, ranging from 50 to 2°C (37,110–113). The milk fat fractions thus obtained can be used as such or the fractions can be blended in various proportions for use as ingredients in various food-fat formulations.

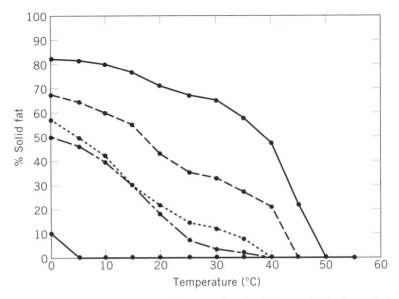

Figure 1.17 Solid fat content profiles of the control and anhydrous milk fat (–·–·–); the low melting, high melting, and 20% very high melting milk fat (·····); and the milk fat fractions used in the low melting, high melting, and 20% very high melting butter: 30S fraction (– – – –), 13L fraction (——, bottom line), and 15S fraction (——, top line). The fraction number includes the fractionation temperature (°C) and its physical form (solid or liquid) (110).

The major shortcoming inherent in this system is the long residence time (8–12 h) for nucleation and crystal growth.

Butter samples made from low melting liquid fractions and from a combination of primarily low melting liquid fractions and a small amount of high melting solid fractions exhibited good spreadability at refrigerator temperature (4°C) but were almost melted at room temperature (21°C). Butters made with a high proportion of low melting liquid fraction, a small proportion of high melting solid, and a small proportion of very high melting solid fractions were spreadable at refrigerator temperature and maintained their physical form at room temperature (Figure 1.17).

4.11 Anhydrous Milk Fat Manufacture

The quality assurance program for manufacture of butter oil, or anhydrous milk fat (AMF), also focuses on the quality of the raw materials. Naturally, many of the same considerations apply to handling raw cream for AMF manufacture that apply to butter, except that vacreation is not used. Because it is stored under ambient conditions, care against oxidation is essential. Oxidation is perhaps the most important mechanism by which milk fat deteriorates

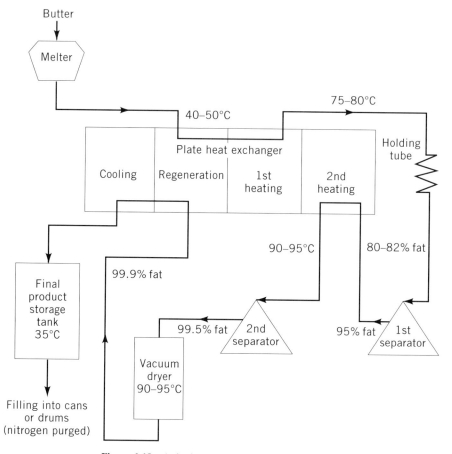

Figure 1.18 Anhydrous milk fat manufacture (119).

in quality. Because the oxidation reaction is autocatalytic (i.e., the products of the reaction act as catalysts to promote further reaction), the normal quality control tests, peroxide value and free fat acidity, could give misleading results when applied to stored butter. Methods of deaeration have been developed that could reduce potential oxidation (74).

Milk fat is present in milk or cream as part of a stable oil-in-water emulsion. The emulsion is stabilized as a result of the protein and phospholipid-rich milk fat globule membrane (MFGM) surrounding the milk fat. During AMF manufacture the aim is to break the emulsion and to separate out all of the nonfat solids and water (Figure 1.18). To achieve this, the MFGM must first be disrupted mechanically or chemically. Homogenization, an example of mechanical treatment, disrupts the membrane, destroying the membrane layer. For chemical destruction of the membrane layer, an acid such as citric acid can be added to lower the pH of milk or cream to about 4.5 (119). The protein will precipitate, removing a component to maintain an intact fat globule. Direct-from-cream AMF plants usually have three separators. The first concentrates cream from 40% to about 75% fat before phase inversion in a homogenizing device. The oil separator then separates the liberated butter oil to about 99% purity. The oil is washed with water before the third (polishing) separator and the final traces of moisture are removed in a dehydrator at 95°C and under a vacuum of 35–50 torr. The dehydrator is usually a simple vessel, and the butter oil is introduced either as a thin film on to the walls or as a spray, to maximize the surface area exposed to the vacuum. Such a device will not remove significant off-flavors, because the vapor flows, temperatures and pressures are inappropriate for flavor stripping. When producing AMF from butter, fresh or block butter is softened just to a pumpable stage (approximately 50°C) and transferred to a plate heat exchanger to increase the temperature (70–80°C). The oil phase is concentrated through separators and dried under vacuum. Some washing is possible before the final separator removes the last traces of nonfat components (see Figure 1.18).

In terms of the preparation of products and their appearance and texture, AMF has several advantages over traditional butter. The latter is in fact subject to seasonal variations, which affect its physical properties. The advantages of AMF are linked to the possibility of standardizing its physical properties (by the selection and mixture of the raw materials used in production) and the possibility of adapting its properties using the fractionation technique. This is particularly important for its use on an industrial level where, given the automation of production and the standardization of the texturization stage (temperature), it is necessary to maintain constant physical (rheological) and organoleptic properties (120).

4.12 Packaging

The objective of any packaging system is to protect the product from deleterious environmental conditions. Many packaging systems have been developed to protect the milk fat from biological, chemical, and mechanical deterioration. Bulk containers such as 5-L cans and 20- and 25-L drums are popular for packaging industrial materials such as butter oil and AMF. These containers usually have a welded side seam and are plain internally. Unlike edible oils, milk fat is corrosive to tinplate, so an internal gold epoxy phenolic lacquer such as International Paint's "IP 180" is required. If the containers have a welded side seam, the internal raw steel edge of the weld needs to be protected by applying a side-strip lacquer (i.e., a two-part epoxy polyamine). All lacquers used on food cans should have FDA status (121). Larger drums (i.e., 210-L nominal and 218-L maximum capacity) need more rigid walls to withstand the greater mechanical stresses in handling. When chilled storage is available, less rigid forms of packaging may be used. Concentrated butter is now being retailed for domestic cooking, aided by a European Community subsidy. This concentrated butter is normally packaged in either parchment or a foil–parchment laminate similar to that used for butter (122).

For packaging in foil, parchment, or other flexible film, the milk fat needs to be in a plastic state, similar to that of freshly churned or reworked butter. The optimum softness of the fat depends on the packaging system being used. Traditionally, the United States has packaged consumer portion sizes of individually wrapped four 0.25-lb sticks in 1-lb units. Multiple sizes from 5- to 25-g units of foil-wrapped "continentals" or polyethylene molded cups, have been used in the food service sector. For bulk handling, the plastic fat or butter can be packed into polyethylene-lined fiberboard boxes. In all cases, the aim is to fill the desired quantity of fat with the minimum giveaway. Aeration of the milk fat should also be avoided as oxidative rancidity is the principal factor limiting the shelf life.

4.13 Storage and Transport

Storage conditions depend on the end use for the product, the packaging system, and the desired storage time. Milk fat in hermetically sealed cans and drums is the least affected by its storage environment. Ambient storage is commonly used and must be to the same standards as used for other food stuffs (121). In the European community and the United States, temperature is not a major factor but it can be in tropical countries, where the temperature may rise to 35–40°C. At temperatures in excess of 30°C, the milk fat deteriorates significantly more rapidly and there is an increased risk that the stored fat will have a stale, oxidized off-flavor.

Transport of drums and cans may be at ambient temperature, though the storage life may benefit from refrigeration. High humidity and wet conditions should be avoided to minimize the risk of corrosion or mold growth on the packaging that would entail an additional cleanup operation. The drums should be stowed away from excessive heat and noxious chemicals. For most journeys, standard freight containers may be used.

Milk fat in polyethylene-lined fiberboard boxes is at far higher risk from its storage environment. Since the packaging is permeable to oxygen, the product is more prone to oxidation. Chilled storage (10°, preferably 5°C) must be used both to reduce the rate of oxidation and to maintain the rigidity of the pack. The humidity of the storage area must also be controlled to prevent mold growth on the fiberboard. Because the pack is also permeable to odors, the storage area should not be shared with other food stuffs with strong odors that could cause off-flavor absorption (e.g., fish, onion, garlic).

Flavor transmission and oxidation is less of a problem at temperatures below −20°C, which is preferable for long-term storage. When conserving and preserving stocks of butter for extended periods (5 years or more), a process has been developed by which butter is placed in the refrigerated chamber or warehouse, which has been sealed airtight. The air is evacuated and replaced with nitrogen or other inert gas mixture so that the pressure in the chamber is equal to the exterior atmospheric pressure. This process allows for extended storage without mold growth and development of rancidity (123).

When butter has been frozen, textural characteristics may have been deleteriously affected. An invention to improve texture has been described in which large, deep-frozen blocks of butter with the desired moisture content are reworked by chipping them in a butter chipper while adding measured quantities of water. The butter chips are then fed through a vacuum chamber into a butter churn designed as a continuous kneading mill. The butter chips are continuously conveyed under pressure through the kneading mill by means of a high-pressure butter pump (124).

It is not possible to set rigid standards for the shelf life of milk fat. Shelf life depends primarily on the acceptance quality criteria of the user and will be affected by (*1*) the quality of the feedstock, (*2*) the packaging system, and (*3*) temperature. With increasing storage time, the flavor defects are more likely to become noticeable. Flavor does not correlate easily with peroxide value (125). At a peroxide value of <0.6, oxidized flavors are unlikely, but if the peroxide value is >1, then some oxidized flavor may be expected. It must be pointed out that these figures are based on the International Dairy Federation (IDF) method (126) for measuring peroxide value and that other methods are likely to give different results. Other grades of milk fat defined in the IDF standard are anhydrous butter oil and butter oil. Anhydrous butter oil is the product obtained from butter for cream; it may be of different ages and should have no pronounced, unclean, or other objectionable taste or odor. Butter oil is the product obtained from butter or cream; it may also be of different ages and should have no pronounced, unclean, or other objectionable taste or odor.

BUTTERFAT PRODUCTS

5.1 Butterfat–Vegetable Oil Blends

The first commercial development of a spread made from a combination of butterfat and vegetable oil was in Sweden in 1963. The product, Bregott, contains 80% fat, of which 80% is milk fat and 20% is soybean oil (127). Bregott is a margarine according to Swedish and American food standards. Swedish scientists also developed and successfully commercialized the first reduced-fat spread in Europe in 1974. The product, Latt and Lagom, contains 39–41% fat, of which 60% is milk fat and 40% is soybean oil. It is considered to be a low calorie margarine. Bregott is exported to Australia, and Latt and Lagom to Japan and France (89). In Finland, where Bregott is popular, oil from rapeseed is used (83).

Other products (under license from the Bregott patent) are Voimariini in Finland, Bremykt in Norway, Smjorvi in Iceland, and Dairy Soft in Australia.

Similar products are Clover in the U.K.; and Dairy Gold, Kerry Gold, and Gold'n Soft in Ireland. This list is, however, not complete. The latter blends are high fat products (75–82% fat), and the amount of butterfat of the total fat is about 50% (93). Most of these products are manufactured in a churning process in a churn or a continuous butter machine.

The first steps in the manufacture of Bregott are pasteurization of the cream, followed by cooling and temperature treatment. The cultures are the same as those used in buttermaking. Measured quantities of cream and soybean oil are mixed in the churn or the oil is continuously injected before churning in a continuous butter machine. The by-product is sour buttermilk.

The most commonly used vegetable oil is soybean oil. Products with a low percentage of butterfat will contain not only vegetable oil but also hydrogenated vegetable fats to achieve a good plasticity. If the minor part of the total fat is butterfat, as in Golden Churn from the U.K., the manufacturing process is completely different from modern butter production. In this case, the technology is analogous with normal margarine manufacture, where some part of the fat is replaced with butterfat. The emulsion is cooled in scraped-surface coolers (Figure 1.19).

Very low fat spreads have recently been developed. The first European commercialized product was made by St. Ivel and is called St. Ivel's Lowest. It contains 25% butterfat and has a lower saturated fat content than sunflower margarine (128).

In the early 1980s blends of butter and vegetable oil products appeared in the U.S. market. The U.S. market leader was Country Morning Blend made

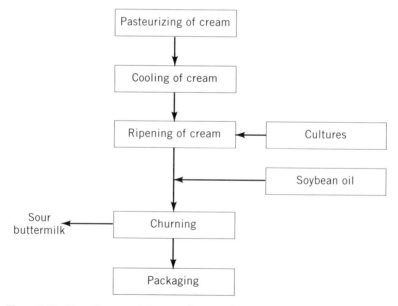

Figure 1.19 Flow diagram of the manufacture of butter–vegetable oil mixtures (93).

Table 1.16 Market share for butter, margarine, blends, and spreads for 1980–1990[a]

Year	Butter	Margarine	Blends	Spreads
1980	25	74	1	–
1985	22	73	3	2
1990	17	48	2	33

[a] Some data from Ref. 88.

by Land O'Lakes. These blends generally were 40% butter and 60% vegetable oil for a total fat content of 80%, within the margarine Standard of Identity and designation. With the increasing popularity of reduced-fat (less than 80%), spreads, starting in the mid-1980s, other blends with butterfat contents of 2–25% were introduced. As noted in Table 1.16, the 80% fat margarine blends are losing market shares to lower fat spreads and blends (129).

A number of processes have been developed using continuous churns (97) and alternative systems similar to the Cherry-Burrell Gold'n Flow process (98). The major disadvantage to churning, either batch or continuous, is that the resultant buttermilk is adulterated with some vegetable fat and is less valuable than standard buttermilk. An advantage of alternative processing systems is their ability to accommodate easily the manufacture of reduced-fat spread blends. All butterfat–vegetable oil blends provided alternatives to butter for the consumer when concerns of health (e.g., fat saturation) and spreadability are desired.

5.2 Ghee

By definition, ghee is a product obtained exclusively from milk and/or fat-enriched milk products of various animal species by means of processes that result in the near total removal of water and nonfat solids (similar to anhydrous milk fat) and in the development of a characteristic flavor and texture. Even so, most ghee contains some nonfat solids to enhance the flavor.

Typically, ghee is manufactured by heating butter to temperatures well above those used during AMF manufacture. The high temperature treatment of the nonfat milk solids and milk fat leads to the development of a strong buttery flavor. However, traditional ghee, as produced in the Middle East and Asia, has a more rancid taste due to less sophisticated methods of preparation and storage. Manufacturers in the European Community are also producing ghee by adding ethyl butyrate to anhydrous milk fat (119,130). Alternate synthetic flavors have been developed to add ghee flavor notes to butter oil.

A synthetic mixture consists of $\delta\text{-}G_{10}$ lactone (3 ppm), $\delta\text{-}G_{12}$ lactone (15 ppm), decanoic acid (5 ppm) and kenanone-2 (10 ppm). This technique is simpler, less time-consuming, and more economical that the technique that

52 Butter

uses powders. The shelf life of this flavored butter oil is 2.5 months. However, the addition of synthetic antioxidants, butylated hydroxyanisole (BHA) at the 0.02% level, enhances its shelf life so it can compete well with conventional ghee (130).

5.3 Butterfat as an Ingredient

Recombined Products. For recombination of milk and dairy products, the two primary ingredients required are AMF and nonfat milk solids. A range of fat sources is available for the recombining industry, but only a few of these are in widespread use. In most countries, anhydrous milk fat is usually the sole fat source. Numerous products can be made by putting together the correct proportions of water, flavor, and other ingredients as desired (e.g., sugar, emulsifier, and stabilizers) to make sweetened condensed milk, ice cream, recombined butter, or milk (1%, 2%, or full fat). For these applications, AMF of the highest quality should be used to avoid off-flavor development. The recombination of butter has been reported in detail (131).

In response to the problems associated with the handling of unsalted butter, a new milk fat product was developed in New Zealand to combine the superior flavor of butter with the ease of handling of AMF. Initial shipments of this product, fresh frozen milk fat for recombining (FFMR) were favorably received, and FFMR quickly became established as the preferred alternative to unsalted butter (72).

Pastry, Cake, and Biscuit Products. In general, fats play several essential nutritional, technological, functional, and organoleptic roles in most all bakery applications. Due to its physical properties, fat plays a major part in the production of the majority of items in the pastry, cake, biscuit, and chocolate confectionery sector; for example, in the preparation of pastry cream and in the desired appearance and texture of the end product. These physical properties include, above all, the rheological properties (consistency, plasticity, texture, etc.), and the properties of fusion and crystallization depend on the type of fat, the temperature, and the working conditions of the product.

The fats used in pastry and biscuit confectionery have different functions, which are determined by their rheological properties (plasticity and texture). In pastry these principal functions are (*1*) an increase in the plasticity of the pastry (e.g., hard pastry with a low level of hydration) and (*2*) a break in the body of the pastry (i.e., the fat makes the gluten structure discontinuous, which gives the desired crumbliness in, for instance, biscuits) (120).

Confectionery–Liquors and Liqueur. In chocolate confectionery and for pastry creams, it is the physical properties linked to the fusion and the crystallization of the fat that are essential. For milk chocolate, for coating or in bars, AMF can be used in proportions that depend on its compatibility with cocoa butter, whose properties of hardness and rapid fusion at 35°C cannot be altered. Thus it is currently accepted that AMF with high fusion levels obtained by the fractionation technique can be used. In general, milk fat has an interesting characteristic: it inhibits the appearance of fat bloom (132).

For pastry creams, the ideal is an AMF, which causes rapid melting in the mouth. Depending on the type of pastry cream, a wider choice of AMF can also be offered, thanks to the fractionation technique.

Due to the low level of milk fat in dark chocolate, fat bloom is a problem with this product. The hard fraction of milk fat (milk fat stearin) has been reported to act as an antibloom in dark chocolate, giving the chocolate an increased shelf life. However, the use of hardened milk fat is limited in several major chocolate producing countries (132).

Regulations for the amount of milk fat, nonfat milk solids, whole milk solids, and total milk solids allowed in milk chocolate vary among countries. Fats other than milk fat are allowed in milk chocolate in some countries, although different flavors and textures may result in the chocolate.

While the preferred source of milk fat in cream liqueurs and associated beverages is undoubtedly double cream, its use may lead to problems. In particular, cream contains calcium and other ionic materials. One solution is to wash the cream to remove all ionic materials, but this approach is cumbersome in practice. The preferred approach is to use anhydrous butterfat as the starting material.

Other Uses. The use of butter or anhydrous milk fat requires more added emulsifiers in ice creams and ice milks, because the naturally occurring milk fat emulsion will have been destroyed in the manufacturing process. Milk fat is also used in fresh cream, frozen cream, dry cream, and plastic cream. Ice creams contain a high level of milk fat, and its manufacture uses substantial quantities of milk fat worldwide.

5.4 Butterfat Powders

The main purposes of transforming fats into powdered forms are to increase their microbial stability and to enhance their handling and functional properties. Powdered fats are available in two principal forms. The first form is feasible only if the fat contains sufficiently large amounts of high melting compounds. If a fine jet of molten fat is sprayed into an ambient atmosphere, it will set as fine discrete particles. The second form embodies the use of a carrier. This form is necessary for oils and fats, such as milk fat, that contain appreciable amounts of low melting triglycerides. The carrier can be added to the fat before or after spraying.

Dry creams are commonly produced as an ingredient for many applications. They consist of at least 40% butterfat, but can range up to 70% fat, 22–57% nonfat milk solids, and 0.5–5% moisture.

The Commission for Dried and Condensed Milk of the International Dairy Federation (1962) proposed to designate cream powders of 60% or more milk fat as "butter powder." Later, it was suggested that a minimum of 80% milk fat be used for butter powder (133). Butter powder composition can vary in the amounts and types of the nonfat constituents, depending on the application. Powders can include nonfat milk solids, sodium caseinate, sodium citrate, sodium aluminum silicate, sucrose, and lactose (133). Other possible minor constituents are antioxidants, emulsifiers, flow agents, and stabilizers. The objective is to create a stable, fluffy, free-flowing, nongreasy, and loose creamy powder. Butter powder containing 80–85% fat could not possibly meet the legal and organoleptic texture requirements of butter when reconstituted with water; therefore, it is not in market competition for conventional butter use. The potential applications for high fat powdered products include butterlike spreads, coffee or beverage whiteners, soup, sauce and dessert creamers, convenience ice cream, scrambled eggs and omelettes, pudding and pancake mixes, and aerating fats.

5.5 Specialty Butterfat Products

Flavored Butter. Flavored butters (garlic, onion, pepper, lemon, etc.) have been successfully mass marketed in Europe (134), but this success has not translated to the U.S. market. There exists a limited specialty market in upscale stores and delicatessens and certain food service applications. Manufacture is quite simple. Creating the desired flavor blends remains an art and skill.

Hypoallergenic Butter. A U.S. patent was granted in 1992 for the manufacture of hypoallergenic butter (135). The patent has limited claims. The product is a sterile butterlike product made from anhydrous milk fat; it contains no nonfat solids (99.9% free of NFS).

Butter Flavors. Technologies for the hydrolysis of butterfat to produce and concentrate the free fatty acids to enhance the butter flavor of products have been available for decades. More recently, biotechnologists have developed methods for producing a variety of fairly pure enzymes, economically and in large quantities. The increased availability of lipases (glycerol ester hydrolases) from microbial sources has made it possible for researchers to employ the catalytic properties of these enzymes in innovative ways. One application in which the use of lipases has become well established is the production of lipolyzed flavors from feedstocks of natural origin.

Immobilization of lipases on hydrophobic supports has the potential to (*1*) preserve, and in some cases enhance, the activity of lipases over their free counterparts; (*2*) increase their thermal stability; (*3*) avoid contamination of the lipase-modified product with residual activity; (*4*) increase system productivity per unit of lipase employed; and (*5*) permit the development of continuous processes. Because the affinity of lipases for hydrophobic interfaces constitutes an essential element of the mechanism by which these enzymes act, a promising reactor configuration for the use of immobilized lipases consists of a bundle of hollow fibers made from a microporous hydrophobic polymer (136).

Extended-life Creams. Extended-life creams are produced using normal separation techniques but involve a high temperature single or double heat treatment (95–135°C). The temperatures employed render the product almost sterile. Any surviving bacteria tend to be spore forming types. Packaging is usually carried out on aseptic machines or nonaseptic machines modified with, for example, H_2O_2 spray and ultraviolet lights (76).

Short-life Creams. For short-life creams, the shelf-life depends on a low bacteriological count milk with good plant hygiene. Heat treatment tends to be in the region of 75–90°C with 3–30 s hold, followed by cooling to below 10°C. Final cooling to below 5°C is normally carried out in aging tanks or in the retail container in the cold store. Shelf life can be up to 12 days (76).

Ultrahigh Temperature Creams. Heat treatment for ultrahigh temperature (UHT) creams is produced by indirect or direct heating to 135–140°C with 1–4 s hold before cooling to ambient temperature. Aseptic packaging is essential. As this product is designed for long shelf life (3–4 months), formation of a cream plug or fat rise in the container must be avoided. Hence all UHT creams, including whipping cream, must be homogenized. Homogenization can be carried out either upstream or downstream. If carried out downstream, an aseptic homogenizer must be employed (76).

Decholesterolized Milk Fat. In the 1980s, there was significant research and market activity in developing decholesterolized milk fat. All this activity was for naught, for the hypothesis of creating a "healthier" fat (for butter or milk or other dairy product) was not sound. The nutrition community had long recognized that the link between dietary cholesterol and serum cholesterol was weak and that the ratio of total fat-saturated fat had a greater impact on health. In addition, the FDA issued new standards in 1993 (63) which effectively negated the value of decholestering milk fat. The new law required that to be called low cholesterol, the fat must contain no more than 2 g of saturation fat per serving. Butterfat is approximately 65% saturated. Since the technology to desaturate milk fat is not cost-effective, decholesterization has no economic value.

Desaturated Milk Fat. In addition to chemical and enzymatic means of desaturation, there have been extensive studies on feeding cows specific diets to change butterfat saturation as well as increasing the ratio of potentially desirable fatty acids (38,115,116). In general, good progress in the understanding of rumen physiology, digestion, and function has occurred, but economic potential remains unacceptable. The most promising technologies are the use of protected fats in a feeding regimen. These fats are protected in a way that they pass through the rumen (point of fat hydrogenation) into the remaining digestive system for absorption and subsequently into the mammary glands. Unsaturated fats that are fed to cows have a great opportunity to remain unsaturated as they are synthesized into milk fat. Biotechnology may offer alternatives in the modification of edible fats and milk fat. Research has led to new methods of lipolysis and esterification, but the developments are still at the laboratory level. Nevertheless, commercial application may emerge from these interesting areas of research.

Nutraceauticals and Healthy Fats. Almost all dairy products tend to be excellent carriers of specialized nutrients (vitamins, minerals, specialized cultures, and micronutrients), thus there is potential for fortification to enhance the natural nutritive properties of dairy products, creating nutraceauticals or functional foods. The use of specialized cultures such as *Lactobacillus acidophilus* and *Lactobacillus bifidus,* which have generally recognized nutritive characteristics, is ideal for cultured butter. A butter spread product using these cultures (Fittisport) has been launched in the French market; another with lower fat content has been introduced in the German market (137). In addition, it has been shown that the free fatty acids of milk fat have inhibitory effects against certain pathogens (e.g., *Listeria monocytogenes*) (138).

A structured lipid containing dairy fat is covered by a U.S. patent (95). The invention relates to a transesterification product of a mixture of fatty acids and triglycerides, including milkfat, in the form of cream or butter as the main component. The product has nutritional applications and may also be used as an enteral or parenteral supplement.

Nonfood Applications. The nonfood use of milk fat has been insignificant. Milk fat and milk fat fractions could, however, have some potential possibilities for profitable use, for example, in manufacture of pharmaceutical or technochemical products. Land O'Lakes, in association with Amerchol, has pioneered the use of milk fat fractions in cosmetics. The first product, called Cremoral, was launched into the marketplace in 1993 (139). The major factor that has stimulated renewed interest in using milk fat for technochemicals and other nonfood applications in the United States has been the significant decline in price, especially relative to alternative fats.

It is doubtful if any single market for milk fat can be found that will compensate for the decrease in butter sales. Rather, it will be necessary to look for a large number of relatively small outlets. If this is to be done, the dairy industry must understand the detailed structure of milk fat and establish the functional properties of its constituent fractions. This approach has been applied with considerable success to the protein fraction of milk. It may be just as rewarding when applied to milk fat.

MARKETING

Production and consumption of butter continue a long-declining trend (Figure 1.20). A dramatic shift occurred, starting in 1985, to the table spreads category of products (less than 80% fat) from full-fat butter, margarine, and blends (88). The spreads category encompasses all non–Standard of Identity table spreads (i.e., 0–79.9% fat).

In a 1984 survey, the most important barriers to increased butter sales were listed in the following order (88):

   Price (opinion of an overwhelming majority when butter is compared with margarine).
   Health (negative consumer attitudes toward cholesterol and saturated fats are increasing).
   Poor spreadability.
   Inadequate promotional spending.
   Product innovation in margarine and spreads.
   Legislation and regulatory restrictions.

Butter manufacture continues to serve as the safety valve for the dairy industry. It absorbs surplus milk supply above market requirements for other dairy products. Milk not required by the demand for these products overflows into the creamery, is skimmed, and the cream is converted to butter. When the milk supply for other products runs short of their demand, milk normally intended for buttermaking is diverted into the channels where needed. Even though consumption patterns have dramatically changed over the years, the butter industry never fails to take up the slack in the relationship of supply and demand for all other dairy products.

Butter is both an intervention and a market product. To counteract growing stocks, special uses are created within the scope of the milk market organization, which contribute to not having too much butter in common storage. For

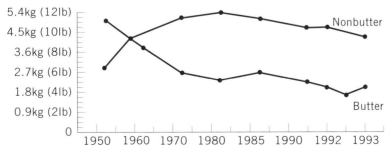

Figure 1.20 Per capita consumption of butter and margarine in the United States, 1968–1993 (140).

some years now, between 300,000 and 400,000 tons of butter have been sold annually at reduced prices to the pastry, ice cream, and chocolate industries in competition with vegetable fats (140).

Despite this and other measures, it was not possible after the milk market organization took effect to prevent an imbalance between butter production and consumption. Even the U.K.'s accession to the EC in 1973 did not bring about a turn in the overall development. Although imports from New Zealand into the new member country were cut from nearly 165,000 to 55,000 tons and although the U.K. turned mainly to France, the Netherlands, and Germany as new supplier countries, the rise of butter prices caused restricted consumption on the English market. In the course of a few years, per capita consumption dropped from 8.8 kg in 1970 to 3.3 kg in 1991 (141).

Between 1977 and 1987, there was an overall decline in the per capita consumption of butter of 16% in 14 European countries (Table 1.17), a similar trend was noted in the United States. By 1993, the per capita butter consumption increased in the United States (Figure 1.20). Price is a strong purchase determinant, and the price of butter has significantly decreased in the United States due to USDA price support policy shifts (89).

A peak in production surplus in the EC was reached in 1986 (Figure 1.21). This was due not only to increasing supplies but also to a notable drop in the consumption of milk fat. The consumer turned to products with a reduced fat content. This trend applies to almost all milk products and has substantially increased the availability of milk fat for butter production (67,142).

Table 1.17 Per capita butter consumption of selected European countries in the International Dairy Federation (kg/per capita) (89)

Country	1977	1987
Ireland	11.9	5.8
United Kingdom	7.8	4.5
France	9.5	6.9
Federal Republic of Germany	6.6	8.1
the Netherlands	3.0	3.8
Luxembourg	7.1	9.1
Belgium	8.1	7.9
Denmark	8.0	7.7
Spain	0.4	0.5
Norway	5.1	4.5
Sweden	3.6	3.1
Finland	12.2	8.2
Austria	5.0	4.7
Switzerland	7.2	4.9
United States	2.3	2.0
New Zealand	14.4	–

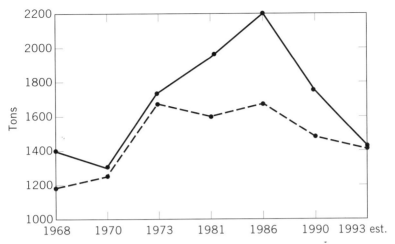

Figure 1.21 EC butter, production, and consumption (141). ——, Production; ----, consumption.

All products sell on a combination of price, perception, and performance. Unfortunately, butter is easily the most expensive of the yellow fats. In terms of perception, all fats are under pressure because of their caloric density. Butter suffers further because it was labeled saturated and has a high cholesterol content; both properties have been the subjects of adverse comment by the nutritional and medical community. As noted in Figure 1.22 (142) the rise in concern for fat and cholesterol in the U.S. market has overshadowed the concern for chemicals and preservatives. Also note that in the 1991–1993 the concern for cholesterol levels decreased significantly while the concern for fat content continued to rise.

The flavor and mouth feel of butter are greatly superior to any other yellow fat, but its physical and rheological properties, particularly its poor spreadability at refrigerated temperature, make butter less attractive to many

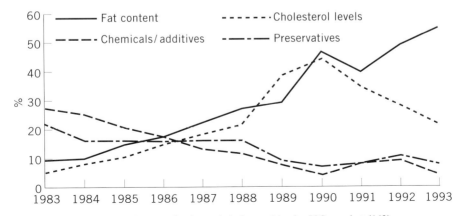

Figure 1.22 Concern for fat and cholesterol in the U.S. market (142).

consumers. The margarine and spread industry can tailor its product in terms of spreadability. As noted earlier, many advances in the ability to alter the texture and rheology of butter have been made, but costs tend to deter manufacturers from applying the technologies for marketplace consumption. Apparently, the consumer demand for spreadability characteristics is inhibited by an unwillingness to pay for the convenience. To those who demand a natural food and appreciate its delicate flavor, butter will remain the preferred product.

REFERENCES

1. O.F. Hunziker, *The Butter Industry,* 3rd ed., Printing Products Corp., Chicago, 1940, pp. 1–23
2. D. Swern, ed., *Bailey's Industrial Oil and Fat Products,* Vol. 1, 4th ed., John Wiley & Sons, Inc., New York, 1979, pp. 392–411.
3. R.G. Jensen and R.W. Clark in N.P. Wong, ed., *Fundamentals of Dairy Chemistry,* 3rd ed., Van Nostrand Reinhold Co., Inc., New York, 1988, pp. 171–213.
4. W.R. Morrison in F.D. Gunstone, ed., *Topics in Liquid Chemistry,* Logos Press, London, 1970, p. 51.
5. R.G. Jensen, *J. Am. Oil Chem. Soc.* **50,** 186–192 (1973).
6. J.R. Brunner in E. Tria and A.M. Scann, eds., *Structural and Functional Aspects of Lipoproteins in Living Systems,* Academic Press, Inc., London, 1969, p. 545.
7. U. Bracco, J. Hidalgo, and H. Bohren, *J. Dairy Sci.* **55,** 165 (1972).
8. J.L. Henderson and E.L. Jack, *Oil Soap* **21,** 90–92 (1944).
9. E.L. Jack and J.L. Henderson, *J. Dairy Sci.* **28,** 65–78 (1945).
10. E.L. Jack and L.M. Smith, *J. Dairy Sci.* **39,** 1–25 (1956).
11. G.A. Gartan and W.R. Duncan, *J. Dairy Sci. Food Agri.* **7,** 734–739 (1956).
12. Anonymous, *Composition and Constants of Natural Fats and Oil by Gas-Liquid Chromatography,* Archer-Daniels-Midland, Minneapolis, Minn., 1961.
13. R.G. Jensen, G.W. Gander, and J. Sampugna, *J. Dairy Sci.* **45,** 329–331 (1962).
14. J.L. Moore, T. Richardson, and C.H. Amundson, *J. Am. Oil Chem. Soc.* **42,** 796–799 (1965).
15. M.F. White and J.B. Brown, *J. Am. Oil Chem. Soc.* **26,** 385–388 (1949).
16. J.D. Hay and W.R. Morrison, *Biochem. Biophys. Acta* 202, 237 (1970).
17. J.L. Iverson, J. Eisner, and D. Firestone, *J. Am. Oil Chem. Soc.* **42,** 1063–1068 (1965).
18. A. Kuksis, *Can. J. Biochem.* **49,** 1245 (1972).
19. A. Kuksis, *J. Chromatog. Sci.* **10,** 53 (1972).
20. L.J. Nutter and O.S. Privett, *J. Dairy Sci.* **50,** 1194 (1967).
21. T.W. Scott, L.J. Cook, and S.C. Mills, *J. Am. Oil Chem. Soc.* **48,** 358–364 (1972)
22. L.F. Edmondson, R.A. Yoncoskie, N.H. Rainey, F.W. Douglas Jr., and J. Bitman, *J. Am. Oil Chem. Soc.* **51,** 72–76 (1974).
23. R.D. Plowman and co-workers, *J. Dairy Sci.* **55,** 204 (1972).
24. G.D. Elsdon and P. Smith, *Analyst,* **52,** 63–66 (1927).
25. G.D. Elsdon and P. Smith, *Oil Fat Ind.* **4,** 103–105, 111 (1927).
26. K.A. Williams, *Analyst* **74,** 504–510 (1949).
27. L. Ahren, L. Björck, T. Raxnikiewictz, and O. Claesson, *J. Dairy Sci.* **63,** 741–745 (1980).
28. B.W. Tharp and S. Patton, *J. Dairy Sci.* **43,** 475 (1960).
29. J.E. Kinsella, S. Patton, and P.S. Dimick, *J. Am. Oil Chem. Soc.* **44,** 202–205 (1967).

References

30. J.E. Kinsella, S. Patton, and P.S. Dimick, *J. Am. Oil Chem. Soc.* **44,** 449–454 (1967).
31. R.C. Lawrence, *J. Dairy Sci.* **30,** 161 (1963).
32. J.E. Langler and E.A. Day, *J. Dairy Sci.* **47,** 1291 (1964).
33. C.R. Brewington, E.A. Caress, and D.P. Schwartz, *J. Lipid Res.* **11,** 355 (1970).
34. V. Antila in *Proceedings of the XXI International Dairy Congress, Moscow 2,* 1982, p. 159.
35. P.W. Parodi, *J. Dairy Res.* **48,** 131 (1981).
36. J. Kisza, K. Batura, B. Staniewski, and W. Zatorski, *Brief Commun.* **1**(1), 352 (1982).
37. A. Boudreau and J. Arul, *Int. Dairy Fed. Bull.* **260,** 7–10, 13–16 (1991).
38. W. Banks, *Int. Dairy Fed. Bull.* **260,** 3–6, 13, 22 (1991).
39. A. Huyghebaert, H. DeMoor, and J. Decatelle in K.K. Rajah and K.J. Burgess, eds., *Milk Fat,* Society of Dairy Technology, Huntingdon, U.K., 1991, pp. 44–61.
40. E. Frede, *Int. Dairy Fed. Bull.* **260,** 12 (1991).
41. S.S. Marschner and J.B. Fine (to General Mills, Inc.), U.S. Pat. 4,804,555 (1989).
42. Anonymous, *R&D Applications, Cholesterol Reduced Fat,* Prepared Foods, Chicago, Ill. 1989, p. 99.
43. Anonymous, *Short Path Molecular Distribution,* CVC Products, Inc., Rochester, N.Y., 1988.
44. U. Bracco, Brit. Pat. 1,559,064 (1980).
45. D.H. Hettinga, *Altering Fat Composition of Dairy Products,* Council Agri. Sci. Technol., Ames, Iowa, 1991.
46. J. Arul, A. Boudreau, J. Makhlouf, R. Tardif, and B. Grenier, *J. Dairy Res.* **55,** 361 (1988).
47. A.R. Keen, Eur. Pat. 0318326A2 (1989).
48. D. Oakenfull and G.S. Sidhu (to Sidoak), Austral. Pat. Appl. 54768/90 and 55112/90 (1990).
49. F. Morel, *Food Process.* **1056,** 19–21 (1990).
50. T.J. Micich, *J. Agr. Food Chem.* **38,** 1839 (1990).
51. Anonymous, *Research Review Special Report 2,* Wisconsin Milk Marketing Board, Madison, Wis., 1989.
52. Anonymous, *Dairy Ind. Int.* **55**(6), 37–38 (1990).
53. Anonymous, *Food Eng.* **61**(2), 83 (1989).
54. H. Chen, S.J. Schwartz, and G.A. Spanos, *J. Dairy Sci.* **75**(10), 2659–2669 (1992).
55. D.H. Hettinga, *Designing Foods,* National Academy Press, Washington, D.C., 1988, p. 292.
56. J. Arul, A. Boudreau, J. Makhlouf, R. Tardif, and T. Bellavia, *J. Am. Oil Chem. Soc.* **65,** 1642 (1988).
57. S. Harlander, *J. Am. Oil Chem. Soc.* **65**(11), 1727 (1988).
58. D. Beitz, personal communication, 1991.
59. P.A.E. Cant in Ref. 39, pp. 122–135.
60. *Code of Federal Regulations,* Title 7, Part, 58, Subpart P, U.S. Government Printing Office, Washington, D.C., 1983.
61. D.H. Hettinga in Y.H Hui ed., *Encylopedia of Food Technology,* Vol. 1, J. Wiley & Sons, Inc., New York 1993, pp. 231–237.
62. M.M. Chrysam in T.H. Applewhite, ed., *Bailey's Industrial Oil and Fat Products,* Vol. 3, John Wiley & Sons, Inc., New York, 1985.
63. R. Wood, K. Kubena, B. O'Brien, S. Tseng, and G. Martin, *J. Lipid Res.* **34**(1), 1–11 (1993).
64. *International Dairy Federation Standard 68A,* Brussels, IDF, 1977.
65. *FAO/WHO Codex Standard A2,* Codex Alimentarius, Rome, 1986.
66. M. Collomb and M. Spahni, *Mitteil. Gebiete Lebensmit. Untersuch. Hyg.* **82**(6), 615–662 (1991).

67. T. Sato, S. Kawano, and M. Iwanoto, *J. Dairy Sci.* **73**(12), 3408–3413 (1990).
68. M. Lucisano, E. Casiraghi, and C. Pompei, *J. Texture Stud.* **20**(3), 301–315 (1989).
69. C.J. Strugnell, *J. Food Sci. Technol.* **14**(2), 128–129 (1990).
70. L. Forman, P. Stern, and E. Matouskova, *Milchwissenschaft* **44**, 761–764 (1989).
71. B. Schaffer and S. Szakaly, *Brief Commun.* **1**(1), 337 (1982).
72. R. Norris, *Int. Dairy Fed. Bull.* **260**(5), 23–25 (1991).
73. T. Richardson and M. Korycka-Dahl in P.F. Fox, ed., *Developments in Dairy Chemistry*, Vol. 2, *Elsevier Applied Science*, London, 1983.
74. M. Schroder, *Ger. Pat. DF 36 23 313 A1* (1988).
75. F. Hulsemeyer, *Kieler Milchwirtschaft. Forschung.* **42**(2), 115–117 (1990).
76. J. Bird in Ref. 39, pp. 30–36.
77. B.D. Dixon, *Int. Dairy Fed. Bull.* **204**, 3–5, (1986).
78. M. Kimenai, *Int. Dairy Fed. Bull.* **204**, 16 (1986).
79. L.H. Wiedermann, *J. Am. Oil Chem. Soc.* **55**(11), 823 (1978).
80. M. Schweizer, *Int. Dairy Fed. Bull.* **204**, 10 (1986).
81. P. Bjerre in Ref. 39, pp. 63–73.
82. F.H. McDowall, *The Butter Maker's Manual*, New Zealand University Press, Wellington, 1953.
83. P. Friis, *Brief Commun.* **1**(1), 326 (1982).
84. Anonymous, *Gold'n Flow Continuous Buttermaking Method*, Bulletin **G-493**, Cherry-Burrell Corp., Chicago, 1954.
85. M. Kawanari, K. Okamoto, Y. Honda, and M. Jukushima, *J. Jpn. Soc. Food Sci. Technol.* **39**(3), 227–232 (1992).
86. D.H. Kleyn, *Food Technol.* **46**(1), 118–121 (1992).
87. J. Toma, J.D. Curtis, and C. Sobotor, *Food Technol.* **42**, 93–95 (1988).
88. Anonymous, *Int. Dairy Fed. Bull.* **170** (1984).
89. W.P. Charteris and M.K. Keogh, *J. Soc. Dairy Technol.* **44**(1), 3–8 (1991).
90. I. Macnab, *Modern Dairy* **68**(3), 23 (1989).
91. S. Gulyayev-Zaitsev, V. Dobronos, J. Berezansky, and N. Zhirnaya, *Brief Commun.* **1**(1), (1982).
92. O. Gerstenberg, *Dairy Ind. Int.* **53**(11), 28 (1988).
93. K. Anderson, *Int. Dairy Fed. Bull.* **260**, 17–18 (1991).
94. A. Pedersen, *Scand. Dairy Inform.* **4**(3), 74–75 (1990).
95. V.K. Babayan, G.L. Blackbrun, and B.R. Bistrian (to N.E. Decaness Hospital) U.S. Pat. 4,952,606 (1990).
96. G. Rosenbaum, *Dairy Foods* **93**(7), 13 (1992).
97. F. Graves (to Land O'Lakes, Inc.) U.S. Pat. 4,425,370 (Jan. 10, 1984).
98. D. Antenore, D. Schmadeke, and R. Steward (to Land O'Lakes, Inc.) U.S. Pat. 4,447,463 (May 8, 1984).
99. B. Schaffer and S. Szakaly, *Brief Commun.* **1**(1), 363 (1982).
100. A. Giles, *J. Am Oil Chem. Soc.* **65**, 1708–1712 (1988).
101. G. Van den Berg in *Proceedings of the XXI International Dairy Congress, Moscow 2*, IDF, Brussels, 1982, p. 153.
102. T. Siek and J. Lindsay, *J. Dairy Sci.* **53**, 700 (1970).
103. J.E. Kinsella, *Food Technol.* **29**(5), 82–98 (1975).
104. N. Cullinane and D. Eason, *Brief Commun.* **1**(1), 349 (1982).

105. B.K. Mortensen and K. Jansen, *Brief Commun.* **1**(1), 334 (1982).
106. S. Berntsen, Int. Pat. Appl. WO 92/19111 A1 (1992).
107. A.M. Fearson and D.E. Johnston, *Dairy Ind. Int.* **53**(10), 25–29 (1988).
108. A. Petersen, *Int. Dairy Fed. Bull.* **260,** 12 (1991).
109. E. DeJong, G. vanden Berg, *Voedingsmiddelenttechnologie* **19**(5), 35–38 (1986).
110. K.E. Kaylegian and R.C. Lindsay, *J. Dairy Sci.* **75**(12), 3307–3317 (1992).
111. J. Bumbalough (to Wisconsin Milk Marketing Board, Inc.) U.S. Pat. 4,839,190 (June 13, 1989).
112. F.M. Fouad, F.R. Vandevoort, W.D. Marshall, and P.G. Farrell, *J. Am. Oil Chem. Soc.* **67**(12), 981–988 (1990).
113. J. Makhlouf, J. Arul, A. Bordreau, P. Verret, and M.R. Sahasrabudhe, *J. Can. Inst. Food Sci. Technol.* **20**(4), 263–245 (1987).
114. J.B. Mickle, *Tech. Bull. T-83,* Oklahoma Agricultural Experiment Station, Stillwater, 1960.
115. W. Banks and W.W. Christie, *Outlook Agr.* **19**(1), 43–47 (1990).
116. R.P. Middaugh, R.J. Baer, D.P. Casper, D.J. Schingoethe, and S.W. Seas, *J. Dairy Sci.* **71**(12), 3179–3187 (1988).
117. G.A. Stegeman, R.J. Baer, D.J. Schingoethe, and D.P. Casper, *J. Dairy Sci.* **75**(4), 962–970 (1992).
118. H. Rohm and S. Raaber, *J. Food Sci* **57**(3), 647–650 (1992).
119. K.K. Rajah and K.J. Burgess, in Ref. 39, pp. 37–43.
120. R. Barts, *Int. Dairy Fed. Bull.* **260,** 19–20 (1991).
121. R.A. Wibley in Ref. 39, pp. 136–147 (1991).
122. Anonymous, *Technical Guide for the Packaging of Milk and Milk Products,* Document 143, 2nd ed., IDF, Brussels, 1982.
123. R.G. Mansour, Int. Pat. WO 9202 144 (Feb. 20, 1992).
124. R. Egli (to Egli AG), *Int. Pat. WO 89/10689 A1* (1989).
125. Proceedings of IDF Seminar, Singapore, *Int. Dairy Fed. Doc.* **142** (1982).
126. Anonymous, IDF Study 74, Brussels, 1974.
127. M. Zillen, *Document 116,* IDF, Brussels, 1977, pp. 1–4.
128. W.P. Carteris and M.K. Keogh, *J. Soc. Dairy Technol.* **44,** 3 (1991).
129. G. Behrens, *Document 224,* IDF Brussels, 1988, pp. 8–9.
130. B.K. Wadhwa and M.K. Jain, *Indian J. Dairy Sci.* **44**(6), 372–374 (1992).
131. R. Spieler, *Document 149,* IDF, Brussels, 1980, pp. 132–135.
132. M. Gordon, *Int. Dairy Fed. Bull.* **260,** 20 (1991).
133. P. Sitaram and S.K. Gupta, *Agr. Rev.* **9**(2), 81–104 (1988).
134. B. Guenault, *Dairy Ind. Int.* **57**(6), 32 (1992).
135. L.S. Girsh (to Immunopath Profile, Inc.) U.S. Pat. 5,112,636 (May 12, 1992).
136. F.X. Malcata, C.G. Hill and C.H. Amundson, *Biotechnol. Bioeng.* **39**(11), 1097–1111 (1992).
137. G. Comi, S. D'Aubert and M. Valenti, *Ind. Aliment.* **29**(281), 358–361 (1990).
138. E.J. Mann, *Dairy Ind. Int.* **2,** 19 (1992).
139. L. Kintish, *J. Soap Cosmetics Chem. Special.* **69**(2), 51 (1993).
140. U.S. Department of Commerce, Bureau of the Census, *Statistical Abstracts of the United States,* 113th ed., U.S. Government Printing Office, Washington, D.C., 1993.
141. E. Hetzner and P. Jacknik, *Int. Dairy Fed. Bull.* **293,** 19 (1993).
142. Anonymous, *Trends in the U.S. Consumer Attitudes and the Supermarket,* Food Marketing Institute, Washington, D.C., 1993.

2
Margarines and Spreads

This chapter will discuss margarine as well as vegetable oil margarine and butter substitutes containing less than 80% fat. These products will be generically referred to as table spreads. The table spread category in the United States has undergone great changes in this century and continues to change rapidly even today. One hundred years ago, annual butter consumption was 8.6 kg (19 lb) per person. Now it is less than 2.2 kg (5 lb), and per capita table spread consumption is over 4.5 kg (10 lb). Historical production and consumption data for table spreads and butter are shown in Table 2.1. Although consumption appears to have peaked around 1980 and the average amount of fat in these products has declined in recent years, table spreads still make a significant contribution to the fat content of the American diet. Approximately 75% of production is sold at retail, with 19% going to food service and 6% for bakery and other industrial uses.

The market for table spreads, which was originally driven primarily by cost relative to butter, has, in the past 40 years, been greatly impacted by health claims relative to content of cholesterol, polyunsaturated fat, saturated fat, and also including most recently, total fat and trans fatty acids. Therefore, since composition, as well as price and taste, is of major importance to the consumer, regulations regarding labeling and advertising have become a driving force in formulation of these products and the oil blends used therein. The resulting proliferation of table spreads available today is somewhat overwhelming and often confusing to the consumer.

HISTORICAL DEVELOPMENT OF MARGARINE (1–4)

Margarine was patented and first manufactured in 1869 by Hippolyte Mege Mouries, a French chemist. The product was developed to meet butter shortages caused by the increasing urban population during the Industrial Revolution, as well as the need for a table spread with satisfactory keeping quality

Table 2.1 **Production and consumption of margarine and butter**[a]

Year	Annual Production, million kg (million lb)		Average Per Capita Consumption, kg (lb)	
	Margarine[b]	Butter	Margarine[b]	Butter
1930	147.7 (325.7)	737.5 (1,625.9)	1.2 (2.6)	8.0 (17.6)
1935	173.1 (381.6)	758.2 (1,671.5)	1.4 (3.0)	8.0 (17.6)
1940	145.3 (320.4)	833.2 (1,836.8)	1.1 (2.4)	7.7 (17.0)
1945	278.5 (614.0)	618.6 (1,363.7)	1.9 (4.1)	4.9 (10.9)
1950	425.0 (937.0)	628.9 (1,386.4)	2.8 (6.1)	4.8 (10.7)
1955	596.4 (1,314.9)	628.8 (1,386.2)	3.7 (8.2)	4.1 (9.0)
1960	768.9 (1,695.2)	622.8 (1,372.9)	4.3 (9.4)	3.4 (7.5)
1965	863.6 (1,903.9)	599.1 (1,320.7)	4.5 (9.9)	3.0 (6.5)
1970	1011.7 (2,230.5)	517.1 (1,140.0)	5.0 (11.0)	2.4 (5.3)
1975	1093.1 (2,409.8)	446.3 (984.0)	5.0 (11.0)	2.1 (4.7)
1980	1158.1 (2,553.2)	519.4 (1,145.0)	5.1 (11.2)	2.0 (4.5)
1985	1180.9 (2,603.3)	566.1 (1,248.0)	4.9 (10.8)	2.2 (4.9)
1990	1255.2 (2,767.1)	590.6 (1,302.0)	4.9 (10.9)	2.0 (4.4)
1991	1266.3 (2,791.7)	606.0 (1,336.0)	4.8 (10.6)	1.9 (4.2)
1992	1249.8 (2,755.2)	629.2 (1,365.0)	5.0 (11.0)	1.9 (4.2)
1993	1247.4 (2,750.0)	598.3[c] (1,319.0[c])	n.a.[d]	2.1[c] (4.6[c])

[a] Data courtesy of the National Association of Margarine Manufacturers. One kg equals 2.2 lb.
[b] Includes spread products beginning in 1975.
[c] Preliminary figures.
[d] Not available.

for the armed forces. The original process was designed to imitate production of butterfat by the cow. Fresh tallow was subjected to a low-temperature rendering with artificial gastric juice and slowly cooled to approximately 26°C (80°F) to partially crystallize the fat. The olein, a soft yellow semifluid fraction obtained in about 60% yield, was then dispersed in skim milk along with cow udder tissue. The emulsion was agitated for several hours and cold water added to the churn, causing the fat to solidify. The water was drained, and the granular mass that remained was kneaded and salted.

During the late nineteenth century, some margarines were prepared from lard or unfractionated beef suet to which liquid oils such as cottonseed or peanut were added to reduce the melting point of the blend. In the early 1900s some 100% vegetable oil margarines were formulated with coconut and palm kernel oils. Examples of these blends and the early margarine manufacturing processes are described by Clayton (5). Although hydrogenation came into practice around 1910, particularly in Europe, it was not used extensively in U.S. margarine manufacture until the 1930s. At that time the use of lauric oils became politically unpopular and tariffs were imposed.

In addition to the use of new oils, significant advances were made in the manufacturing process. Peptic digestion of fat and the use of udder extracts soon were abandoned. Pasteurization allowed the milk to be cultured in order to develop a more buttery flavor. The invention of dry chilling using a metal drum containing circulating brine resulted in improved cleanliness and reduced fat and milk losses. Regardless of whether ice or a chill roll was used, however, the product still had to be worked subsequently to achieve consistency. It was not until about 1940 that the first closed continuous systems were used for margarine manufacture.

Legislation has played an important role in the evolution of margarine. When the product was introduced in the United States in 1874, dairy farming was expanding more rapidly than the needs of the population. Hostility from the large number of dairy farmers set the tone for almost a century of antimargarine legislation and taxation of the lower priced spread. Adulteration of butter with margarine, and later the use of imported tropical oils, also added to the call for government intervention. The first tax was levied in 1886, and in 1902 a much more severe tax and producer-licensing fees were imposed on colored margarine. Many states also passed laws prohibiting the sale of colored margarine. In 1941 margarine gained recognition as a food in itself with the adoption of a Federal Standard of Identity that defined the product and provided for vitamin fortification. In response to both public pressure generated by high butter prices following World War II and expanding soybean and cottonseed oil farming interests, federal taxes were abolished by the Margarine Act of 1950. This legislation also stipulated that the product could not be sold at retail in units greater than 0.4536 kg (1 lb). Most states were quick to follow the lead of the federal government. However, it was 1967 before Wisconsin became the last state to repeal a law prohibiting colored margarine, and the final state margarine tax was repealed by Minnesota in 1975.

RECENT U.S. TRENDS

In the early 1950s almost all consumer margarine was a stick variety usually sold in packages of 113.4 g (quarter-pound) prints. With most restrictions on colored margarine removed, sales continued to rise and per capita consumption surpassed that of butter in 1957. Sales increases stimulated important technological progress and new product development (Table 2.2). Spreadable margarine, polyunsaturated margarine, and low-fat products were developed to satisfy the desire for convenience, nutritional awareness, and weight consciousness of the consumer.

The first major departure from the past was the introduction of soft margarine spreadable from the refrigerator and packaged in tubs. This soft, full-fat

68 Margarines and Spreads

Table 2.2 New table spread products introduced in the United States since 1950

Product	Year Introduced
Spreadable stick margarine	1952
Whipped margarine	1957
Corn oil margarine (high polyunsaturates)	1958
Soft margarine	1962
Liquid margarine	1963
Diet margarine (40% fat)	1964
Spread (60% fat)	1975
Whipped spread	1978
Butter blends (widespread distribution)	1981
Improved 40% fat spreads[a]	1986
Lower fat spreads (20% fat)	1989
Non-fat spread	1993

[a] Containing gelling agents.

product captured one-fourth of the margarine market by 1973. Today stick table spreads account for slightly less than half of the total table spread market. The most significant recent trend is away from margarine (80% fat) to spreads containing lower fat levels. Originally introduced at 60% fat, since 1980 spreads containing from 75% to less than 5% fat have appeared in the marketplace. The products are sold in soft, stick, liquid, and soft whipped forms. Market share increased from under 5% in 1976 to more than 15% in 1983, and, in the early 1990s, this trend has accelerated to the point that there are very few full-fat margarine products available in the United States. Reasons for the growth in spread production are (*1*) low price, (*2*) availability in sizes greater than 0.4536 kg (1 lb), and (*3*) fewer calories and lower fat than margarine. Much of the recent rapid shift from margarine to spread production is attributable to manufacturers' cost reduction efforts in times of rising vegetable oil prices.

The table spread market is composed of several segments differentiated by image and price. At the high end of the price range are products perceived either as being healthy or as having a superior buttery taste. The healthy image products generally contain liquid oils such as corn, sunflower, or canola and have been promoted with statements relating to cholesterol, total fat, or saturated fat contents. For the last 10 years this segment has held at 16–17% of the market. Unsalted products are not popular and account for less than 1.5% of table spread volume. The buttery image or blend segment has sharply increased during the last decade and consists of spreads that, by their name or positioning, merely convey a strong dairy taste perception to the consumer, as well as spreads that actually contain various amounts of butter. The branded value grouping, which accounts for about 50% of the retail market, includes brands that are nationally advertised and priced lower than the premium

Table 2.3 Categorization of the 1993 U.S. margarine/spread market[a]

Form	Segment				Total Form
	Healthy Image	Buttery Image	Branded Value	Other Value	
Stick	7.6	6.2	26.9	8.5	49.2
Soft	9.1	9.3	22.0	7.9	48.3
Liquid	0.3	0.6	1.5	0.1	2.5
Total segment	17.0	16.1	50.4	16.5	—

[a] Data derived from Information Resources, Inc. InfoScan syndicated scanner databases representing scanner data collected from IRI's sample of grocery stores for the 52-week period ending December 26, 1993, and projected for the total United States for that period.

segments discussed. The remaining products, which usually have the lowest everyday price, consist of regional and private label store brands. Market share of the various forms is shown in Table 2.3. The 1993 retail table spread market (including butter blends, but not butter) was just over 907 million kg (2 billion lb). Food service volumes totaled 250 million kg (546 million lb), while bakery and industrial uses other than food service accounted for about 86 million kg (190 million lb).

REGULATORY STATUS IN THE UNITED STATES

The following section contains a description of the legal requirements for composition and labeling of margarine and table spreads. Such information is essential for formulating products to meet specific claims; however, because it contains considerable detail, the reader wanting only an overview may wish to skip all or part of this section.

3.1 Standard of Identity

There are two standards of identity for margarine in the United States. Vegetable oil margarines are regulated by the Food and Drug Administration (FDA), and animal fat and animal–vegetable margarines are subject to federal meat inspection regulations of the United States Department of Agriculture (USDA). The standards are similar but not identical. The FDA standard was revised in 1973 (6) and the USDA standard in 1983 (7) to conform more closely to the international standard as adopted by the Food and Agriculture Organization/World Health Organization Codex Alimentarius Commission. In 1993, a new Codex standard for fat spreads was proposed by the commission.

70   Margarines and Spreads

Finalization of this standard, which includes margarine, butter, and lower fat spreads, is not expected for several years. The following is the current FDA standard (8) including recent revisions allowing the use of marine oils (9) and removing previous emulsifier restrictions (10):

(a) Margarine (or oleomargarine) is the food in plastic form or liquid emulsion, containing not less than 80 percent fat determined by the method prescribed in "Official Methods of Analysis of the Association of Official Analytical Chemists," 13th Ed. (1980), section 16.206, "Indirect Method," under the heading "Fat [47]—Official Final Action," which is incorporated by reference. Copies may be obtained from the Association of Official Analytical Chemists, 2200 Wilson Blvd., Suite 400, Arlington, VA 22201-3301, or may be examined at the Office of the Federal Register, 800 North Capitol Street, NW., suite 700, Washington, DC 20001. Margarine contains only safe and suitable ingredients as defined in section 130.3(d) of this chapter. It is produced from one or more of the optional ingredients in paragraph (a)(1) of this section, and one or more of the optional ingredients in paragraph (a)(2) of this section, to which may be added one or more of the optional ingredients in paragraph (b) of this section. Margarine contains vitamin A as provided for in paragraph (a)(3) of this section.

(1) Edible fats and/or oils, or mixtures of these, whose origin is vegetable or rendered animal carcass fats or any form of oil from a marine species that has been affirmed as GRAS or listed as a food additive for this use, any or all of which may have been subjected to an accepted process of physico-chemical modification. They may contain small amounts of other lipids such as phosphatides, or unsaponifiable constituents and of free fatty acids naturally present in the fat or oil.

(2) One or more of the following aqueous phase ingredients:
   (i) Water and/or milk and/or milk products.
   (ii) Suitable edible protein including, but not limited to, the liquid, condensed, or dry form of whey, whey modified by the reduction of lactose and/or minerals, nonlactose containing whey components, albumin, casein, caseinate, vegetable proteins, or soy protein isolate, in amounts not greater than reasonably required to accomplish the desired effect.
   (iii) Any mixture of two or more of the articles named under paragraphs (a)(2) (i) and (ii) of this section.
   (iv) The ingredients in paragraphs (a)(2) (i), (ii), and (iii) of this section shall be pasteurized and then may be subjected to the action of harmless bacterial starters. One or more of the articles designated in paragraphs (a)(2) (i), (ii), and (iii) of this section is intimately mixed with the edible fat and/or ingredients to form a solidified or liquid emulsion.

(3) Vitamin A in such quantity that the finished margarine contains not less than 15,000 international units per pound.

(b) Optional ingredients: (1) Vitamin D in such quantity that the finished oleomargarine contains not less than 1,500 international units of vitamin D per pound.
(2) Salt (sodium chloride); potassium chloride for dietary margarine or oleomargarine.
(3) Nutritive carbohydrate sweeteners.
(4) Emulsifiers.
(5) Preservatives including but not limited to the following within these maximum amounts in percent by weight of the finished food: Sorbic acid, benzoic acid and their sodium, potassium, and calcium salts, individually, 0.1 percent, or in combination, 0.2 percent, expressed as the acids; calcium disodium EDTA, 0.0075 percent; propyl, octyl, and dodecyl gallates, BHT, BHA, ascorbyl palmitate, ascorbyl stearate, all individually or in combination, 0.02 percent; stearyl citrate, 0.15 percent; isopropyl citrate mixture, 0.02 percent.

(6) Color additives. For the purpose of this subparagraph, provitamin A (beta-carotene) shall be deemed to be a color additive.
(7) Flavoring substances. If the flavoring ingredients impart to the food a flavor other than in semblance of butter, the characterizing flavor shall be declared as part of the name of the food in accordance with section 101.22 of this chapter.
(8) Acidulants.
(9) Alkalizers.
(c) The name of the food for which a definition and standard of identity are prescribed in this section is "margarine", or "oleomargarine".
(d) Label declaration. Each of the ingredients used in the food shall be declared on the label as required by the applicable sections of parts 101 and 130 of this chapter. For the purposes of this section the use of the term "milk" unqualified means milk from cows. If any milk other than cow's milk is used in whole or in part, the animal source shall be identified in conjunction with the word milk in the ingredient statement. Colored margarine shall be subject to the provisions of section 407 of the Federal Food, Drug, and Cosmetic Act as amended.

3.2 Label Requirements

Detailed labeling regulations for margarine are set forth in the Code of Federal Regulations (11). In general, the label must include the product name, net weight, name and address of the manufacturer or distributor, an ingredient statement, serving size, number of servings per package, and nutritional information. The name *margarine* or *oleomargarine* must appear in lettering at least as large as any other on the label. For products resembling margarine but that contain less than 80% fat, the product name should include the term *spread*, the total percentage of fat, and a listing of each fat ingredient in order of predominance or a generic term such as *vegetable oil*, e.g., "60% vegetable oil spread." This nomenclature, proposed in 1976 (12), generally has been followed by the industry and a final regulation was not published. In January 1993 the FDA promulgated extensive labeling regulations implementing the Nutrition Labeling and Education Act of 1990 (13). Clerical corrections (14) and technical amendments (15) to these regulations have also been published and all of these will be incorporated in subsequent editions of the Code of Federal Regulations. All food products labeled on or after May 8, 1994, must be in compliance. Included in these regulations are provisions for use of a standardized term such as *margarine* to describe foods that do not comply with the standard of identity because of a deviation described by an expressed nutrient content claim that has been defined by FDA regulation (e.g., reduced fat margarine, see next section). Other than the deviation described by the claim, the product must comply with the margarine standard in all respects with the exception that safe and suitable ingredients not provided for in the standard may be added to improve functional characteristics so that these are not inferior to those of margarine.

A complete listing of all ingredients in order of predominance is required. If nonstandard ingredients are contained in a product labeled as margarine described by a nutrient claim, these ingredients must be so noted in the ingredient statement. Fats must be listed according to type and must be declared as hydrogenated or partially hydrogenated if applicable. The individual fats may be placed in order of predominance in the listing or they may be grouped together in parenthesis and prefixed by "vegetable oil blend." The listing of fats may not be preceded by "and/or" or "may contain" if, by weight, fat constitutes the predominant ingredient. Dry or condensed dairy ingredients (e.g., whey powder, dry milk, or buttermilk) may be declared as such or as the reconstituted ingredient. Any water in excess of that required for standard reconstitution also must be declared at the appropriate place in the ingredient statement. The specific form of added vitamins must be listed. If colors or flavors are listed by specific name such as "colored with beta-carotene," the term "artificial" is not required. Flavors usually are not labeled specifically, and "artificially flavored" appears in the ingredient statement. Preservatives must be declared by their common names together with a statement of intended use, for example, "sodium benzoate as a preservative" or "potassium sorbate as a mold inhibitor."

All products must be labeled with a nutrition panel. Details of the content and format are prescribed (13,15). All information in the nutrition panel is based on a single serving of 15 mL (one tablespoon) rounded to the nearest gram. For unwhipped 80% fat margarine this amount is 14 g; however, this weight may vary for whipped products and some low-fat spreads if the density is significantly different than that of margarine.

As regards fat labeling, the nutrition panel must include total fat, calories from fat, saturated fat, and cholesterol. Declaration of polyunsaturated or monounsaturated fat content are mandatory only if the other is declared or if a claim about fatty acids or cholesterol is made, unless the product meets the criteria for a *fat-free* claim. All fat amounts are expressed to the nearest 0.5 g increment below 5 g, to nearest gram above 5 g, and as zero if a serving contains less than 0.5 g. Saturated fat includes all fatty acids that do not contain a double bond. Polyunsaturated fat includes only *cis, cis*-methylene-interrupted fatty acids, and monounsaturated fat is defined as cis-monounsaturated fatty acids. Saturated, polyunsaturated, or monounsaturated fat is declared as weight of the fatty acids, while fat is expressed as the weight of triglyceride based on the total amount of lipid fatty acids.

3.3 Label Claims

Nutrient Content Claims. The following is a brief description of nutrient content claims of most relevance to the positioning of margarine and spread products. For a more complete discussion the reader is referred to the regulations (13,15). Use of the terms *fat free, low fat,* and *reduced fat* or their equivalents have been fully defined by the new regulations. A product may be called *fat free* or *no fat* if it contains less than 0.5 g of fat per serving. If the food contains an ingredient generally understood by consumers to contain fat, that ingredient must be asterisked in the ingredient statement with a following explanation that the ingredient contributes a dietarily insignificant amount of fat. A claim of 100% fat free may only be made if the product contains no added fat. FDA has indicated (15) that the terms "fat free" and "margarine" cannot be used together since standardized foods must contain a significant level of characterizing ingredients, in this case, fat. Such a product can be called a fat-free spread. A margarine or spread may be called *low fat* if it contains 3 g or less of fat per serving and per 50 g, thus limiting its use to spreads containing 6% or less fat. The term *saturated fat free* may be used if the margarine or spread contains less than 0.5 g of saturated fatty acids and less than 0.5 g of trans fatty acids per serving. To make the claim *low in saturated fat* the food must contain 1 g or less of saturated fat per serving and derive not more than 15% of its calories from saturated fat. A *reduced fat* or *reduced saturated fat* claim may be made provided the reduction is at least 25%. The term *light* or *lite* may be used only if the fat content is reduced by at least 50% compared to an appropriate reference product if the spread derives 50% or more of its calories from fat. The minimum reduction is 33% if the spread derives less than 50% of its calories from fat. For *reduced, light,* or comparative claims, the identity of the reference food and the percent reduction must be declared in immediate proximity to the most prominent claim. Quantitative information comparing the grams of fat or saturated fat per serving with that of an appropriate reference food that it replaces also must be declared in proximity to the claim or on the information panel.

A *no cholesterol* or equivalent claim may be made if the product contains less than 2 mg of cholesterol and 2 g or less of saturated fat per serving. If the spread contains more than 13 g of fat per 50 g, the amount of total fat per serving must be stated in immediate proximity to the most prominent claim on each panel except the panel that bears nutrition labeling. If the food contains an ingredient understood by consumers to contain cholesterol, this ingredient must be asterisked in the ingredient list followed by a statement that this ingredient contributes a negligible amount of cholesterol. The requirements for a *low-cholesterol* claim are similar with the exception that the spread must contain less than 20 mg of cholesterol and less than 2 g of saturated fat per serving. *Cholesterol reduced* or comparative claims require declaration of the identity of the reference food, the percent reduction, and quantitative information regarding the reduction in immediate proximity to the most prominent label claim, e.g.: "Cholesterol free margarine, contains 100% less cholesterol than butter, no cholesterol compared with 30 mg in one serving of butter. Contains 11 g of fat per serving."

Health Claims. Approved health claims with potential applicability to margarines or spreads are the relationships between fat and cancer, saturated fat and cholesterol and risk of heart disease, and sodium and hypertension. A food that contains more than 13 g of fat, 4 g of saturated fat, 60 mg of cholesterol, or 480 mg of sodium per serving or per 50 g is disqualified from making any health claim. In addition the product must qualify as "low fat" for the cancer claim, "low fat, low saturated fat," and "low cholesterol" for the coronary heart disease claim, and "low sodium" for the hypertension claim. The first two claims therefore would be limited to spreads containing 6% or less fat, and the hypertension claim would require the spread to contain less than 140 mg of sodium per 50 g (approximately 0.7% salt).

PRODUCT CHARACTERISTICS

The consumer-directed functional aspects of spreads and margarines, which primarily depend on fat level, type of fat, and stability of the emulsion, are spreadability, oiliness, and melting properties. Spattering, which is a concern for products intended for pan frying, is discussed in Section 6.2.

4.1 Spreadability

Spreadability is one of the most highly regarded attributes of margarine, perhaps second only to flavor. Products with a solid fat index (SFI) of 10–20 at serving temperature were found to be optimal on a consumer panel (16). The standard method (17) for evaluation of hardness of fatty materials utilizes a cone penetrometer. For some products, hardness measurements may not correlate well with SFI since, in addition to the amount of solid fat, the fat crystal network, which is processing dependent, is also important rheologically (18). The penetration is the distance, in units of 0.1 mm, traveled by a standard cone in 5 s after its release on the surface of the product. The values may be converted mathematically to a hardness index (19) or to a yield value (20) that are independent of cone weight. In an assessment of hedonic spreadability preferences of butter, margarines, and spreads as a function of temperature, correlation with penetration measurements indicated optimum spreadability in the yield value range of 30–60 kPa (21). Dynamic techniques involving motor-driven penetrometers (22–24), extrusion devices (25,26), viscometers (27,28), and a mechanical spectrometer (29) also have been used to evaluate margarine consistency. Although the action of a penetrometer may seem quite different from assessment of the spreadability with a knife, both techniques evaluate the force required to bring about a significant deformation, and cone penetrometer readings have been found, in general, to correlate with spreadability (30). This method has the advantage of low equipment cost, minimal sample preparation, and reproducible results. Smoothness and brittleness, which depend on the crystal network and are not measured with the cone penetrometer, were measured by means of an Instron compression test (31). In a study of North American stick margarines (32) compression of cylindrical samples was found to be the most sensitive method for detecting differences in textural attributes. Constant speed penetration was the next most sensitive, while the cone penetrometer was least sensitive. These methods were also used to evaluate a series of North American soft margarines (33). Consistency measurements have been reviewed by deMan (19).

Instrumental methods are used by manufacturers to ensure product uniformity and for comparison to competitive products. The sample is generally evaluated at a single temperature, most often in the 4.4–10°C (40–50°F) range. The results are meaningful only at the measurement temperature. For example, two margarines with the same solids content at 7.2°C (45°F) may have much different solids–temperature slopes at that point; one has significantly greater solids at 4.4°C (40°F) and less solids at 10°C (50°F) than the other. One must consider this when attempting to relate spreadability measurements to consumer perception in the home, where storage and use conditions are variable.

4.2 Oil Separation

Oil-off occurs in margarine when the matrix of fine fat crystals is no longer of sufficient size or character to be able to enmesh all of the liquid oil. The problem is most serious for stick products as the outside of the inner wrappers may become oil soaked and, if severe, oil will leak out of the package. Prints are also the most susceptible to oil-off because of the pressures to which the product is subjected when pallets or individual packages are stacked. If the margarine is to be merchandised outside of dairy cases, the 21.1°C (70°F) SFI should be specified as high as possible consistent with 10 and 33.3°C (50 and 92°F) requirements. Soft margarines in bowls are not as much a problem because the package supports and contains the product. Emulsion damage, that is, coalescence and settling of the milk phase, seldom will be significant as long as 3–4% solid fat remains. Oil-off testing is most often conducted by placing a margarine sample of defined geometry and weight on a wire screen (34) or on filter paper (35) at a temperature of 26.7°C (80°F), or sometimes greater, for a period of 24–48 h. The oil exuded into the filter paper or through the screen is then measured. Another test used for determining structural stability, most often for stick margarine, is slump. This involves placement of a standard size cube of margarine at 23.9–29.4°C (75–85°F) for several hours. Deformation of the cube is graded according to visual reference to a standard slump chart (36). Results of these types of test are only directional; products are more appropriately evaluated in the package under actual distribution conditions. The harshest of such conditions would generally be a stacked, out-of-dairy-case display as can be observed in some markets.

4.3 Melting

A high-quality table margarine melts quickly with a cooling sensation on the palate. Flavor and salt components of the aqueous phase are immediately perceptible by the taste buds, and there is no lingering greasiness or waxiness.

The factors affecting these qualities are the melting profile of the fat, the tightness of the emulsion, and the storage conditions of the finished product. In order for a margarine to melt cleanly without seeming gummy or waxy, it should be completely melted at body temperature and contain less than 3.5% solid fat at 33.3°C (92°F). A cool tasting product, typified by butter, results from the almost instantaneous absorption of the heat of crystallization due to a steep melting profile between 10 and 26.7°C (50 and 80°F). The cooling sensation, as measured by differential scanning calorimetry, is significant only for butter and high-fat stick margarines and spreads (37). When margarine is produced, quick chilling results in solid solutions of high- and low-melting glycerides. If the product subsequently is stored at higher temperatures for several days, the recrystallization of the higher melting portions of the melted solid solutions may result in a product with greater waxiness and a slower flavor release (38).

Emulsion tightness is a function of processing, emulsifier content, and formulation of the aqueous phase. If the aqueous droplets are uniformly small or heavily stabilized with emulsifiers, the flavor and salt release will be delayed. A margarine in which about 95% of the droplets have a diameter of 1–5 μm, 4% of 5–10 μm, and 1% of 10–20 μm is described as light on the palate (1). Droplet size also affects the microbiological susceptibility of the product and, to some extent, the consistency. A pulsed nuclear magnetic resonance (NMR) method for determining droplet size based on restricted diffusion in droplets has been reported (39). Zschaler (40) has described a method for microscopically evaluating the size distribution of the aqueous phase in margarine. For low-fat spreads the rheological changes generally associated with melting may be more a function of the stabilizers employed and the degree of emulsification than the melting behavior of the fat blend (37). This is also true of some of the oil-in-water emulsion-type products containing very little fat or no solid fat that have recently appeared in the marketplace.

Melting quality usually is assessed by oral response. Such evaluations are run under standard conditions using an established rating scale (41). Empirical methods also have been used in attempts to quantify melt-in-the-mouth properties that depend on both emulsion tightness and the melting profile of the fat blend. Moran (42) describes a "phase instability temperature," which is the temperature at which the product shows a marked increase in electrical conductivity under shear conditions similar to those that occur on the palate. A viscometer is used in the determination, which is intended for low-calorie products. Cooling sensation on the palate has been estimated by recording the temperature drop in a 35°C (95°F) metal sensing head when placed in contact with the product for 6 s (36). The rate of salt release has been evaluated by determining the chloride ion increase in water at 36°C (97°F) in which a margarine sample is suspended (43). Softening point, which may relate to how a product will melt in hot food applications, can be measured using a Mettler dropping point apparatus. The softening points of a series of North American soft (33) and stick (44) margarines have been determined.

OILS USED IN VEGETABLE OIL MARGARINES AND SPREADS

5.1 Source Oils

Usage of fats and oils in table spreads in the United States is shown in Table 2.4. Most table spreads are formulated with soybean oil with the exception of the healthy image category in which corn, sunflower, or safflower oil is usually the primary ingredient. Canola oil (low-erucic-acid rapeseed oil, erucic acid content <2%) was approved for use in foods in the United States in 1985 (46) and recently has been used in some table spreads that are primarily positioned in the healthy image category. Lard and tallow, of which lard constitutes the predominant portion, are blended mostly in very low cost products. In Europe use of palm oil, lauric fats, and hydrogenated marine oils is common. The tropical oils, palm, palm kernel, and coconut, are not used currently in table spreads in the United States, in part due to their high saturate content and the intense negative public opinion generated by consumer advocate groups in the late 1980s. Marine oils, particularly California sardine oil, were at one time used in American margarines, usage reaching a peak of 18 million kg (40 million lb) in 1936 (47). By 1951, these fish had almost completely disappeared, and no provision had been made for the use of marine oils in food. Hydrogenated and partially hydrogenated menhaden oil (iodine value 10–85) was affirmed as a GRAS (generally recognized as safe) food ingredient in 1989 (48), but to date it has not been utilized for table spreads in the United States. The GRAS status has not yet been affirmed for liquid (unhydrogenated) marine oil. A European study of margarine containing 10–20% liquid fish oil indicated that there are shelf-life problems that would need to be overcome (49). Several methods for purification of fish oil to remove flavors (50–52) or to stabilize the flavor in foods (53–56) have been patented. Today, fish oil in the United States, predominantly menhaden oil, is used for manufacture of protective coatings and fatty acids and derivatives. Much of the oil is exported for use in margarines and shortenings abroad (57). In the future other oils may be used in table spreads as a result of positive comments in the popular press based on epidemiological and dietary studies reported in the scientific literature. Spreads containing olive oil are being marketed in Europe and Canada. Rice bran oil is expected to become available in commercial quantities in the United States in 1995 (58). Margarines prepared from rice bran oil have been described (59).

5.2 Fat Crystallization

Solid Fat. The consistency and the emulsion stability of margarine and most other table spreads depends on crystallized fat. Freeze-fracture electron microscopy of deoiled margarine shows the crystalline nature of the water

Table 2.4 Oils used in U.S. margarine and spreads (45)

Oil	1950	1955	1960	1965	1970	1975	1980	1985[a]	1990[a]
					Millions of kg (lb)				
Soybean	142 (312)	338 (746)	501 (1105)	504 (1112)	640 (1410)	711 (1568)	749 (1651)	739 (1628)	812 (1790)
Cottonseed	190 (418)	126 (278)	62 (136)	52 (114)	31 (68)	21 (46)	11 (25)	nr[c]	10 (21)
Corn	0.5 (1)	0	25 (55)	73 (161)	84 (185)	85 (188)	101 (222)	95 (210)	99 (218)
Peanut	3 (7)	1 (2)	2 (4)	2 (4)	0 (0)	0.5 (1)	nr	0	0
Coconut	0	3 (6)	2 (4)	2 (5)	5 (10)	9 (20)	nr	nr	nr
Safflower	—	—	—	4 (10)	10 (22)	3 (7)	nr	nr	0
Other vegetable[b]	6 (13)	8 (17)	0.5 (1)	6 (14)	0.5 (1)	18 (39)	15 (34)	20 (43)	44 (97)
Lard	2 (4)	6 (13)	25 (56)	45 (100)	41 (90)	20 (45)	47 (104)	30 (65)	15 (33)
Beef fats	4 (9)	4 (9)	3 (6)	6 (14)	4 (8)	3 (7)	—[d]	—[d]	—[d]
Total	347 (764)	486 (1071)	620 (1367)	696 (1535)	814 (1794)	871 (1921)	923 (2036)	883 (1946)	979 (2159)

[a] Data courtesy of the National Association of Margarine Manufacturers (NAMM).
[b] Figure includes unidentified vegetable stearines in 1950 and 1955. Later figures include palm oil, sunflower oil, and oils not reported separately.
[c] Not reported separately to avoid disclosure of individual manufacturer's figures.
[d] Included under lard.

Table 2.5 **Typical solid fat indices of U.S. table spreads**

Product	Solid Fat Index				
	10°C (50°F)	21.1°C (70°F)	26.7°C (80°F)	33.3°C (92°F)	37.7°C (100°F)
Stick	28	16	10	2	0
Soft stick	20	13	9	2.5	0
Soft tub	11	7	5	2	0.5
Liquid	3	2.5	2.5	2	1.5
Butter	32	12	7	2	0

droplet interface as well as a continuous fat matrix that appears to be an interconnected network structure composed of single crystals and sheetlike crystal aggregates (60). The microstructure of margarine and shortenings has been reviewed (61). Two factors predominantly determine the influence of the margarine oil on the textural properties of the finished product: the amount of solid fat present and the conditions under which the margarine is processed. For identical processing conditions of products containing oil blends with similar crystallization characteristics, there is a direct relationship between solids content and consistency (62,63). There are two primary methods in use today for measuring the amount of solid fat in an oil blend. The SFI (64) is an empirical method based on density variations between solid and liquid fat. The more recent NMR procedure (65) yields an absolute percent solids and depends on differences in the magnetic environment of protons in the liquid and solid phases. The results of the NMR determination, referred to as solid fat content (SFC), are similar but not directly comparable to the SFI; however, the two methods can be correlated (66). In the United States, SFI is still in wide use for commercial specification, whereas in Europe SFC has found greater acceptance.

Typical SFI values of U.S. margarine oils are given in Table 2.5. The numbers are representative of commercial products. For a given type of margarine, however, specifications may vary considerably between manufacturers depending on (*1*) the organoleptic characteristics desired, (*2*) compositional requirements to meet nutrient content claims or other information on the nutrition panel, (*3*) whether the product will be marketed using unrefrigerated display, and (*4*) the type of packaging equipment available. The values at 10, 21.1, and 33.3°C (50, 70, and 92°F) are used by most, if not all manufacturers in the United States and are considered to be critical specifications. These solids values are indicative of finished product spreadability at refrigerator temperatures, resistance to oil-off at room temperature, and melt-in-the-mouth qualities, respectively.

Polymorphism. The structural stability of margarine is influenced by the properties of the crystal lattice and by the actual amount of solid fat present.

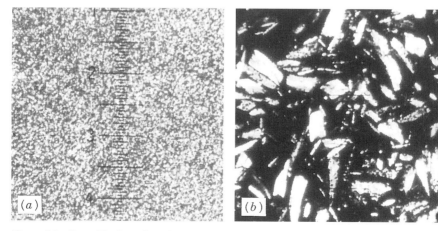

Figure 2.1 Crystallization of sunflower oil margarine. (*a*) Normal β' form, (*b*) sandy β form. (Courtesy of Grindsted Products, Inc., Industrial Airport, Kansas.)

Many organic compounds or mixtures such as fats can solidify in more than one crystalline pattern. The primary crystal forms of triglycerides are designated α, β', and β, which correspond to three principal cross-sectional arrangements of the fatty acid chains (67). These may be differentiated by characteristic x-ray diffraction patterns (68) and heats of transition observed in calorimetric studies (69). Phase behavior of triglycerides is not completely understood. Studies using Raman spectroscopy, x-ray crystallography, and differential scanning calorimetry (DSC) indicate that there are two closely related β' polymorphs of both triundecanoin (70) and tristearin (71). However, the data of Kellens and Reynaers (72) do not support the existence of two distinct β' forms. The observed differences are attributed to the degree of crystal perfection of a single β' polymorph.

The α phase, the least stable and lowest melting polymorph, is initially formed during the rapid chilling conditions used in margarine manufacture; however, it quickly transforms to the β' state (38). Riiner (73) studied production of margarines containing hydrogenated marine oil and found no significant influence of the α phase on consistency. The β' structure of margarine, which may be relatively stable, consists of a very fine network that, because of its great surface area, is capable of immobilizing a large amount of liquid oil and aqueous-phase droplets. Although produced in the β' form, if a margarine oil has strong β tendencies, it may, under certain storage conditions, transform to β, the highest melting and most stable crystalline state. This is usually accompanied by development of a coarse, sandy texture consisting of large crystals (Figure 2.1). A storage study of commercial canola oil margarines indicates the crystal size threshold for consumer detection of sandiness to be about 22 μm (74). In severe cases, transformation to the β form also may result in exudation of liquid oil from the product and partial coalescence of the aqueous phase, which increases the microbiological susceptibility.

Table 2.6 Classification of fats and oils according to crystal habit (75)

β Type	β' Type
Soybean	Cottonseed
Safflower	Palm
Sunflower	Tallow
Sesame	Herring
Peanut	Menhaden
Corn	Whale
Canola	Rapeseed (high erucic)
Olive	Milk fat (butter oil)
Coconut	Modified lard (interesterified)
Palm kernel	
Lard	
Cocoa butter	

Source: Reprinted from the *Journal of the American Oil Chemists' Society,* with permission.

Some triglycerides such as 1,3-dipalmitoyl-2-stearoylglycerol are very stable in the β' form, but tristearin has strong β-crystal forming tendencies (67). Similarly, fats, which are complex triglyceride mixtures, have characteristic polymorphic preferences. Wiedermann (75) has grouped some common fats according to their crystal habits (Table 2.6). In general, the more diverse the triglyceride structure of the highest melting portion of a fat, the lower the β forming tendencies. Therefore, fats such as sunflower, safflower, and canola (low-erucic rapeseed) are most likely to undergo this transformation because the palmitic acid contents are very low, and when hydrogenated, the solid component consists of a series of closely related homologues. In a study of the 1:1 interesterification products of completely hydrogenated soybean oil with each of nine common vegetable oils, it was found that the β' form was favored by higher palmitic acid content in the liquid oil (76). The effect of some surfactants on the kinetics of polymorphic transitions in saturated monoacid triglycerides has been studied by Aronhime and co-workers (77). Heertje (60) has used scanning electron microscopy and light microscopy to detail the microstructural features of graininess.

Problems with sandiness in 100% canola oil margarines have been reported (78). In a study of low- and medium-erucic-acid content hydrogenated rapeseed oils, the $\beta'-\beta$ transition occurred more readily at low-erucic-acid content and with a higher degree of hydrogenation (79). The stability of the β' polymorph of hydrogenated canola oil as it relates to iodine value and trans fatty acid content has been studied by deMan (80). The fat crystals of commercial soft (33) and stick (44) canola oil margarines were found to be in the β form. Hernqvist and Anjou (81) found that in canola oil margarines the relatively rapid development of graininess could be significantly retarded by the addition of 0.5–5% diglycerides. The transition is much more retarded by 1,2-diglycerides than by 1,3-diglycerides, and it appears that diglycerides containing saturated fatty acids of the same chain length as the fat to be stabilized are most effective (82). The effect of diglycerides on the crystallization properties of fats has been reviewed (83). Interesterification has been employed to reduce the development of sandiness in sunflower oil margarine (84,85). Sorbitan tristearate has been extensively studied and has been found to retard the β'–β transformation (86–89) and is permitted by the Canadian margarine standard for use as a crystal inhibitor. Addition of 10% of a hard palm fraction or 15% hydrogenated palm oil was effective in delaying polymorphic transformation in a canola oil stick margarine formulation (90). Interesterified palm stearine and hydrogenated canola oil, when used as a hard stock in soft margarine oil blends, was found to reduce the tendency of the blend to crystallize in the β form relative to blends containing the uninteresterified hard stocks (91). Processing techniques to minimize problems with stick canola margarines have been suggested by Ward (92). Graininess was at one time found to be a problem in 100% soybean oil margarines, and cottonseed oil was added to modify the product (38). Today, with the development of more spreadable stick margarines and the use of multiple base stocks, formulation of 100% soybean or corn oil margarines with stable textural characteristics is achieved readily.

Sandiness develops rapidly when margarine is stored under conditions where melting and recrystallization of the fat can occur. By reducing the storage temperature to 0°C (32°F) undesirable polymorphic transitions can be decreased substantially (93). Accelerated procedures for evaluation of sandiness development involve cycling the margarine between 7.2 and 23.8°C (45 and 75°F) or storage at 21.1–23.8°C (70–75°F) for several weeks. Madsen (94) found that a sunflower oil margarine transformed to the β form after 4 weeks at 18.3°C (65°F) or after 2 weeks at 25°C (77°F). Van der Hock (95) has described a rapid method for estimating structural stability using the solid fat that is isolated from margarine by use of a high-pressure press. The high-melting fraction thus obtained is evaluated both by differential thermal analysis and by observing recrystallization microscopically. The polymorphic behavior of some commercial margarines has been related to the composition of the high-melting glyceride fractions obtained by crystallization from acetone (96,97) or the solid fat obtained by extraction with isobutanol (97).

Crystallization Rate. The rate at which crystals form and undergo polymorphic transitions is of critical importance to the processing of margarine. The α–β' transition rate is accelerated by agitation. Haighton (62) has used metal dilatometers with scraper blades to study crystallization rates under dynamic conditions. Blanc (98) has evaluated the speed of crystallization of several fats under static, supercooled conditions. Crystallization times range from 3 min for coconut oil to 27 min for palm oil and 45 min for shea butter. Riiner (99) has studied polymorphic transition times and found that palm oil has a long α lifetime, that is, the transition for α to β' occurs very slowly. It was also found that the degree of hydrolysis of shea nut oil influences the rate of polymorphic transition (100). Berger (101) observed that the α lifetime of a palm oil sample was reduced from 30 to 4 min after partial glycerides and other nontriglyceride components were removed by column chromatography. Palm oil from unbruised fruit has been found to contain almost 6% diglycerides (102). Okiy (103) demonstrated that the α lifetime of pure palm oil triglycerides increases when diglycerides isolated from palm oil are added back in concentrations from 2 to 15%. The melting point of the β' form and the heat of fusion of the palm oil also decreased with increasing amounts of added diglyceride. Yap and co-workers (104) have studied the crystallization rates of palm oil, palm oil fractions, and hydrogenated palm oil.

Monoglycerides have been found to hasten the onset of crystallization of various fats under agitation in the presence of 16% water (105), but lecithin had the opposite effect (106). In a study of mixtures of margarine oil components, Chikany (107) found that the crystallization rate is accelerated by increasing the amount of coconut oil or decreasing the amount of palm oil in the blend. Crystallization problems associated with blends containing predominantly palm oil can be alleviated by interesterifying a portion of the palm oil with other fats (108). The crystallization rate of a margarine-type oil blend was found to be increased by the addition of palmitate sucrose polyester (SPE, average degree of esterification ca. 5.4), stearate SPE, and sorbitan tristearate (109). The rate was decreased by laurate SPE, and oleate SPE had little effect. Van Meeteren and Wesdrop (110) have found that 1,3-saturated-2-trans-unsaturated glycerides, particularly 1,3-dipalmitoyl-2-elaidoylglycerol, accelerate fat crystallization from a supercooled melt, increasing the processability of spreads formulated with a slowly crystallizing fat.

5.3 Oil Blending

In the United States, most margarines are formulated using hydrogenation as the only means of oil modification. For the most part soybean and corn oils are used, although some healthy image products contain sunflower, canola, or safflower oil as the liquid portion of the blend. High-quality margarines can be prepared using selectively hydrogenated base stocks that are characterized by steep SFI slopes resulting from high trans isomer development with minimal saturate formation. Although interesterification can be used in preparation of soybean margarine oils (111,112), there has been no necessity or incentive for the use of either fractionated or interesterified components unless such components are by-products of another operation such as winterization. The type of margarine oil blends used in the preponderance of table spread products in the United States have changed very little in the last 15 years with the notable exception that in many soybean oil products, the manufacturer has replaced the softest (lightly hydrogenated) base stocks with liquid soybean

Margarines and Spreads

Table 2.7 Typical base stocks used in soybean oil margarine (113)

Stock Number	1	2	3	4
Hydrogenation conditions				
Initial temperature, °C (°F)	148.8 (300)	148.8 (300)	148.8 (300)	148.8 (300)
Hydrogenation temperature, °C (°F)	165.5 (330)	176.6 (350)	218.3 (425)	218.3 (425)
Pressure, kPa (psig)	103 (15)	103 (15)	103 (15)	35 (5)
Nickel (%)	0.02	0.02	0.02	0.02
Characteristics				
Iodine value	80–82	106–108	73–76	64–68
Congeal point, °C (°F)	—	—	23.8–25 (75–77)	33–33.5 (91.4–92.3)
SFI at 10°C (50°F)	19–21	4 max.	36–38	58–61
SFI at 21.1°C (70°F)	11–13	2 max.	19–21	42–46
SFI at 33.3°C (92°F)	0	0	2 max.	2 max.

Source: Adapted from the *Journal of the American Oil Chemists' Society*, with permission.

oil. However, public awareness of the results of recent studies on the effects of various fatty acid classes on serum lipids and possible changes in fatty acid labeling regulations may lead to significant changes in the near future.

Latondress (113) has described some typical soybean oil base stocks and the margarine oils prepared therefrom (Tables 2.7 and 2.8). The percentages of each stock may be adjusted somewhat to take into account batch variability in hydrogenation. The greater the number of base stocks, the greater the flexibility in allowing a slightly out-of-specification stock to be used to meet final blend requirements. The 66-IV stock (number 4 in Table 2.7) is the most

Table 2.8 Typical soybean margarine oil formulas (113)

Type	Soft Stick	Stick	Stick	Soft Tub
Composition (%)				
Stock 1 (Table 2.7)	—	—	60	—
Stock 2	—	42	—	80
Stock 3	—	20	25	—
Stock 4	50	38	15	20
Liquid soybean oil	50	—	—	—
SFI at 10°C (50°F)	20–24	27–30	28–32	10–14
SFI at 21.1°C (70°F)	12–15	17.5 min	16–18	6–9
SFI at 33.3°C (92°F)	2–4	2.5–3.5	1–2	2–4

Source: Adapted from the *Journal of the American Oil Chemists' Society*, with permission.

difficult to control because, at the endpoint of hydrogenation, the saturate content is increasing rapidly. Cottonseed or corn oil can be used to prepare base stocks similar to those shown in Table 2.7. A refiner may have as many as eight base stocks with which to formulate blends. Since different refiners may have different specifications for their base stocks, the margarine producer must realize that the composition of blends received from alternate suppliers may vary somewhat in crystallization characteristics even though the blends meet the same solid fat profile. Moustafa (114) has described the composition, solid fat characteristics, and polyunsaturate content of oil blends found in soft margarines available in the United States. Low-calorie soft and stick spreads containing 40–75% fat are usually formulated from the same oil blends as those used for the manufacture of soft and stick margarines, respectively.

Oil blends for liquid margarine consist of a liquid oil and a highly hydrogenated fat (e.g., liquid and 5-IV soybean oil). Pichel (115) used 0.75–7.5% hard fat depending on the melting point. Too much of the hard fraction gives too viscous a product, whereas too little results in oil-off and water separation. Melnick and Josefowicz (116) describe the SFI ranges of a similar liquid product to be 1.5–4.0 at 10°C (50°F) and 1.0–3.0 at 33.3°C (92°F). If an oil such as base stock 2 from Table 2.7 is used in place of liquid oil, a product with improved oxidative flavor stability and greater temperature dependence of the viscosity will result.

Because of the greater variety of competitively priced oils and because labeling of source oils in order of predominance is not generally legislated in Europe, oil blends are often more complex and interesterified oils are common. When prices permit, low percentages of lauric fats sometimes are added for their sharp melting characteristics (117). Frequently, oil formulation changes may be dictated by economics and availability. Blends that are optimum in terms of solid content and cost can be calculated according to the method of statistical dilatation equivalents (62). These calculations, which are based on measurements taken on a large number of mixtures of available components, assume that each component yields a linear contribution to the solids content of the blend. The contribution factors, which are calculated by multiple regression analysis, must be determined at each temperature of interest. For example, coconut oil would have a strong positive solids contribution at 10°C (50°F) and a negative coefficient at 21.1°C (70°F). The validity of the relationships are bound by the range of mixtures studied and the linearity of contributions in that range. Cho, deMan, and Allen (118) used an experimental design to identify suitable margarine and shortening blends formulated from four component oils, two of which were interesterified. Besides solids requirements, least-cost computer programs can be written to take into account other constraints such as minimum essential fatty acid content required and limits on individual components due to cost, crystallization behavior, availability, or production capacity (119).

Wieske (119) also has described a more general calculation method in which fatty acids are assigned to one of four classes: high-melting saturated

C_{16} or greater; trans monoenes and dienes; medium melting (C_{12} and C_{14}) saturated; and low melting saturated and cis unsaturated. If triglyceride structures based on these four fatty acid types are subgrouped into types with similar melting characteristics, a model can be generated containing 16 glyceride classes that contribute to the solids content. The triglyceride types in any oil blend are the sums of the values found by analysis of the triglyceride structures of the components. The effect of interesterification of any part or all of the blend can then be calculated mathematically. A margarine oil patent specification based on these fatty acid and triglyceride classifications has been issued (120). As more knowledge is gained of mutual solubility relationships and melting behavior of triglyceride classes, and as economical methods of achieving specific rearrangement of triglyceride structures are developed, it may become possible to formulate optimum oil blends with knowledge of only the fatty acid composition of the components available.

High-Liquid-Oil Blends. From the late 1950s to the 1980s considerable effort was devoted to the development of margarines containing high levels of linoleic acid. Although polyunsaturate content is less emphasized in the United States today, oils such as corn, sunflower, and safflower continue to be perceived as healthy by the consumer, as are olive oil with its high monounsaturates and canola oil with its low saturated fat content. Therefore, it appears advantageous to market a product containing 100% of one of these oils. When the selectively hydrogenated liquid oils are used as hard stocks, the upper limit on the liquid oil content of such a stick margarine blend is about 60–65% for formation of a satisfactory print. For a high-liquid-oil blend, the SFI at 10°C (50°F) is limited on the high side by the percentage of trans fatty acids that can be developed in the hard stock without increasing the saturate content to such an extent that the 33.3°C (92°F) solids are too high. These high-trans hard stocks can be achieved most consistently by use of a sulfur-poisoned catalyst (121). Soft margarine blends containing as much as 85% liquid oil are feasible. As discussed in Section 5.2, 100% sunflower, safflower, or canola margarines may become grainy during storage. Randomization of the hard component (122) or interesterification of the hard stock with a portion of the liquid oil (123) has been used to alleviate texture problems. Soft margarines containing more than 80% of a highly polyunsaturated liquid oil have been found to develop a granular, lumpy appearance accompanied by texture breakdown when stored at less than −23.3°C (−10°F) McNaught (124) discovered that randomization of at least part of the liquid oil alleviates this problem.

Low-Trans Oil Blends. Based on recently published studies (125–133) that indicate that trans fatty acids may not be neutral in their effect on serum lipids, negative press has heightened consumer awareness of these fatty acids, particularly in North America. A consumer advocate group has petitioned the U.S. FDA that trans acids be labeled (134). At the present time it is unknown whether these acids will eventually be classed as cholesterol-raising fatty acids like saturates or whether health implications more or less negative than saturates will be attributed to them. Over the last decade such concerns have led to considerable research in the development of low- and no-trans oil blends for margarines and spreads. Most of these formulations also contain a high level of liquid oil. Of course, for very low fat spreads the trans content of the oil blend is less significant, since only the amount of trans fatty acids per serving would be labeled. The approaches that some companies are taking to reduce trans levels and new product introductions in the area of no-trans margarines and spreads has been discussed by Haumann (135).

Margarines high in polyunsaturates and low in trans fatty acids are available in Canada and Europe. These products can be formulated from tropical oils, with no hydrogenated fats or with fats that are completely hydrogenated. Tropical oils are viewed negatively by the consumer in the United States, and therefore are not used to any significant extent. Palm oil and lauric fats such as coconut, palm kernel, and babassu oils are rich in saturated fatty acids. However, because of eutectic formation and steep melting profiles, satisfactory stick margarine oil blends cannot be achieved with these oils as the sole hard stock without some additional modification. Ward has found that interesterification of a completely hydrogenated mixture of palm and babassu oil yields hard stocks that can be used in very high polyunsaturated soft (136) and stick (137) margarines capable of being processed on standard equipment. Similar soft margarines containing approximately 90% liquid oil have been formulated from fully hydrogenated interesterified palm oil and palm kernel oil (138) or palm kernel olein (139). Soft and stick margarines were prepared with an unhydrogenated hard fraction consisting of an interesterified mixture of coconut oil, palm oil, and palm stearine (140). By using fractionation or cofractionation of palm and lauric oils, soft (141) and stick (142) margarines were prepared without addition of hydrogenated or interesterified fats. A process for producing hard stocks from high-palmitic oils, high-lauric oils, and hydrogenated oils having a high-behenic-acid content by 1,3-specific enzymatic interesterification has been described (143). Spreads have been prepared from a blend containing liquid oil and 3–10% of a fully hydrogenated fat composed of at least 25% palmitic acid, less than 3% trans acids, and less than 10% glycerol tristearate (144). Suitable hard stocks are fully hydrogenated fish oils or palm oils. Hard stocks produced by interesterification of high palmitic and/or stearic acid fat and high behenic acid fat were used at low levels to prepare a margarine containing low saturated and trans fatty acids (145). The characteristics of hard stocks obtained by interesterification of fully hardened soybean oil and nine common vegetable oils (1:1) have been described (76). Margarine oils containing very low levels of trans and C_8–C_{16} fatty acids can be prepared by enzymatic transesterification of a stearic acid source and a liquid oil (146).

Formulation of zero-trans margarines from single-source high-linoleic oils also has been reported. The random interesterification of an 80/20 blend of liquid soybean oil and completely hydrogenated soybean oil resulted in SFI values of 8, 3.5, and 2.2 at 10, 21.1, and 33.3°C (50, 70 and 92°F) respectively. The margarine had a stable crystal structure and good oxidative stability (147). The enzymatic interesterification of blends of liquid and 10–15% fully hydrogenated soybean oil has been reported to result in a fat base suitable for preparing table spreads (148). Low-trans margarines also were prepared from 100% sunflower, soybean, or corn oils by fractionating an interesterified mixture of liquid and completely hydrogenated oil (149). The olein thus obtained was blended with additional liquid oil to obtain fats suitable for stick margarine. A spread formulated using a single unmodified fat has also been characterized (150). The fat is preferably high-stearic soybean oil containing at least 30% disaturated triglycerides.

An alternate method for preparing low-trans margarines makes use of directed interesterification to prepare fats from liquid oil without the aid of hydrogenation. Sreenivasan (151) prepared a 100% sunflower oil with solids of 10.7 at 0°C (32°F), 6.0 at 21.1°C (70°F), 5.2 at 26.7°C (80°F), and 2.1 at 33.3°C (92°F) by directed interesterification in an aprotic solvent for 6 days at -9.4–0°C (15–32°F). At these temperatures the di- and trisaturated glycerides are precipitated from the reaction mixture as they are formed. The reaction is then stopped at low temperature and the oil blend has the same fatty acid composition as the original liquid oil. Directed interesterification of liquid corn oil using temperature cycling between 0.6°C (33°F) and 10°C (50°F) in the absence of solvent was accomplished in 6 h (152). The resulting solids were 13.1, 10.3, 7.2, 4, 8, and 2.2 at 0, 10, 20, 30, and 40°C (32, 50, 68, 86, and 104°F), respectively. In order to increase the amount of solids attainable by directed interesterification, the liquid oil can first be enriched with a small percentage of the completely hydrogenated oil (153). Alternatively, the solids may be enriched without hydrogenation by removing a portion of the saturate-reduced liquid oil during the reaction and adding additional starting oil (154). Although soft zero-trans margarines can be prepared from 100% liquid oils, it is unlikely that satisfactory stick margarine oils will be obtained through directed interesterification alone. The limited melting ranges and high melting points of triglycerides composed of stearic and palmitic acid would appear to preclude this possibility.

A novel alternative to the use of hydrogenated or saturated fats for structural stability in oil-continuous emulsions is the addition of oil-soluble polymers as thickening or texturizing agents (155). These polymers are condensation products of hydroxyacids or polyhydric alcohols and polybasic acids. Currently they are not approved for food use. Another option to hydrogenated oils is to base the product on an oil-in-water emulsion. Such a product, which contains 80% liquid canola oil, has been introduced in the United States (135).

Special Dietary Oils. Some reports have been published regarding special health margarines that contain nontraditional ingredients. Stahl (156) has described a low-calorie margarine in which the oleaginous phases consist of monoglycerides with no appreciable quantities of di- or triglycerides. The product may be used for nutritional supplementation where certain digestive disorders exist. Margarines also have been prepared using medium-chain triglycerides based on caprylic and capric acids (157–159). Such fats are rapidly absorbed and are useful where fat metabolism is impaired. Formulation of low-calorie margarines and spreads using sucrose octaoleate as a fat replacer has been reported (160,161). Improved margarine hard stocks composed of sucrose polyesters of short- and long-chain fatty acids (162) as well as palm oil fatty acids (163) also have been described. These sucrose polyesters are nonabsorbable fat substitutes that have been found to be useful in reducing calorie intake and plasma cholesterol in clinical trials (164). In the future there may be a market for spreads formulated from other oils perceived to be healthy such as rice bran oil (165) or oils containing long-chain highly unsaturated ω-3 acids (166) or γ-linolenic acid (167). The use of the latter two types of oil in the United States will depend not only on their approval for food use but also on whether sufficient scientific evidence is generated to support approval of a health claim. The food uses and properties of fish oils, which contain high levels of ω-3 fatty acids, have been discussed (49,168,169). Plant sources of γ-linolenic acid such as borage, black currant, and evening primrose oils, as well as microbial sources have been reviewed by Gunstone (170).

5.4 Oil Specifications

Patel (171) has listed the parameters usually included in specifications for a margarine oil blend (Table 2.9). In addition to those listed, specifications sometimes include anisidine value, totox value, and limits on heavy metals and microorganisms. The oil used in margarine should be of the highest quality and as bland as possible. Flavors are evaluated organoleptically upon receipt of the oil using a standard scoring system such as that recommended by the American Oil Chemists' Society (172). Some manufacturers may confirm organoleptic results using volatiles analysis. Physical properties are described by SFI. Some specifications also include SFI at 26.7 and 37.8°C (80 and 100°F) as well as a melting point or Mettler dropping point. Characteristics of the base stocks used to formulate the blend generally are left to the discretion of the oil supplier. As Wiedermann (173) has pointed out, this may, at times, lead to difficulties in the manufacture of a margarine product for which process control is critical. The SFI profile should not be viewed merely as three separate specification ranges. In the example in Table 2.9, oils with SFIs of 26-15-1.5 and 22-16-3 would both meet specification but could result in noticeably different finished product characteristics. If the SFI–temperature curve for a given oil blend is routinely steeper or flatter than the slope through the midpoints of the SFI ranges, the base stocks being used may not be suited to the specifications or the specifications may not be realistic. The specification of polyunsaturates or saturates is usually included only if these are necessary

Table 2.9 Typical margarine oil specification

Parameter (139)	Example
Composition	
Source oil(s)	100% corn
Blend	50% liquid oil, min.
Additives (permitted/required)	Citric acid permitted
Quality	
Flavor (organoleptic score)	7, min.
Color (Lovibond red)	3, max.
Peroxide value (mEq/kg)	1.0, max.
Free fatty acids (% as oleic)	0.05, max.
Moisture (%)	0.05, max.
Stability (active oxygen method), 8-h AOM	Peroxide value less than 10
Physical	
SFI at 10°C (50°F)	22–26
SFI at 21.1°C (70°F)	14–17
SFI at 33.3°C (92°F)	1.5–3
Nutritional	
Percent polyunsaturates (enzymatic)	28, min.
Percent saturates	19, max.
Shipping	
Mode	Truck
Nitrogen blanketing	Yes
Loading temperature	57.2°C (135°F) max.
Arrival temperature	48.8 ± 15°C (120 ± 5°F)

to meet specific label claims on the margarine product. If labeling of trans fatty acids becomes mandatory in the future, the trans content may become part of many manufacturers' specifications.

As many margarine plants in the United States are located at considerable distance from their suppliers, truck or rail shipment is required. Loading temperatures and whether the oil must be nitrogen blanketed are specified. For truck deliveries, the temperature on arrival often is specified to ensure that the oil is completely liquid. In order to minimize the possibilities for oil degradation, it is optimal to process and deodorize the oil blend at the margarine production location.

OTHER COMMON INGREDIENTS

The following section discusses the functionality of some of the ingredients, other than the fat blend, that are commonly used in table spreads. There is considerable latitude in the choice of which ingredients to utilize and the levels of these ingredients. The recipe is generally decided by consumer prefer-

Table 2.10 Typical margarine and spread formulations

Ingredient	Percent in Finished Product		
	80% Fat	60% Fat	40% Fat
Oil phase			
Liquid and partially hydrogenated soybean oil blend	79.884	59.584	39.384
Soybean lecithin	0.100	0.100	0.100
Soybean oil mono- and diglycerides (IV 5, max.)	0.200	0.300	—
Soybean oil monoglyceride (IV 60)	—	—	0.500
Vitamin A palmitate-β-carotene blend[a]	0.001	0.001	0.001
Oil-soluble flavor	0.015	0.015	0.015
Aqueous phase			
Water	16.200	37.360	54.860
Gelatin (250 bloom)	—	—	2.500
Spray-dried whey	1.600	1.000	1.000
Salt	2.000	1.500	1.500
Sodium benzoate	0.090	—	—
Potassium sorbate	—	0.130	0.130
Lactic acid	—	to pH 5	to pH 4.8
Water-soluble flavor	0.010	0.010	0.010

[a] Custom blended for correct vitamin content and color; suspended in corn oil.

ence testing and process considerations. Some typical formulations are listed in Table 2.10.

6.1 Milk Products and Protein

In the margarine standard of identity, milk products have been interpreted (174) by the FDA to include butter or butterfat in any percentage as long as some vegetable oil is used to meet the 80% minimum fat requirement. Therefore butter blends fall under the FDA regulations governing margarines and spreads. In the past cultured milk was used in almost all margarines. Because of the time and space required for culturing, this practice was largely abandoned in favor of using skim milk together with starter distillate, diacetyl, or other flavors. Skim milk, although still used by some manufacturers in the United States, for the most part has now been replaced by spray-dried whey, which is sometimes supplemented with potassium caseinate to a standardized protein content. Soy protein can be used for products where dietary considerations preclude dairy ingredients.

Protein affects margarine products in several ways. In addition to flavor, dairy solids undergo the Maillard reaction and brown during frying. Low-lactose milk or whey proteins can be used to control or eliminate the browning effect while retaining desirable flavor characteristics (175). Milk solids also act as a preservative by sequestering metals that promote oil oxidation (176). Protein exerts a destabilizing effect on water-in-oil emulsions. If protein is removed from a margarine formulation without changing the processing or the fat/emulsifier system, flavor and salt release will be impaired because the aqueous-phase droplets are smaller and the emulsion is more resistant to breaking. Linteris (177) found that the addition of 0.01–0.1% sodium caseinate enhanced the salt sensation of milk-free margarine. In very low fat spreads where the aqueous phase contained only water, preservative, salt, acid, and flavor, the incorporation of 5–10 ppm of protein caused a significant enhancement of flavor release (178). The emulsion instability caused by the presence of protein is a particularly important concern in formulating products that contain less than 50% fat. In the past all 40% fat diet margarines in the United States contained no protein. Today, however, many 40% and lower fat spreads contain gelling agents or other water-binding ingredients that afford sufficient stability to allow addition of milk protein.

6.2 Emulsifiers

Emulsifiers are multifunctional in margarine. They reduce surface tension between the aqueous and oil phases so that the emulsion forms with minimal work. Emulsifiers stabilize the finished product during storage to prevent leakage or coalescence of the aqueous phase. They also act as antispattering agents by preventing coalescence and violent eruption of steam during frying. Common emulsifiers and their uses are listed by Dziezak (179). The role of emulsifiers in spreads and shortenings has been discussed by Madsen (180). The effects of emulsifiers and their interactions are complex and become more critical at lower fat levels. Gaonkar and Borwankar (181) have reported the influence of lecithin, monoglyceride, and surface-active impurities present in the oil on the vegetable oil–water interface. Using microscopy, Heertje (60) found that saturated monoglycerides appear to be more effective in displacing proteins at the interface than unsaturated monoglycerides, and that phospholipids are much more surface active than monoacylglycerols.

Crude lecithin is used at levels of 0.1–0.5% in almost all margarines because of its antispattering properties; however, in very low fat spreads it may lead to decreased emulsion stability and increased tendency to oil-off. The production, properties, and food uses of lecithin have been reported by Schneider (182). Lecithin may be unbleached or single or double bleached with hydrogen peroxide. In addition to producing an even, stable foam during frying, lecithin contributes to a fine dispersion of the protein sediment, interacts with protein to form a brown gravy, and results in a quicker salt release (183). Enzymatic hydrolysis of lecithin yields α-monoacylglycerophosphatides that improve frying performance and resistance to oil-off in liquid margarines (184). Fractionation and partial hydrolysis of the alcohol-insoluble fraction is claimed to afford improved antispattering effects in margarine (185). Antispattering properties have also been claimed for proteose-peptone-enriched milk protein (186), finely divided metal or metalloid oxides (187), citric acid esters of monoglycerides (188), sodium sulfoacetate derivatives of monoglycerides (189), and polyglycerol esters (190). Incorporation of finely dispersed gases such as nitrogen, carbon dioxide, or air also reduces spattering (191). Nitrogen is preferred and is most often used in whipped margarines and spreads. The use of helium or its admixtures with other gases is reported to result in a very fine dispersion that can be attained relatively easily (192).

In stick margarines where the high percentage of solid fat is capable of stabilizing the crystallized emulsion, only lecithin is necessary. However, most margarines also contain mono- and diglycerides of low or intermediate iodine value (IV) for added protection against weeping. Very high IV monoglycerides such as those produced from sunflower or safflower oil have been found to function well in low-fat products (193). The rate of coalescence of a 50% water-in-vegetable-oil emulsion as a function of concentration of monolinolein has been studied (194). Some very low fat spreads contain polyglycerol esters. A combination of monoglycerides and polyglycerol esters is effective in allowing production of very low fat spreads containing a significant amount of milk protein, especially if the spread also contains a gelling agent (195). Although polyglycerol polyricinoleate is said to be particularly effective (196,197), and is used in Europe, this specific polyglycerol has not been approved in the United States. Erucic acid esters of polyglycerols also were found to stabilize high internal phase water-in-oil emulsions (198). The use of sucrose esters to form a stable liquid margarine with only unhydrogenated liquid oil has been described (199). It has also been reported (200) that a high liquid oil containing fluid margarine with low viscosity, and which is not susceptible to oiling-out, can be formulated using a distilled behenic acid monoglyceride.

In addition to emulsifiers, low-fat spreads may contain aqueous-phase gelling and/or thickening agents such as gelatin, pectin, carrageenans, agar, xanthan, gellan, starch or starch derivatives, alginates, or methylcellulose derivatives. Commercially, the most important of these is gelatin. High-quality gelatin is a costly ingredient and a process for removal of off-flavors from the less expensive grades of gelatin by membrane filtration has been patented (201). The functionality of gelling agents is discussed later in this chapter. The microstructural nature of the emulsion in margarine and low-fat spreads has been elucidated by Heertje (60) using microscopical techniques.

6.3 Preservatives

The preservatives that can be added to margarine fall into three categories: antioxidants, metal scavengers, and antimicrobial agents. Because of hygienic manufacture of both raw materials and the finished table spreads, lipolytic microorganisms are not found and hydrolytic rancidity is not a problem, even in products formulated with lauric oils. Antioxidants may be necessary for keeping quality of spreads formulated with significant amounts of animal fat, but they are not added to most vegetable oil margarines. However, in the future, these may become necessary in order to incorporate even low levels of highly unsaturated fats such as unhydrogenated fish oils. Vegetable oil margarines containing milk protein have been shown to be stable for 6 months at 4.4°C (40°F) with no added antioxidant (202). Residual tocopherol levels in vegetable oils are said to be near optimum for protection (203); however, excess tocopherol may have a prooxidant effect (204). Lecithin and ascorbic acid act as antioxidant synergists (205). The addition of nonlipolytic, nonproteolytic, oxygen-consuming yeasts has been suggested as a means of preventing autoxidation of the finished product (206). Salt-tolerant lactobacillus that convert aldehydes to alcohols are said to improve margarine shelf life by removing the oxidation products responsible for off-flavors (207).

The presence of heavy metals can cause serious metallic off-flavors to develop in margarine within days. Copper has the strongest prooxidant effect. The maximum amount of copper that can be tolerated is indicated to be on the order of 0.02 ppm (208). Citric acid, citrates, and salts of ethylenediaminetetraacetic acid (EDTA) act as sequestering agents to inactivate metals that may be present. EDTA has been found to be very effective in preventing off-flavors due to copper-induced degradation and often is added to milk-free margarines in the form of calcium disodium salt. Melnick (209) has patented a process of crystallizing salt in the presence of EDTA to reduce the heavy-metal content. Such high-purity salt is available for use in margarine.

Microbiologically, water-in-oil emulsions are more stable than the aqueous phase itself because only a small fraction of the droplets are occupied by microorganisms. The droplet size limits growth provided that the organism does not excrete lipases (210). Since droplet size depends to a great extent on processing parameters, process control is critical, particularly in the manufacture of low-fat spreads. The median and largest diameters of the droplets, pH, available nutrients, and degree of inoculation play essential roles in determining the fate of contaminating microorganisms (211). The concentration of salt in the aqueous phase of a 2% salt, 80% fat margarine is also very effective; however, in the absence of additional preservatives or acidulents, molds occasionally do proliferate. Sorbic acid, benzoic acid, and their salts are used as preservatives, particularly in low-fat and low-salt products. The undissociated acids are primarily responsible for the preservative effect, and therefore, the lower the pH, the greater their effectiveness. However, the acids are more soluble in the oil than in the aqueous phase, where protection is needed, and sorbic acid has the more favorable partition coefficient. Although salt acts synergistically with these preservatives in aqueous solution, in an emulsion it has a negative effect on the partition coefficient, driving more of the free acid into the oil (212). Studies have been conducted on the effectiveness of benzoic acid (213) and sorbic acid (214) in margarine as a function of concentration. Castenon and Inigo (215) recommend the addition of 0.05% sorbic or benzoic acid and a pH of 4–5 for unsalted margarine and a pH of 5–6 for salted margarine. Lactic acid is said to be the most effective acidulent for use as a preservative as long as at least 0.2% is present (211); however, citric and phosphoric acids are also used. Demineralization of acidified milk can reduce its sour taste by lessening its buffering capacity and hence the acidity required to achieve a given pH (216). The use of two aqueous phases formulated such that the preservatives and the nutrients are concentrated in the same phase has been suggested (217). Klapwijk (218) has discussed the hygienic production of low-fat spreads and has outlined a predictive modeling approach to microbiological hazard analysis.

6.4 Flavors

Many synthetic butter flavors are available for use in margarine. These are based generally on mixtures of compounds that have been identified as contributing to the flavor of butter, such as lactones, ethyl esters of short-chain fatty acids, ketones, and aldehydes (219). Diacetyl is a primary volatile constituent of many margarine flavorings and contributes significantly to a buttery aroma. The concentration in butter varies from 1 to 4 ppm (220). It is formed from citric acid present in milk during the culturing process. If the milk is not cultured, synthetic diacetyl or starter distillate can be added. Flavors obtained by lipolysis of butterfat also are available, and the use of a combination of starter distillate and heat-treated butterfat has been described (221). The tightness of the emulsion and the melting characteristics of the fat will affect the rate and the order in which flavors are perceived. Salt concentration and pH also affect flavor balance since they may influence the partition coefficients of various flavor components. As the fat content of spreads is reduced to very low levels, the challenge of formulating flavors whose oral response is similar to that of high-fat spreads is increased considerably.

6.5 Vitamins and Colors

The mandatory fortification of margarine with vitamin A is accomplished by the addition of β-carotene (provitamin A) and/or vitamin A esters. The carotene level is adjusted for the desired color and the colorless esters (acetate, palmitate, etc.) are used to standardize the vitamin content. Addition of vitamin D is optional. Fortification with vitamin E is not permitted by the U.S. margarine standard, but recently some spreads fortified with vitamin E have appeared in the marketplace in the United States, and fortification of both margarines and spreads has recently been done in Europe. The naturally occurring vitamin E content of vegetable oil margarines available in the United States has been reported (222).

Margarine is colored with carotenoids, and synthetic β-carotene is by far the most widely used. Carotene dissolves very slowly in oil. Therefore, the compound is pulverized to a particle size of 2–5 μm and the microcrystals are suspended in oil to retard oxidation (223). Natural extracts containing carotenoids, for example, annatto (bixin), carrot oil, and red palm oil also have been utilized. Annatto, which is used in butter, is somewhat sensitive to light and may have an orange or slightly pink hue, particularly when the aqueous phase is acid (1). Mixtures of annatto and turmeric extracts result in a more typical color than annatto alone (224). Many margarine manufacturers purchase blends of colors and vitamins customized for their specific products.

PROCESSING

There is a wide range of formulations used in the margarine industry today. Just as the fat must be tailored to suit the product, in order to attain the desired finished product characteristics, processing parameters must be established that are appropriate for level, solids content, and crystallization rate of the fat used in the formulation. The basic process consists of five operations: emulsification, cooling, working, resting, and packaging. The following is a brief general description of these operations and their usage for different product types together with some published process modifications. Detailed descriptions of the equipment, process, packaging, and plant layout, as well as the production of puff pastry margarine, are given elsewhere in this work. Some aspects of low-fat spread processing will be discussed later in this Chapter.

7.1 Processing Operations

Emulsification. The formation of the coarse initial emulsion may be a strict batch process in which the warm oil and the oil-soluble ingredients are individually weighed or melted into an agitated tank, which, in the past, was referred to as a churn. The pasteurized, aqueous phase is then weighed in with agitation. In many plants the aqueous phase is held at 4.4–10°C (40–50°F) after pasteurization, so that heat may be necessary to maintain the emulsion at a temperature above the melting point of the fat. Usually the emulsion is held at about 43.3–42.8°C (110–120°F). If the temperature is not sufficiently high, crystal nuclei and precrystalline structures may be formed that can affect the consistency of the finished product (225). Spattering properties of the margarine may be affected by the aqueous-phase temperature (226). The emulsion at this point is very unstable. Without agitation the milk-phase droplets would immediately begin to coalesce and settle out. After the emulsion is well mixed to ensure uniformity, it is pumped to an agitated holding tank that feeds the processing line.

Alternatively, the emulsion may be formed on a continuous basis. If all the oil-phase ingredients are added to the primary oil storage tank, the two phases can be metered into the line feed tank or simply mixed in-line using metering pumps or mass flow meters. Whether this is feasible depends on the oil holding tanks available and the formulation and production scheduling of products in which the base oil is used. The ultimate in flexibility consists of a multihead pump capable of metering individual components such as brine, water, whey concentrate, flavor, and preservative solutions to compose the aqueous phase, and also the individual oil base stocks, emulsifiers, and colors to form the oil phase. In-line static mixers are used to blend the separate phases, which are then joined in-line and emulsified through another static mixer.

Prechilling the emulsion just enough to form a low level of solid fat (0.5–2.5%) followed by homogenization prior to normal processing is said to result in a very fine dispersion of the aqueous phase that obviates the need for preservatives (227). A novel process for forming stable low-fat emulsions of uniform droplet size has been recently described (228). The method consists of dispersing an aqueous-phase or an oil-in-water emulsion at low pressure into the oil phase through a hydrophilic microporous membrane that has been pretreated with the oil phase. The process is said to be applicable for fat levels as low as 20 and 25% to water-in-oil and oil-in-water-in-oil emulsions, respectively. Recently, a 25% fat spread, which is believed to have been manufactured by this process, and which contains no preservatives and has a 6-month shelf-life, has been introduced in Japan.

Chilling. When the emulsion is formed, it is fed via a high-pressure positive pump to a tubular swept-surface heat exchanger, usually referred to as an A unit. Examples of the tube chillers available are the Votator (Cherry-Burrell, United States), the Chemetator (Crown Chemtech Limited, United Kingdom), the Perfector (Gerstenberg and Agger, Denmark), and the Kombinator (Shroeder, Germany). In these devices the product passes through the annulus between a rotating shaft and an insulated outer jacket containing a refrigerant, usually liquid ammonia. Temperature control is achieved by regulating the suction pressure on the refrigerant. The tube often is fabricated from chrome-plated nickel or steel, which have high heat transfer coefficients. Free-floating blades attached to the mutator shaft are caused by centrifugal force to scrape the jacket inner wall continuously to achieve maximum cooling. In general, shaft speeds range from 300 to 700 rpm, scraping the surface clean as many as 1500 times per minute (229). The high internal pressures and shearing forces generated by the blades and their holding pins cause rapid crystal nucleation. High rotational speeds in the A unit result in a much finer emulsion than low speeds (230). Warm water is circulated through the shaft to prevent buildup of solid fat. The A units are available in a range of sizes and commonly are used in series, which allows for processing flexibility.

Working. When the margarine emerges from the cooling tubes, it is only partially crystallized. In many processes it then goes to a working unit or blender; sometimes called a B unit. In some patent literature it is referred to as a C unit or crystallizer; however, an eccentric swept-surface heat exchanger following a blender has also been described as a C unit. The working unit has pins arranged in a helical pattern on a variable-speed shaft. These pins intermesh with stationary pins positioned on the cylinder wall. In the blender, crystallization is intense and a temperature rise results both from the heat of crystallization and mechanical work. Agitation in the crystallizer facilitates free diffusion of crystals to the surface of the aqueous-phase droplets, forming a crystalline shell (the so-called Pickering stabilization). Due to the increased viscosity, some coarsening of the emulsion may occur during passage through the working unit (230). Intermediate crystallizers, which are driven by the mutator shaft of the A unit are available from some manufacturers. The temperature rise across the blender is an indication of the amount of crystallization achieved. The degree of crystallization in the working unit depends on the residence time (volume and throughput) and the rotational speed of the shaft as well as the crystallization rate of the fat. An experimentally derived relationship between these variables and an automated process for controlling the degree of crystallization in the crystallizer has been described (231).

Resting. If the product requires a stiffer consistency for packaging, this is accomplished through the use of a static B unit or quiescent tube. This is a warm-water-jacketed cylinder that sometimes contains baffles or perforated plates to keep product from channeling through the middle of the cylinder. The static B unit usually consists of flanged sections so that the length can be varied to suit the product. Additional resting time is often achieved with two quiescent tubes in parallel, with the use of a timed, rotary valve that alternates flow to the two units.

Packaging. Two basic types of stick packaging machinery are in use in the United States. The first of these forms a molded print that is then wrapped. This may be an open or closed system. With an open system the product exits the quiescent tube through a perforated plate, forming "noodles" that drop into a hopper. Screw impellers feed the margarine to a mold where the print is formed prior to wrapping and cartoning. The closed system is similar except that the margarine is not extruded but filled directly into the mold cavity with line pressure. On the second type of machine, the filled print, margarine from the holding tube is still in a semifluid state. It is filled directly into a cavity that is prelined with the inner wrap. The wrapper is then folded and the print ejected from the mold. This equipment is more suitable for filling a soft stick because the product is not molded before wrapping. Soft tub margarines are packaged in a fluid or semifluid state on straight-line or rotary-head fillers.

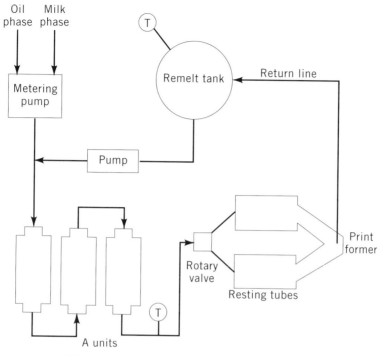

Figure 2.2 Continuous stick margarine production.

7.2 Specific Processes

Stick Margarine. A typical equipment configuration for closed-system stick margarine production is illustrated in Figure 2.2. The emulsion is proportioned together, cooled and allowed crystallization time prior to packaging. For *molded print* equipment, the product passes from the A units directly to the quiescent tubes. For *filled print* packaging, a small blender may be used prior to the resting tubes to achieve the proper consistency for packaging and a slightly softer finished product. The remelt line is necessary because, in a closed system, some overfeeding must be present in order to maintain sufficient product to the filler for adequate weight control. The excess product is returned to a tank where it is remelted and pumped back into the product stream to be reprocessed. In the event of a packaging equipment malfunction, all the product goes back to the return tank. If an emulsion tank is used to feed the line, the excess product can be remelted in-line and returned to the emulsion tank.

The crystallization rate of the fat blend affects the equipment requirements and the processing parameters necessary to achieve a satisfactory product. If the fat blend is prone to extreme supercooling, the crystal lattice will not be formed in sufficient time for the blender to be effective. In this case the product may set up too firm in the quiescent tubes, or severe postpackaging hardening may occur. In order to process a blend containing a high level of palm oil, Kattenberg and Verburg (232) used a slowly agitated working unit between the A units to provide a residence time of 2–3 min for crystallization to occur.

Several processing variations have been claimed to improve the organoleptic quality of margarines. Controlled precrystallization is used to achieve early crystallization and crystal enlargement of the higher melting glycerides. This is achieved by feeding all or part of the oil blend to a precooler prior to mixing it with the aqueous phases. Alternatively, part of the emulsion leaving the A units is recycled and mixed with the emulsion at the inlet of the cooling tubes. The greater distribution of crystal sizes is said to improve consistency and flavor release (233). In another modification of the process, the oil blend and 25% of the aqueous phase are chilled and the remaining 75% of the cold milk added to the emulsion just prior to the blender (234). This milk injection results in a greater range of droplet sizes, some quite large, which improves flavor and salt release. Part of the milk is added prior to cooling so as to obtain a uniform opaqueness, with no regions of pure fat. Flavor improvement also is claimed for margarine composed of a double emulsion (oil-in-water-in-oil). The emulsion is formed by first preparing an emulsion of a portion of the fat in the aqueous phase and then emulsifying this into the remainder of the oil blend (235–237). Advantages of this process are said to include (*1*) improved flavor release due to the greater surface area of the aqueous phase, (*2*) the potential for increase of the liquid oil content by using this as the internal oil phase while the consistency depends on the external phase, and (*3*) the ability to incorporate high flavor levels for baking and cooking into the internal oil phase and at the same time retain acceptable flavor levels if the product is used as a spread. Phase inversion processing of high-fat table spreads is claimed to impart some of the textural and flavor release properties of butter. The margarine is prepared by churning a filled vegetable oil cream (238,239). Injection of some of the fat, preferably a liquid oil, into the cream as it is being churned is reported to facilitate the process (240).

Whipped Margarine. Whipped margarines usually contain 33% nitrogen by volume (50% overrun). The gas is introduced in-line through a flow meter before or between the A units. A backpressure valve is placed after the A units to maintain a constant line pressure against which the nitrogen is injected. The nitrogen also may be introduced at low pressure between two pumps with the second operating at a higher throughput rate. A high-speed whipper is used just prior to the filler to ensure a fine dispersion of the nitrogen. In order to control overrun, a heat exchanger in the return line completely melts the product to eliminate nitrogen from the recycled emulsion. Closed-system packaging equipment needs some modification to run whipped sticks because of the expansion when line pressure is released at the fill head. Line pressures are critical in producing a product of uniform texture and appearance (241).

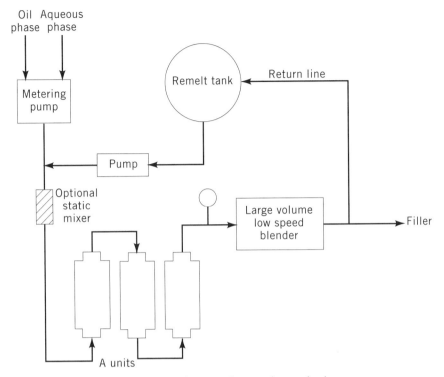

Figure 2.3 Continuous soft margarine production.

Equipment especially designed for packaging six sticks to a 1-lb carton generally is used.

Soft Margarine. In order to fill the container properly, soft margarine must be quite fluid. Usually there are no quiescent tubes. A large blender is used for work softening so that the product does not become brittle as a result of excess crystallization in the bowl (Figure 2.3). For a low-solids oil blend the blender can be placed between the A units or replaced with an extrusion valve to avoid overworking the margarine. Bouffard (242) has described precrystallization techniques for controlling the consistency of spreadable margarines. Faur (243) has demonstrated that, for an A unit–blender–A unit configuration, by increasing the relative amount of cooling in the first A unit or by adding another blender after the second A unit, the spreadability of products packed at the same temperature is enhanced significantly.

Liquid Margarine. Liquid or "squeeze" margarine can be prepared by use of the same equipment to chill and work the product as that used for soft margarine (115). Improved resistance to oil separation is claimed for a process in which the oil blend is cooled and held in a quiescent state for at least 5 h prior to adding the aqueous phase, rechilling, and whipping (116). Stability also is said to be improved by finely dispersing about 5% nitrogen in the emulsion (244). Improved resistance to oil-off may also be attainable by use of suitable emulsifiers (see Section 6.2).

LOW-CALORIE SPREADS

Low-fat spreads, as originally introduced in the 1960s, contained 40% fat and only water, salt, and preservatives in the aqueous phase. The spreads were labeled "diet margarine" and usually consisted of very tight water-in-oil emulsions that had poor melting and flavor release properties. Some such products continue to be marketed today. The first 40% fat products containing dairy solids to be marketed in the United States in the early 1980s were in fact stabilized by the water binding ability of the very high level of milk protein employed. These spreads exhibited poor microbiological stability, did not have the 4- to 6-month shelf life of other spreads, and were withdrawn from the marketplace. Today, lower fat, higher quality spreads, usually containing milk protein and various stabilizing agents, are being produced. Formulation and processing of modern low-fat spreads has been summarized by Moran (245), and production of low-calorie spreads and butters in Europe has been discussed by Madsen (246). In Europe these products are understood by the consumer to be spreads and not as replacements for full-fat margarine. In the United States there remains considerable consumer confusion about the distinction between margarine and low-fat spreads, perhaps because of the multitude of fat levels available between 20 and 75% in the spread category. However, products containing 40% or less fat usually have a statement on the package that the spread is not intended for baking or frying.

Oil-continuous products are favored because of their lubricity, release of oil-soluble buttery flavors, microbiological stability, and reduced tendency for moisture loss. It is preferred that the product melts in the mouth and on hot foods, and that it does not exude water when being spread. Common tests for emulsion stability are electrical resistance, absence of staining with a water-soluble dye on a microscope slide, failure to disperse easily when stirred in water at room temperature, and negative or only slight reaction with paper impregnated with a water-indicating dye, such as bromophenol-blue, after spreading.

The poor organoleptic response of the original diet margarines, which contain only fat, water, salt, emulsifier, flavor, vitamins, and preservatives, is due to the fact that the products are highly emulsified in order to achieve the stability required. Some of these products consist of such tight emulsions that even when held at temperatures high enough to melt all of the fat, phase separation does not occur. This results in a greasy sensation on the palate and little flavor and salt delivery. Processing is similar to that described in the preceding section, except that the lower the fat content, the more critical the manufacturing parameters as a result of emulsion instability. In forming the emulsion, the aqueous phase and oil should be similar in temperature and should be combined slowly. Stronger agitation is required to ensure homogeneity because of the inherent viscosity of a high internal phase emulsion. Care is also taken to avoid incorporation of air during emulsification. To prepare a low-fat spread with protein, Altrock and Ritums (247) used deaerated water and circulated the fat phase through the system to precoat the equipment and displace air. Then the phases were fed to the chiller in concentric streams with the oil blend on the outside. Low-fat emulsions have been found to be sensitive to line pressures and cooling rate (248). Fill temperatures are higher than corresponding 80% fat products because the emulsion is more viscous. If the fill temperature is too low, the product will mound in the bowl with excessive lid contact and may become crumbly with water leakage as it is packaged (249). If too much crystallization takes place early in the process, the shearing forces of processing become so great that they break down the emulsion (245). Therefore, low-fat products are more easily prepared by use of high-liquid-oil and low-SFI blends (250). Norton (251) has prepared oil-continuous spreads with less than 14% fat utilizing a high-melting mono-, di-, or triglyceride seeding component. The seeding component allows the use of higher crystallization temperatures, where the crystallization in the α form is avoided and crystallization directly into the β or β' form is slower and more controllable.

This poor oral response and processing difficulties encountered with these low-fat spreads can be overcome by utilizing gelling agents. Gelatin is particularly suitable since the gel melts in the mouth allowing dairy solids, or other oil-in-water-promoting ingredients incorporated in the oil or in the aqueous phase, to disrupt the water-in-oil emulsion. Thus the spread melts cleanly with good flavor and salt release. The aqueous phase can be cooled to below the gelation temperature and worked to form gelled beads that are then mixed with the fat. Alternatively, the fat can be dispersed in the heated aqueous phase as an oil-in-water emulsion, which upon cooling to below gelation temperature and working, inverts to a water-in-oil emulsion (252). For very low fat spreads the use of a combination of monoglyceride and polyglycerol ester emulsifiers is claimed to result in wider processing latitude and enhanced finished product stability (195). In practice the oil-in-water emulsion can be cooled by standard swept-surface heat exchangers, or even by static heat exchangers (253), and inversion accomplished in a high-shear working unit (254). It is critical that the cooling times and temperatures employed allow gelation to occur before exiting the working unit. Inclusion of low levels of starch or starch derivatives was found to be beneficial for increasing the gelation rate of the gelatin. Spreads containing less than 27% fat have been prepared in this manner (255). It can be difficult to maintain inversion at high throughput rates or if fats with a high solid fat content such as butter oil are used. These difficulties may be overcome by injecting a small amount of the molten fat into the process stream just prior to the inversion unit after the bulk of the emulsion has been cooled (256). Spreads containing 20% fat and high levels of milk protein and gelling agents have been reported (257). Stability was enhanced by homogenization of the aqueous phase prior to forming a water-in-oil emulsion.

Some of the alternatives to gelatin that have been proposed are low-melting carrageenan gels (258), agar and/or pectin (259), gellan gum blends (260), and gelling starch derivatives (261). Oil-continuous spreads containing 15–35% fat and high levels of a nongelling, low-dextrose equivalent starch hydrolyzate have also been prepared by an inversion process (262). The use of nongelling proteinaceous aqueous phases, which contain starch and/or sodium alginate and which have specific viscosity characteristics, has been reported (263). It has been suggested that finished product stability problems may be encountered if the aqueous phase contains amino acid residues in excess of 200 ppm, and such amino acid residues may be components of commercial gums (264).

Very low fat spreads have also been prepared by homogenizing and chilling oil-in-water emulsions containing substantial amounts of gelling and/or thickening agents (265–268). The finished products may, in a sense, be described as "bicontinuous" because the nature of the fat phase, although essentially dispersed, contributes significantly to the plasticity and organoleptic properties. Water-continuous spreads containing as little as 5% or less fat have also been disclosed. These products consist of at least two gelled aqueous phases, one or more of which is continuous, and at least one of which contains an aggregate-forming gelling material such as modified starch, denatured protein, or microcrystalline cellulose (269). Some quick-setting purified starches have been found to be particularly applicable in terms of flavor, texture, and stability in this application (270). Preferably, besides the aggregate gel, the dispersion also includes a network-forming gelling agent such as gelatin, carrageenan, alginate, agar, gellan, pectin, or mixtures of these. The properties of such a spread were enhanced by the addition of high levels of a casein source (271). Preferably the dairy ingredient is demineralized to reduce the acidity required in order to attain the pH necessary for microbiological preservation. Extremely low fat spreads have also been prepared from a gelled or thickened aqueous phase and a mesomorphic phase consisting of edible surfactants and water (272).

DETERIORATION AND SHELF LIFE

In the United States, the "sell by" or "best when used by" date stamped on most margarines is 6–12 months from the date of manufacture. Some lower fat products are given only a 4-month shelf life. Table spreads generally are distributed and merchandised under refrigeration. At higher temperatures the product is more susceptible to oiling out, discoloration, off-flavor development, flavor loss, texture problems, and mold. In a sensory evaluation of polyunsatu-

Deterioration and Shelf Life 105

rated margarines, McBride and Richardson (273) found the high-quality shelf life to be approximately 8 months with storage at 5°C (41°F) and 6 months at 10°C (50°F). Naudet and Biasini (274) evaluated the organoleptic and chemical characteristics of three types of margarine over a 12-month period. Animal–vegetable margarine was acceptable for 20 weeks at 5°C (41°F) but became unacceptable after 12 weeks at ambient temperature. Vegetable oil stick and soft margarines were both acceptable for 6 months in the refrigerator or for 3 months at ambient temperature. Deterioration in flavor score was accompanied by increases in peroxide value and Kreis number. Shelf life of two commercial margarine samples was estimated by calculating induction times based on peroxide value measurements during long-term storage (275). The temperature at which margarine is stored may influence its texture. When fresh production samples of commercial margarine were stored at either 4 or 13°C (39.2 or 55.4°F) these products became softer and harder, respectively, when rheological measurements were conducted at 15°C (59°F) (276). In the United States it is not uncommon that margarine products are "end-aisle" displayed outside of the refrigerator case during promotional events. This can sometimes lead to recrystallization of higher melting glycerides and an increase in melting point. Organoleptically this results in a heavier sensation on the palate and slower flavor and salt release. If an oil has a significant tendency to form β crystals, this type of abuse will accelerate the development of sandiness in the margarine.

Packaging plays a significant role in maintaining the quality of the product. If margarine is packed in plastic tubs or bottles that transmit light, storage under fluorescent lighting in the dairy case can cause development of off-flavors within days. Products containing liquid soybean oil, liquid canola oil, or other oils with a high linolenic acid content are particularly susceptible to developing a fishy, "light-struck" flavor and aroma. Care also must be exercised in the selection of resins, pigments, adhesives, and other additives used in the manufacture of plastic containers and laminated wrappers for margarine. The package must allow no unapproved components to enter the food and should be free of solventlike odors and metals such as copper. Margarine readily picks up off-flavors and odors from its surroundings, so that care must be exercised during shipping and storage. In a study of volatile halocarbons in margarines, Entz and Diachenko (277) generally found less than 100 ppb, probably contributed by the oil and water as well as packaging materials. In those instances where a high level (1–5 ppm) was found, the products had been obtained from a supermarket adjacent to a dry cleaning establishment.

Discoloration of the margarine surface is another package-related defect observed frequently, particularly in oil-continuous spreads containing low fat and/or salt. Surface darkening occurs when moisture evaporates, leaving an outer layer enriched in the fat phase that contains the coloring matter. This reportedly is alleviated by dispersing a material such as titanium dioxide in the product, which minimizes the change in optical properties as the moisture evaporates (278). The use of a thin, imperceptible, edible fat layer on the surface of tub-type spreads has been suggested as a means of reducing mold susceptibility and discoloration (279). Poot and Verburg (280) have derived a mathematical equation correlating increased water content and lower solid fat content of a margarine to greater moisture loss. The drying rate also is affected by temperature and relative humidity. Discoloration of margarine packed in bowls is not a problem if the container is heat sealed or has a tight-fitting lid. Stick margarine wrapped in laminated foil sometimes is observed to form dark yellow streaks on the surface in the region of overlapping flaps or on product surfaces not in contact with the foil due to irregularities on the surface of the product. Moisture loss is considerably greater for parchment wrapped prints; however, discoloration may be less apparent than when foil is used, because evaporation occurs from the whole surface rather than from highly localized areas (281). Moisture loss during shelf life must also be considered in determining the fill weight required to ensure that the product remains in compliance with existing net weight regulations. One advantage of moisture loss is that it provides a surface that is less prone to proliferation of microorganisms.

Margarine has an excellent health record in regard to food-borne disease of microbial origin. An implication in the British press in 1989 that a case of listeriosis was caused by contaminated margarine was later concluded to be false (282). Mossel (283) has reported the results of an extensive study of microorganisms found in margarine and has recommended quality control procedures and microbiological specifications. Molds and yeasts, especially lipolytic yeasts, occasionally cause spoilage of margarine. The use of salt and preservatives and the control of the aqueous-phase droplet size can prevent the growth of organisms in an 80% fat product. However, increased vigilance must be exercised in product formulation and inspection of raw materials, packaging, and equipment when lower salt and higher moisture spreads are produced.

The effect of adverse storage and transportation conditions has been reviewed by Faur (284), and Muys (211) has discussed the microbiological storage qualities of margarine. Dritschel (285) has detailed a rating system for evaluation of finished product quality based on appearance, texture, flavor, and packaging.

REFERENCES

1. A.J.C. Andersen and P.N. Williams, *Margarine,* 2nd ed., Pergamon Press, New York, 1965.
2. S.F. Riepma, *The Story of Margarine,* Public Affairs Press, Washington, D.C., 1970.
3. S.C. Miksta, *J. Am. Oil Chem. Soc.* **48,** 169A (1971).
4. K. Snodgrass, *Margarine as a Butter Substitute,* Food Research Institute, Stanford University, 1930.
5. W. Clayton, *Margarine,* Longmans Green, New York, 1920.
6. *Fed. Reg.* **38,** 25672 (1973).

References 107

7. *Fed. Reg.* **48,** 52692 (1983).
8. Code of Federal Regulations, Title 21, Sec. 166.10, 1993.
9. *Fed. Reg.* **58,** 43580 (1993).
10. *Fed. Reg.* **58,** 21648 (1993).
11. Code of Federal Regulations, Title 21, Sec. 166.40, 1993.
12. *Fed. Reg.* **41,** 36509 (1976).
13. *Fed. Reg.* **58,** 2066 (1993).
14. *Fed. Reg.* **58,** 17096, 17171, 17328 (1993).
15. *Fed. Reg.* **58,** 44020 (1993).
16. J.M. deMan, J.E. Dobbs, and P. Sherman in P. Sherman, ed., *Food Texture and Rheology*, Academic Press, New York. 1979, p. 43.
17. American Oil Chemists' Society, *Official Methods and Recommended Practices*, 4th ed., AOCS Press, Champaign, Ill., Method Cc 16–60, Reapproved 1989.
18. J.M. deMan and A.M. Beers, *J. Texture Stud.* **18,** 303 (1987).
19. J.M. deMan, *J. Am. Oil Chem. Soc.* **60** (1983).
20. A.J. Haighton, *J. Am. Oil Chem. Soc.* **36,** 345 (1959).
21. H. Rohm and S. Raaber, *J. Sensory Stud.* **8,** 81 (1991).
22. B.S. Kamel and J.M. deMan, *Can. Inst. Food Sci. Technol. J.* **8,** 117 (1975).
23. P.W. Board, K. Aichen, and A. Kuskis, *J. Food Technol.* **15,** 277 (1980).
24. M. Tanaka, J.M. deMan, and P.W. Voisey, *J. Texture Stud.* **2,** 306 (1971).
25. J.H. Prentice, *Lab. Pract.* **3,** 186 (1954).
26. I. Vasic and J.M. deMan, *J. Am. Oil Chem. Soc.* **44,** 225 (1967).
27. F. Shama and P. Sherman, *J. Texture Stud.* **1,** 196 (1970).
28. P. Stern and J. Cmolik, *J. Am. Oil Chem. Soc.* **53,** 644 (1976).
29. J.L. Kokini and A. Dickie, *J. Texture Stud.* **13,** 211 (1982).
30. A.J. Haighton, *J. Am. Oil Chem. Soc.* **46,** 570 (1969).
31. L. deMan. J.M. deMan, and B. Blackman, *J. Am. Oil Chem. Soc.* **66,** 128 (1989).
32. L. deMan, E. Postmus, and J.M. deMan, *J. Am. Oil Chem. Soc.* **67,** 323 (1990).
33. L. deMan, C.F. Chen, and J.M. deMan, *J. Am. Oil Chem. Soc.* **68,** 70 (1991).
34. P. Seiden (to Procter and Gamble Ltd.), Can. Pat. 718,372 (1965).
35. J.M. deMan and F.W. Wood, *J. Dairy Sci.* **41,** 369 (1958).
36. R.B. Lomneth, D.R. Blair, G.L. Parnell, and B.Y. Tao (to Procter and Gamble), U.S. Pat. 4,388,339 (1983).
37. R.P. Borwankar, L.A. Frye, A.E. Blaurock, and F.J. Sasevich, *J. Food Eng.* **16,** 55 (1992).
38. E.S. Lutton, *5th Margarine Research Symposium*, October 17, 1958.
39. B. Balinov, O. Söderman, and T. Wärnheim, *J. Am. Oil Chem. Soc.* **71,** 513 (1994); P.T. Callaghan, K.W. Jolley, and R.S. Humphrey, *J. Coll. Interface Sci.* **93,** 521 (1983); K.J. Packer and C. Rees, *J. Coll. Interface Sci.,* **40,** 206 (1972).
40. R. Zschaler, *Fette Seifen Anstrichm.* **79,** 107 (1977).
41. C. Poot, C. Verburg, D. Kirton, and A. MacNeill (to Lever Brothers), U.S. Pat. 4,087,564 (1978).
42. D.P. Moran (to Lever Brothers), U.S. Pat. 4,115,598 (1978).
43. J. Henricus, M. Rek, and E. Baas (to Unilever), Brit. Pat. 1,358,260 (1974).
44. E. Postmus, L. deMan, and J.M. deMan, *Can. Inst. Food Sci. Technol. J.* **22,** 481 (1989).
45. U.S.D.A., *Agricultural Statistics*, U.S. Government Printing Office, Washington, D.C., 1972, 1977, 1983.

46. *Fed. Reg.* **50,** 3745 (1985).
47. M.E. Stansby, *J. Am. Oil Chem. Soc.* **50,** 220A, 222A, 224A (1973).
48. *Fed. Reg.* **54,** 38223 (1989).
49. V. Young, *Lipid Technol.,* **2,** 7 (1990).
50. S.S. Chang (to Vitrum AB), U.S. Pat. 4,101,673 (1978).
51. M. Takso (to Q.P. Corporation), U.S. Pat. 4,623,488 (1986).
52. I.P. Freeman, G.J. van Lookeren, F.B. Padley, and R.G. Polman (to Unilever), Eur. Pat. Appl. 0,304,115 A2 (1989).
53. L.R. Schroeder and D.J. Muffett (to General Mills), U.S. Pat. 4,913,921 (1990).
54. R.L. Antrim and J.B. Taylor (to Nabisco Brands), U.S. Pat. 4,961,939 (1990).
55. R.L. Antrim, N.E. Lloyd, and J.B. Taylor (to Nabisco Brands), U.S. Pat. 4,963,368 (1990).
56. R.L. Antrim and J.B. Taylor (to Nabisco Brands), U.S. Pat. 4,963,385 (1990).
57. M.E. Stansby, *J. Am. Oil Chem. Soc.* **55,** 238 (1978).
58. F. Orthoefer, private communication, 1994.
59. A. Joshi, V.V.R. Subrahmanyam, and S.A. Momin, *J. Oil Technol. Assoc. India* **25,** 7 (1993).
60. I. Heertje, *Food Struct.* **12,** 77 (1993).
61. A.C. Juriaanse and I. Heertje, *Food Microstruct.* **7,** 181 (1988).
62. A.J. Haighton, *J. Am. Oil Chem. Soc.* **53,** 397 (1976).
63. J. Lefebvre, *J. Am. Oil Chem. Soc.* **60,** 295 (1983).
64. American Oil Chemists' Society in Ref. 17, Method Cd 10–57, Reapproved 1989.
65. American Oil Chemists' Society in Ref. 17, Method Cd 16–81, Reapproved 1989.
66. J.C. van den Enden, A.J. Haighton, K. van Putte, L.F. Vermaas, and D. Waddington, *Fette Seifen Anstrichm.* **80,** 180 (1978).
67. E.S. Lutton, *J. Am. Oil Chem. Soc.* **49,** 1 (1972).
68. C.W. Hoerr, *J. Am. Oil Chem. Soc.* **41,** 4, 22, 32, 34 (1964).
69. I. Wilton and G. Wode, *J. Am. Oil Chem. Soc.* **40,** 707 (1963).
70. T.D. Simpson and J.W. Hagemann, *J. Am. Oil Chem. Soc.* **59,** 169 (1982).
71. L. Hernqvist and K. Larsson, *Fette Seifen Anstrichm.* **84,** 349 (1982).
72. M. Kellens and H. Reynaers, *Fett Wiss. Technol.* **94,** 94 (1992).
73. U. Riiner, *Lebensm. Wiss. Technol.* **4,** 175 (1971).
74. M. Vaisey-Genser, B.K. Vane, and S. Johnson, *J. Texture Stud.* **20,** 347 (1989).
75. L.H. Wiedermann, *J. Am. Oil Chem. Soc.* **55,** 823 (1978).
76. M.A.M. Zeitoun, W.E. Neff, G.R. List, and T.L. Mounts, *J. Am. Oil Chem. Soc.* **70,** 467 (1993).
77. J.S. Aronhime, S. Sarig, and N. Garti, *J. Am. Oil Chem. Soc.* **65,** 1144 (1988).
78. B. Weinberg, *Can. Inst. Food Sci. Technol. J.* **5,** A57 (1972).
79. V. Persmark and L. Bengtsson, *Riv. Ital. Sostanze Grasse* **3,** 307 (1976).
80. J.M. deMan, *J. Inst. Can. Sci. Technol. Aliment.* **11,** 194 (1978).
81. L. Hernqvist and K. Anjou, *Fette Seifen Anstrichm.* **85,** 64 (1983).
82. L. Hernqvist, B. Herslöf, K. Larsson, and O. Podlaha, *J. Sci. Food Agric.* **32,** 1197 (1981).
83. V.S. Wähnelt, D. Meusel, and M. Tülsner, *Fat Sci. Technol.* **93,** 117 (1991).
84. K.F. Gander, J. Hannewijk, and A.J. Haighton (to Unilever), Can. Pat. 830,938 (1969).
85. B. Freier, O. Popescu, A.M. Ille, and H. Antoni, *Ind. Aliment.* **24,** 61, 75 (1973).
86. N. Krog, *J. Am. Oil Chem. Soc.* **54,** 124 (1977).
87. S. Lee and J.M. deMan, *Fette Seifen Anstrichm.* **86,** 460 (1984).
88. J. Hojerová, S. Schmidt, and J. Krempasky, *Food Struc.* **11,** 147 (1992).

References

89. K. Sato and T. Kuroda, *J. Am. Oil Chem. Soc.* **64,** 124 (1987).
90. C.F. Shen, L. deMan, and J.M. deMan, *Elaeis* **2,** 143 (1990).
91. F. Cho and J.M. deMan, *J. Food Lipids* **1,** 53 (1993).
92. J. Ward, *J. Am. Oil Chem. Soc.* **65,** 1731 (1988).
93. G.K. Belyaeva, N.I. Kozin, V.S. Golosava, and E. Fal'k, *Maslozhirovaya Promyshlennost* **38,** 15–16 (1969); through *Food Sci. Technol. Abstr.*, No. 3, N0096 (1970).
94. J. Madsen, *Zesz. Probl. Postepow Nauk Roln.*, No. 136, 147 (1973).
95. W. van der Hoek, *Rev. Fr. Corps Gras* **16,** 761 (1969).
96. V. D'Souza, L. deMan, and J.M. deMan, *J. Am. Oil Chem. Soc.* **68,** 153 (1991).
97. V. D'Souza, L. deMan, and J.M. deMan, *J. Am. Oil Chem. Soc.* **69,** 1198 (1992).
98. I. Blanc, *Rev. Fr. Corps Gras* **16,** 457 (1969).
99. U. Riiner, *Lebensm. Wiss. Technol.* **3,** 101 (1970).
100. U. Riiner, *Lebensm. Wiss. Technol.* **4,** 76 (1971).
101. K.G. Berger, *Chem. Ind. (London)* **1975,** 910–913.
102. B. Jacobsberg and Oh Chuan Ho, *J. Am. Oil Chem. Soc.* **53,** 609 (1976).
103. D.A. Okiy, *Oleagineaux* **33,** 625 (1978).
104. P.H. Yap, J.M. deMan, and L. deMan, *Fat Sci. Technol.* **91,** 178 (1989).
105. E. Sambuc, Z. Dirik, G. Reymond, and M. Naudet, *Rev. Fr. Corps Gras* **27,** 505 (1980).
106. E. Sambuc, Z. Dirik, G. Reymond, and M. Naudet, *Rev. Fr. Corps Gras* **28,** 13 (1981).
107. B. Chikany, *Olaj Szappan Kosmet.* **31,** 107 (1982); through *Food Sci. Technol. Abstr.*, No. 10. N0472 (1983).
108. H.R. Kattenberg and C. Poot (to Lever Brothers), U.S. Pat. 4,016,302 (1977).
109. A. Yuki, K. Matsuda, and A. Nishimura, *J. Jpn. Oil Chem. Soc.* **39,** 236 (1990).
110. J.A. Van Meeteren and L.H. Wesdorp (to Unilever), Eur. Pat. Appl. 0,498,487 A1 (1992).
111. W. Heimann and J. Baltes, *Fette Seifen Anstrichm.* **73,** 113 (1971).
112. A. Yaron, B. Turzynski, C. Shmulinzon, and A. Letan, *Fette Seifen Anstrichm.* **75,** 533 (1973).
113. E.G. Latondress, *J. Am. Oil Chem. Soc.* **58,** 185 (1981).
114. A. Moustafa, *Rev. Fr. Corps Gras* **26,** 485 (1979).
115. M.J. Pichel (to Swift), U.S. Pat. 3,338,720 (1967).
116. D. Melnick and E.L. Josefowicz (to Corn Products Company), U.S. Pat. 3,472,661 (1969).
117. F.V.K. Young, *J. Am. Oil Chem. Soc.* **60,** 374 (1983).
118. F. Cho, J.M. deMan, and O.B. Allen, *J. Food Lipids* **1,** 25 (1993).
119. T. Wieske, in *7th European Symposium*, 1977, pp. 214–224.
120. T. Wieske (to Unilever), Brit. Pat. 1,479,287 (1977).
121. J. Baltes (to Harburger Oelwerke Brinckman & Mergell), Can. Pat. 846,842 (1970).
122. T.A. Pelloso and L. Kogan (to Nabisco Brands), U.S. Pat. 4,316,919 (1982).
123. H. Heider and T. Wieske (to Lever Brothers), U.S. Pat. 4,230,737 (1980).
124. J.P. McNaught (to Lever Brothers), U.S. Pat. 3,900,503 (1975).
125. R.P. Mensink and M.B. Katan, *N. Engl. J. Med.* **323,** 439 (1990).
126. P.L. Zock and M.B. Katan, *J. Lipid Res.* **33,** 399 (1992).
127. R.P. Mensink, P.L. Zock, M.B. Katan, and G. Hornstra, *J. Lipid Res.* **33,** 1493 (1992).
128. A.H. Lichtenstein, L.M. Ausman, W. Carrasco, J.L. Jenner, J.M. Ordovas, and E.J. Schaefer, *Arteriosclerosis Thrombosis* **13,** 154 (1993).
129. R. Wood, K. Kubena, B. O'Brien, S. Tseng, and G. Martin, *J. Lipid Res.* **34,** 1 (1993).

130. E.N. Siguel and R.H. Lerman, *Am. J. Cardiol.* **71,** 916 (1993).
131. R. Troisi, W.C. Willett, and S.T. Weiss, *Am. J. Clin. Nutr.* **56,** 1019 (1992).
132. W.C. Willett and co-workers, *Lancet* **341,** 581 (1993).
133. J.T. Judd and co-workers, *Am. J. Clin. Nutr.* **59,** 861 (1994).
134. *Food Labeling News* **2**(21), 15 (1994).
135. B.F. Haumann, *INFORM* **5,** 346, 350, 352, 354, 356, 358 (1994).
136. J. Ward (to Nabisco Brands), U.S. Pat. 4,341,812 (1982).
137. J. Ward (to Nabisco Brands), U.S. Pat. 4,341,813 (1982).
138. H.A. Graffelman (to Lever Brothers), U.S. Pat. 3,617,308 (1971).
139. J. Van Heteren, J.N. Pronk, W.J. Smeenk, and L.F. Vermaas (to Lever Brothers), U.S. Pat. 4,386,111 (1983).
140. M. Fondu and M. Willens (to Lever Brothers), U.S. Pat. 3,634,100 (1972).
141. R. Keuning, A.J. Haighton, W. Dijkshoorn, and H. Huizinga (to Lever Brothers), U.S. Pat. 4,360,536 (1982).
142. W. Dijkshoorn, H. Huizinga, and J. Pronk (to Lever Brothers), U.S. Pat. 4,366,181 (1982).
143. A. Nagoh, O. Kaizuka-shi, and M. Miyabe (to Fuji Oil Company), Eur. Pat. Appl. 0,526,980 A1 (1993).
144. R. Schijf and V.K. Muller (to Unilever), Eur. Pat. Appl. 0,470,658 A1 (1992).
145. A.J. Lansbergen and R. Schijf (to Unilever), Eur. Pat. Appl. 0,455,278 A2 (1991).
146. D.K. Yayashi, R.C. Dinwoodie, M.T. Dueber, R.G. Krishnamurthy, and J.J. Myrick (to Kraft General Foods, Inc.), U.K. Pat. Appl. 2,239,256 A (1991).
147. G.R. List, E.A. Emken, W.F. Kwolek, T.D. Simpson, and H.J. Dutton, *J. Am. Oil Chem. Soc.* **54,** 408 (1977).
148. M.A.M. Zeitoun, W.E. Neff, and T.L. Mounts, *Rev. Fr. Corps Gras* **39,** 85 (1992).
149. W. Stratmann and L.F. Vermaas (to Lever Brothers), U.S. Pat. 4,425,371 (1984).
150. P.W. Eiilott, M.R.J. Greep, J.A. Van Meeteren, and L.H. Wesdorp (to Unilever), Eur. Pat. Appl. 0,369,519 A2 (1990).
151. B. Sreenivasan (to Lever Brothers), U.S. Pat. 3,859,447 (1975).
152. R. DeLathauwer, M. Van Opstal, and A.J. Dijkstra (to Safinco), U.S. Pat. 4,419,291 (1983).
153. J. Boot, A. Rozendaal, and R. Schijf (to Unilever), Eur. Pat. Appl. 0,060,139 A2 (1982).
154. L. Kogan and T. Pelloso (to Nabisco Brands), U.S. Pat. 4,335,156 (1982).
155. A. Zaks, R.D. Feeney, and A. Gross (to Opta Food Ingredients, Inc.), Internat. Pat. Appl. WO 92/03937 (1992).
156. M. Stahl, Ger. Fed. Rep. Pat. Appl. 2,935,572 (1981).
157. H. Menz, J. Rost, and T. Wieske (to Lever Brothers), U.S. Pat. 3,658,555 (1972).
158. T. Weiske and H. Menz, *Fette Seifen Anstrichm.* **74,** 133 (1972).
159. G. Von Rappard and W. Kretschner (to Walter Rau), Ger. Fed. Rep. Pat. Appl. 2,832,636 (1980).
160. B.A. Roberts (to Procter and Gamble), U.S. Pat. 4,446,165 (1984).
161. P.D. Orphanos and co-workers, (to Proctor and Gamble), U.S. Pat. 4,940,601 (1990).
162. R.J. Jandacek and J.C. Letton (to Proctor and Gamble), U.S. Pat. 5,017,398 (1991).
163. F.W. Cain, F.R. De Jong, A.J. Lanting Marijs, and J.J. Verschuren (to Unilever), Eur. Pat. Appl. 0,415,468 A2 (1991).
164. C.J. Glueck and co-workers, *Am. J. Clin. Nutr.* **35,** 1352 (1982).
165. R. Nicolosi, *INFORM* **1,** 831 (1990).
166. A.P. Simopoulos, *Am. J. Clin. Nutr.* **54,** 438 (1991).

167. D.F. Horrobin, *Prog. Lipid Res.* **31,** 163 (1992).
168. A.P. Bimbo, *J. Am. Oil Chem. Soc.,* **66,** 1717 (1989).
169. F.V.K. Young, "The Chemical and Physical Properties of Crude Fish Oil for Refiners and Hydrogenators," *International Association of Fish Meal Manufacturers Fish Oil Bull., No. 18,* Hertfordshire, UK, 1986.
170. F.D. Gunstone, *Prog. Lipid Res.* **31,** 145 (1992).
171. S. Patel, "Quality Considerations in Margarine Oil Manufacturing," presented at ISF AOCS World Congress, New York, 1980.
172. American Oil Chemists' Society in Ref. 17, Method Cg 2-83, Reapproved 1989.
173. L.H. Wiedermann, *J. Am. Oil Chem. Soc.* **49,** 478 (1972).
174. *Fed. Reg.* **38,** 25671 (1973).
175. T.H. Smouse, J.K. Maines, and R.R. Allen (to Anderson Clayton), U.S. Pat. 4,038,436 (1977).
176. C.E. Eriksson, *Food Chem.* **9,** 3 (1982).
177. L.L. Linteris (to Lever Brothers), U.S. Pat. 3,721,570 (1973).
178. C.F. Cain, I.J. Day, M.G. Jones, and I.T. Norton (to Unilever), Eur. Pat. Appl. 0,279,499 A2 (1988).
179. J.D. Dziezak, *Food Technol.* **42,** 172, 174 (1988).
180. J. Madsen, *Fett Wiss. Technol.* **89,** 165 (1987).
181. A.G. Gaonkar and R.P. Borwankar, *Colloids and Surfaces* **59,** 331 (1991).
182. M. Schneider, *Fett Wiss. Technol.* **94,** 524 (1992).
183. W. van Nieuwenhuyzen, *J. Am. Oil Chem. Soc.* **58,** 886 (1981).
184. H.W. Lincklaen and J.H.M. Rek (to Lever Brothers), U.S. Pat. 3,796,815 (1974).
185. T. Wieske, K.H. Todt, J.A. De Feÿter, and W.A. Castenmiller (to Van den Bergh Foods), U.S. Pat. 5,079,028 (1992).
186. H.J. Duin, A.F. van Dam, and J.H.M. Rek (to Lever Brothers), U.S. Pat. 4,148,930 (1979).
187. J.H.M. Rek and P.M.J. Holemans (to Lever Brothers), U.S. Pat. 4,325,980 (1982).
188. J. Madsen, *Res Discl.* No. 238, 91 (1984).
189. B.R. Harris, U.S. Pat. 1,917,255 (1933).
190. M.F. Stewart and E.J. Hughes, *Process Biochem.* **7,** 27 (1972).
191. W.A. Gorman, R.G. Christie, and G.H. Kraft (to National Dairy Products Corporation), U.S. Pat. 2,937,093 (1960).
192. W.A.M. Castenmiller, A.K. Chesters, and P.B. Ernsting (to Lever Brothers), U.S. Pat. 4,874,626 (1989).
193. K. Brammer and T. Wieske (to Lever Brothers), U.S. Pat. 3,889,005 (1975).
194. F. Groeneweg, F. van Voorst Vader, and W.G.M. Agterof, *Chem. Eng. Sci.* **48,** 229 (1993).
195. J. Van Heteren, T.R. Kelly, R.M. Livingston, and A.B. MacNeill (to Unilever), Eur. Pat. Appl., 0,420,314 A2 (1991).
196. Anon., *Res. Discl.* No. 329, 689 (1991).
197. Anon., *Res. Discl.* No. 352, 512 (1993).
198. K. Matsuda and M. Kitao (to Mitsubishi Kasei Corporation), Eur. Pat. Appl. 0,430,180 A2 (1991).
199. K. Terada, S. Fujita, and N. Yoshida (to Asahi Denka), U.S. Pat. 3,914,458 (1975).
200. P.F. Pedersen, *Res. Discl.* No. 321, 64 (1991).
201. B. Barmentlo and N.K. Slater (to Van den Bergh Foods), U.S. Pat. 5,145,704 (1992).
202. H. Kanematsu, E. Morise, I. Niiya, M. Imamura, A. Matsumoto, and G. Katsui, *J. Jpn. Soc. Food Nutr.* **25,** 343 (1972); through *Food Sci. Technol. Abstr.* No. 12, N0647 (1973).

203. E.R. Sherwin, *J. Am. Oil Chem. Soc.* **53,** 430 (1976).
204. J. Cillard, P. Cillard, and M. Cormier, *J. Am. Oil Chem. Soc.* **57,** 255 (1980).
205. K. Klaui, *Flavours* **7,** 165 (1976).
206. G.T. Muys, C.T. Verrips, and R.T.S. van Gorp (to Lever Brothers), U.S. Pat. 3,995,066 (1976).
207. C.T. Verrips and H. Vonkeman (to Lever Brothers), U.S. Pat. 3,904,767 (1975).
208. W.G. Mertens, C.E. Swindells, and B.F. Teasdale, *J. Am. Oil Chem. Soc.* **48,** 544 (1971).
209. D. Melnick (to Corn Products), U.S. Pat. 3,243,302 (1966).
210. C.T. Verrips and J. Zaalberg, *Eur. J. Appl. Microbiol. Biotechnol.* **10,** 187 (1980).
211. G.T. Muys, *Process Biochem.* **4,** 31 (1969).
212. E. Lueck, *Antimicrobial Food Additives,* Springer-Verlag, New York, 1980.
213. F. Kapp and B. Mittag, *Lebensmittelindustrie* **29,** 160 (1982).
214. N.E. Harris and D. Rosenfield, *Food Process. Ind.* **43,** 23 (1974).
215. M. Castenon and B. Inigo, *Lebensmittel Wiss. Technol.* **6,** 70 (1973).
216. J. Bodor, A.W. Schoenmakers, and W.M. Verhue (to Van den Bergh Foods), U.S. Pat. 5,013,573 (1991).
217. Anon. (to Unilever), Austral. Pat. Appl. 52102 (1990).
218. P.M. Klapwijk, *Food Control* **3,** 183 (1992).
219. T.J. Siek and R.C. Lindsay, *J. Dairy Sci.* **53,** 700 (1970).
220. J.G. Keppler, *J. Agric. Food Chem.* **18,** 998 (1970).
221. P.G.M. Haring, J.G. van Pelt, and C.F. Andreae (to Unilever), Eur. Pat. Appl. 0,478,036 A2 (1992).
222. H.T. Slover, R.H. Thompson Jr., C.S. Davis, and G.V. Merola, *J. Am. Oil Chem. Soc.* **62,** 775 (1985).
223. H. Klaui and O. Raunhardt, *Alimenta* **15,** 37 (1976).
224. P.H. Todd (to Kalamazoo Spice Extraction Company), U.S. Pat. 3,162,538 (1964).
225. E. Sambuc and M. Naudet, *Rev. Fr. Corps Gras* **12,** 239 (1965).
226. R. Presse, H. Quendt, and H. Raeuber, *Lebensmittelindustrie* **22,** 34 (1975).
227. G. Gabriel, M. Havenstein, P.M.J. Holemans, and B.E. Kapellen (to Unilever), Eur. Pat. Appl., 0,422,712 A2, 0,422,713 A2, and 0,422,714 A2 (1991).
228. S. Okonogi and co-workers, (to Morinaga Milk Industry Company, Ltd.), U.S. Pat 5,279,847 (1994).
229. N.T. Joyner, *J. Am. Oil Chem. Soc.* **30,** 526 (1953).
230. I. Heertje, J. Van Eendenburg, J.M. Cornelissen, and A.C. Juriaanse, *Food Microstruct.* **7,** 189 (1988).
231. P. De Bruijne, J. Van Eendenburg, and H.J. Human (to Unilever), Eur. Pat. Appl. 0,341,771 A2 (1989).
232. H.R. Kattenberg and C.C. Verburg (to Lever Brothers), U.S. Pat. 4,055,679 (1977).
233. A.D. Wilson, H.B. Oakley, and J. Rourke (to Lever Brothers), U.S. Pat. 2,592,224 (1952).
234. B.D. Miller, P. Phelps, and H.W. Bevarly (to Girdler Corp.), U.S. Pat. 2,330,986 (1943).
235. K. Terada, S. Fujita, H. Kohno, and H. Sugiyama (to Asahi Denka), U.S. Pat. 3,917,859 (1975).
236. K.F. Gander (to Lever Brothers), U.S. Pat. 3,488,199 (1970).
237. D.P.J. Moran (to Lever Brothers), U.S. Pat. 3,490,919 (1970).
238. G.C. Cramer (to Madison Creamery), U.S. Pat. 4,315,955 (1982).
239. R.D. Price and W.L. Sledzieski (to Nabisco Brands), Eur. Pat. Spec. 0,139,398 B1 (1988).
240. J.J. Brockhus, D. Schnell, and K.T. Vermaat (to Unilever), Eur. Pat. Appl. 0,505,007 A2 (1992).

References 113

241. W.A. Gorman, R.G. Christie, and G.H. Kraft (to National Dairy Products Corporation), U.S. Pat. 2,937,093 (1960).
242. C. Bouffard, *Rev. Fr. Corps Gras* **21,** 351 (1974).
243. L. Faur, *Rev. Fr. Corps Gras* **27,** 319 (1980).
244. I. Wilton and K. Bauren (to Margarinbolaget), U.S. Pat. 3,682,656 (1972).
245. D.P.J. Moran, *PORIM Technology,* Publication Number 15, Palm Oil Research Institute of Malaysia, 1993.
246. J. Madsen, in D. Erickson, ed., *World Conference Proceedings, Edible Fats and Oils Processing: Basic Principles and Modern Practices,* American Oil Chemists' Society, Champaign, Ill., 1990, pp. 221–227.
247. W. Altrock and J.A. Ritums (to Lever Brothers), U.S. Pat. 4,366,180 (1982).
248. E.L. Josefowicz and D. Melnick (to Corn Products Company), U.S. Pat. 3,457,086 (1969).
249. J.G. Spritzer, J.J. Kearns, and O. Cooper, U.S. Pat. 3,360,377 (1967).
250. L.L. Linteris (to Unilever), Can. Pat. 871,647 (1971).
251. I.T. Norton (to Van den Bergh Foods), U.S. Pat. 5,244,688 (1993).
252. I.T. Norton and J. Underdown (to Van den Bergh Foods), U.S. Pat. 5,306,517 (1994).
253. P.B. Ernsting (to Lever Brothers), U.S. Pat. 4,883,681 (1989).
254. B. Sreenivasan (to Lever Brothers), U.S. Pat. 4,849,243 (1989).
255. I.T. Norton and J. Underdown (to Van den Bergh Foods), U.S. Pat. 5,151,290 (1992).
256. B. Milo and R. Ochmann (to Unilever), Eur. Pat. Appl. 0,396,170 A2 (1990).
257. S. Madsen, *Res. Discl.* No. 330, 774 (1991).
258. I.T. Norton and C.R.T. Brown (to Unilever), Eur. Pat. Appl. 0,271,132 A2 (1987).
259. I.T. Norton (to Unilever), Eur. Pat. Appl. 0,474,299 A1 (1991).
260. D.J. Pettitt, W. Gibson, and I.A. Challen, *Res. Discl.* No. 301, 338 (1989).
261. F.W. Cain, M.G. Jones, and I.T. Norton (to Lever Brothers), U.S. Pat. 4,917,915 (1990).
262. A.L. Morehouse and C.J. Lewis (to Grain Processing Corporation), U.S. Pat. 4,536,408 (1985).
263. I.T. Norton and R.M. Livingston (to Unilever), Eur. Pat. Appl. 0,496,466 A2 (1992).
264. F.W. Cain, M.G. Jones, and I.T. Norton (to Unilever), Eur. Pat. Appl. 0,279,499 A2 (1988).
265. M.G. Jones and I.T. Norton (to Van den Bergh Foods), U.S. Pat. 5,217,742 (1993).
266. P.M. Bosco and W.L. Sledzieski (to Standard Brands), U.S. Pat. 4,279,941 and 4,292,333 (1981).
267. D.E. Miller and C.E. Werstak (to SCM Corporation), U.S. Pat. 4,238,520 (1980).
268. D.F. Darling (to Lever Brothers), U.S. Pat 4,443,487 (1984).
269. F.W. Cain and co-workers, (to Lever Brothers), U.S. Pat. 4,956,193 (1990).
270. L.H. Wesdorp, R.A. Madsen, J. Kasica, and M. Kowblansky (to Van den Bergh Foods), U.S. Pat. 5,279,844 (1994).
271. G. Banach, L.H. Wesdorp, and F.S. Fiori (to Van den Bergh Foods), U.S. Pat. 5,252,352 (1993).
272. I. Heertje and L.H. Wesdorp (to Unilever), Eur. Pat. Appl. 0,547,647 A1 (1993).
273. R.L. McBride and K.C. Richardson, *Lebensm. Wiss. Technol.* **16,** 198 (1983).
274. M. Naudet and S. Biasini, *Rev. Fr. Corps Gras* **23,** 337 (1976).
275. M. Maskan, M.D. Öner, and A.K. Aya, *J. Food Qual.* **16,** 175 (1993).
276. J.A. Segura, M.L. Herrera, and M.C. Añón, *J. Am. Oil Chemists' Soc.* **67,** 989 (1990).
277. R.C. Entz and G.W. Diachenko, *Food Additives and Contaminants* **5,** 267 (1988).
278. H.M. Princen and M.P. Aronson (to Lever Brothers), U.S. Pat. 4,176,200 (1979).

279. A.G. Havenstein, W. Kahle, and D. Schnell (to Unilever), Eur. Pat. Appl. 0,240,089 A1 (1987).
280. C. Poot and €.C. Verburg, *Fette Seifen Anstrichm.* **76,** 178 (1974).
281. E.L. Josefowicz and D. Melnick (to Corn Products Company), U.S. Pat. 3,148,993 (1964).
282. P. Barnes, *Lipid Technol.* **1,** 46 (1989).
283. D.A.A. Mossel in *Margarine Today: Technological and Nutritional Aspects,* symposium held at Dijon, France, March 21, 1969, pp. 104–125.
284. L. Faur, *Rev. Fr. Corps Gras* **27,** 371 (1980).
285. M.E. Dritschel, *Food Eng.* **42**(10), 90 (1970).

3

Shortening: Science and Technology

INTRODUCTION

1.1 Definition and Characteristics

Shortening is a commercially prepared edible fat used in frying, cooking, baking, and as an ingredient in fillings, icings, and other confectionery items. It may have been so named because, when dough is mixed, water-insoluble fat prevents cohesion of gluten strands literally "shortening" them and thus generating tender baked goods. Shortening is a typically 100% fat product formulated with animal and/or vegetable oils that have been carefully processed for functionality and to remove undesirable flavor and aroma. Overall, shortening improves the texture and palatability of food products while its calories provide heat and energy to fuel the body.

In its most recognized form, household shortening, it is a white, relatively soft, plastic solid with a bland flavor and no detectable odor. Some types have a butterlike color and flavor added. Household and industrial all-purpose shortenings are products formulated with properties permitting their use in both frying and baking. Pourable types include clear liquid or fluid (opaque) shortenings. Liquid shortenings are typically used as cooking or salad oils. Fluid or opaque shortenings are pourable products with a small amount of solid fat or emulsifier suspended in oil. Because they are convenient to use, pourable shortenings are increasing in popularity especially for frying and baking. Shortening is also available in dry form as powder, pellets, or flakes encapsulated in a water-soluble material. Skim milk, cheese whey, corn syrup, soy protein isolate, and cellulose compounds have proven feasible as encapsulating materials (1).

1.2 Products with Characteristics Similar to Shortening

Lard, tallow, and ghee are traditional animal fats that have existed for centuries. Like most shortenings, all of these products are 100% fat. Vanaspati, another all-fat product now primarily vegetable-based, is popular in all Eastern countries but especially India and Pakistan. Other commercial shortening-like products are available with fat contents from 5 to 90%. Most of these contain an aqueous phase emulsified in the oil phase. Butter and margarine are water-in-oil emulsions manufactured worldwide, and most areas have legal labeling stipulations fixing their fat content at 80% minimum. Table spreads are formulated with intermediate fat levels generally from 40 to 80% and many popular brands fall within the 50–70% range. New low-fat or dietary products, spreads with fat levels of 5–40%, have recently been developed and are just entering the marketplace.

Animal fats were once the primary source oils in both North America and Europe; however, shortening, margarine, spread, and low-fat, dietary table products are now usually formulated from vegetable oils. Blends containing animal fats are still available and popular in certain areas. In fact, new "blends" of butter and vegetable oil are gaining acceptance. Animal fats and marine oils are important fat sources in many areas of the world, and quality products are available in Latin America, Australia, and Asia based exclusively on or containing significant levels of these oils.

1.3 Worldwide Production of Shortening and Margarine

Shortening is generally considered an American invention and the data in Table 3.1 (2) certainly confirm that North America is the world's leading producer of both shortening and margarine. As these data indicate, compound fat and shortening production has been stable at nearly 4 million tons per year since 1990, with North America responsible for about half of this. Margarine production is more than double that of shortening and compound oils and currently exceeds 9 million tons per year. The margarine production figures include at least 2 million tons of Vanaspati, a product more like shortening than margarine.

The focus of this chapter is shortening; brief information regarding margarine and other similar products is offered due to similarities in raw materials, usage, production methods, and equipment.

1.4 Functionality

Functionality is a term food technologists employ to describe how well a product performs in a specific application. Shortening and margarine are often characterized as highly functional products. In baking, margarine and

Table 3.1 Production of secondary oils and fats in major countries (2)

Country	Production (1000 ton)				
	1989	1990	1991	1992	1993[a]
Compound fats and shortening					
Australia	104	108	106	108	56
Canada	228	242	296	320	155
Eastern Europe	146	155	148	155	77
Germany	113	114	114	105	52
Japan	216	249	253	255	127
Netherlands	263	263	258	245	115
United Kingdom	121	121	140	137	62
United States	2411	2536	2596	2594	1299
Other	48	49	51	50	25
World total	3650	3837	3962	3969	1968
Margarine					
Australia	162	156	165	161	72
Canada	153	151	156	148	74
Eastern Europe	428	382	382	429	225
Germany	650	685	705	728	363
India[b]	950	853	828	857	428
Japan	214	176	171	177	91
Netherlands	225	233	241	244	130
Pakistan[b]	990	1040	1120	1200	605
Turkey[c]	486	481	554	565	265
United Kingdom	466	451	446	457	230
United States	1148	1256	1224	1278	629
Other	3387	3292	3090	3064	1565
World total	9259	9156	9082	9308	4677

[a] Production January through June 1993.
[b] Vanaspati.
[c] Includes vanaspati.

shortening contribute to the quality of the finished product by imparting a creamy texture and rich flavor, tenderness, and uniform aeration for moisture retention and size expansion. Liquid and fluid shortenings are used in salad oils and for restaurant and industrial deep-fat and pan frying. In frying, shortening functions as more than a heat transfer medium; it also reacts with components in the food to develop unique, savory flavors and odors. Dry shortenings are convenient to store and use. Grease will not soak packaging materials and, although expensive, it can be used in prepackaged cake, biscuit, and pie crust mixes, which are free-flowing at room temperature. Those fat-based products formulated and processed for plasticity spread readily and disperse thoroughly and uniformly in dough, batter, icing, etc. over a wide temperature range.

1.5 Solid Fat Profiles for Margarine

The fat in shortening and margarine products exists in both liquid and solid form. The solid fat index (SFI) is an analytical measure approximating the solid fat content. It is always less than the actual solid content and, to be meaningful, must be determined at several standard temperatures, usually 10°C (50°F), 21.1°C (70°F), 26.7°C (80°F), 33.3°C (92°F), 37.8°C (100°F), and sometimes 40°C (104°F).

The SFI measurements for table margarine are usually determined at 10°C (50°F) as an indication of consistency during crystallization and refrigeration, at 21.1°C (70°F) to simulate room conditions during use, and at 33.3°C (92°F) to approximate "mouth feel" or eating quality. If the 33.3°C (92°F) SFI level is too high, the margarine will melt slowly in the mouth, often creating a "waxy" sensation. SFI curves for stick table-grade margarine are generally steep with solids levels from about 30% at 10°C (50°F) to less than 5% at 33.3°C (92°F) (3). Soft tub margarine oils have less steep SFI curves for a smooth, more plastic consistency. The SFI curves in Figure 3.1 are typical for U.S. tub- and stick-type margarine.

1.6 Solid Fat Profiles for Shortening

The SFI profile is a good indicator of the plastic range of a fat formulated for shortening. *High-stability* shortenings have a steep SFI profile and a narrow plastic range. Typical all-purpose plastic shortenings retain much of their solid fat content over a wider temperature range than high-stability types and consequently possess much flatter profiles. *Liquid* pourable shortenings in-

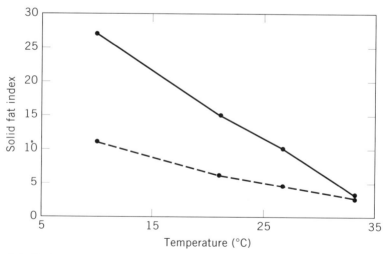

Figure 3.1 Typical solid fat indices for margarine oils. (Courtesy of Kraft Food Ingredients, Memphis, Tennessee.) ———, Table-grade stick; ----, soft tub.

clude clear oils as well as *fluid* or *opaque* types. Pourable shortenings contain low solids levels with very flat SFI profiles. Specialty shortenings have been formulated for specific applications including cakes, dry cake mixes, bread, Danish and puff pastry, pie crust, cookies, crackers, icing, creams and fillings, coating fat, nondairy products, and frying. Specialty shortenings may be of any general type depending on the requirements for that specific application, and their SFI profiles will be characteristic of that particular type.

High-Stability Shortenings. A steep SFI profile is indicative of a narrow plastic range. Products with this type of profile are often referred to as high-stability shortenings. The SFI values may be 50 or greater at 10°C (50°F) but usually less than 10 at 40°C (104°F). As their SFI profiles indicate, these shortenings are not intended to be workable over a wide temperature range. They tend to be hard and brittle below 18.3°C (65°F) and soft above 32.2°C (90°F). High-stability shortenings are used for deep frying, as center fat for confectionery and bakery items, replacements for butter and coating fats, in vegetable/dairy systems, and for crackers and hard cookies.

All-Purpose Shortenings. All-purpose shortenings were developed for household use and to allow production of a wide variety of baked goods by firms that cannot stock individual types formulated for every specialty item. Typical all-purpose plastic shortenings contain 15–30% solid (crystalline) fat and retain many of these solids over their intended temperature usage range of 16–32°C (60–90°F) (4). A wide plastic range is essential as these products must resist breakdown during creaming and are subject to wide temperature variations in the workplace and during shipping and storage. All-purpose shortenings for baking contain emulsifiers to enhance creaming ability and to improve air retention. Since they reduce the smoke point, emulsifiers should be omitted from the formulation when this type shortening is used for deep frying. All-purpose shortenings are formulated as a compromise of individual properties but yet to possess capabilities making them suitable for frying, baking, and confectionery uses.

Pourable Shortenings. It is not unusual to classify an edible fat or oil as shortening simply to differentiate it from products such as margarine that contain moisture and other nonfat materials. Liquid shortenings include clear oils as well as fluid, opaque pourable products. The SFI profiles for clear oils are very flat as they normally contain very low levels of oil-soluble emulsifiers or hard fat. Clear oils can be used in household grilling and frying and in institutional deep frying provided the turnover rate is high enough (15–25%) so that stability is not a concern. The flavor and oxidative stability of such oils is greatly improved by partially hydrogenating soybean, safflower, corn, sunflower, or other source oils. Following hydrogenation, the oil is fractionated and the clear liquid oil separated from solid portion. Above 16°C (60°F), these oils are usually free of suspended solids.

Fluid shortenings can be distinguished from liquid shortenings by their opacity resulting from the suspension of high-melting emulsifiers or fully hydrogenated fats. The total amount of suspended solids ranges between 5 and 15%. These products are usually fluid between 18.3°C (65°F) and 32.2°C (90°F); outside of this range, opaque shortenings may loose their pourability or become more fluid depending on how temperature has altered their solid content. Fluid shortenings are widely used in commercial frying but have also been formulated and marketed successfully for baking cakes, bread, buns, rolls, and pie crust.

Specialty Shortenings. *Roll-in* shortenings are specialty products used almost exclusively for baking. Their primary use is as an ingredient in puff pastry. Puff pastry is prepared by placing a layer of shortening on a layer of dough. This is folded and sheeted until there are more than 700 fat–dough layers. When baked, the shortening melts, liberating moisture that becomes steam "puffing" the thin dough layers into a very delicate flaky structure. The SFI profile for puff pastry is fairly flat with solid levels of 40% or higher at 10°C (50°F) to about 20% at 33.3°C (92°F).

Dry shortenings are fats that have been encapsulated in a water-soluble coating material; their fat content is generally between 75 and 80%. Dry shortenings are used in ready-to-use mixes where only water is added to form a batter ready for baking.

Ambient temperatures, source oils, performance requirements and storage conditions change throughout the world, requiring SFI profile adjustment to meet varying product needs. Typical SFI shortening profiles are shown in Figure 3.2.

PLASTIC THEORY

2.1 Plastic Solids

Modern edible fats are blends of one or more of about a dozen common oils. Those with higher fat levels such as shortening, margarine, and spreads are formulated to possess special physical characteristics. These products appear to be solid yet, when subjected to a shearing force great enough to cause a permanent deformation, all assume the rheological flow characteristics of a viscous liquid. Such solids are referred to as *plastic solids*. Their plastic nature enables them to spread readily and combine thoroughly with other solids or liquids without cracking, breaking, or liquid oil separating from the crystalline fat. These solids are usually relatively soft at ambient temperature and may actually contain as little as 5% solidified fat; assuming that fat crystals are uniform spheres packed in a close cubical pattern, the theoretical maximum solid content is slightly more than 52%.

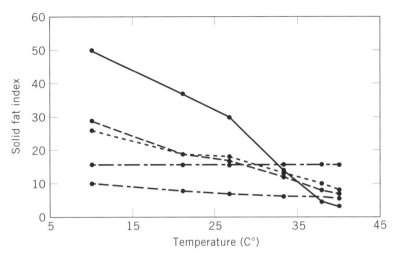

Figure 3.2 Typical solid fat indices for shortening. (Courtesy of Kraft Food Ingredients, Memphis, Tennessee.) ———, High stability frying; – – – –, all-purpose; · · · ·, all-purpose emulsified; — · — · —, bread; — — · — — ·, liquid (fluid) frying.

2.2 Process Definition for Shortening

Shortening is a classic example of a plastic solid. In fact, from a process view, shortening can be defined as a highly functional plastic solid commercially prepared by carefully cooling, plasticizing, and tempering correctly formulated blends of molten, edible fats and oils.

2.3 Conditions Essential for Plasticity

Plastic solids derive their functionality from their unique plastic nature. Three conditions are essential for plasticity (5): (*1*) both liquid and solid phases must be present; (*2*) the solid phase must be so finely dispersed that the entire solid–liquid matrix can be effectively bound together by internal cohesive forces; and (*3*) proper proportions must exist between the phases. Incorrect phase ratios adversely influence product rheology. For example, deficient solids content may result in oil separation while excessive solids can cause hardness or brittleness instead of the desired viscous flow.

2.4 Influence of Crystal Size

Crystal size has a major influence on the rheological properties of plastic solids and is therefore a critical factor that must be considered in their formulation. Fats exist as a three-dimensional liquid–solid matrix in which the liquid oil must be contained. Statically cooled molten fats always form large crystals; crystal population is low and, when the total surface area is insufficient to bind the liquid phase within the crystalline matrix, oil separation occurs. Products become progressively firmer as the crystal size decreases. Rapidly chilling the same fat produces many more and much smaller crystals with a far greater combined surface area enabling the liquid phase to adhere more effectively to the crystal surface. A rapidly chilled shortening formulation will be more stable, much firmer, and possess a greater plastic range than a statically or slowly cooled fat. Typical commercially prepared shortenings, especially those formulated from vegetable or marine oils, usually have mean crystal sizes ranging from 5 to 9 μm (6).

FORMULATION

3.1 Crystalline Nature

Commercial fats solidify in several crystalline polymorphic forms. Two desirable stable forms are commonly designated by the Greek letters beta (β) and beta-prime (β'). Table 3.2 (7) lists many common fats and oils and their most commonly exhibited polymorphic crystalline form.

When the higher melting portion of a solidifying fat crystallizes in a stable β' form, the entire fat will crystallize in this same β' configuration. Plastic shortenings in this polymorphic form consist of small, uniform needlelike crystals, exhibit a smooth texture, aerate well, have excellent creaming properties, and make good cake and icing shortenings. Palm or cottonseed oils in their liquid and partially hydrogenated forms are often included in shortening and margarine formulations to promote β' crystallization ensuring these qualities.

Fats that crystallize in the β polymorphic form tend to be more coarse

Table 3.2 Classification of fats and oils according to crystal habit (7)

β Type	β' Type
Soybean	Cottonseed
Safflower	Palm
Sunflower	Tallow (beef)
Sesame	Herring
Peanut	Menhaden
Corn	Whale
Canola	Rapeseed (high erucic acid)
Olive	—
Coconut	Milk fat (butter oil)
Palm kernel	—
Lard	Modified lard
Cocoa butter	—

textured with large granular crystals. They are poor aerators yet function well in pie crust applications. Lard crystals tend to be large and grainy, but pie crusts formulated from it have earned wide acceptance because of their flaky texture.

The degree to which these crystallization tendencies will be exhibited is influenced somewhat by blending but more significantly by the hardness of the fat in the formulation. For example, fluid margarine, stick products with 75–80% oil, and soft polyunsaturated tub types can be formulated by blending a high level of liquid oil with a low level of a high-melting-point fat. Stick and soft tub products containing 50% oil can be blends comprised of an intermediate level of a liquid oil plus an intermediate level of a moderately hard fat. Base stock blending is the basis for the successful formulation of stick margarine using 100% soybean oil—an oil with a definite predisposition toward β crystallization (8).

3.2 Fatty Acid Distribution

Fats and oils are essentially triglycerides—glycerol molecules to which three fatty acids are attached. The symbolic representation of this chemical structure is shown in Figure 3.3.

Three different fatty acid chains are represented in Figure 3.3 by R_1, R_2, and R_3. In actuality, either two or all three may be the same. However, each source oil has characteristic fatty acid compositions and distributions within their triglyceride molecules that influence the melting point and crystalline structure of the solidified fat in ways that are not always beneficial or desirable. For example, if all of the fatty acid chains represented by the letter R are different or R_1 and R_2 the same but different than R_3, the triglyceride is asymmetrical. Fat blends with a high proportion of asymmetrical triglycerides tend to develop a granular consistency on cooling that is objectionable for most shortening and margarine compositions. Palm or cottonseed oil is often included in blends because their β' behavior promotes small needlelike crystals that result in smooth textured products. One reason for this behavior may be that palm and cottonseed oils have a high ratio of symmetrical triglycerides (9).

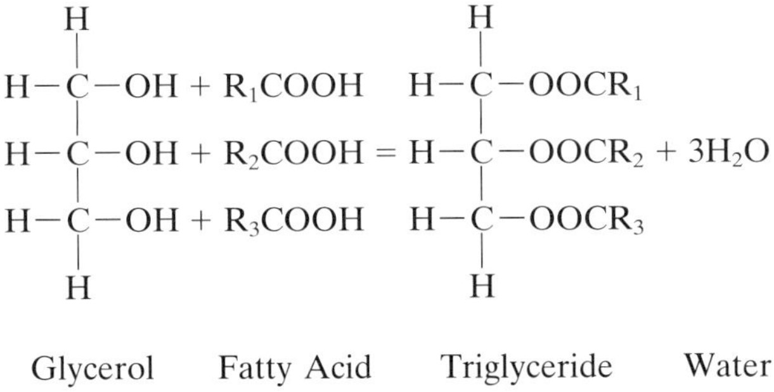

$$\begin{array}{c}
\text{H} \\
| \\
\text{H}-\text{C}-\text{OH} \\
| \\
\text{H}-\text{C}-\text{OH} \\
| \\
\text{H}-\text{C}-\text{OH} \\
| \\
\text{H}
\end{array}
+ \begin{array}{c} R_1\text{COOH} \\ R_2\text{COOH} \\ R_3\text{COOH} \end{array}
= \begin{array}{c}
\text{H} \\
| \\
\text{H}-\text{C}-\text{OOCR}_1 \\
| \\
\text{H}-\text{C}-\text{OOCR}_2 \\
| \\
\text{H}-\text{C}-\text{OOCR}_3 \\
| \\
\text{H}
\end{array}
+ 3\text{H}_2\text{O}$$

Glycerol   Fatty Acid   Triglyceride   Water

Figure 3.3 Structure of a triglyceride.

Fat and *oil* are interchangeable terms. The distinguishing criterion is their physical state at ambient temperature; oils are usually thought of as liquids while fats are considered as solids. Table 3.3 (10) lists the melting point, titer, and iodine value (IV) of many oils commonly used to formulate shortening. Melting points for liquid oils are usually not measured, but titer and iodine value are general indicators of the relative fluidity of these oils. Titer is an analytical method for determining the congeal point of fats and for measuring the melting point of fatty acids. Although the melting point of a triglyceride is not the same as its titer, it generally approximates within a few degrees the actual melting point of harder fats. A high IV is a characteristic of a liquid oil while a low IV is indicative of a solid fat. Some fats such as lard, palm, and cocoa butter are solids only at the highest ambient temperatures while sesame, soybean, and peanut are clearly liquid except in cold climates.

To achieve shortening products with desirable physical and functional end properties, the melting and crystallization habits of commercial fats are manipulated through hydrogenation, fractionation, interesterification, or combinations of these processes. Each process is a legitimate topic for study. Without these processes, or more specifically, without the modified oil products obtained from them, margarine and shortening manufacture would be much more difficult. This chapter deals briefly and only in a general way with these processes and the role of each in the formulation of edible fat products. It is assumed that properly modified oil products are available and used correctly.

Table 3.3 Melting point, titer, and iodine value of selected oils (10)

Fat or Oil	Melting Point (°C)	Titer (°C)	Iodine Value
Coconut	24–27	20–24	7.5–10.5
Palm kernel	24–26	21–27	14–22
American lard	36.5	36–42	46–70
Tallow	—	40–46	35–48
Butter oil	38	34	33–43
Palm oil	38–45	43–47	48–56
Sunflower	—	16–20	125–136
Sesame	—	20–25	103–116
Corn	—	14–20	103–128
Safflower	—	—	140–150
Rapeseed	—	11.5–15	97–108
Soybean	—	—	120–141
Cottonseed	—	30–37	99–113
Peanut	—	26–32	84–100
Cocoa butter	—	45–50	35–40
Herring	—	25	115–160
Menhaden	—	32	150–165
Whale	—	—	110–135
Olive	—	17–26	80–88

Producers normally rely on only a few source oils indigenous to their geographic area or that can be imported economically. Soybean is the primary oil used in the United States while very little palm is consumed and none is produced. Canada's major oil is canola (low-erucic-acid rapeseed). Malaysia, Indonesia, and Central America are the largest producers and users of palm oil. Eastern Europe, like Canada, relies on low-erucic-acid rapeseed (LEAR), sunflower, and soybean oils. It is apparent from Table 3.3 that it is virtually impossible to formulate products with controlled melting and crystalline properties using only one of these oils. Even in areas where conditions and economics justify a variety of types, modification methods other than blending are essential to adequately control rheologic properties.

3.3 Fractionation

Gasoline, lubricating oil, fuel oil, diesel fuel, and various solvents are all familiar products obtained from petroleum. Several processes are employed in a petroleum oil refinery to separate and recover these useful "fractions" from the base oil.

Edible oils also contain liquid and solid fractions that can be separated by a fractionation process. Dry fractional crystallization is a process in which two or more components with different melting points are cooled and separated based on their solubility or crystallization at different temperatures. Fractional crystallization is frequently applied to palm oil to separate liquid palm olein from solid palm stearin.

Solvent fraction is a process in which the various fractions are separated by dissolving the triglyceride in a solvent. This solution is then carefully cooled until the desired fraction precipitates. The precipitate is recovered by filtration. Solvent fractionation can be applied to virtually any edible oil (11).

3.4 Hydrogenation

Hydrogenation is a chemical process in which hydrogen gas is reacted with oils to increase their oxidative and thermal stability by converting liquid components to semisolid fractions. The melting and crystalline characteristics developed are essential for formulating shortenings with specific desirable physical and functional properties.

It is a catalyzed reaction dependent on catalyst type, temperature, time, pressure, agitation, and the starting oil. Platinum and palladium catalysts have been used but nickel, supported on an inert carrier, is now much more common. The catalyst must be removed after hydrogenation usually by filtration.

Each carbon atom in a fatty acid chain can be bonded to as many as four other atoms—two hydrogen and two carbon. When four bonds are present, they are referred to as single bonds and the fatty acid chain is saturated with hydrogen atoms. Naturally occurring triglycerides contain unsaturated fatty acid chains with carbon atoms interconnected by double bonds. In the hydrogenation reaction, hydrogen gas reacts with triglycerides at these selective points of unsaturation in their fatty acid chains.

Hydrogenation can be conducted in batch converters or continuous reactors. The reaction is controlled by stopping the flow of hydrogen gas. As hydrogen is added at the double-bond sites, the melting point of the original oil or fat gradually increases. If only a small amount of hydrogen is added to liquid oils such as soybean or cottonseed, the end product can still remain liquid. As more hydrogen reacts, more saturation is achieved, and soft base fats suitable for shortening formulations will be obtained. Hydrogenation can be continued until all of the double bonds have been saturated with hydrogen and the oil "fully" hardened. Fully hardened products are solids at room temperature and, although generally hard and brittle, are still useful formulation tools.

Since the hydrogenation reaction is exothermic, it affords interesting possibilities for energy conservation. This heat of reaction can be used to preheat the feed oil, which in turn cools the hydrogenated fat. Systems are also available that use this heat to produce steam.

Hydrogenation is the most widely used and practical method of preparing fats and oils capable of imparting essential physical and functional properties to shortening. It is presently used to modify and stabilize marine, animal, and all types of vegetable oils. An excellent synopsis of batch and continuous hydrogenation processes is presented by Edvardsson and Irandoust (12).

3.5 Interesterification

Interesterification is an effective tool for raising and/or lowering the melting points of edible oils. Like hydrogenation, it is a catalyzed chemical reaction; however, it alters fats by rearranging the fatty acid distributions in their triglyceride molecules. This rearrangement can be effected in a random or directed manner. Total randomization is the most widely used practice but either randomization process results in profoundly different triglyceride compositions, which follow the laws of probability based on the composition of the starting triglycerides.

The random rearrangement reaction can be conducted in continuous or batch reactors. The batch reaction vessel is agitated and fitted with a nitrogen sparger and coils for heating and cooling. Moisture, which poisons the alkaline catalyst, is removed by heating the fat or oil blend under vacuum. After drying and cooling to the reaction temperature, the catalyst is added to the reaction vessel and the oil–catalyst mixture vigorously agitated for 30–60 min. In the continuous reactor, the fats are flash dried and the fat–oil slurry is formed continuously by adding catalyst to the oil as it passes through coils sized to provide adequate time for randomization. When the reaction is completed, the catalyst is neutralized with water or acid and the salts formed or removed by filtration or centrifugation.

In directed rearrangement, the randomization process is interrupted by selectively removing one or more of the reaction products through continuous distillation or fractional crystallization. The remaining reactants continue to randomize promoting the formation of specific glycerides.

Palm oil production has increased significantly in the last 10 years and may soon exceed that of all other edible oils. Palm oil is also the only β' type whose crystal habit is not changed by interesterification. Interesterification has no effect on the oils in the β classification either; however, randomization in the presence of another oil can moderate their β tendency (13).

Interesterification is regularly used to process palm, palm kernel, and coconut oils for use in various types of confectionery, margarine, cooking and frying fats, and as blends with lauric oils in reduced-calorie spreads. These three oils crystallize slowly, are often difficult to chill and package, and tend to become hard and grainy during storage. Interesterification often reduces or eliminates these undesirable charactcristics (14).

MANUFACTURING PROCESSES AND EQUIPMENT

4.1 General

The ultimate consistency attainable depends on the fats and oils in the formulation, the processes to which these have been subjected, the equipment and conditions used to solidify them, and the conditions under which these products are stored prior to utilization. Properly formulated liquid blends can be converted to true plastic solids only when the apparatus employed provides controlled cooling, crystallization, and working techniques. The manner in which these plasticity and crystallization theories have been applied and employed in practice can be discerned by examining commercial production apparatus.

4.2 Anco Cooling Roll

The cooling roll is one of the earliest unions of equipment and theory actually applied to solidifying lard and shortening. Anco, with a commercial installation in 1881 in Chicago, was a pioneer supplier of this type of apparatus. As shown in the sketch in Figure 3.4, the device consisted of a hollow, internally refrigerated cast-iron cylinder rotating at 7–11 rpm in a trough containing molten fat slightly above its melting point. As this cylinder revolved, a thin film of fat solidified on its surface and was continuously removed by a doctor blade. The solidified fat film dropped into a special screw conveyor called a picker box. The flights of the conveyor were interspersed with blades that incorporated air while beating and working the fat. High-pressure pumps then

128 Shortening: Science and Technology

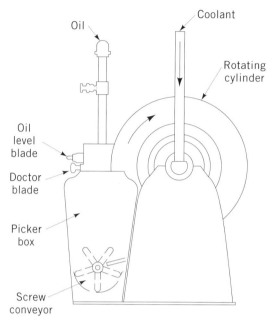

Figure 3.4 Anco cooling roll sketch. (Courtesy of Cherry-Burrell, Louisville, Kentucky.)

forced the fat through orifices, slots, screens, and valves to break crystal aggregates and further disperse the entrapped air.

Figure 3.5 is a photograph of a roll with its ancillary equipment. Cylinder sizes ranged from 610 mm (2 ft) in diameter by 762 mm (2.5 ft) long to 1219 mm (4 ft) in diameter by 2743 mm (9 ft) long. Capacities for lard ranged from 454–6350 kg/h (1000–14,000 lb/h) and for shortening from 227–3175 kg/h

Figure 3.5 Anco cooling roll. (Courtesy of Cherry-Burrell, Louisville, Kentucky.)

(500–7000 lb/h). Although still employed to produce fat flakes, the cooling roll is now virtually extinct having been replaced almost entirely by the Votator process for crystallizing and plasticizing shortening and margarine.

4.3 Votator Process

More than 60 years ago, Votator invented what is still the world's only closed continuous process for cooling, crystallizing, and plasticizing edible fat. A simplified diagram for this process is shown in Figure 3.6. This system employs Votator scraped-surface heat exchangers (A units) for cooling and agitated holding units (B units) for working and plasticizing the product as it crystallizes. Positive-displacement pumps develop high internal product pressure within the cooling and working units that, when combined with special extrusion valves, ensure that the crystallized shortening will be free of crystal aggregates, uniformly aerated, and possess the desired texture and plastic structure.

4.4 Votator Scraped-Surface Heat Exchanger

Scraped-surface heat exchangers are the most commonly used devices for chilling edible fats. Votator manufactured the first such heat exchanger in the early 1920s. The name has since become synonymous with the device and many scraped-surface heat exchangers are now commonly referred to as "Votators."

Figure 3.7 is a photograph of a two-cylinder Votator with a gravity ammonia refrigeration system. Its basic construction is shown in Figure 3.8, which is a cross-sectional view of a Votator scraped-surface heat exchanger. Each cylinder consists of a hollow cylindrical tube usually 152 mm (6 in.) in diameter by 1829 mm (72 in.) long. This tube is externally jacketed for cooling using brine or direct expansion refrigerants such as ammonia. As the molten oil formulation passes through the tube and cools, an electric motor rotates a shaft centrally located inside the product tube. This "mutator" shaft is fitted with mechanical seals at each end and floating blades that, as the shaft spins, constantly clean the heat transfer surface by scraping and removing the product film from the tube wall. Each mutator shaft has two effective rows of 152-mm (6-in.) long blades staggered along its entire length. This staggered blade arrangement provides improved mixing over the older conventional in-line blade mounting system. All margarine and shortening cooling units are equipped with a hollow mutator shaft and a rotary joint through which hot water can be circulated to prevent solids buildup on the shaft body. Standard shafts are 119 mm (4.6875 in.) in diameter and are rotated at about 400 rpm. Larger and smaller diameter shafts with three and four rows of in-line or staggered blades are available for special applications.

Shortening units are constructed of carbon steel. The water phase in margarine is corrosive, and sanitation procedures require that all of the equipment used to manufacture it must be chemically cleaned. Margarine processing

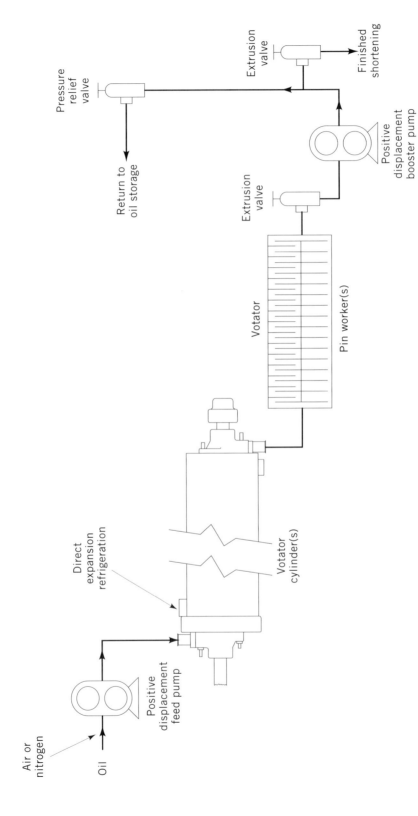

Figure 3.6 Votator shortening process—block flow diagram. (Courtesy of Cherry-Burrell, Louisville, Kentucky.)

Manufacturing Processes and Equipment 131

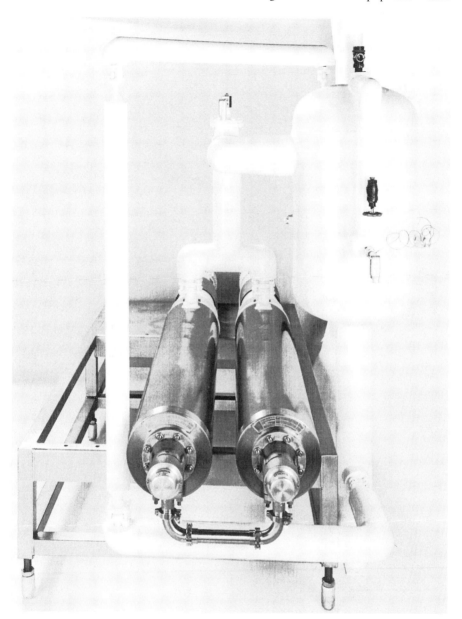

Figure 3.7 Votator scraped-surface heat exchanger. (Courtesy of Cherry-Burrell, Louisville, Kentucky.)

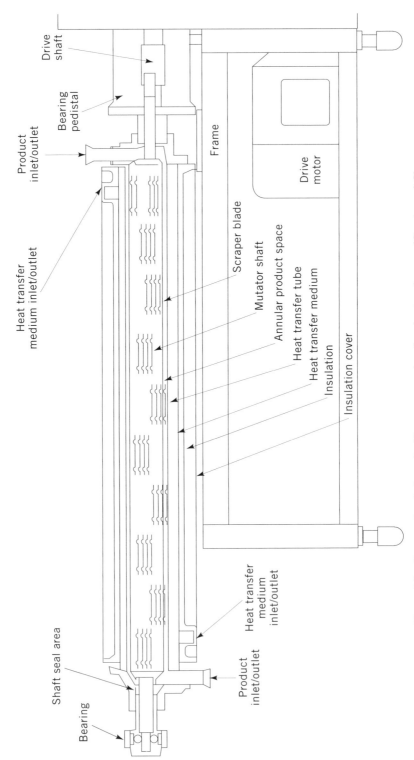

Figure 3.8 Cross-sectional view of a Votator scraped-surface heat exchanger. (Courtesy of Cherry-Burrell, Louisville, Kentucky.)

units contain chromium-plated commercially pure nickel heat transfer tubes and stainless steel for all product-contacted metal surfaces.

4.5 Supercooling and Direct Expansion Refrigeration

In order to form small crystals, shortening must be cooled very rapidly; so rapidly in fact that the flow at the exit of the Votator must contain virtually no crystals although cold enough for significant levels of solid fat to exist. In reality, the heat exchanger must be capable of supercooling the molten shortening formulation. Fortunately, all triglycerides exhibit a definite propensity for supercooling but, to achieve it, the residence time within the cooler must be limited to invariably less than 20 s. This mandates highly efficient heat exchangers and the use of refrigerants with favorable physical properties and effective heat transfer characteristics. Ammonia and chlorofluorocarbons fulfill these demands and are widely used as refrigerants in heat exchangers for cooling shortening.

This cooling concept is commonly called *direct expansion refrigeration*. Currently there are four basic variations employed: gravity, forced circulation, pool boiling, and liquid overfeed.

Gravity Refrigeration System. The gravity refrigeration system shown in Figure 3.9 is the simplest to comprehend. Liquid refrigerant flows from its receiver in the compressor plant to a surge drum installed above the freezing cylinder(s). A modulating thermostatically controlled expansion valve automatically maintains the correct refrigerant level in this vessel. The temperature of the liquid refrigerant, even if subcooled, has no detrimental influence on the operation of this type of system. Gravity forces the refrigerant into the cooling jacket where product heat vaporizes a portion of the liquid while reducing the bulk density of the remainder. The flow of vapor and this density difference combine to create the classic *thermosyphon effect,* which forces liquid refrigerant to circulate from the surge drum to the cooler. Baffles in the surge drum effectively separate the vapor from the liquid. A pressure-regulating valve controls the pressure in the surge drum and, consequently, the temperature of the refrigerant in the cooling cylinder. Makeup liquid enters through the level control and the entire cycle continues. Individual surge drums can be provided for each cooling cylinder or a single drum can be used for as many as three cylinders.

The gravity system protects against freeze-up through an instantaneous current relay system. The current drawn by the heat exchanger drive motor is continuously monitored and, at a preset but higher than desired operating level, it closes a solenoid valve in the line from the cylinder to the surge drum. Refrigerant circulation is stopped and residual heat from hot metal surfaces and the product vaporizes enough refrigerant to build sufficient pressure to instantly force any remaining liquid refrigerant to return to the surge drum.

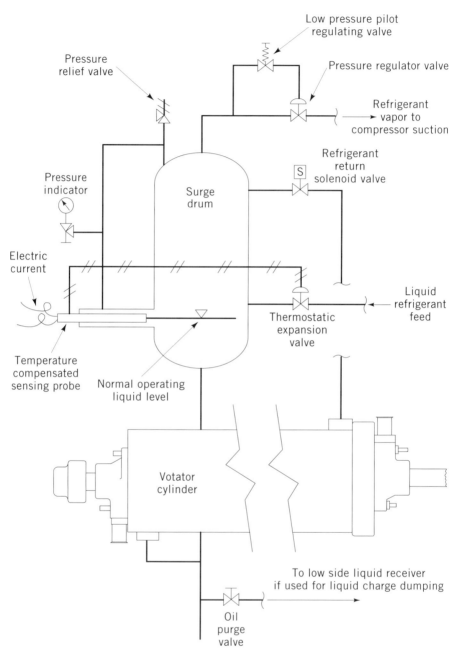

Figure 3.9 Direct expansion gravity refrigeration system. (Courtesy of Cherry-Burrell, Louisville, Kentucky.)

When the current draw returns to normal, the solenoid valve opens and the heat exchanger automatically resumes cooling. Optional hot gas controls can be provided to assist in removing refrigerant for pump down.

Forced Circulation Refrigeration Systems. Forced circulation direct expansion refrigeration systems employ principles similar to gravity systems. Refrigerant must be physically transported to the cooling cylinder since the surge drum is generally located below the heat exchanger. Liquid level can be regulated with thermostatic controls or float valves. Some designs use a mechanical pump while others depend on pressure differential and gas pressure to force the liquid to flood the heat transfer tube. Injectors, based on Bernoulli's theorem, take advantage of the liquid supply pressure to transport refrigerant from the surge drum to the cylinder. Freeze-up protection devices sense motor load and either turn off the pump or stop the motive refrigerant flow to the injector. Since the surge drum is physically below the cylinder, refrigerant naturally drains to it. Hot gas systems can also be provided.

Pool Boiling Refrigeration Systems. In a pool boiling system, the jacket area surrounding the heat exchanger tube serves as the refrigerant surge drum. The heat exchanger tube is flooded with liquid refrigerant, and a level control device maintains this flooded condition by simply replacing the liquid as it vaporizes. A valve in the line returning vapor to the compressor plant regulates the pressure and temperature of the refrigerant in the jacket. This type of system is often provided with a "drop tank" to which the refrigerant charge is dumped during pump down or to prevent freeze-up. Hot gas systems are available.

Liquid Overfeed Refrigeration Systems. One commonality of all of the previous systems is that all of the liquid refrigerant entering is intended to be returned as vapor. Liquid overfeed (LOF) is a proven direct expansion concept in which only 25–35% of the liquid flowing to the heat exchanger is actually vaporized. A large low-pressure receiver replaces individual surge drums. This receiver, normally located in the compressor plant, is designed to separate the vapor from the circulated liquid (15). Overload protection against freeze-up is included and hot gas systems can be provided.

Where LOF is already employed or when new installations require four or more surge drums, economics favor LOF. Advantages include:

1. One low-pressure receiver replaces multiple surge drums.
2. Surge drum controls and safety devices are eliminated.
3. Refrigerant volume in the operating area is reduced.
4. Oil does not accumulate in heat exchanger jackets.
5. The low-pressure receiver also serves as a suction trap.
6. Subcooled liquid can be used.

7. Liquid from one receiver can be returned to receivers at higher or lower pressures.
8. Has extremely rapid boil-off.
9. Refrigerant is immediately evacuated to the low-pressure receiver during pump down.

Only Votator is known to have designed scraped-surface heat exchangers for LOF refrigeration and to have plants actually operating successfully using this principle.

4.6 Crystallization

With efficient heat exchangers cooling through direct expansion refrigeration, the product delivered is supercooled significantly below its equilibrium temperature and primed for crystallization. A supercooled fat composition allowed to solidify without agitation and mechanical work will solidify to form an extremely strong crystal lattice and exhibit a narrow plastic range. This is a desirable characteristic for stick margarine formulations but, where specific body and plasticity is necessary, the plastic range can be altered and extended by mechanically working the fat while it is crystallizing from the supercooled state (16). In general, these fats require crystallization times with mechanical working of about 2–5 min. Votator developed a special device for this purpose—the agitated working unit.

4.7 Votator Agitated Working Unit

Figure 3.10 is a cross-sectional, cutaway view of a Votator working unit often referred to as a B unit. Depending on the product and the required residence time, B units vary in size from 76 mm (3 in.) in diameter by 305 mm (12 in.) long to 457 mm (18 in.) in diameter by 1372 mm (54 in.) long. All sizes contain relatively small diameter shafts with pins fixed throughout the length of the shaft. A photograph of a B unit is shown in Figure 3.11. Pins welded into the product cylinder intermesh with the shaft pins as the shaft rotates. The mechanical working accomplished by this device during the primary crystallization period distributes the latent heat evolved uniformly and forms fine discrete crystals throughout the crystallizing mass. For shortening, the residence time is normally 2–3 min with standard shaft speeds of 100–125 rpm. Margarine processing generally requires less residence time but variable and often higher agitation speeds. Although the product temperature rises within, B units are normally not jacketed for external cooling, but it is beneficial to jacket margarine B units with hot water to aid in melting and cleaning. Shortening worker units can be constructed of carbon steel, but stainless steel is required for margarine units.

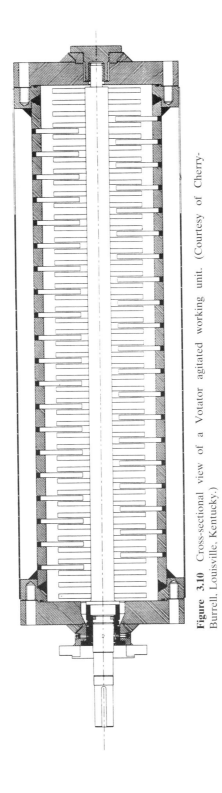

Figure 3.10 Cross-sectional view of a Votator agitated working unit. (Courtesy of Cherry-Burrell, Louisville, Kentucky.)

138 Shortening: Science and Technology

Figure 3.11 Votator agitated working unit. (Courtesy of Cherry-Burrell, Louisville, Kentucky.)

4.8 Tempering

Definition. Except for table margarine, plasticized edible fats are usually packaged and immediately transported to and stored in a constant temperature room for 24–72 h. This quiescent heat treatment is referred to as *tempering*. It is generally agreed that tempering should be conducted at about 27°C (80°F) and that it improves plasticity and creaming properties and a plastic fat's ability to maintain a uniform consistency despite reasonable temperature fluctuations during storage.

Theory. Considerable confusion exists and theories abound as to exactly what happens during tempering. One contends that low-melting-point triglycerides dissolve and recrystallize to produce a stronger more homogeneous structure (17). Another proposes that all crystalline forms exist initially and that conversion to a stable polymorphic form for slowly transforming formulations requires quiescent holding at a temperature just below the melting point of the lowest melting polymorph with 29°C (85°F) being an acceptable temperature compromise (18). These products have complex triglyceride structures that can and do crystallize in multiple polymorphic forms. Since they pass from molten oils to packaged plastic solids in normally less than 5 min, all possible crystalline forms will be present. Most of these, however, are not stable but, given time and sufficient energy input, all will revert to stable structures. Whatever the mechanism, it is generally agreed that the rheologic properties most affected by crystal transformation are substantially stabilized during the first 48 h of tempering.

Fluid shortenings are finding increased acceptance in all segments of the bakery industry. While fluid products based on β' hard fats can be stabilized by normal quiescent tempering; products based on β hard fats require tempering under agitation to form a stable suspension. Votator's current recommendation for bakery shortening systems and fluid shortenings in general is that agitated tempering for at least 4 and preferably 6 h is necessary to form a crystalline dispersion with a viscosity low enough for pumping but high enough for prolonged suspension.

Constant–temperature rooms for tempering shortening are expensive to operate and create logistic problems maintaining and rotating inventories. While it cannot be eliminated, tempering time can be reduced in many cases by 50% or more. Obviously, the material leaving the B unit is not completely crystallized nor has it crystallized into a completely stable polymorphic form. Nevertheless, it must still conform to the laws of physical chemistry. Phase equilibria dictate that, if cooled further, additional liquid must solidify. Further cooling of this viscous crystalline mass is possible through the use of a special scraped-surface heat exchanger or "postcooler" often called a C unit.

The Votator C Unit. The Votator C unit is essentially an A unit with the mutator shaft installed 6 mm ($\frac{1}{4}$ in.) off the centerline of the heat transfer tube. This "eccentric" design forces the scraper blades into a cam-type motion with each revolution of the mutator, and this continuous oscillation gently kneads the product while cleaning the tube surface. Consequently, sufficient mixing is developed to maintain efficient heat transfer even at very low shaft speeds. Since high shaft speeds are not required, mechanical power input is minimized. The net result is that the viscous crystalline material from the B unit can be cooled back to the temperature achieved in the A unit. Liquid fat, which in the quiescent state would ordinarily crystallize onto existing crystals increasing their size, is forced to crystallize creating more stable individual crystals. The C unit can reduce tempering time while also providing a means of controlling both viscosity and temperature at the filling station.

SHORTENING PRODUCTION SYSTEMS

5.1 Votator Carbon Steel Lard and Shortening Systems

Because shortening is a 100% oil product, expensive stainless steel equipment for a shortening is not required, and the Votator model LS182 shortening system is constructed entirely of mild steel. All of the equipment is cleaned simply by circulating oil until all of the fat has been melted and then purging with air or inert gas. No chemical cleaning is necessary.

The basic system components are an accumulator for storage of the direct expansion refrigerant and its necessary valves and controls, two 6 × 72 Votators located directly below the accumulator and an 18 × 54 agitated working unit. The accumulator is shown on the top right in Figure 3.12 immediately above the two cylinder scraped-surface heat exchanger. The working unit is installed on the same frame beneath the cooling cylinders. The pumping system on the left includes a raw material tank and two high-speed rotary gear pumps driven by a single, double-shafted motor. Special extrusion and backpressure valves are included in the interconnected product piping provided and installed at the factory.

Figure 3.12 is the process flow diagram for the Votator model LS182 shortening system described above. Prepared shortening formulations are normally stored at between 49°C (120°F) and 60°C (140°F). From storage, the molten oil is either pumped or gravity flows to a raw material tank located at the lower right-hand corner of the diagram. A float valve maintains a constant level in this tank as a high-speed positive displacement rotary gear pump draws oil from it. Air or inert gas is usually injected into the oil at the suction of this pump. The backpressure control valve maintains a constant pressure of approximately 24 bar (350 psig) at the pump discharge. With a controlled level in the raw material tank and a constant-speed positive-displacement pump, a fixed ratio of oil and air is maintained ensuring a constant product density. Shortening normally contains 10–15% air by volume although considerably more can be uniformly incorporated in proper formulations.

A water-jacketed shell-and-tube-type heat exchanger precools the molten oil to just above its melting point, usually 43–46°C (110–115°F). From this precooler, the oil flows directly to the Votator two-cylinder A unit. The primary function of the precooler is to reduce the heat load on the A units, thus maximizing their cooling capacity and ensuring that the greatest number of crystal nuclei are developed as the fat is supercooled. The A units chill the oil from the precooler to about 18°C (65°F) or to a previously determined temperature required to produce the desired plasticity. Very little crystallization actually occurs in the A unit although SFI profiles indicate that 25% or more solidification is possible at the A-unit exit temperature. The fat leaves the A unit in a semifluid supercooled state primed for crystallization and ideally prepared for plasticizing via the B unit.

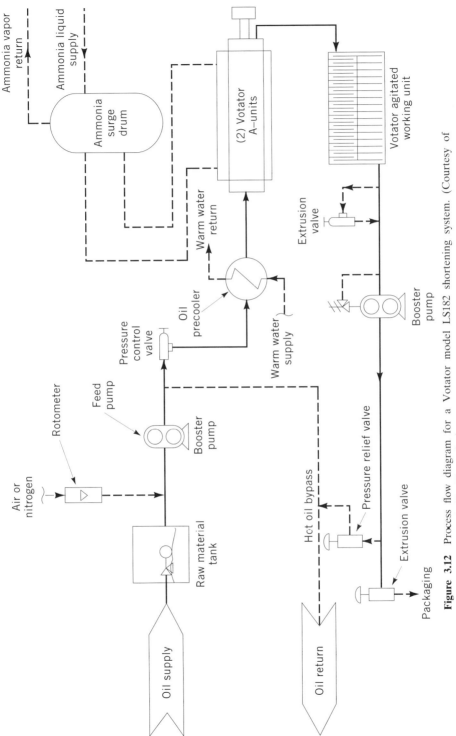

Figure 3.12 Process flow diagram for a Votator model LS182 shortening system. (Courtesy of Cherry-Burrell, Louisville, Kentucky.)

The supercooled stream from the A unit flows directly to the worker unit. There is normally a 5–8°C (10–15°F) temperature rise across the B unit most of which results from latent heat of crystallization; mechanical power does not add significantly to the total heat input. The plasticized fat from the B unit is forced through a special extrusion valve that also maintains an internal pressure of 17–20 bar (250–300 psig) inside all the A and B units. This pressure combines with the action of the extrusion valve to ensure thorough gas dispersion and breaks any crystal aggregates that form. A second rotary gear pump develops the pressure necessary to transport the viscous crystalline mass to distant filling points while producing enough pressure for final texturization by a second extrusion valve located just prior to filling.

By varying the size and number of the coolers and working units, systems with virtually any desired capacity can be provided. Completely engineered and prepackaged shortening systems are available in capacities from 1361–9072 kg/h (3000–20,000 lb/h). Pilot plants for 91 kg/h (200 lb/h) are also available.

5.2 Votator Stainless Steel Margarine/Shortening Systems

While many manufacturers produce only shortening, there is an increasing demand for combination systems capable of processing margarine as well. In 1993, Votator introduced the stainless steel lard/shortening (SLS) series. Completely engineered, prepackaged, assembled, and factory-tested SLS combination margarine/shortening systems are now available with capacities from 1361 to 9072 kg/h (3000–20,000 lb/h) of shortening or 1043 to 6350 kg/h (2300–14,000 lb/h) of stick or tub-type margarine.

Figure 3.13 is a photograph of a system that has a capacity of 4536 kg/h (10,000 lb/h) for shortening and 3175 kg/h (7000 lb/h) for margarine. The metal surfaces in the feed pump, B unit, and all interconnecting valves and piping are stainless steel as is the mounting frame. All metal product-contacted surfaces in the Votator A units are also stainless steel except for the heat transfer product tube, which is commercially pure nickel with a hard chromium plating inside. Independent gear motor drives are provided for all A and B units.

The process flow diagram for the Votator model SLS182 margarine/shortening system in Figure 3.13 is shown in Figure 3.14. A units supercool the molten oil blend while B units work and texturize it as it crystallizes. A high-pressure stainless steel feed pump generates sufficient pressure to overcome the resistance created during supercooling and plasticizing and to transport this viscous product to the filling station. A units and B units experience higher internal pressures than with the two-pump arrangement and must be designed for at least 41 bar (600 psig). Newer designs are currently capable of operating at 68–102 bar (1000–1500 psig). The only option required to convert from shortening to soft margarine production are variable-speed drives for the feed

Shortening Production Systems   **143**

Figure 3.13   Votator model SLS182 margarine/shortening system. (Courtesy of Cherry-Burrell, Louisville, Kentucky.)

pump and B units. Because of the elimination of the raw material tank, one feed pump and several control valves, a Votator SLS system is only about 20% more expensive than an all-carbon-steel Votator carbon steel lard/shortening (LS) system despite the upgrade to all stainless steel construction and individual gear motor drives.

A completely redesigned manual control panel is shown in Figure 3.15. The operator panel, now a stainless steel washdown enclosure, also contains the necessary start/stop push buttons and running lights as well as digital indicators for product and refrigeration temperatures. The refrigeration temperature is adjustable from the operator panel. Circuit breakers, conforming to the International Electrical Code (IEC), have replaced fuse-protected motor starters. Motor current is sensed and indicated on an ammeter with a digital display. Freeze-up is prevented by stopping cooling at a factory preset current level. This power level can easily be reset in the field.

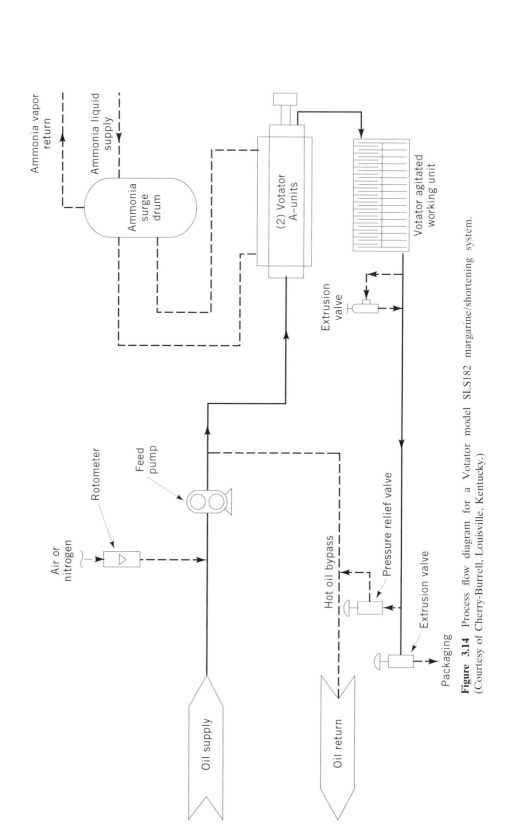

Figure 3.14 Process flow diagram for a Votator model SLS182 margarine/shortening system. (Courtesy of Cherry-Burrell, Louisville, Kentucky.)

Shortening Production Systems **145**

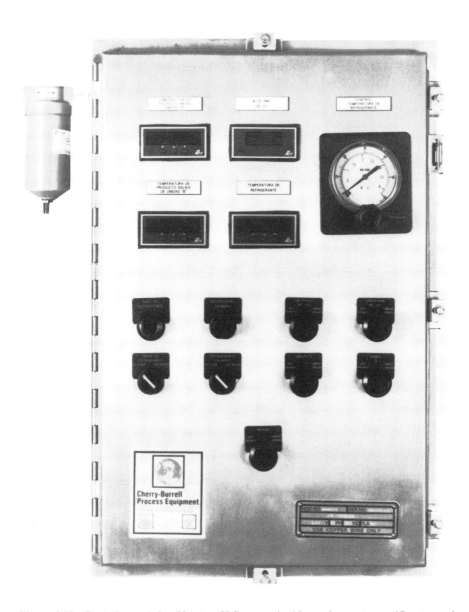

Figure 3.15 Control panel for Votator SLS margarine/shortening systems. (Courtesy of Cherry Burrell, Louisville, Kentucky.)

5.3 Chemetator Scraped-Surface Heat Exchanger

In 1993, Crown Iron Works Company of Roseville, Minnesota, acquired controlling interest in a new U.K.-based joint-venture operation. The new company, Crown Chemtech Ltd., provides processing systems for the manufacture of margarine and shortening. Figure 3.16 is a photograph of a Chemetator model 246-A4M with four cooling cylinders. Each cylinder has an individual ammonia refrigeration system and a separate drive motor for each mutator.

5.4 Gerstenberg & Agger Margarine/Shortening Systems

Gerstenberg & Agger A/S also provides processing systems to manufacture shortening and margarine. Figure 3.17 is a photograph of a shortening system with a production capacity of 10,000 kg/h (22,000 lb/h). It has four chilling tubes, two independent cooling systems, and is also suitable for production of soft margarine. Special features include a unique drop tank system to ensure

Figure 3.16 Chemoreactor scraped-surface heat exchanger unit. (Courtesy of Crown Chemtech Ltd., Reading, U.K.)

Figure 3.17 Gerstenberg & Agger Perfector type (2 + 2) × 92R. (Courtesy of Gerstenberg & Agger A/S, Copenhagen, Denmark.)

product does not freeze inside the chilling tubes in case of short production breaks, a maximum allowable product pressure rating of 80 bar (1176 psig), high-efficiency outside corrugated chilling tubes with chrome plating inside, tungsten carbide mechanical product seals, and optional separate cooling systems for each tube. The heat exchangers feature special floating knives as shown in Figure 3.18.

Figure 3.19 is a photograph of a puff pastry margarine system with a production capacity of 4000 kg/h (8800 lb/h). It has six chilling tubes, three independent cooling systems, a maximum product pressure rating of 180 bar (2646 psig) and a "bulldog" knife system. This unit also contains the special

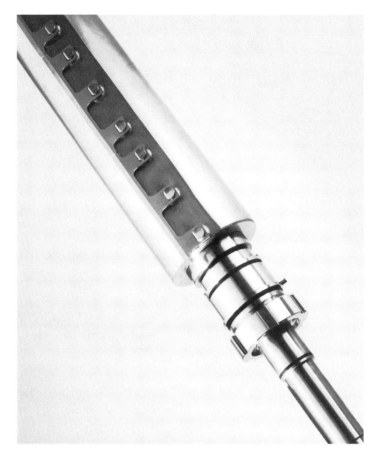

Figure 3.18 Gerstenberg & Agger floating knife mutator shaft system for Perfector type (2 + 2) × 92R. (Courtesy of Gerstenberg & Agger A/S, Copenhagen, Denmark.)

features described for the shortening Perfector type (2 + 2) × 92R shown in Figure 3.18.

Figure 3.20 is a photograph of an intermediate crystallizer that can be mounted directly on each chilling tube to ensure correct plasticity and crystal formation in all types of products such as table margarine, soft margarine, shortening, and cake and cream margarine. Intermediate crystallizers can be equipped with T pins that are especially suited for production of puff pastry.

ANALYTICAL EVALUATION AND QUALITY CONTROL

Control of the quality and consistency of the final product depends on understanding, adhering to, and applying the principles underlying cooling, crystalliz-

Figure 3.19 Gerstenberg & Agger Perfector type (2 + 2) × 180. (Courtesy of Gerstenberg & Agger A/S, Copenhagen, Denmark.)

ing, and texturizing to production and selecting and controlling the ingredients that constitute the formulation. The *Official Methods and Recommended Practices of the American Oil Chemists' Society* (19) is the definitive analytical reference for evaluating raw materials and finished products in the fats and oils industry. Referred to in the industry as simply the AOCS Methods, it is an essential tool for every analytical laboratory. The fourth edition contains approximately 1200 pages in two heavy-duty loose-leaf ring binders, details more than 400 collaboratively tested and verified methods, and is currently used in more than 40 countries. Procedures and the apparatus required are described for all analyses required to properly characterize shortening. Typically, analytical methods are conducted to measure crystal size, color, solid fat index or content (SFI or SFC), iodine value, refractive index, Wiley melting point, dropping and softening point, oxidative stability, peroxide value, viscosity, penetration, consistency, and texture and color.

Sophisticated instruments have been developed to reduce the time needed to measure critical factors influencing shortening's rheologic properties. These include nuclear magnetic resonance (NMR), polarized light microscopes, x-ray scatterers, revolving laser beams, gas–liquid chromatography (GLC), high-

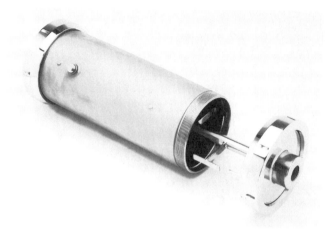

Figure 3.20 Gerstenberg & Agger intermediate crystallizer type 15 Ltr. for Perfector type 125. (Courtesy of Gerstenberg & Agger A/S, Copenhagen, Denmark.)

performance liquid chromatography (HPLC), thin-layer chromatography (TLC), and differential scanning calorimetry (DSC).

Crystal size and distribution was first measured using polarized light microscopy but the analysis is time consuming and inaccurate. Recently, light or x-ray scattering and sedimentation methods coupled with computer analysis have provided new and improved techniques for measuring mean crystal size and size distribution. A revolving laser beam generated by a laser particle counter can be coupled with a computer using special software to measure and record crystalline information. Crystal structure can be viewed by polarized light microscopy.

The SFC and SFI values can be determined by dilatometric methods or by pulsed NMR. The dilatometric method is still considered the most accurate, but NMR provides reliable information much more quickly.

Until the advent of chromatography, triglyceride and fatty acid analysis was difficult and time consuming. With GLC, HPLC, or TLC these analyses are now routine.

Differential scanning calorimetry is used to determine the melting behavior of shortenings. A small sample is rapidly chilled using liquid nitrogen and then reheated. The cooling and heating rates are accurately controlled at about 5°C per minute. Energy is graphed as a function of temperature. Exothermal crystallization appears as a peak in these curves. Curves for most shortenings contain a single major peak occurring at a temperature 4 or 5°C (39.2 or 41°F) higher than the softening point. Products crystallizing in the β' polymorphic form exhibit a single sharp melting peak while those in the β configuration have broader peaks. Hydrogenated fat may appear as additional peaks at higher melting temperatures. Differential scanning calorimetry analyses can be used to determine the polymorphic form of pure components, but these melting curves are not absolute indicators of the crystalline structure of compound shortenings (20).

Devices have also been developed to measure hardness and plasticity by compression. A curve is obtained comparing the compressing force applied to the resulting deformation. Initially, the curve is straight with the degree of deformation dependant on the formulation and the force applied. If the compressing force is increased further, a break will appear in most shortenings. Samples that are hard and brittle have a narrow plastic range and the breaking point occurs after very little deformation. Viscous flow and plasticity is indicated by curves that round off with long flat sections.

A cone penetrometer can also be used to determine hardness (AOCS Method Ce 16–60) (21). A cone of specified mass and dimensions is dropped into a prepared sample. The relative hardness of the sample is determined by dividing the mass of the cone by the depth of penetration. The cone penetrates farther into soft products and produces subsequently lower relative hardness values. A single temperature penetration value is not a true indication of overall relative plasticity. Narrow differences in penetration values at low and high temperatures indicate a wide plastic range while huge differences indicate a narrow range. Some products are formulated to be naturally firmer than others, depending on geographic area and intended usage.

Management support is the key to any quality assurance program. High product quality standards must be established and line production supervision must have the technical training, authority, and support of upper management to enforce those controls and ensure the production of quality shortenings.

PACKAGING AND STORAGE

Semisolid plastic shortenings are usually packaged in 0.5- and 1.5-kg (1- and 3-lb) cans, cubes, 50-kg (110-lb) cartons, and in drums containing about 175 kg (380 lb). This type of plastic shortening is also available in "chubs," prints, and sheets ready for direct end-use application.

Cubes, comprised of cardboard cartons with plastic liners, are probably the most popular foodservice and food processor packaging form. Systems equipped with optical sensors assure that an empty carton is in position on a scale before filling is initiated, and automatic controls then fill it to the correct weight. Most systems use two scales and switch to the second carton slightly before the first finishes filling. Cube sizes range from 10 to 25 kg (22 to 55 lb).

Chub packages are especially popular in Latin America. They are produced by machines that form plastic into cylindrical sheets and fill the cylinders with from 0.11 to 5 kg (0.25 to 10 lb) of shortening. The chubs are sealed by crimping both ends with metal clips.

Prints can be formed by extruding product directly into sticks or bricks with weights ranging between approximately 0.5 and 5 kg (1 and 10 lb). Prints can also be produced using rotary machines to fill plasticized shortening in paper-lined containers. The quick chilled shortening is fluid enough at filling to assume the shape of the container as it is filled but rapidly crystallizes into a rigid form.

Puff pastry shortening can be extruded in flat sheets or printed in 2- or 4.5-kg (5- or 10-lb) blocks. It is usually provided in corrugated boxes containing several sheets or slabs with paper sheets between the slabs to prevent them from sticking together.

Fluid and liquid shortenings are usually sold in 4-, 8-, and 20-liter (1-, 2-, and 5-gal) containers or in bulk 200-liter (50-gal) drums and 40,000-liter (10,000-gal) or larger tanks. The 20-liter (5-gal) size is available as a rigid package consisting of a soft plastic container inserted in a corrugated paperboard box. It is best to package liquid oil in dark containers since light catalyzes oxidation reactions in fats and oils. However, for household use, the consumer seems to prefer clear containers and some brands are packaged in clear glass or plastic. Polyethylene is permeable to oxygen and, because of low turnover rates, household oil stored in clear polyethylene containers frequently becomes rancid before it can be consumed. Other oxygen-impermeable plastics such as polyvinyl chloride, opaque polyethylene, and saran-coated polyethylene seem to be satisfactory packaging materials.

Pockets of free oil sometimes form in the package if the shortening has poor gas dispersion, a weak plastic structure, insufficient hard fat in the formulation, or the storage temperature is too high. If exposed to a sufficiently high temperature long enough for the lower melting fat fractions to liquify, plastic shortenings will recrystallize improperly resulting in a loss of functionality. Products that have suffered heat damage during storage must be remelted and completely reprocessed to restore their plastic properties. In general, solid shortenings need not be refrigerated during storage but, since they will absorb odor, the storage area should be cool, dry, and free of odoriferous materials. Plastic shortenings that have been correctly formulated and properly prepared are stable and will tolerate considerable abuse during storage and transportation.

Clear liquid cooking and salad oils require no special storage considerations. They have no crystal structure or suspended solids and, if the storage temperature is low enough to cause some solids to form, these usually melt when the normal storage temperature is resumed. Should these solids not remelt, it is advisable to mix and suspend them before using the affected containers.

Fluid (opaque) shortenings are frequently used in applications where plastic crystallographic properties are not as important as pourability, homogeneity, and stability. Because they contain suspended solids, storage temperature is important. Most fluid products in the United States are formulated to be stored between 18.3 and 35°C (65 and 95°F). Storage below this temperature will cause the shortening to set up or lose fluidity. Warming normally reverses this condition. Storage at too high of a temperature results in partial or complete melting of the suspended solids. This situation is not reversible since reducing the storage temperature will result in the formation of large crystals that may settle to the bottom of the container. The loss of these solids may be of little consequence or it may have a disastrous effect on the functionality of the product.

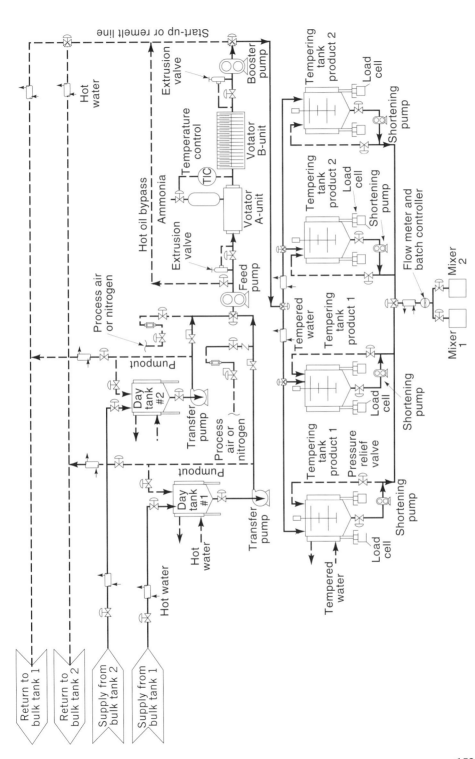

Figure 3.21 Process flow diagram for an automatic Votator bakery shortening system. (Courtesy of Cherry-Burrell, Louisville, Kentucky.)

153

154 Shortening: Science and Technology

Dry or powdered products are packaged in multiwalled paper bags or fiber drums. The bags usually contain 25 kg (50 lb) while drum capacities range from 45 to 90 kg (100–200 lb). As is recommended for semisolid plastic products, the storage area for these shortenings should also be cool, dry, and free of odoriferous materials.

All types of shortening have been formulated and prepared to possess essential functional properties. It is apparent that these properties can be impaired or totally destroyed by even limited exposure to excessively high temperatures during packaging, storage, or shipment.

RECENT INNOVATIONS

8.1 Automation and Computer Control

With a proven process and time-tested reliable equipment, recent innovations have been directed toward improved process control and automation. Shortening lines are now available with semiautomated or completely automated screen-based control systems.

Figure 3.22 Control panel for an automatic Votator bakery shortening system. (Courtesy of Cherry-Burrell, Louisville, Kentucky.)

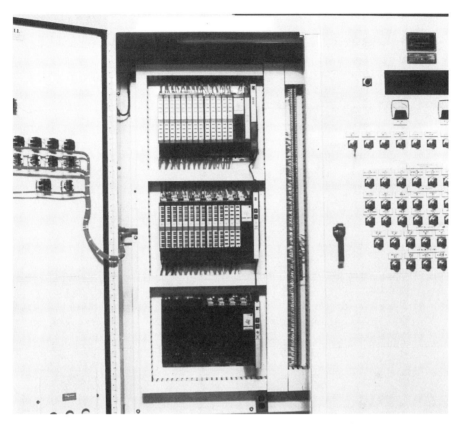

Figure 3.23 PLC for an automatic Votator bakery shortening system. (Courtesy of Cherry-Burrell, Louisville, Kentucky.)

The process flow diagram for an automatic bakery shortening system is shown in Figure 3.21. The heart of the processing system is a standard A and B unit combination. A C unit can be added after the B unit for additional temperature control, crystal stabilization, and product flexibility. Shortening from the B unit is transferred to and held in agitated, jacketed tempering tanks until properly tempered and ready for use. Shortening is metered, on demand, from these tanks to end users. A programmable logic controller (PLC) continuously monitors and maintains product levels in the tempering tanks, filling each with the proper formulation as required. For each formulation change, the system is automatically purged to prevent intermixing. A user-friendly message center reports any anomaly and a graphic panel displays the current system status.

The next three figures are photographs of a typical control panel, PLC, and graphic display for the Votator bakery shortening system in Figure 3.21. Figure 3.22 contains all of the high-voltage switch gear, motor starters, as well as an operator control section with push button stations and running lights. The PLC is shown in the photograph in Figure 3.23. Figure 3.24 depicts the

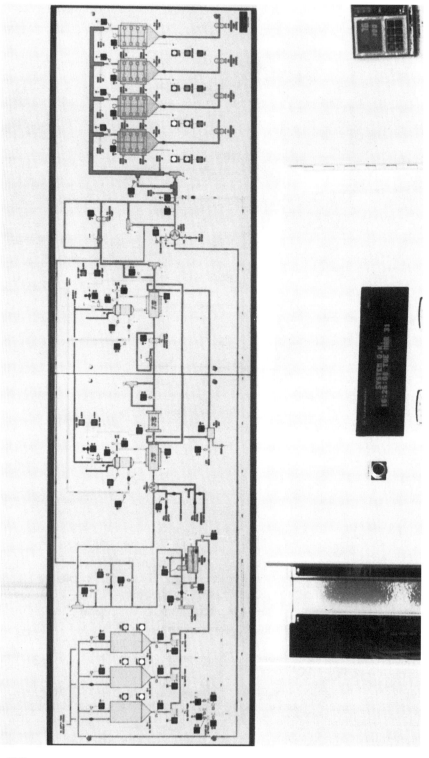

Figure 3.24 Graphic display for an automatic Votator bakery shortening system. (Courtesy of Cherry-Burrell, Louisville, Kentucky.)

Recent Innovations 157

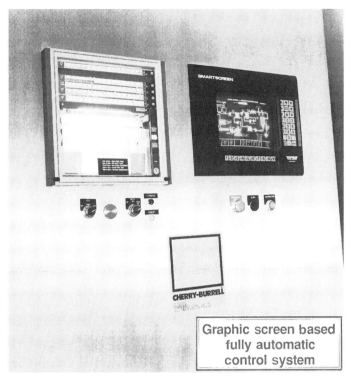

Figure 3.25 Graphic screen-based fully automatic control system. (Courtesy of Cherry-Burrell, Louisville, Kentucky.)

process flow diagram in Figure 3.22 and provides a visual indication of the process and equipment status.

Automation and control devices have been integrated into the system to pass information developed in the semiautomatic system to a full-color screen-based interface. This interface allows remote access to view, while in operation, process flow diagrams indicating product and utility status, valve positions, motor loading conditions, alarm status, and all other critical parameters with the ability to reset control variables simply by touching the display screen. A fully automatic screen-based control system is shown in Figure 3.25. Figure 3.26 is a typical screen display of a process and instrumentation diagram.

State-of-the-art control systems are also capable of accumulating, storing, and presenting production records. Many feature trending with records offered as options for formatted tables and histograms. Of course, all include screens for indicating and resetting control values and displaying alarm messages.

8.2 Future

Table 3.4 is a summary of world production of 12 major fats and oils from 1960 through 1985 with forecasts for 1990 and 1995. This projection was very

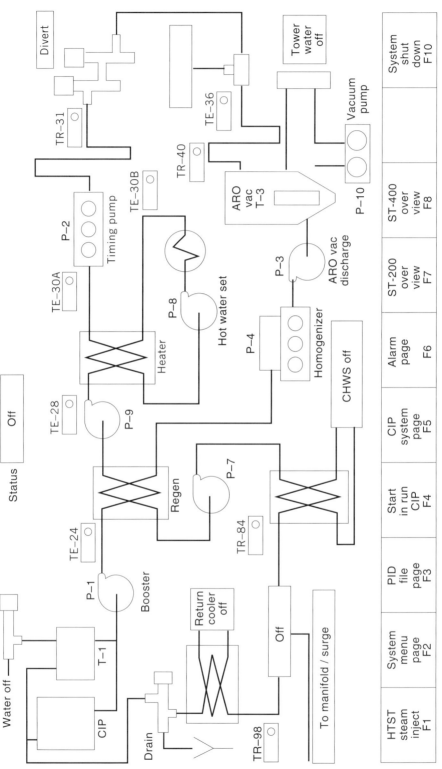

Figure 3.26 Schematic of a process flow diagram graphic screen display. (Courtesy of Cherry-Burrell, Louisville, Kentucky.)

Table 3.4 World production of 12 major oils and fats (22)

Fat or Oil	Production (1000 ton)							
	1995	1990	1985	1980	1975	1970	1965	1960
Soybean	18,200	15,800	13,939	13,358	8,025	6,381	3,999	3,355
Cottonseed	4,100	3,700	3,807	3,030	2,929	2,511	2,733	2,324
Groundnut	4,100	3,600	3,144	2,801	2,672	2,730	2,806	2,282
Sunflower	9,260	7,950	6,549	5,044	3,900	3,491	3,077	1,788
Rapeseed	8,550	7,500	5,988	3,484	2,392	1,778	1,500	1,164
Corn	1,470	1,230	1,036	765	580	475	436	370
Coconut	3,800	3,350	2,630	2,722	2,593	2,019	2,032	1,949
Palm kernel	2,270	1,520	935	633	474	393	412	432
Palm	17,800	11,210	6,898	4,621	2,858	1,796	1,446	1,306
Lard	6,170	5,700	5,278	5,025	4,268	3,901	3,613	3,142
Fish	1,600	1,540	1,474	1,211	1,059	1,078	783	470
Tallow	7,080	6,800	6,558	6,345	5,270	4,907	4,238	3,436
Total	84,400	69,900	58,236	49,039	37,020	31,460	27,075	22,018

a Forecast for 1990 and 1995.

close to the actual 1990 production and appears to be realistic for 1995 as well. The data in Table 3.4 indicate that production has been increasing an average of about 20% for every 5-year period from 1960 through 1990. This forecast extrapolates that for 1995 world production will exceed 84 million tons. Considering that the 26 million ton increase from 1985 to 1995 represents nearly the entire oil production in 1970, the challenge and opportunity for the fats and oils industry is unmistakable.

REFERENCES

1. T.J. Weiss, *Food Oils and Their Uses*, AVI Publishing Co., Westport, Conn., 1983, p. 129.
2. *Oil World Weekly* **14,** 7 (1993).
3. E.G. Latondress, *J. Am. Oil Chem. Soc.* **58,** 187 (1981).
4. A.E. Thomas III, *J. Am. Oil Chem. Soc.* **55,** 831 (1978).
5. A.E. Bailey, *Industrial Oil and Fat Products,* 2nd ed., Wiley-Interscience, New York, 1951, pp. 211–212.
6. P. Chawla and J.M. deMan, *J. Am. Oil Chem. Soc.* **67,** 329 (1990).
7. L.H. Wiedermann, *J. Am. Oil Chem. Soc.* **55,** 825 (1978).
8. p. 826 in Ref. 7.
9. M.S.A. Kheiri, *J. Am. Oil Chem. Soc.* **62,** 414 (1985).
10. N.O.V. Sonntag in *Bailey's Industrial Oil and Fat Products,* Vol. 1, 4th ed., Wiley-Interscience, New York, 1979, pp. 292–448.
11. Institute of Shortening and Edible Oils, Inc., *Food Fats and Oils,* 13 (1988).
12. J. Edvardsson and S. Irandoust, *J. Am. Oil Chem. Soc.* **71**(3), 235 (1994).

13. p. 825 in Ref. 7.
14. S.J. Laning, *J. Am. Oil Chem. Soc.* **62,** 403 (1985).
15. American Society of Heating, Refrigeration and Air-conditioning Engineers, *Refrigeration Handbook,* Vol. 2, 1990, pp. 2.1–2.9.
16. p. 827 in Ref. 7.
17. N.T. Joyner, *J. Am. Oil Chem. Soc.* **30,** 531 (1953).
18. p. 832 in Ref. 4.
19. *Official Methods and Recommended Practices of the American Oil Chemists' Society,* Champaign, Ill.
20. L. deMan, J.M. deMan, and B. Blackman, *J. Am. Oil Chem. Soc.* **68**(2), 64 (1991).
21. W.E. Link, ed., *Official and Tentative Methods of the American Oil Chemists' Society,* American Oil Chemists' Society, Champaign, Ill., 1974.
22. S. Mielke, *J. Am. Oil Chem. Soc.* **64**(3), 298 (1987).

4
Shortening: Types and Formulations

INTRODUCTION

Shortening once was the reference to naturally occurring fats that are solid at room temperature and were used to "shorten" or tenderize baked foods. The material makeup of shortening has changed from a natural fat to blends of oils with hard fats to hydrogenated liquid oils to blends with additives like emulsifers, antioxidants, antifoamers, metal scavengers, antispattering agents, etc. With all these changes, shortenings are still intended to tenderize, or shorten, baked products as well as to provide many other functional attributes for baked and other food preparations. Today's shortenings are essential ingredients in virtually every type of prepared food product. Shortenings affect the structure, stability, flavor, storage quality, eating characteristics, and the eye appeal of the foods prepared.

1.1 Historical

The first fat and oil products used by humans were probably rendered from wild animal carcasses. Then, as animals were domesticated, their body fat became an important food source and was used for other things such as lubricants, illuminates, and soap. Lard or hog fat became the preferred animal fat for edible purposes, while the other animal fats were utilized for nonedible applications. The more pleasing flavor of lard may have been one reason for choosing it for edible purposes. However, the principal reason undoubtedly was the plastic consistency of this fat. At room temperature, lard had a good consistency for incorporation into breads, cakes, pastries, and other baked products. The tallows were too firm for this purpose and the available marine oils were too fluid (1).

Plastic fat usage has been more prevalent in the Western Hemisphere, largely because of the meat-eating proclivities of the people and the availability of fats from the animals eaten. Vegetable shortening with all the physical characteristics of a plastic animal fat was an American invention; it was created by the cotton raising industry and perfected for soybean oil utilization. Cotton acreage expansion between the American Civil War and the close of the nineteenth century resulted in large quantities of cottonseed oil. At the same time, the pork industry could not satisfy all of the plasticized shortening requirements. Initially, cottonseed oil was blended with animal fats and called *lard compounds* or simply *compounds* and were marketed as substitutes for lard. Therefore, meatpackers had a prominent role in the development of the shortening industry due to the essential ingredient of the product: hard animal fat (2).

The introduction of the hydrogenation process into the United States about 1910 split the development of vegetable shortenings into two divergent courses. The meatpackers continued to process compounds or blended shortenings, employing the hydrogenation process only for the production of hardened oils to serve as occasional substitutes for oleostearine. Other manufacturers abandoned the lard substitute concept to create products with properties better than the traditional properties of lard. The terms *lard compound* and *compound* were replaced with proprietary names or brands not suggestive of meat fats.

The vegetable shortening manufacturers had to devise methods for improving their product to gain market share. It was necessary to more carefully refine and bleach the vegetable oils than had been the custom with the compounds. New deodorization methods with high temperatures and better vacuums were developed to remove all traces of odor and flavor to produce "bland" shortenings. Further improvements in the techniques of hydrogenation enabled the production of shortening products with increased oxidative stability, improved uniformity, and enhanced performance characteristics. Additionally, better processes for solidification, filling, packaging, and crystal development were devised to enhance the appearance and improve the keeping qualities and performance of the shortenings (3).

Pure vegetable shortening quickly assumed a position of preeminence and was accepted as a premium product both by the housewife and the commercial user or baker. The bland flavor, uniform white color, and smooth texture were probably the prime factors influencing acceptance by the consumer. Undoubtedly, these factors also influenced the bakers, but the deciding factors were the increased stability and the improved creaming properties (4).

The chemistry of natural fats was virtually an unexplored field; it did not attract many research chemists. However, a rapid expansion of interest in lipid chemistry began to be evident in the late 1920s. The obvious opportunities presented by the advancements in the hydrogenation of liquid oils attracted new investigators to this field. This increased activity led toward the development of many new analytical tools and techniques plus new commercially important products to dispel the previous academic theory of the simplicity of natural fats and oils. It was found that these substances were capable of undergoing many of the classic organic chemistry reactions: isomerization, polymerization, oxidation, esterification, interesterification, condensation, addition, and substitution (5).

High-ratio shortenings, introduced about 1933, brought about significant changes for the baker and the shortening industry. These shortenings contained mono- and diglycerides that contributed to a finer dispersion of fat particles causing a greater number of smaller sized fat globules, which strengthened cake batters. Emulsified shortenings allowed the baker to produce cakes with additional liquids, which permitted higher sugar levels. Additionally, the high-ratio shortenings had excellent creaming properties or the ability to incorporate air. Altogether, the superglycerinated shortenings produced more moist, higher volume cakes with a fine grain and a more even texture. Then, as a bonus, it was determined that lighter icings with higher moisture levels could be produced with these shortenings (6).

Development of emulsifiers and their ability to significantly improve bakery products added a new dimension to the fats and oils industry; it ushered in the era of tailor-made shortenings. New shortenings, specifically designed for a single application, such as layer cakes, pound cakes, cake mixes, creme fillings, icings, whipped toppings, breads, sweet doughs, and other baked products were developed and introduced at a rapid pace after World War II (7). Specialty shortening development fostered further improvements in all aspects of the fats and oils industry: processing techniques and equipment, quality control procedures and tools, processing aids, additives, analytical equipment and methods, packaging containers and equipment, plus others too numerous to list.

The advances in technology have increased the storehouse of fats and oils information allowing the introduction of even more sophisticated products for all aspects of the food industry: retail consumer, foodservice, and food processor. The specialty shortenings have been responsible for many of the recent advances within the foodservice and food processor industries. Likewise, new food developments have created the need for totally new shortening products. Currently, plasticized shortenings have been joined by three other shortening forms: liquid shortenings, shortening chips, and powdered shortenings. These additional shortening products were introduced to fill needs expressed by the food industry.

1.2 Source Oils

Raw material selection for shortenings has been influenced by availability and economics. These factors were the main reasons for the development of lard substitutes, introduction of pure vegetable shortenings, major processing improvements, the development of additives to fortify specific performance char-

Table 4.1 Fats and oils used in U.S. shortenings (in millions of pounds) (8–11)[a]

Source Oil	1940	1950	1960	1970	1980–1981	1990–1991	1992–1993
Soybean	212	841	1169	2182	2675	4090	4465
Cottonseed	823	549	365	276	132	272	238
Coconut	18	NR	10	45	103	NR	NR
Palm	NR	NR	NR	90	215	NR	NR
Peanut	23	12	2	16	NR	NR	NR
Corn	1	1	4	12	W	304	241
Lard	17	177	480	430	328	277	274
Tallow	58	31	268	522	730	498	405
Other	44	27	3	7	NR	NR	NR
Total	1196	1727	2302	3580	4224	5793	5960

[a] NR = none recorded; W = withheld.

acteristics, and most of the other changes. Table 4.1 reviews the source oils and quantities used to produce shortenings in the United States since 1940.

Lard usage in shortening decreased to a low point in 1940 while cottonseed oil became the major source oil used in shortenings. Interestingly, soybean oil accounted for almost 18% of the oil used for shortenings at this early date. The next decade was a time of change as illustrated by the shortening source oils utilization: (*1*) soybean oil replaced cottonseed oil as the highest volume oil to attain 49% of the total oil used and (*2*) the introduction of crystal-modified lard-based shortenings in 1950 returned this raw material to shortening production. Interesterified lard shortenings were comparable to good-quality pure vegetable plasticized shortenings in appearance and creaming properties.

Soybean oil began its rise from a minor, little known, problem oil before 1940 due to the scarcity of other oils during World War II. After the hostilities ended, these gains were in jeopardy unless the technology could be developed to improve the flavor stability of soybean oil products. The technologies developed included metals deactivators, hydrogenation techniques and more selective catalysts, antioxidants, surface active agents, other surfactants, etc. These and other considerations helped soybean oil reach and maintain a dominant shortening source oil position in 1960 with over 50% of the total requirement. The soybean oil share has risen to 75% for the 1992–1993 crop year.

Palm oil threatened to become a major source oil for U.S. shortenings in the mid-1960s. It grew to command better than 15% of the total shortening requirement in less than 10 years. Palm oil was found to be an excellent plasticizing agent to force a shortening's crystal habit to β'. Therefore, the growth of palm oil usage in shortenings was primarily at the expense of cottonseed oil and tallow. Palm oil's decline, which began in the late 1980s, was due to unfavorable publicity highlighting nutritional concerns with saturated fatty acids. Palm oil dropped to an estimated 2–3% of the total shortening source oil requirement after this time.

Coconut oil and the other lauric oils are not among the more desirable materials for shortening manufacture because of their short plastic range and tendency to foam in deep-fat frying when mixed with other source oils. Nevertheless, coconut oil was a popular frying media for Mexican foods, which probably accounts for the high usage in 1980–1981. This use also suffered from unfavorable publicity that convinced the foodservice industry to change to frying shortenings with a more healthy image. This change probably accounted for the rise in corn oil utilization in 1990.

Lard and tallow were both important shortening raw materials after the meat fat shortenings regained popularity during the 1950s: interestified lard in bakery shortenings and tallow in frying shortenings and as a hardening/plasticizing agent for shortenings. Later, technology was developed to replace the crystal-modified lard with tallow. The meat fat usage continued to grow until cholesterol concerns brought pressure on the major users to provide products that had a better nutritional image. Meat fats account for less than 11% of the shortening source oil requirements in 1992, after controlling slightly more than 25% for many years.

The U.S. shortening industry has been able to develop processing technologies for the utilization of many different source oils. The nutritional challenges of the 1990s may require that the technology effort be directed toward a change in the source oil's composition. Plant biotechnology research has developed the potential for genetic variation in the fatty acid composition of oils. This means that performance may be grown into the oil rather than relying on processing to provide these characteristics. Table 4.2 illustrates some of the potential changes in the fatty acid composition of vegetable oils with biotechnology.

SHORTENING ATTRIBUTES

The development of a shortening for a food application is dependent on many interlaced factors. These requirements may differ from customer to customer

Table 4.2 Genetically modified vegetable oils (12)

Oil Seed	Type	Fatty Acid Composition (%)					
		C-16:0	C-18:0	C-18:1	C-18:2	C-18:3	C-22:1
Soybean	Normal	11.0	3.0	22.0	56.0	8.0	—
	N-87-2122-4	5.3	3.2	48.0	38.9	4.6	—
	N85-2176	9.5	3.3	44.4	39.5	3.3	—
	A-6	8.4	28.1	19.8	35.5	6.6	—
	C-1727	17.3	2.9	16.8	54.5	8.3	—
Sunflower	Normal	7.0	5.0	20.0	68.0	—	—
	G-8	3.3	8.2	84.2	3.5	0.8	—
Rapeseed	Normal	4.0	2.0	18.0	14.0	9.0	53.0
	M-30	2.4	1.0	91.6	1.5	3.3	0.2

depending on the equipment, processing limitations, product preference, customer base, and many other contributors. Fat and oil products are now being designed to satisfy individual specific requirements as well as the general-purpose products with a broader application potential. The design criteria for the general-purpose products must be of a less exacting nature than those developed for a specific product or process.

The important attributes of a shortening in different food products vary considerably. In some food items, the flavor contributed by the shortening is of minor importance; however, it does contribute a beneficial effect to the eating quality of the finished product. This fact has been relearned by the developers and experienced by the initial consumers, to their dissatisfaction, with many of the fat-free products recently introduced. A characteristic failing of the fat-free products was a lack of the eating characteristics normally contributed by shortening. In many products, such as cakes, pie crusts, icings, cookies, and other pastries, shortening is the major contributor to the product structure and eating character as well as contributing other significant effects upon the finished product's quality. Satisfactory shortening performance is dependent on many factors. Five of the most important considerations, which affect most applications are: (*1*) flavor, (*2*) physical characteristics, (*3*) crystal structure, (*4*) emulsification, and (*5*) additives.

2.1 Flavor

Generally the flavor of shortening should be as completely bland as possible so it can enhance a food product's flavor rather than contribute a flavor. In some specific cases the desired shortening flavor is typical of the original flavor of the source oil, for example, a lardy flavor is somewhat desirable in some products. Also, artificial flavors are added to some shortening products to enhance the functionality. Both the bland or typical flavor must be stable throughout the life of the food product. Therefore, the oxidative stability requirements of the finished product must be established to determine the minimum requirements for the shortening. The oxidative reversion rate of a shortening is directly related to the type and amount of unsaturated fatty acids available. The expected oxidation rates for the three most common unsaturated fatty acids are as follows (13):

Fatty Acid	Oxidation Rate
Oleic C-18:1	1
Linoleic C-18:2	10
Linolenic C-18:3	25

Reversion of deodorized shortenings brings back the flavor characteristics of the original crude oil. Shortening must be designed with a flavor stability suitable for the finished product requirements. Reduction of unsaturated fatty acids can be accomplished with hydrogenation or fractionation to increase flavor stability.

2.2 Physical Characteristics

The characteristics of the fats and oils utilized for a shortening are of primary importance in the design of a shortening for a specific use. Oils can be modified through various processes to produce the desired properties. Hydrogenation has been the primary process used to change the physical characteristics of oils. Melting points or hardness of an oil can be completely altered with this process and the changes controlled by the conditions used to hydrogenate the oil. In the hardening process, hydrogen gas is reacted with oil at a suitable temperature and pressure in the presence of a catalyst with agitation. Control of these conditions and the end point enables the operator to better meet the desired physical characteristics of the shortening products. A range of typical shortening solids fat index curves are shown in Figure 4.1. This chart illustrates how the hydrogenation process can be utilized to produce physical characteristics suitable for the performance desired.

2.3 Crystal Structure

Each source oil has an inherent crystallization tendency, either β or β'. The small, uniform, tightly knit β' crystals produce smooth textured shortenings with good plasticity, heat resistance, and creaming properties. The large β crystals can produce sandy, brittle consistency shortenings that result in poor baking performance where creaming properties are important. However, the large β crystals are desirable for some applications such as pie crusts or frying. Crystal habit is controlled by the oil source selection and complemented by the plasticization conditions employed and tempering for slow crystallization after packaging.

2.4 Emulsification

Shortening emulsifying properties are accomplished with adjustment of the fat structure and with the addition of surface-active agents. The typical food emulsifiers supplement and improve the functionality of a properly developed shortening, i.e., act as lubricants, emulsify fat in batters, build structure, aerate, improve eating qualities, extend shelf life, crystal modifier, antisticking, dispersant, moisture retention, etc. Obviously, no single emulsifier or emulsifier system can perform all of these different functions. In selection of the proper emulsifier or system, the developer must consider the usage application, the preparation method, emulsion type, effects of the other ingredients, econom-

168 Shortening: Types and Formulations

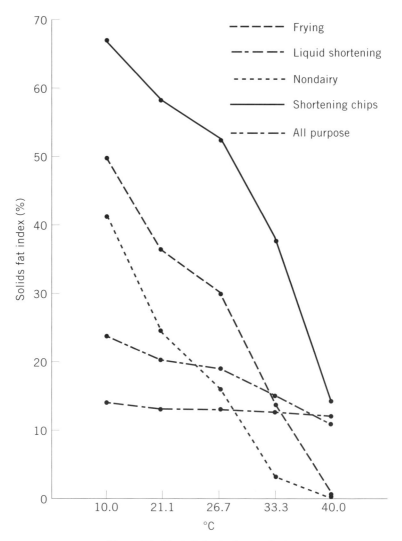

Figure 4.1 Typical shortening products.

ics, and any other applicable criteria for the finished product. Most of the emulsifiers used in shortenings are listed in Table 4.3 with Code of Federal Regulations (CFR) Title 21 Part numbers and a suggested application guide.

2.5 Chemical Adjuncts

In addition to emulsifiers a number of other chemical compounds provide a specific function for certain types of shortenings. These additives can be classified into the following categories:

Table 4.3 Emulsifiers used in shortening

Emulsifier	CFR 21	Applications[a]
Mono- and diglyceride	182.4505	All
Lecithin	182.1400	All
Lactylated monoglyceride	172.852	C and M
Calcium stearyl lactylate	172.844	B and S
Sodium stearoyl lactylate	172.846	B and S
Propylene glycol mono esters	172.856	C,M,B, and S
Diacetyl tartaric monoglycerides	182.4101	B
Ethoxylated monoglycerides	178.834	B
Sorbitan monostearate	172.842	C and M
Polysorbate 60	172.836	All
Polyglycerol esters	172.854	All
Succinylated monoglycerides	172.830	B
Sodium stearoyl fumarate	172.826	B and S
Sucrose esters	172.859	C,B, and S
Stearoyl lactylate	172.848	C,B, and M

[a] Applications code: B, bread; C, cake; F, fillings; I, icings; M, cake mix; S, sweet doughs; All, all applications.

Antifoamers. Dimethylpolysiloxane serves as an antifoaming agent that forms a monomolecular layer on the surface of a heated frying shortening to retard oxidation and foaming. The silicone compounds are added to frying shortenings at levels of 0.5–3.0 ppm. Higher concentrations do not inhibit foaming any more effectively and can cause immediate foaming at levels in excess of 10 ppm, the maximum allowed by U.S. federal regulations. Other potential problem areas with antifoamer use are: (*1*) unintentional addition to bakery shortenings can cause cake failures, (*2*) glazes may not adhere to donuts fried with an antifoamer present, and (*3*) potato chips may lack crispness (14).

Antioxidants. Antioxidants are materials that can retard the development of off-flavors and odors by inhibiting oxidation. Vegetable oils contain natural antioxidants, tocopherols, that can survive most processing. Several phenolic compounds have been identified that can also provide oxidative stability. Table 4.4 presents active oxygen method (AOM) stability data for select source oils treated with four different phenolic compounds (14). This data should only be used as a guideline to distinguish effectiveness of the antioxidants. In general, tertiary butylhydroquinone (TBHQ) is the most effective synthetic antioxidant in unhardened vegetable oils, followed by propyl gallate. TBHQ also appears to be effective for the meat fats, but protection is also afforded by butylated hydroxytoluene (BHT) and butylated hydroxyanisole (BHA).

Metal Inactivators. Fats and oils obtain metal contents from the soils where the plants are grown and later from contact during crushing, processing,

Table 4.4 Antioxidants effect upon AOM stability

	AOM Stability (h)			
	70 PV End Point		20 PV End Point	
Antioxidant[a]	Soybean	Cottonseed	Lard	Tallow
None	11	9	4	16
TBHQ	41	34	55	133
BHA	10	9	42	95
BHT	13	11	33	138[b]
Propyl gallate	26	30	42	[c]

[a] 200 ppm added.
[b] 100 each of BHA and BHT.
[c] Not analyzed.

and storage. Many of the metals promote autoxidation that results in off-flavors and odors accompanied by color development in the finished shortening. Studies have identified copper as the most harmful metal with iron, manganese, chromium, and nickel following. The results of a lard study is presented in Table 4.5. Metal scavengers, added at low levels during or immediately following deodorization, facilitate removal of the harmful metals. The most widely used chelating agent is citric acid at 50–100 ppm. Phosphoric acid at 10 ppm and lecithin at 5 ppm have also been used to inactivate metals.

Colorants. Pigments used in shortenings are usually the oil-soluble carotenoids in the yellow to reddish orange range. The carotenoids with U.S. Food and Drug Administration (FDA) approval include the carotenes, bixin, and apo-6-carotenal. Carotenoids are heat sensitive but can be stabilized with

Table 4.5 Trace metals concentration required to reduce lard AOM by 50% (15)

Trace Metal	ppm
Copper	0.1
Manganese	0.6
Iron	0.6
Chromium	1.2
Nickel	2.2
Vanadium	3.0
Zinc	19.6
Aluminum	50.0

BHA and BHT for greater heat stability. Lakes are FDA approved and heat stable but violent agitation is required to keep these oil-insoluble colorants in suspension. Apocarotenal, a synthetic, FDA-approved pigment is used primarily as a color intensifier for β-carotene (16).

Flavors. Most of the flavors used in shortenings are butterlike. Diacetyl was the major butter flavor used in fat and oil products until improved analytical techniques identified other flavor components in butter. Today, the U.S. FDA regulations allow safe compounds that impart a suitable flavor to the finished product (17). The choice of a particular flavor or blend of flavors depends on the expertise and the taste preference of the product developer.

BASE STOCK SYSTEM

In the past, prepared food products were formulated with the ingredients available. Today, most prepared foods are formulated with ingredients designed for their applications or, in many cases, specifically for the particular product and/or processing technique employed by the producer. These customer specialty products have expanded the product base for fat and oil processors from a few basic products to literally hundreds.

A shortening manufacturer could formulate products so that a special base fat or series of base fats is required for each different product. This practice, with the ever-increasing number of finished products, would result in a scheduling nightmare with a large number of product heels tying up tank space and inventory. A base stock system utilizing a limited number of hydrogenated stock products for blending to meet the finished product requirements is practiced by many shortening producers.

3.1 Base Stock System Advantages

The advantages provided by a base stock system are twofold (18).

Control.

1. Blending of hydrogenated oil batches to average minor variations.
2. Improved uniformity by producing the same hydrogenated product more often.
3. Cross-product contamination reduction by scheduling more of the same product consecutively.
4. Reduction of product deviations from attempts to utilize finished product heels.
5. Eliminate rework created by heel deterioration before use.

Efficiency.

1. The hydrogenation process scheduled to maintain base stock inventories rather than reacting to customer orders.
2. Hydrogenation of full batches instead of some smaller batches to meet demands without creating excessive heels.
3. Better order reaction time to improve customer service.

3.2 Base Stock Uniformity Control

Base stock consistency control is the secret to first pass yield, i.e., meeting finished product specification limits with the specified blend of base stocks. Most finished shortenings' physical characteristics are controlled with solids fat index (SFI) iodine value (IV), and/or a melting point analysis. However, time restraints during hydrogenation do not allow the long elapsed times required for these results. More rapid techniques to control base stock end points are:

1. Soft base stocks with IV of 90+ are adequately controlled by refractometer (RI or RN) determinations performed within the hydrogenation department. Mettler dropping points and most other melting point analysis have poor reproducibility at this level or are too time consuming.
2. Intermediate base stocks (55–89 IV) preferred controls are a combination of RI or RN readings and Mettler dropping point (AOCS Method Cc 18–80) (19). The oil is hydrogenated to the predetermined refractometer end point and then the reaction is stopped until a Mettler dropping point result can be determined. The melting point is the controlling analysis. If the analysis indicates that the base stock is softer than desired, hydrogenation is continued and the process repeated until the internally specified end point has been attained.
3. Low iodine value hard stocks can be controlled with "quick titer" determinations. There are many versions of this evaluation, but basically, the test involves dipping the bulb of a glass thermometer into liquid fat heated above the melting point, then rotating the thermometer stem to cool the fat. The end point is the temperature when the fat on the bulb clouds. Constants can be determined to add to the quick titer result to approximate "real titer" determined by AOCS Method Cc 12–59 (20).

The time-consuming but more accurate analytical methods should be performed as an audit function to ensure that the correlation with the quicker controls are still valid.

3.3 Soybean Oil Base Stocks

Base stock requirements will vary dependent on the customer base served, which obviously dictates the finished product mix. A wide variety of source oils are available that could be used for shortening base stocks, as illustrated by Table 4.1. The choices are narrowed by factors such as customer specifications, costs, religious prohibitions, traditional preferences, crop economics, legislation, availability, transportation, and others. These factors have favored soybean oil in the United States for several decades. Therefore, most U.S. shortening processors have base stock systems dominated by soybean oil with only minor representation by the other source oils. The minor oils in most base stock systems serve as β' promoters for plasticity, like cottonseed and palm oils, or those source oils required for specialty product preparation.

Table 4.6 outlines a soybean-oil-based system with seven hydrogenated base stocks; ranging from a partially hydrogenated 108 IV base stock to a saturated hard stock with a 5 max IV. An eighth base stock is refined and bleached liquid soybean oil, the same oil that serves as the feedstock for the hydrogenated bases. Utilization of a similar base stock system should enable the processor to meet most shortening specifications by blending two or more base stocks, except for some specialty products that can only be made with special hydrogenation conditions.

Hydrogenation has two principal aims: to produce solid or semisolid products with certain plastic properties and to increase the stability of the oil. A huge variety of products can be produced with the hydrogenation process depending on the conditions used and the degree of saturation. The fatty acid composition and resultant characteristics of the hydrogenated products depend on the controllable factors: temperature, pressure, agitation, catalyst type and activity, and catalyst concentration. Agitation is fixed on most converters by design and, therefore, cannot be considered an operator controllable factor.

Table 4.6 Soybean oil base stock

Iodine Value	Solid Fat Index					Mettler Dropping Point (°C)	Quick Titer (°C)
	10.0°C	21.2°C	26.7°C	33.3°C	40.0°C		
108	4.0 max.	2.0 max.	—	—	—	—	—
85	15–21	6–10	2–4	—	—	28–32	—
80	22–28	9–15	4–6	—	—	31–35	—
75	38–44	21–27	13–19	3.5 max.	—	34–37	—
65	59–65	47–53	42–48	23–29	3–9	41–45	—
60	65–71	56–62	51–57	37–43	14–18	45–48	—
>5	a	a	a	a	a	a	50–54

Hydrogenation Conditions	Temperature	Pressure	Catalyst Concentration	Agitation
Nonselective	Low	High	Low	Fixed
Selective	High	Low	High	Fixed
Selectivity unimportant	High	High	High	Fixed

^a Too hard to analyze.

Changes in the reaction conditions affect the selectivity of the hydrogenated base stock; e.g., saturation of linoleate over oleate and the rate of trans acids formation. Selectivity affects the slope of the SFI curve; steep SFI slopes are produced with selective hydrogenation conditions while flat SFI slopes are the result of nonselective hydrogenation conditions.

The soybean oil base stock system in Table 4.6 utilizes both selective and nonselective conditions for the partially hydrogenated bases. Selectivity is not important for low-IV hard stocks because these reactions are continued to almost complete saturation. The main objective for these products is to reach maximum saturation as quickly as possible. Actual values for the hydrogenation conditions were not identified because results vary from one converter to another due to design and other variables within each plant (21). It is necessary to develop conditions for each installation separately to meet the SFI, IV, and melting point relationships.

SHORTENING FORMULATION

Most shortenings are identified and formulated according to usage. Figure 4.1 illustrates the diverse SFI and melting point relationships among five different shortening products. This figure indicates the differences in plastic range necessary to perform the desired function in the finished products. Shortenings with the flattest SFI curves have the widest plastic range for workability at cool temperatures as well as elevated temperatures. All-purpose shortenings have the widest plastic range. Nondairy and solid frying shortenings have relatively steep SFI curves, which will provide a firm, brittle consistency at room temperature but will be almost fluid at only slightly elevated temperatures.

The product with the very flat SFI slope is a fluid opaque or pumpable liquid shortening that has become popular due to the convenience offered, handling cost savings in some situations, and lower saturated fatty acid levels. In these systems, the β-crystal form is necessary to produce and maintain fluidity.

Shortening chips are a somewhat recent development as a specialty ingredient for addition to doughs, biscuits, cookies, and other baked products. These specialty products are modifications of fat flakes, which formerly indicated only the saturated oil products or hard fats. The chips are formulated with selectively hydrogenated base stocks with melting points high enough to flake but low enough for good eating characteristics. This product type now includes the traditional hard fat flakes, shortening chips, and stabilizers for icings and glazes.

4.1 Wide Plastic Range Shortenings

The basic all-purpose shortening has been the building block for shortenings where creaming properties, a wide working range, and heat tolerance are important. The functionality of an all-purpose shortening at any temperature is largely a function of the solids content at that temperature. The all-purpose shortenings are formulated to be not too firm at 10–16°C (50–60°F) and not too soft at 32–38°C (90–100°F). Initially, a liquid oil was blended with a hard fat to make a compound shortening that had a very flat SFI curve, which provided an excellent plastic range. However, the low oxidative stability of these shortenings preclude their use today for most products. Currently, most of these products are formulated with a partially hydrogenated soybean oil base stock and a low-IV cottonseed or palm oil hard stock. Hard fats are added to shortenings both to extend the plastic range, which improves the tolerance to high temperatures, and for crystal type and stability. The β'-crystal-forming cottonseed oil hard stock functions as a plasticizer for improved creaming properties.

Hydrogenation of a shortening base increases the oxidative stability. As a rule, the lower the base IV the longer the AOM stability. However, as base hardness is increased, the level of hard stock required to reach a desired consistency decreases. Hard stock reduction reduces the plastic range and heat tolerance. Therefore, oxidative stability improvements are achieved at the expense of plasticity. The extent that one attribute can be compromised to improve another must be determined by the requirements of the intended food product.

Shortening plastic range is important for bakery shortenings intended for roll-in and creaming applications alike because of the consistency changes with temperature. Shortenings become brittle above the plasticity range and soft below the range; both conditions adversely affect creaming and workability alike. Shortenings are normally plastic and workable at SFI values between 15 and 25%. Therefore, shortenings with flatter SFI slopes fall within the plasticity window for a much greater temperature range than products with steep SFI slopes. The all-purpose shortening in Figure 4.1 has a plastic range of 23°C (73.4°F) while the frying shortening has only a 4°C (39.2°F) plastic range. Theoretically, the frying shortening should perform equally as well as the all-purpose shortening for baking applications if it is used within the 4°C (39.2°F) range from 29 to 33°C (85 to 92°F). The frying shortening's use for the baking application would require very strict controlled temperatures probably not available in most bakeries. The 23°C (73.4°F) range from 10 to 33°C (50 to 92°F) for the all-purpose shortening is decidedly more practical.

Two of the basestocks outlined in Table 4.6 are designed for shortenings requiring a wide plastic range, e.g., the nonselectively hydrogenated 80 and 85 IV base stocks. Even these two base stocks with only a 5 IV difference provide measurable differences in plastic range and stability when cottonseed oil hard fat is added to produce equivalent consistencies at 26.7°C (80°F). The softer 85 IV base stock required 2.5% more hard fat to achieve the targeted 20% SFI at 26.7°C (80°F). The higher hard fat level indicates greater heat resistance and a wider plastic range but a lower AOM stability from the lower IV value. The firmer 80 IV base stock required 2.5% less hard fat to attain the 20% SFI at 26.7°C (80°F), which helped reduce the plastic range by 4.5°C (40.1°F) but increased the AOM stability to 100 h verses the 65 h for the shortening with the 85 IV base. Figure 4.2 graphically illustrates these effects.

The effect of hard stock upon the SFI slope and plastic range is illustrated by Figure 4.3. Cottonseed oil hard fat was added to the 85 IV soybean oil base stock to demonstrate the plasticizers effect; as the hard fat level is increased the shortening becomes firmer with a flatter slope. The higher hard fat levels are used to formulate roll-in, puff pastry, and other shortenings where a plastic but firm consistency is desired.

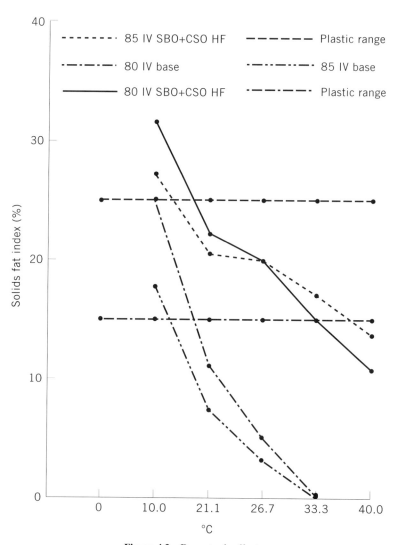

Figure 4.2 Base stock effect.

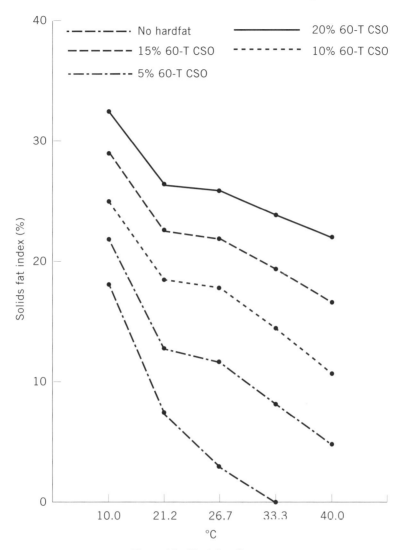

Figure 4.3 Hard fat effect.

The use of a partially hydrogenated base plus hard fat to produce a wide plastic range with good creaming properties has been expanded into a whole family of specialized shortenings. The development of these products has involved the selection of the proper hydrogenated base stock and hard fat to produce the desired plastic range and AOM stability. These developments have taken two directions: (*1*) the addition of an emulsifier or an emulsifier system to an all-purpose shortening base or (*2*) formulating nonemulsified products for a specific functionality. Table 4.7 outlines shortening types in these two categories.

Table 4.7   Wide plastic range

Nonemulsified	Emulsified
All-purpose	Household
Danish roll-in	Cake and icing
Puff pastry	Icing and filling
Cookie	Cake mix
Pie crust	Specialty cakes
Donut frying	Yeast raised

4.2   Narrow Plastic Range Shortenings

Plasticity is of minor importance and can be a detriment for products requiring high oxidative stability and/or sharp melting characteristics. Shortenings specially designed for specific frying situations, nondairy systems, cookie fillers, and confectionery fats require flavor stability's and eating characteristics not possible with blends of moderately hydrogenated oils with hard fats. These products require lower IV shortenings for oxidative stability with steeper SFI slopes and a melting point lower than body temperature for good eating characteristics. The frying and nondairy shortenings plotted on Figure 4.1 illustrate the steep SFI slopes.

The frying shortening meets the restaurant industry requirements for a stable heat transfer media that becomes a part of the food to supply texture, mouth feel, and to enhance the food flavor. The nondairy shortening requirements are similar to those of the frying shortening. Shortening effects both texture and mouth feel of the finished products. A shortening with a flat SFI will mask the desirable flavors and coat the mouth rather than melt cleanly. Flavor stability is also a concern because reverted or oxidized flavors change the flavor profile of dairy-type products.

The steep SFI products are generally produced with selectively hydrogenated bases that have high solids at the lower temperature readings but fall off rapidly to a low melting point for the desired eating or mouth feel characteristics and excellent oxidative stability. These products are usually composed of a single straight hardened base or possibly two selectively hydrogenated bases for a slightly different slope than available with a single base. For example, the nondairy shortening in Figure 4.1 is composed of one straight hydrogenated base; it is the 75 IV selectively hydrogenated soybean oil base stock. The frying shortening can be produced with a blend of the 65 and 75 IV base stocks.

4.3   Liquid Shortenings

Liquid or pumpable shortenings are flowable suspensions of solid fat in liquid oil. The liquid oil phase may or may not be hydrogenated depending on the finished product's consistency and oxidative stability required. Low-IV soybean oil hard fat seeds crystallization. It can vary from as low as 1% to higher levels as required to produce the desired finished product viscosity. The ease in which soybean oil converts to the stable β crystal form makes it ideal for liquid shortenings (22).

Liquid opaque shortenings have been developed for food products where pourability or pumpability at room temperature and below is important. The major uses identified for these products are:

1. *Frying.* Liquid frying shortenings provide a convenience to the operator; it can be poured directly into the fryer and quickly heated to frying temperature.
2. *Bread and Cake.* Bakery liquid shortening systems usually contain high melting emulsifiers, which provide optimum performance, combined with the oil phase, which allows bulk handling without heating while still maintaining the desired specialty product performance.
3. *Nondairy.* Liquid shortening systems offer the room temperature pumpable convenience with a stable shortening high in polyunsaturates, which are attractive for some nondairy applications such as creamers, filled milks, toppings, etc.

Typical compositions and SFI results for four different types of liquid shortenings are compared in Table 4.8. These products must contain high levels of liquid or only lightly hydrogenated oils to maintain fluidity. Therefore, additives must contribute heavily to the product performance in all liquid shortenings. The major additives in liquid shortenings, as indicated with the compositions in Table 4.8, are antifoaming agents for frying applications and emulsifiers for the cake and nondairy products. Bread shortenings are the only

Table 4.8 Typical liquid opaque shortening formulations

Soybean Oil Basestock (%)	Frying	Bread	Cake	Nondairy
135 IV	—	90	—	—
108 IV	98	—	99	98
5 IV	2	10	1	2
Base total	100	100	100	100
Additives				
Dimethylpolysiloxane	1.0 ppm	None	None	None
Emulsifiers (%)	None	None	Yes	Yes
Analytical limits:				
% SFI at 10.0°C	4.0 max	13.0	7.0	5.5
% SFI at 26.7°C	2.0 max	12.0	5.5	2.5
% SFI at 40.0°C	0.5 max	10.0	1.0	1.0
α monoglycerides (%)	—	—	1.5	4.0
PGME (%)	—	—	2.8	—

product where an additive is not mandatory for liquid shortening performance. Lubrication is the primary function of a bread shortening. However, the emulsifiers that improve bread shelf life are usually produced with fully saturated fats that are difficult for the baker to add to doughs. Liquid shortening is an ideal carrier for these emulsifiers, providing another convenience for the user.

Other additives used with liquid shortenings are color and flavor. Butterlike flavors and yellow colors have been added to the nondairy liquid shortenings to make them more complete systems. In the foodservice area, a modification of the liquid frying shortening has become another product type—*pan and grill* liquid shortening. Additions of butterlike flavors, yellow colors, and lecithin to the liquid frying shortening are the usual procedure employed for making the pan and grill liquid products. Lecithin, a natural emulsifier, functions as an antisticking agent in this application.

4.4 Shortening Flakes and Chips

Shortening flake describes the high melting edible oil products solidified into a thin flake form for ease in handling, quicker remelting, or for a specific function in a food product. The traditional flaked products have been the saturated oil products known as low IV hard fats or stearines. The hard fat flakes have been joined by products formulated for special uses where the flake form is desirable. These specialty flaked products are shortening chips and stabilizers for icings and glazes.

Hydrogenation selectivity is not important for the hard fats or stearines because the reactions are carried to almost complete saturation. However, selectivity is very important for the proper functionality of both the shortening chips and icing stabilizers. The shortening chips are selectively hydrogenated to attain a steep SFI slope with a melting point as low as possible to still allow the product to maintain the chip form after packaging, during shipment to the food processor, while processing into the finished product, and until heated by the homemaker during baking to produce a flaky, tender product. The chips can be formulated from the base stock system with a blend of the 60 and 65 IV base stocks depending on the desired melting point and SFI requirements. Shortening chips, made with nonlauric oils, melting points usually range from 43 to 48°C (110 to 118°F). Shortening chips can be made with hydrogenated lauric oils (palm kernel or coconut) as well. The lauric-oil-based chips offer a sharper melt with a lower melting point than the chips made with soybean or cottonseed oils. The melting points for the lauric oil chips normally range from 38 to 42°C (101 to 108°F).

Flavors, colors, and/or spices can be encapsulated in the shortening chips to leave pockets of color and flavor where the chips have melted during baking. Shortening chips have been commercially available in flavors such as plain, butter, cinnamon sugar, and blueberry.

Icing stabilizers are also selectively hydrogenated with melting points centered at 45, 47, and 52°C (113, 116.6, and 125.6°F) and formulated with and without lecithin dependent upon the desired degree of fluidity in the finished icing or glaze. Lecithin addition increases fluidity up to 0.5% but restricts this function at higher levels. The icing stabilizers must be hydrogenated specially to meet the SFI–melting point requirements, except for the 45°C (113°F) melting point product; it can be blended like the shortening chips. In either case, low-IV hard fats should not be used to adjust melting points, due to the effect upon mouth feel.

4.5 Powdered Shortenings

Two types of powdered shortenings are produced: (*1*) spray dried fat emulsions with a carrier and (*2*) spray chilled or beaded hard fat blends. The spray dried powdered shortenings are partially hydrogenated shortenings encapsulated in a water-soluble material. The fats can be homogenized in solution with a variety of carriers; i.e., skim milk, corn syrup solids, sodium caseinate soy protein isolate, and others. Emulsifiers may be included with the fat for finished product functionality. Fat contents range from 50 to 82% depending on the original emulsion composition (23). The spray dried powdered shortenings are used in some prepared mixes for their ease in blending with the other dry ingredients.

Hard fats can be powdered or beaded without the aid of a carrier. Spray chilling in a tower or solidification on a chill roll followed by grinding and screening for particle sizing are two methods used for powdering or beading the hard fats (24). Some of the uses for these products are as peanut butter stabilizers, specialty prepared mix products, and in place of flaked hard fats where more rapid melting is desirable.

SHORTENING CRYSTALLIZATION

The functionality of solidified edible oil products is influenced by three basic processes: (*1*) formulation, which includes the choice of source oils and the hydrogenation techniques used for the base stocks; (*2*) chilling, which initiates the crystallization process; and (*3*) tempering, where the desirable crystal nuclei are developed and stabilized. Formulation, the first requirement for consistency control, has been reviewed in the preceding section. Chilling and tempering, which develop, mature, and stabilize the desired crystal structure introduced by the product composition, is discussed in this section.

Chilling and tempering processes are employed when shortenings are packaged. The physical form of edible oil products is important for the proper handling and performance in food products. The importance of shortening consistency cannot be overemphasized, many applications depend on the physical properties peculiar to each packaged product, such as softness, firmness, oiliness, creaming properties, melting behavior, surface activity, workability, solubility, aeration potential, pourability, and others. In the case of plastic shortenings, consistency is important from the standpoint of usage and performance in bakery and related products. And the importance of the consistency for a liquid shortening is obvious; it must be maintained as a uniform, homogeneous suspension to perform properly (25).

5.1 Fat Plasticity

Edible fat products appear to be soft homogenous solids; however, microscopic examination shows a mass of very small crystals in which a liquid oil is enmeshed. The crystals are separate discrete particles capable of moving independently of each other. Therefore, shortenings possess the three conditions essential for plasticity in a material:

1. A solid and a liquid phase.
2. The solid phase dispersed finely enough to hold the mass together by internal cohesive forces.
3. Proper proportions of the two phases.

Plasticity, or consistency, of an edible oil product depends on the amount of solid material, the size, shape, and distribution of the solid material, and the development of crystal nuclei capable of surviving high-temperature abuse and serving as starting points for new desirable crystal growth. The factor most directly and obviously influencing the consistency of a plastic shortening is the proportion of the solid phase. As the solids contents increase, an edible oil product becomes firmer. The proportion of the solid phase is influenced by the extent of hydrogenation and attendant isomerization. Shortening products are also firmer with smaller crystal sizes due to an increased opportunity for the solids particles to touch and resist flow. Stiffening is also increased more with the interlacing of long needlelike crystals than with more compact crystals of the same size. The crystal nuclei ("memory") are developed by the further treatment of a newly solidified product. Exposure immediately after chilling to 29°C (85°F) for 24+ h before cooling to 21°C (70°F) provides a softer consistency with the ability to withstand widely fluctuating temperatures and still revert back to the original consistency at room temperature (26).

5.2 Fat Crystal Habit

Fat crystals represent lower energy states of molecular configuration. At elevated temperatures, fats retain enough molecular motion to preclude organization into stable crystal structures. Edible oils go through a series of increasingly organized crystal phases with cooling until a final stable crystal form is achieved. This process can occur in fractions of a second or over a number of months. The crystal types formed define the textural and functional properties of most fat-based products.

The crystal structure of a shortening or other fat-based product is determined by: (*1*) source oil composition, (*2*) processing, and (*3*) tempering or maturing. Crystallization is induced when melted fat is cooled rapidly to initiate the formation of crystal nuclei or seed. The seeds form templates upon which crystals grow. Formulation, cooling rate, heat of crystallization, and agitation levels affect the number and type of crystals formed.

Each source oil exhibits inherent crystallization tendencies. A fat may pass through one or more unstable crystalline stages before assuming either β or β' crystal forms. The differences among the three crystal forms are (27):

1. α-crystal forms are unstable and will convert into the more stable β or β' crystal forms.
2. β crystals are large, coarse, and self-occluding.
3. β' crystals are small and needlelike and can pack together into dense fine-grained structures.

Each common fat or oil has a definite crystal habit that is determined by four factors: (*1*) palmitic fatty acid content, (*2*) distribution and position of palmitic and stearic fatty acids on the triglyceride molecule, (*3*) degree of hydrogenation, and (*4*) the degree of randomization. Table 4.9 identifies the crystal habit of hydrogenated edible oils (28).

Many edible oil products contain various combinations of β and β' tending components. The ratio of β–β' crystals helps to determine the dominant crystal habit, but the higher melting triglyceride portions of a solidified fat product

Table 4.9 Hydrogenated oils crystal structure

β	β'
Canola oil	Cottonseed oil
Cocoa butter	Butter oil
Coconut oil	Herring oil
Corn oil	Menhaden oil
Olive oil	Modified lard
Lard	Palm oil
Palm kernel	Rapeseed oil
Peanut oil	Tallow
Safflower oil	Whale oil
Sesame oil	
Soybean oil	
Sunflower oil	

usually force the fat to assume that crystal form. The crystal form of the solidified fat product has a major influence upon the textural properties. Fats exhibiting a stable β' form appear smooth, provide good aeration, and have excellent creaming properties for the production of cakes, icings, and other bakery-type products. Conversely, the β'-polymorphic form tends to produce large granular crystals for products that are waxy, grainy, and provide poor aeration. These β formers perform well in applications such as pie crusts where a grainy texture is desirable (29) and opaque liquid shortenings where the large granular crystals are preferred for stability and maintenance of fluidity.

PLASTICIZED SHORTENING CONSISTENCY

The consistency of a shortening is the culmination of all the factors influencing crystallization and plasticity: chilling, working, creaming gas, pressure, and tempering. Each process is individually and collectively important; a shortening's performance can be adversely affected if any of the consistency factors do not conform to the standards established for each individual shortening product during development. Table 4.10 summarizes the effects of the crystallization processes upon shortening consistency covered in greater detail in the following paragraphs.

6.1 Chilling

One of the earliest methods of solidifying shortenings involved the use of a chill roll that was internally refrigerated by circulating cold brine. The melted fat was fed into a trough from which a thin film of the liquid was picked up as the roll revolved and crystallized into a semisolid state during a revolution. A blade removed the solidified fat film, which fell into a trough equipped with a rotating shaft with metal fingers called a picker box, which worked the product to make it homogeneous and to incorporate air. Pumps then picked up the shortening for transfer to filling machines. This process has now become obsolete in the United States except for some very special products. The reason

Table 4.10 Factors influencing shortening consistency

Soft	Process	Firm
Cold	Chilling	Warm
More	Working	Less
High	Creaming gas	None
High	Pressure	Low
Warm	Tempering	Cold

for obsolescence for shortening processing was the difficulty in controlling the variables to produce a uniform product (30), poor thermal efficiency, and moisture condensation on the chilled roll surface.

Most shortenings are now quick-chilled in closed thin-film scraped-surface heat exchangers. The principal of operation for most of these units is a combination of chilling the fat in a very thin film, which is continually removed by scraper blades with concise temperature control. The residence time within the heat exchanger tubes is very short, almost always less than 20 s. Since all triglycerides exhibit a high propensity for supercooling, the product exits from the heat exchanger cooled well below the equilibrium crystallization temperature, usually within a range of 15.5–26.7°C (60–80°F) depending on the shortening type.

The chilling unit temperature control limits are determined by fill tests to identify the condition necessary to produce the consistency and plasticity requirements for each shortening product. The initial guidelines are usually identified by trial and error to determine the temperature and pressure where the shortening forms a slight rounded or mounded surface upon filling plus evaluation of the product characteristics after tempering. The finished shortening's consistency becomes softer as the temperature is decreased and firmer more brittle shortenings develop with higher temperatures. After the operating limits are identified for a product, the processing control should be within 1°C (33.8°F) for uniformity. Thereafter, if the plasticization controls and the formulation specifications are met, the original product should be duplicated.

6.2 Working and Filler Pressure

If the product is allowed to solidify without agitation at this point, it will form an extremely strong crystal lattice and exhibit a narrow plastic range. This consistency may be desirable for stick margarine but is harmful for products requiring a plasticlike consistency. Therefore, processing after the initial quick chilling cycle has been adapted to the product consistency or form desired (31,32). Enclosed worker units with speed controls have replaced the picker boxes that followed the chill roll in the early plasticization process. The heat of crystallization is dissipated rapidly in these units while working the product to provide fine crystals. Extrusion valves are employed in most systems to deliver a homogeneous smooth product to the filler at 17–27 atm (250–400 psig) (30).

6.3 Creaming Gas

Air incorporation, while agitating in the open picker box, has been replaced with nitrogen injection into the inlet side of the chiller, in precisely controlled quantities, normally 12–14% by volume for standard plasticized shortenings, which provides a white creamy appearance and increases the shortening's workability. The correct gas content is important for appearance and stability. Shortenings containing the proper levels are white and creamy with a bright surface sheen. Too low a gas content gives a yellowish greasy appearance and shortenings packed without creaming gas develop a Vaseline-like appearance. High gas levels cause dead-white chalky appearances with a lifeless surface appearance and often large air pockets within the product, which give a puffy feel or consistency. Nonuniform dispersion of the gas gives an unattractive streaked appearance.

Gas levels are varied from none to as high as 30% in the shortenings produced depending on the product requirements. The typical creaming gas additions for the different shortening types are identified in the following: regular plasticized, 12–14%; precreamed, 19–25%; puff pastry, none; liquid shortening, none; shortening chips, none; and speciality shortenings, as needed.

6.4 Tempering

Most technologists agree that a shortening is tempered when the crystal structure of the hard fraction reaches equilibrium by forming a stable crystal matrix. The crystal structure entraps the liquid portion of the shortening. The mixture of low- and high-melting components of the solids undergoes a transformation in which the low-melting fractions remelt and then recrystallize into a higher melting, more stable form. This process can take from 1 to 10 days depending on the shortening formulation and package size. After a shortening takes an initial set, some α crystals are still present. These crystals remelt and slowly recrystallize in the β' form during tempering. β' crystals are preferred for most plastic shortenings, especially those designed for creaming or laminating (32). Therefore, soybean-oil-based shortenings requiring a plastic range are formulated with 5–20% of a β' tending hard fat. The β' hard fat must have a higher melting than the soybean oil base stock in order for the entire shortening to crystallize in the stable β' form.

The effect of tempering on a plastic shortening can be demonstrated best by performance testing. In some cases penetration values undergo some change during tempering, showing a softening of the conditioned shortening versus a nontempered one. The effect of tempering can be identified with the feel or workability of the shortening; a tempered product is more smooth with good plasticity while the nontempered shortening will be more brittle with less plasticity. β'-crystal-forming shortenings transferred to a cool temperature immediately after filling become permanently hard and brittle and attempts to recondition these products by subjecting them to tempering conditions have not been successful.

Immediately after packaging, a shortening requiring a plastic consistency should be stored 40 h or more in a quiescent state at a temperature slightly above the fill temperature. In practice, holding at 29°C (85°F) for 24–72 h or until a stable crystal form is reached is an acceptable compromise (29). The primary purpose of tempering is to condition the solidified shortening so that it will withstand wide temperature variations in subsequent storage and still have a uniform consistency when brought back to 21–24°C (70–75°F), which is the use temperature for a majority of the plasticized shortenings (31).

6.5 Quick Tempering

The expense and logistical problems associated with constant temperature rooms for tempering have led several equipment manufacturers to investigate the development of mechanical systems to eliminate the need for tempering. The majority of the systems developed do not claim complete elimination but rather a 50% or more reduction in tempering time. Most of the so-called quick tempering systems add a postcooling and kneading, or working, unit to the conventional-type chilling and working systems utilized with tempering. The theory postulated for these systems is that liquid fat is forced to crystallize individually and rapidly, thus creating smaller more stable crystals, rather than crystallizing onto existing crystals and increasing their size as happens with normal tempering (30,31).

Performance characteristics equivalent to well-tempered shortenings with only 24 h conditioning have been claimed for the quick temper products (33). These systems have been accepted by some shortening processors, but the standard tempering procedures are still practiced by many for shortenings requiring a wide plastic range.

6.6 Quality Assurance Evaluations

The consistency of a shortening is controlled by two dominating factors: (*1*) the SFI, which depends on the composition of the fat blend and (*2*) the processing conditions used to solidify, package, and temper the product. When the solidification conditions are constant, a strong correlation is expected between the analytical characteristics of the blend and the consistency of the shortening. Solidified shortening's appearance, texture, and consistency characteristics should be evaluated after tempering to identify potential plasticization defects and devise corrective actions for future production. A suggested evaluation method is to inspect a package of the shortening visually and physically. The physical evaluation involves picking up a small portion and kneading it in your hand. A good wide plastic range product will be smooth and will not become sticky or fall apart when worked. A review of shortening faults previously encountered with corrective actions are:

1. *Streaking* can be caused by a number of plasticization deficiencies: (*1*) chilling too low for the operating backpressure; (*2*) erratic creaming gas dispersion; (*3*) channeling, which allows semiliquid oil to pass through the

chilling unit insulated from chilling properly; (*4*) mixing different chiller unit streams operating at different temperatures; and (*5*) erratic chilling unit pressures (34).

2. *Sandy* or small lumps can be caused by too cold chilling unit temperatures or precrystallization prior to the chilling unit.
3. *Ribby* is a condition of alternating layers of hard and soft product caused by (*1*) too cold chilling unit with low pressure, (*2*) mixing streams from two chilling units operating at different temperatures, or (*3*) too cold chilling with excessive mounding when filling.
4. *Puffy* conditions are caused by too high creaming gas content or too low pressure to finely distribute the air cells.
5. *Brittle* feel is contributed by (*1*) high chilling unit temperature, (*2*) absence of low working after chilling, or (*3*) a formulated narrow plastic range.
6. *Oil separation* or Free oil, is caused by (*1*) high chilling unit temperature, (*2*) precrystallization, (*3*) temperature abuse after packaging, or (*4*) formulation.
7. *Chalky* appearance is caused by too high creaming gas content.
8. *Vaseline-like* or a yellowish greasy appearance is typical of shortenings packaged without creaming gas.
9. *Whiteness* may be controlled by the amount and degree of dispersion of the creaming gas (35).

LIQUID OPAQUE SHORTENINGS

Liquid opaque shortenings are distinguished from liquid oils by composition and appearance. Both are pourable, but liquid oils are clear while liquid shortenings are opaque due to their suspended solids. The suspended solids may be hard fats, emulsifiers, or a combination of the two depending on the intended use, i.e., frying, bread, cake, or nondairy products. Simply described, liquid shortenings are flowable suspensions of solid fat and/or emulsifiers in liquid oil.

7.1 Crystallization Processes

1. β tending hard fats, such as the >5-IV soybean oil hard fat identified in the base stock system, serve as quick-forming nuclei that cause solids in the base oil to precipitate in small enough crystals to ensure pourablity and prevent separation. Many different crystallization procedures for liquid opaque shortenings have been patented, some of which are briefly described below (36–39):
2. Gradual cooling of melted product to form large β crystals with gentle agitation. This procedure normally requires 3–4 days elapsed time.

3. Gradual cooling followed by comminution with a homogenizer or colloid mill. An estimated 3–4 days elapsed time required for processing.
4. Rapid cooling with a swept-surface heat exchanger followed by a holding period of at least 16 h with gentle agitation, to fluidize, just prior to packaging.
5. Suspension of finely ground hard fat or emulsifier in a cool liquid oil with subsequent comminution using a homogenizer, colloid mill, or shear pump.
6. Quick chilling a concentrated mixture of hard fat in liquid oil followed by a holding period in an agitated tank to allow β-crystal growth. The stabilized concentrate is then blended with a diluent oil at room temperature and seeded with an oil containing β-hard fat crystals.
7. Quick chilling a solution of β-crystal-forming hard fat in liquid oil to 38°C (100°F) and allowing the heat of crystallization to carry the temperature of the cooled oil to not over 54°C (130°F). Complete crystallization reportedly requires 20–60 min.
8. Recirculating a hot oil solution of hard fat from an agitated holding tank through a scraped-surface heat exchanger and back into the holding tank. The oil in the tank cools slowly and the crystals formed in the early stages melt in the hot oil. Eventually the mass cools to a point where the crystals do not melt but form α, then β', and finally stable β crystals. At this point, further chilling is stopped and the resulting opaque pourable shortening is packaged.
9. Double chilling/tempering system of first cooling from 65 to 43°C (150 to 110°F) followed by a 2-h crystallization period with gentle agitation and subsequent supercooling to 21–24°C (70–75°F) followed by a second crystallization period of approximately 1 h. The opaque liquid shortening is packaged after an expected heat of crystallization rise of approximately 9°C (15°F) has been experienced.

The summarized processing procedures for opaque liquid shortenings agree on several points: (*1*) all are dispersions of solid and liquid fractions, (*2*) β-forming hard fats is the preferred solid fraction to seed crystallization, and (*3*) heat of crystallization must be dissipated before a stable product is achieved. Another agreement is that aeration of the liquid shortenings must be avoided at all stages of processing before, during, and after crystallization. Air incorporation makes the product more viscous or less pourable and promotes product separation. Storage studies indicate an air content of less than 1.0% is required for a stable suspension stability.

7.2 Tempering and Storage

Opaque liquid shortenings do not require any further tempering after packaging, but the storage temperatures are critical. Storage below 18°C (65°F) will cause the liquid shortenings to solidify with a loss of fluidity. Storage above 35°C (95°F) will result in partial or complete melting of the suspended solids. The solidification due to cool temperatures can be reversed by a controlled heating not to exceed the melting point. However, the damage caused by high temperature cannot be remedied except by complete melting and reprocessing.

SHORTENING CHIPS AND FLAKES

As stated previously, shortening flake describes the higher melting edible oil products solidified into a thin flake form for ease of handling, quicker remelting, or for a specific function in a food product. Flaking rolls, utilized for the chilling of shortening and margarine prior to the introduction of scraped-wall heat exchangers, are still used for the production of shortening flakes. Chill rolls have been adapted to produce several different flaked products used to provide distinctive performance characteristics in specialty formulated foods. Consumer demands have created the need for such specialty fat products. Specialty high-melting fat flakes have been developed for specific applications with varied melting points as shown on Table 4.10. Instead of becoming obsolete, chill rolls have fulfilled an equipment requirement to produce specialty-fat-based products.

8.1 Chill Rolls

Chill rolls are available in different sizes, configurations, surface treatments, feeding mechanisms, etc., but most are a hollow metal cylinder with a surface machined and ground smooth to true cylindrical form. Rolls, internally refrigerated with either flooded or spray systems, turn slowly on longitudinal and horizontal axes. Several options exist for feeding the melted oil product to the chill roll: (*1*) a trough arrangement positioned midway between the bottom and top of the roll, (*2*) a dip pan at the bottom of the roll, (*3*) overhead feeding between the chill roll and a smaller applicator roll, and (*4*) a double- or twin-drum arrangement operating together with a very narrow space between them where the fat product is sprayed for application to both rolls. The coating of fat is carried over the roll to solidify and is removed by a doctor blade, positioned ahead of the feed mechanism with all of the designs.

8.2 Flake Crystallization

In the crystallization of hydrogenated edible oil products, the sensible heat of the liquid is removed until the temperature of the product is equal to the melting point. At the melting point, heat must be removed to allow the crystallization of the product. The quantity of heat associated with this phenomenon is called heat of crystallization. Sensible heat (specific heat) of most common hard fat products is equal to about 0.27 cal/g (0.5 Btu/lb) and the heat of crystallization is equal to 27.8 cal/g (50 Btu/lb). The amount of heat that must be removed to crystallize hardened oil is 100 times the amount of heat that must be removed to lower the product temperature (40).

8.3 Flaking Conditions

The desired shortening flake product dictates the chill roll operating conditions and additional treatment necessary before and after packaging. However, some generalizations relative to chill roll operations and product quality can be made.

1. *Crystal structure:* each flaked product has crystallization requirements dependent on source oils, melting points, degree of saturation, and the physical characteristics desired.
2. *Flake thickness:* four controllable variables help determine flake thickness: (*1*) oil temperature to the roll, (*2*) chill roll temperature, (*3*) speed of the chill roll, and (*4*) the feed mechanism.
3. *In package temperature:* heat of crystallization will cause a product temperature rise after packaging if it is not dissipated prior to packaging. The product temperature can increase to the point where partial melting coupled with the pressure from stacking will cause the product to fuse together into a large lump.
4. *Flake condition:* glossy or wet flakes are caused by a film of liquid oil on the flake surface due to incomplete solidification. Either too warm or too low chill roll temperatures can cause this condition. High roll temperatures may not provide sufficient cooling to completely solidify the flake. Low roll temperatures may shock the oil film, causing the flake to pull away from the surface before completely solidified. Wet flakes from either cause will lump in the package.

REFERENCES

1. K.F. Mattil, in D. Swern, *Bailey's Industrial Oil and Fat Products,* 3rd ed. Interscience, New York, 1964, p. 265.
2. W.H. Shearon, Jr., *Ind. Eng. Chem.,* July, 1278 (1950).
3. Ref. 1, p. 267.
4. J.E. Slaughter, Jr., *Proceedings: A Six Day Short Course in Vegetable Oils,* sponsored by The American Oil Chemists' Society, Aug. 16–21, 1948, p. 120.
5. K.S. Markley, *Fatty Acids,* Part I, 2nd ed., Interscience, New York, 1960, pp. 6–13; *Chem. Soc.* **54,** 557 (1977).
6. D.I. Hartnett, *J. Am. Oil Chem. Soc.* **54,** 557 (1977).
7. H.E. Robinson and K.F. Mattel, *J. Am. Oil Chem. Soc.* **36,** 434 (1959).
8. Ref. 1, p. 296.

9. M.M. Chrysam in T.H. Applewhite, ed., *Bailey's Industrial Oil and Fat Products,* Vol. 3, Wiley-Interscience, New York, 1985, p. 89.
10. *USDA Oil Crops Situation and Outlook Yearbook,* July 25 (1986).
11. *USDA Oil Crops Situation and Outlook Report,* January, 40 (1994).
12. R.F. Wilson, *INFORM* **4,** 193 (1993).
13. F.D. Gunstone and T.P. Hiloitch, *J. Am. Oil Chem. Soc.* **22,** 836 (1945).
14. T.J. Weiss, *Food Oils and Their Uses,* AVI Publishing, Westport, Conn., 1983, p. 112.
15. F.J. Flider and F.T. Orthoefer, *J. Am. Oil Chem. Soc.* **58,** 270 (1981).
16. Ref. 14, p. 114.
17. Ref. 14, p. 116.
18. E.G. Latondress, *J. Am. Oil Chem. Soc.* **58,** 185 (1981).
19. *Official Methods and Recommended Practices of American Oil Chemists' Society,* 4th ed. American Oil Chemists' Society, Champaign, Ill., 1994, Method Cc 18–80.
20. Method Cc 12–59 in Ref. 19.
21. E.J. Latondress in D.R. Erickson, E.H. Pryde, O.L. Brekke, T.L. Mounts, and R.A. Falb, eds., *Handbook of Soy Oil Processing and Utilization,* American Soybean Association and AOCS, Champaign, Ill., 1980, pp. 145–152.
22. R.D. O'Brien in R. Hastert, ed., *Hydrogenation: Proceedings of an AOCS Colloquium,* American Oil Chemists' Society, Champaign, Ill., 1987, pp. 157–165.
23. Ref. 14, pp. 129–130.
24. Ref. 9, pp. 97–98.
25. R.J. Bell in P.J. Wan, ed., *Introduction to Fats and Oils Technology,* American Oil Chemists' Society, Champaign, Ill., 1992, pp. 187–188.
26. Ref. 1, pp. 272–281.
27. D. Best, *Prepared Foods,* May, 168 (1988).
28. L.H. Wiedermann, *J. Am. Oil Chem. Soc.* **55,** 825 (1978).
29. A.E. Thomas, *J. Am. Oil Chem. Soc.* **55,** 830 (1978).
30. C.E. McMichael, *J. Am. Oil Chem. Soc.* **33,** 512 (1956).
31. N.T. Joyner, *J. Am. Oil Chem. Soc.* **30,** 526 (1953).
32. C.W. Hoerr and J.V. Ziemba, *Food Engineering,* 90 (May 1965).
33. Ref. 14, pp. 96–97.
34. Ref. 14, p. 125.
35. D.R. Erickson, *J. Am. Oil Chem. Soc.* **44,** 534A (1967).
36. Ref. 14, pp. 126–129.
37. T. Petricca, *Baker's Digest,* Oct., 39 (1976).
38. Ref. 1, pp. 1064–1068.
39. Ref. 9, pp. 100–104.
40. P.J. Wan and A.H. Chen, *J. Am. Oil Chem. Soc.* **60,** 743 (1983).

5
Cooking Oils, Salad Oils, and Oil-Based Dressings

Lipids used in food products have been conventionally divided into two classes based on their consistency at about 25°C (77°F): (*1*) liquid oils, such as soybean, cottonseed, sunflower, safflower, peanut, olive, and rapeseed oils, and (*2*) solids and semisolids, such as lard, tallow, palm oil, coconut oil, palm kernel oil, and cocoa butter. In the preparation of some foods it is of no particular consequence whether the lipid material is liquid or solid, but in others the consistency is important, depending on the amount of the lipid used. For example, in the preparation of dressing for green salads the object is to provide an oily mouthfeel in the form of a stable emulsion; hence a liquid oil must be used. However, when the oil content of such a dressing is low, plastic fats may be desirable. On the other hand, baked food products generally require plastic fats for the incorporation of air needed for leavening. But with the modern development of emulsifiers, in many of these areas it is possible to use liquid oils with a suitable emulsifier (1,2) to achieve this goal.

In the past, for reasons related to both history and climate, there were distinct geographical divisions of people by fat- and oil-consuming groups. Inhabitants of central and northern Europe derived their lipids from domestic animals. Consequently, the food habits and cuisine of these people developed around the use of plastic fats. On the other hand, the main form of lipids used by people from the warmer climates of southern Europe, Asia, and Africa were liquids. In the Western Hemisphere, plastic fats are the most widely used lipids because of the predominantly northern European extraction of the early settlers. However, a significant shift in this distinction has taken place because of the global movement of peoples, the implication of plastic fats in cardiovascular diseases, and the development of new oil compositions through biotechnology. In general, there is greater use of liquid oils except where functionality, such as highly developed dough structure, cream icings, or fillings, dictates use of a plastic fat.

In addition to their household uses, cooking oils are in considerable demand for deep-fat frying of many food products that are consumed immediately after frying. In the past, for reasons of stability during storage, fats were the preferred frying medium for packaged foods, such as potato and corn chips, that remain in storage for considerable periods. In recent times, the use of liquid oils, which are less unsaturated and more stable in such operations, has significantly increased. Products like doughnuts still require plastic fats for frying, because liquid oils tend to give them a greasy appearance.

Most cooking oils are of vegetable origin although such oils can be produced by fractionation (3) of animal fats like butterfat. The only naturally occurring liquid oil from animal sources is of marine origin. But marine oils in their natural form are highly susceptible to oxidation and, therefore, not desirable for use in foods. In various parts of the world, considerable quantities of marine oils are used in food after hydrogenation. However, technologies (4) are being developed for use of liquid marine oils because of the health benefits (5) of the highly unsaturated eicosapentaenoic and docosahexaenoic acids present in them.

NATURAL AND PROCESSED COOKING AND SALAD OILS

Cooking oils are used either in their natural state or after processing, depending on local taste, custom, and nutritional beliefs. In the Old World, oils were used in their natural state. In Occidental countries, the oil is processed to a bland state. This geographical difference in the use of oils is probably due to the different methods of obtaining the oil from its source. However, use of processed oils has become more common in all parts of the world because of the global expansion of the more efficient technologies, such as high pressure expression and solvent extraction to obtain higher yields of oil, which also produce an oil with considerable amounts of nontriglyceride components, such as coloring bodies and gums. These oils, because of the presence of nontriglyceride substances and their strong flavor and pronounced color, must be suitably processed to make them acceptable and edible. In general, bland neutral oil is produced from crude oils by refining, bleaching, and deodorizing them.

An important distinction between salad and cooking oils is the difference in their oxidative and thermal stability. Cooking oils are more stable than salad oils at higher temperatures, such as deep-fat frying. The term *salad oil* is applied to oils that remain substantially liquid in a refrigerator, i.e., about 4.4°C (40°F). In the standard method of evaluating salad oils (AOCS, Cc11-53) is the cold test. The oil sample is put in a sealed 4-oz bottle and placed in an ice bath at 0°C (32°F). If the oil remains clear after 5.5 h, it meets the criterion of a salad oil or suitably winterized oil. Well-winterized oils remain clear for periods much longer than 5.5 h. From a practical standpoint, the number of hours required for the oil to cloud is of less significance than the character and the amount of the crystals deposited after a prolonged storage at 0°C (32°F), or refrigeration temperature.

Crystallization inhibitors such as lecithin and polyglycerol esters have been used for extending the cold test. Other crystal inhibitors, such as disaccharide esters (6), monosaccharide esters (7), and polysaccharide esters of hydroxy fatty acids (8), have been patented. The extent of commercial use of these additives is not known. Oxystearin, which is produced by controlled oxidation of partially hydrogenated cottonseed oil, is no longer used as a crystal inhibitor.

Sunflower, safflower, and corn oils need dewaxing before they can meet the criteria of a salad oil. Soybean, high erucic, and low erucic rapeseed oils conform to the definition of a salad oil. But, because they contain significant amounts of relatively unstable linolenic acid, they are partially hydrogenated to produce cooking and salad oils. However, both soybean and low erucic acid rapeseed (LEAR) oils are used without hydrogenation in the preparation of salad dressings. Cottonseed oil, because of its high content of palmitic acid must be winterized to obtain a salad oil. Oils meeting the criteria of a salad oil cannot be prepared from peanut oil (because of noncrystallinity of its higher melting fraction) or palm oil (because of its high content of palmitic acid). However, liquid cooking oil is manufactured from palm oil by winterization. High oleic safflower and sunflower oils (9,10), created through crop biotechnology, contain about 80% oleic acid and can serve as stable cooking and salad oils. Rice bran oil, because of its stability, is gaining significant commercial status, particularly in Asian countries. Linola (a low linolenic flax seed oil created through crop biotechnology) was approved in Canada for food use.

Olive oil, a high oleic oil, is used as both a cooking oil and a salad oil in many parts of the world. Olive oil is unique because of its natural flavor relished by users; however, its use is restricted by its high cost. Manufacture of naturally flavored salad oils such as olive oil require nothing more than expression of the oil from oil-bearing materials and clarification of the oil by filtration. Commercial olive oil is probably a blend of oils from different sources. The quality and the flavor of olive oil vary considerably from one season and one locality to another. To turn out a product of uniform quality the packer must have access to a variety of oil sources.

There is a substantial market for cold-pressed peanut, sesame, safflower, and sunflower oils in many countries. In the United States, the use of such unprocessed oils has become more popular in recent years because of nutritional considerations. However, there are no reliable data available for the quantities of such oils consumed.

Consumption of all types of food fats and oils has been gradually increasing all over the world. Fats and oils consumed as part of foods can be conveniently divided into two broad categories: *visible* fats that are used in the preparation of foods and *invisible* fats that are present in the food naturally. Per capita consumption of visible fats in the United States grew from 104.9 kg (47.7 lb) in 1965 to 144.5 kg (65.7 lb) in 1992. Included in this figure are per capita consumption of cooking and salad oils, which increased from 33.9 kg (15.4 lb) in 1970 to an estimated 56.3 kg (25.6 lb) in 1992 (11).

For many years, cottonseed oil was the main source of cooking and salad oils in the United States. However, since World War II soybean oil has become the most prominent oil. Other oils, particularly those with high oleic content or low linolenic content, are becoming more important, as they do not require hydrogenation for stability (12,13). Comparisons of some commercially important cooking and salad oils whose fatty acids have been altered are shown in Table 5.1.

STABILITY OF SALAD AND COOKING OILS

By proper processing, most oils can be made bland. Complete removal of the "corny" flavor from corn oil is rather difficult. However, this flavor of corn oil is preferred by many consumers in the preparation of some foods, such as corn chips. During storage and cooking, even bland oils undergo physical and chemical changes, because they are sensitive to heat and light. In addition, trace metals (e.g., copper and iron), chlorophyll, and other prooxidants of unknown nature catalyze such chemical changes in the presence of oxygen.

Table 5.1 Comparison of regular and fatty acid altered salad oils of commerce

Oil Source	Fatty Acid Composition (% total)[a]					
	Palmitic	Stearic	Oleic	Linoleic	Linolenic	Others
Soybean	11.0	4.0	23.0	55.0	7.0	1.0
Soybean (low linolenic)	11.0	4.0	29.0	52.0	3.0	1.0
Canola	4.0	2.0	61.0	20.0	9.0	4.0
Canola (low linolenic)	4.0	2.0	65.0	26.0	3.0	—
Sunflower	7.0	2.0	20.0	70.0	0.0	1.0
Sunflower (high oleic)	3.0	4.0	87.0	3.0	0.0	1.0
Flax	10.0[b]	—	26.0	16.0	48.0	—
Flax (low linolenic)	10.0[b]	—	18.0	70.0	2.0	—
Cottonseed	25.0	2.0	18.0	53.0	0.0	2.0
Corn	11.0	2.0	26.0	59.0	1.0	2.0
Palm	45.0	5.0	39.0	9.0	0.0	2.0
Peanut	10.0	3.0	56.0	24.0	0.0	7.0[c]
Rapeseed	4.0	1.0	15.0	14.0	9.0	55.0[d]

[a] Figures have been rounded to the nearest whole number.
[b] Covers total of all saturates (palmitic + stearic).
[c] Mostly C20–24 acids.
[d] Primarily C22:1 (45%) and C20:1 (10%).

At low levels of oxidation, the degree and type of change known as flavor reversion is characteristic for each oil. For example, soybean oil develops a flavor that is described as beany or grassy. This flavor has been attributed to the formation of 2-pentylfuran (14) and 3-*cis*-hexenal (15). LEAR (canola) develops flavors similar to those of soybean oil. The flavor reversion of sunflower seed and safflower seed oils are described as "seedy." Similarly, corn and palm oils develop flavors of distinct type. These reversion flavors are observed long before other objectionable oxidative off-flavors are formed. With many of these oils, the cause of the reversion flavors is not known. In the case of most well-processed oils the reversion flavor is not a frequent problem.

The initial product of reaction of oils with oxygen is the hydroperoxide, which has no odor or flavor. When it decomposes, it forms diverse compounds, including aldehydes, ketones, acids, alcohols, hydroxy compounds, lactones, hydrocarbons, dienals, epoxides, and polymers. The breakdown products are often the cause of off-flavors observed in oils and their products. Many of these compounds are present only in parts per billion, and not all of these compounds are objectionable. Many of the related esters, alcohols, acids, ketones, and lactones are responsible for the desirable flavors of tomato juice, ripened banana, butter, olive oil, and other foods. Similarly, the flavor of fried foods has been attributed to 2,4-decadienal, which is also a component of off-flavor soybean oil. The proportions of these compounds in the oil are more important in determining whether a given flavor is favorable, rather than the types of compounds (16).

Shelf stability of salad oils depends on many factors, such as initial quality of the crude oil, conditions of processing and subsequent handling, presence and amount of natural antioxidants, type of container, and temperature of storage. In addition, Neff and co-workers have shown that oil's sensitivity to light depends both on composition and on structure (17). However, it can be generally stated that the oils containing linolenic acid—whether hydrogenated or unhydrogenated—are less stable than oils with corresponding iodine value but without linolenic acid. It has been shown that the oxidative stability of polyunsaturated oils can be improved by blending with a less polyunsaturated oil (18).

In the case of cooking oils, both thermal and oxidative degradation are important. Thermally caused changes result in polymer formation and other types of alteration of the oil. These changes have important nutritional and aesthetic aspects. Therefore, they have been extensively studied and discussed (19–27). Unfortunately, many of these studies have been conducted on fats and oils degraded under conditions far different from practical cooking operations. This has caused a considerable amount of uncertainty about the nutritional qualities of cooking oils that are consumed through fried foods. It has been shown that in fast oil turnover frying operations, the changes slow down after initial rapid changes (28). Both volatile and nonvolatile decomposition products of different cooking oils have been reported (31–33).

QUALITY EVALUATION OF SALAD AND COOKING OILS

Handling and storage of finished oils have been reviewed (34). Minimizing the oxidation of oils and fats from harvest to consumption of the finished oil is important to the quality and stability of oils and the products in which they are used. Therefore, many quality-control methods have been used to measure the extent of such oxidation. The most important quality criterion in the case of salad and cooking oils is their blandness, which is changed by any degree of oxidation. A common method of determining blandness is organoleptic evaluation, which is subjective and is influenced by many variables. To minimize variability, a certain amount of expertise is developed by the tasters through training. In addition, several objective methods to evaluate the quality of oils and fats are in use. These methods are generally based on the fact that oxidative degradation produces both volatile and nonvolatile chemical compounds with functional groups that can be measured by chemical and physical methods. However, for any method to be useful, the results must correlate with the organoleptic properties of the oils and fats in question. A second but equally important requirement is that the method should be capable of predicting the storage stability of the oils as well as any products made with them. Kreis test and carbonyl value are rarely used (35–37).

3.1 Peroxide Value

Peroxide value is the most widely used method (AOCS Cd8-53) to determine the degree of oxidation. The primary oxidation products of oils and fats are the hydroperoxides. They can be quantitatively measured by determining the amount of iodine liberated by its reaction with potassium iodide. The peroxide content is expressed in terms of milliequivalents of iodine per kilogram of fat. However, when these hydroperoxides start breaking down to produce off-flavor compounds, correlation to the quality and stability of the oil will no longer be valid. Freshly deodorized oil should have zero peroxide value. In most cases, for the product to have acceptable storage stability the peroxide value of oils used should be less than 1.0 meq/kg fat at the point of use.

3.2 Benzidine or Anisidine Value

The benzidine or anisidine value method uses the reaction of nonvolatile α- and β-unsaturated aldehydes with these reagents (38,39). Absorption readings are made at 350 nm in 1-cm cells. Originally, benzidine was used a reagent. Later, because of the carcinogenicity of benzidine, anisidine was recommended. This method of determining the degree of oxidation is extensively used in Europe and other parts of the world. Results thus far indicate that this method is more useful in determining the quality of crude oils and the efficiency of processing procedures than the quality of the finished oil during shipment and storage.

An extension of this method is the total oxidation (TOTOX) value (40), which is the sum of anisidine value and two times the peroxide value.

3.3 Pentane and Hexanal Values

One of the compounds produced by the oxidation of salad and cooking oils is pentane. The quantity of pentane is measured by gas liquid chromatography (GLC) procedure. There is good correlation of this value with the organoleptic qualities of the oil for the first few weeks of storage life. But correlation at later periods of storage has not been unequivocally established. Similarly, level of hexanal has been used as a quantitative indicator of the quality of oils and products made with them. As the oil or the product degrades during storage, other degradation products seem to be more important than hexanal in determining the organoleptic quality (41,42).

3.4 Thiobarbituric Acid Test

The thiobarbituric acid test, developed initially for measurement of malondialdehyde, is used to measure the oxidative degradation product of polyunsaturated fatty acids (43,44). However, it lacks sensitivity (45).

All the preceding methods serve as useful tools under a given set of conditions, but seem to fail in their universal applicability to measure the quality and predict the stability of oils, fats, and their products.

3.5 Volatile Profile Method

The volatile profile method is a direct gas chromatographic method to examine the volatiles in oils, fats, and their products (46–50). Under control conditions, the method has good reproducibility. When the results are statistically treated, good correlation with organoleptic results is obtained.

3.6 Accelerated Tests

The study of storage stability of salad oils is time-consuming and requires several months to complete. From a commercial standpoint, it is necessary to complete this determination in a shorter period. Therefore, partially successful attempts have been made to develop accelerated tests using light- and temperature-susceptible properties of oils and fats. Moser and co-workers (51) have designed an apparatus using fluorescent light to accelerate the aging of oils and fats. Radtke and co-workers have studied the influence of the intensity and wavelength of light on the oxidative deterioration of salad oils. They observed that photochemical action depended on the wavelength and increased with decreasing wavelength to a much greater extent than could be predicted from energy considerations. Similar studies have been done by others (53,54).

From these light studies it may be possible to reduce the time of the storage stability test by exposing the oils to short wavelengths, followed by an analysis of the volatiles, using volatile profile methods.

Several temperature catalyzed stability tests are used in evaluating the oxidative stability of oils and fats. The oldest method is the Schaal oven test (35). It is inexpensive but subjective, because it uses organoleptic and odor intensities in the procedure and still requires days to obtain the result. In the active oxygen method (AOM) (35), the development of peroxide is measured with time. Because the formation and decomposition of peroxides are dynamic processes, the results obtained by this method do not correlate well to the actual stability of the oils and fats observed under practical application conditions. Other methods that have been based on oxygen absorption are the gravimetric (55) and the headspace oxygen concentration measurement (56,57).

Recently, with the advent of more sophisticated technology, the older thermal methods are being replaced by procedures using automated instruments (rancimat, oil stability index, and oxidograph) for measurement of oil stability. Rancimat plots the change in electrical conductivity of distilled water with time by the dissolution of gaseous products from the oxidizing oil (58,59). The oil stability index (OSI) is similar to rancimat in its operating principle (60). In both of these instrumental procedures, just as in AOM, air is passed through heated oil under defined conditions of temperature and flow rate. The decomposition products are absorbed in distilled water. The end of the induction period is indicated by the rapid increase in the conductance of the distilled water. The oxidograph (61) is a modified version of FIRA Astell apparatus (62) and is based on a plot of the change in oxygen pressure of the system containing the oil with time. The plots from this method resemble the conductivity versus time graphs from rancimat.

In the case of cooking oils, degradation similar to that observed in salad oils occurs but to a lesser degree. However, during cooking, more severe types of breakdown are caused by high temperatures and the presence of moisture, oxygen, and food materials. These changes are reflected by increases in color, free fatty acids, polymers, polar materials, foaming, and so on. Even today there are no reliable methods to predict the performance of a cooking oil and establish a discard point for the used oil. However, several empirical methods have been suggested to determine the extent of damage to the oil caused by the frying process (63).

3.7 Free Fatty Acid

Free fatty acids are probably the most widely used characteristic of oil quality control (AOCS Ca 5a-40). Well-deodorized oils, generally, have a free fatty acid level of less than 0.05%. During use, there is a buildup of free fatty acids. In the initial stages of cooking, free fatty acids are produced by oxidative breakdown, but in later stages, hydrolysis of the fat caused by the presence of moisture in the food causes free fatty acids to increase. This is a dynamic process, as some of the free fatty acids are lost through oxidation and steam distillation from the food. Furthermore, free fatty acids seem to catalyze hydrolysis of the cooking oils. An increase in free fatty acid decreases the smoke point of the oil. When significant amounts of free fatty acids have accumulated in the oil, smoking becomes excessive and the quality of the food decreases. The oil then must be discarded.

3.8 Color

Oils used for cooking darken because of the formation of oxidative materials, including polymers and the presence of oil-soluble products from the fried food. Generally speaking, such a darkening of the oil without corresponding increase in other degradative products will simply make the food look darker but does not significantly affect the taste of the food. Color alone as a measure of the extent of degradation of the cooking oil is highly unreliable (AOCS Cc 13b-45).

3.9 Oxidized Fatty Acids

Many methods have been proposed to measure the amount of oxidized lipids in a cooking oil (64,65). Oxidative degradation will affect both the appearance and the taste of the food.

3.10 Polymers

When a cooking oil oxidizes, it forms polymers that cause the oil to foam. In addition, the viscosity of the oxidized oil increases, making the cooked food look oily due to retention of a higher amount of oil on the surface of the food. Many methods have been proposed for the determination of polymers in the oil (66–69).

ADDITIVES FOR SALAD AND COOKING OILS

Salad and cooking oils have varying amounts of natural antioxidants, mostly tocopherols. With the introduction of synthetic antioxidants, it has been customary to supplement the natural antioxidants with the synthetic ones. Most commonly used synthetic antioxidants are butylated hydroxy anisole (BHA), butylated hydroxy toluene (BHT), propyl gallate (PG), ascorbyl palmitate, and tertiary butyl hydroquinone (TBHQ). However, in recent times tertiary butyl hydroquinone, because of its better performance, has been used more extensively than the other synthetic antioxidants. Application and status of antioxidants have been reviewed (70).

Antioxidants derived from sage, rosemary, and green tea are becoming more popular because they are natural sources (71–73). In addition, the antioxidants from these sources show a greater thermal stability than the synthetic antioxidants, particularly in meat products. However, they do impart flavors and colors to the product in which they are used.

Citric acid is added primarily to chelate trace metals which otherwise accelerate the oxidative process. Methyl silicone (dimethyl polysiloxane) is added in 1–10 ppm quantities to cooking oils to reduce the rate of oxidation at high temperature usage. Many other additives have been suggested or patented, to improve either stability or functionality, but none of them are being used commercially (74–77).

NUTRITION-ORIENTED SALAD AND COOKING OILS

The U.S. diet, as well as diets of many industrialized nations, is rich in calories contributed by oils and fats. Technological innovations are, therefore, directed toward finding low or no calorie substitutes that will perform like normal oils and fats and that do not affect the aesthetic or health values of food products (78–80). The manufacturers of many of these synthetic fats must still establish that the substitutes are fit for human consumption. Even sucrose polyesters, which have been shown to be safe for food use, are still awaiting approval, and their significant use in food products is still in doubt.

NEW SALAD AND COOKING OILS

Because of the health concerns of saturated fatty acids and trans fatty acids produced by hydrogenation there has been significant biotechnological work to develop crop oils with low saturates and oils that will be stable at high frying temperatures. Some of the oils (high oleic sunflower, low erucic acid rapeseed, low linolenic soybean, and low linolenic canola) are already in commerce. Some of the other oils under development are shown in Table 5.2.

OIL-BASED DRESSINGS

Oil-based products are divided into two broad categories by their texture: spoonable dressings (salad dressings, mayonnaise, and their reduced-calorie

Table 5.2 Composition and functionality of some of the cooking and salad oils in development through crop biotechnology

Crop	Perceived Benefit	Approximate Fatty Acid Content (% Total)				
		Palmitic	Stearic	Oleic	Linoleic	Linolenic
Soybean[a]						
Regular		11	4	24	54	7
Low saturates	Health	4	3	25	61	8
Low linolenic	Stability	9	4	49	36	3
High stearic	Functionality	8	20	16	48	8
High palmitic	Functionality	27	4	15	45	9
Sunflower[a]						
Regular		7	5	19	69	0
High oleic	Stability	4	6	82	8	0
High oleic– Low stearate	Health, stability	4	2	88	5	0
Canola[a]						
Regular		4	2	62	22	10
High oleic	Stability, health	4	2	80	6	6
Low linolenic	Stability	4	2	65	26	3
Peanut[b]						
Regular	Stability	10	3	56	24	Others
High oleic	Health	8	3	76	5	Others

[a] Courtesy of Pioneer Hi-Bred International, Inc.
[b] Ref. 81.

counterparts) and pourable dressings (French dressing and various other varieties, including reduced calorie).

VISCOUS OR SPOONABLE DRESSINGS

Production of dressing products from 1988 to 1993 is shown in Table 5.3. This time period represented an increased consumer awareness of the contribution of dietary fat consumption to various debilitating diseases. The consumption of many oil-containing products, including salad products, slowed as consumers began seeking healthier alternatives in their diet. The production and consumption of salad dressings and mayonnaise showed a general decline from 1988 through 1993. Reduced-calorie salad dressings and mayonnaise peaked in volume in 1989, then declined as fat-free versions were introduced in 1989.

Apparently, this reduction reflects the switch by consumers to fat-free products from reduced-calorie salad dressings and mayonnaise. However, the trend for this 5-year period showed an overall decline in production of salad dressing and mayonnaise, because consumption of bound salads (potato salad, egg salad, and pasta salad) and sandwiches declined during this time of health

Table 5.3 U.S. shipments of salad dressings; mayonnaise; and pourable dressings, including reduced calorie, fat free, and totals, 1988–1993 (millions of pounds) (82)

Year	Total[a]	Salad Dressings	Mayonnaise	Reduced-Calorie Salad Dressings and Mayonnaise	Fat-Free Salad Dressings and Mayonnaise	Total Salad Dressings and Mayonnaise	Pourable Dressings	Reduced-Calorie Pourable Dressings	Fat-Free Pourable Dressings	Total Pourable Dressings
1988	1373.7	292.5	494.8	196.2	—	983.5	290.9	99.3	—	390.2
1989	1380.4	260.8	439.0	267.7	2.3	969.8	294.6	115.0	1.0	410.6
1990	1387.2	256.7	435.7	266.9	3.1	962.4	274.6	110.1	40.1	424.8
1991	1371.0	248.8	438.9	209.4	44.6	941.7	266.1	95.3	67.9	429.3
1992	1371.5	255.3	447.5	175.1	48.7	926.6	273.6	79.6	91.7	444.9
1993	1349.8	249.0	439.4	163.3	47.1	898.8	284.6	69.2	97.2	451.0

[a] Includes total of all salad dressings, mayonnaise, and pourable dressing products.

consciousness. These health-oriented U.S. consumers took an active interest in exercising and embracing a more healthy lifestyle, which included fewer calories from oils and fats.

8.1 Mayonnaise

Composition. The official definition adopted in 1950 by the U.S. Food and Drug Administration (FDA) describes mayonnaise as follows: mayonnaise is the semisolid food prepared from edible vegetable oil, egg yolk, or egg whites (fresh or frozen or dried); any vinegar diluted with water to an acidity, calculated as acetic acid, of not less than 2.5% by weight; lemon and/or lime juice; with one or more of the following: salt, sweetener, paprika, monosodium glutamate, and other suitable food seasonings. The finished product contains not less than 65% of edible vegetable oil. This definition is not universal and other types of dressings, with different percentages of oil, are marketed as mayonnaise in other countries around the world. The sweetener used may be sucrose, dextrose, corn syrup, invert sugar syrup, nondiastatic maltose syrup, glucose, honey, or high fructose corn syrup.

Mustard, paprika, other spices, or any spice oil or spice extract can be used, but not turmeric, saffron, spice oil, or spice extract, which imparts to the mayonnaise a color simulating the color imparted by egg yolk. Both mustard and oleoresin paprika impart a yellow color but, because of the small amounts used, do not cause any variance from the standard of identity.

Monosodium glutamate, because of its nutritional implications, has come under the scrutiny of the FDA and various consumer watchdog groups. Although not outlawed, monosodium glutamate is rarely used in mayonnaise. Any other harmless food seasoning or flavoring can be used, provided it is not an imitation and does not impart to the mayonnaise a color simulating the color imparted by egg yolk.

Acidity, calculated as acetic acid, contributed by lemon and/or lime juice, should not be less than 2.5% by weight of the product. When an optional acidifying ingredient such as citric acid is used, the label on the package must bear the statement "citric acid added." As of January 1, 1978, malic acid has been permitted as an acidifying agent by the FDA.

Calcium disodium ethylenediaminetetraacetic acid (EDTA) or disodium EDTA is permitted to protect the flavor. Physically, mayonnaise consists of an internal or discontinuous phase of oil droplets dispersed in an external or continuous aqueous phase of vinegar, egg yolk, and other ingredients. Therefore, it is an oil-in-water emulsion. Consistency of the emulsion depends to a large extent on the ratio of its aqueous and oil phases and the amount and type of egg solids. Stability of the emulsion not only depends on the ingredients but is also greatly affected by the type of equipment and how it is operated during manufacture. Table 5.4 gives a general composition for mayonnaises. The color of mayonnaise is pale yellow, and is obtained mainly from the egg rather than from the oil.

Oil. Edible vegetable oil used in mayonnaise may contain not more than 0.125% by weight of oxystearin to inhibit crystallization of high melting triglycerides (triacylglycerols). Mayonnaise may legally contain as little as 65% oil. In actuality, most mayonnaise manufactured in the United States is 75–82% edible oil, usually soybean oil. Other salad oils, including winterized cottonseed oil, safflower oil, corn oil, and unhydrogenated and partially hydrogenated winterized or nonwinterized soybean oil, can be used in mayonnaise (83,84). Unhydrogenated soybean oil is the most commonly used oil. Oils with large quantities of saturated acids, such as palm oil, or oils that solidify at refrigeration temperatures, such as peanut oil, are seldom used, because they tend to break the emulsion at refrigerator temperatures. Oil levels 1–2% higher than the range given in Table 5.3 are used in mayonnaise when a stiffer body is needed. If the oil exceeds 85%, the emulsion tends to be unstable.

Vinegar. Vinegar has a dual purpose in mayonnaise and in most other dressings. The acetic acid acts as a preservative against microbial spoilage,

Table 5.4 Approximate composition of mayonnaise

Ingredient	Percent by Weight
Vegetable oil	75.0–80.0
Vinegar (4.5% acetic acid)	10.8–9.4
Egg yolk	9.0–7.0
Sugar	2.5–1.5
Salt	1.5
Mustard	1.0–0.5
White pepper	0.2–0.1

including some types of yeasts (85). The theory of action of acetic acid on microbes has been reviewed (86). Acetic acid also serves as a flavor ingredient when used at the proper level. However, excessive quantities of the acid will impair the flavor. Water in vinegar forms the continuous phase in the emulsion. Generally, the concentration of acetic acid is about 3.5% of the volatiles present in vinegar. Vinegar may contain high quantities of trace metals that can be detrimental to the stability of the mayonnaise. Therefore, the quality of the vinegar used is important.

Egg. The main function of the egg solids in mayonnaise is to serve as an emulsifying agent. The mayonnaise emulsion is built around egg yolk, which is one of nature's perfect emulsions. One of the factors influencing the stiffness and the stability of the emulsion is the amount and kind of egg yolk. Egg yolk is a complicated chemical entity with a high lipid content. The lipids of the egg yolk form about 65% of the total solids. The composition (87) of the lipids in egg yolk is as follows: glycerides 40.3%, phospholipids 21.3%, and cholesterol 3.6%, based on the total solids. Of the fatty acids, 30% are saturated and the rest are unsaturated. Lipid components of egg yolk have a profound effect on flavor, color, and stability of mayonnaise and other dressings. Egg white is a complex system of proteins that aids in emulsification by forming a gel structure on coagulation by the acid component (88). More detailed information on egg and egg protein composition is available (89).

Mustard. Mustard is generally used in mayonnaise in the form of a flour. There are two varieties of mustard: white and brown. White mustard is hot to taste but practically odorless. The brown variety, on the other hand, has a sharp odor. Therefore, the two varieties are blended in various proportions to achieve a desired level of flavor and pungency (90). Mustard contains a glucoside that, when hydrolyzed, releases the pungent oil of mustard, allyl isothiocyanate. This compound imparts the bite to mayonnaise. According to Corran (91), in addition to its flavor contribution, the nature, the origin, and the method of addition of mustard influence the emulsion. He stresses that mustard is an efficient emulsifier and is effective if incorporated with egg yolk.

Because of the variation in the essential oil content and immiscibility of the mustard flour with the vegetable oil, Cummings (92) recommends the use of oil of mustard in place of mustard flour. If oil of mustard is used, the advantage of mustard flour as an emulsifying adjunct and as a possible contributor of color in mayonnaise may be lost.

Other Spices and Seasonings. Of the other optional ingredients permitted, paprika adds to the flavor and salt and sugar contribute flavor and impart some microbial and physical stability to the mayonnaise.

8.2 Other Spoonable Salad Dressings

Salad dressings were first developed by C. E. Nelson as an alternative to mayonnaise during the Depression years. Kraft Miracle Whip became the common everyday product that many people living in the north-central section of the United States used in place of mayonnaise.

Composition. These products are commonly called salad dressings and have a standard of identity as defined by the FDA. The composition of salad dressings as defined in the standard of identity is as follows. Salad dressing is the emulsified semisolid food prepared from not less than 30% vegetable oil; acidifying ingredients (usually vinegar); not less than 4% egg yolks; and a starchy paste component prepared from one or more of a combination of tapioca starch, wheat starch, rye starch, or corn starch (dent or amioca). The starches can be native or chemically modified to impart physical stability against syneresis, or acid breakdown. The starch paste is prepared, usually with vinegar and spices, and then blended with a modified mayonnaise base to yield the finished spoonable salad dressing. The following optional ingredients may also be included: salt, nonnutritive sweeteners, spices, monosodium glutamate, thickeners and stabilizers, sequestrants, and crystallization inhibitors.

The oils are selected for use using the same criteria as for mayonnaise. However, a process for making stable salad dressings containing mixtures of liquid triglycerides that have an iodine value greater than 75 and solid triglycerides that have an iodine value not exceeding 12 has been patented (93).

8.3 Quality Control and Stability Measurements

Mayonnaise may be classified as a semiperishable product. It is sufficiently stable to keep for reasonable lengths of time without refrigeration. Mayonnaise gradually becomes thinner with age. Thinning and separation of the phases of mayonnaise are greatly accelerated by mechanical shock. Separation of phases can also be caused by exposure to low temperatures. Gray (94,95) studied the breakdown of mayonnaise and concluded that the higher the water content, the greater the amount of egg solids needed to stabilize the emulsion. Other causes of breakdown are rapid addition of the oil, unregulated agitation during emulsifying, high temperature storage, and agitation during transit.

Other common forms of spoilage in mayonnaise are due to oxidative degradation of various components, especially vegetable oil and egg lipids. Microbial spoilage is a rare occurrence as the product is well protected by the high acid content. But molds and yeast and, to a lesser extent, *Lactobacilli* may sometimes find conditions favorable for growth (100).

Vegetable oils used must be of good quality. Poor oils will produce inferior-quality products with shortened shelf life. Similarly, the quality of the egg, both microbiologically and chemically, should be high; otherwise both flavor and emulsion stability problems can be encountered. Mustard flour must be free from impurities. Strict microbiological control of raw materials is important. During processing, control must be exercised in maintaining correct temperature and emulsification procedures. Proper sequential addition of ingredients and mixing times are important for emulsion stability (97). For good emulsification, the oil globules should be even and should be 1–4 μ in size (98). It is reported that freshness of mayonnaise can be retained substantially longer by processing in the absence of oxygen (99). Processing under nitrogen increased the shelf life of mayonnaise from 48 to more than 240 days (100). Specific gravity of the product with proper amount of inert gas is about 0.90–0.92, and that of the ungassed product is about 0.94. Amount of entrapped gas must be controlled below 6%; otherwise, shrinkage during transit will cause deterioration of the product (100). Tryptophan or tryptophan derivatives (101), as well as L-cystine (102), have been found to improve the storage stability of mayonnaise.

Various instruments have been proposed for testing the important properties of consistency, stability, and flavor quality of mayonnaise and other dressings. Consistency is measured by a device that is a modified form of the Gardner mobilometer (103), which was originally designed for testing paints, varnishes, and enamels. The mobilometer is in reality a special form of viscometer in which a weighted, perforated plunger is allowed to fall through the sample. Consistency of the product is reported in seconds. Kilgore (104) devised a simple method that is particularly suitable for control work, since it involves nothing more than dropping a pointed rod or a "plummet" into a sample from a definite height and noting the depth of its penetration. Penetration correlates inversely with viscosity. The Brookfield Helipath viscometer can be used to measure viscosities. These consistency measurements do not totally describe the attributes of body and texture of mayonnaise or salad dressing. The ability of mayonnaise emulsion to resist mechanical shock is commonly evaluated by testing samples under simulated shipping conditions.

Because the quality of the oil plays a major role in the flavor stability of mayonnaise and other dressings, only the best grade of oil should be used. The quality of flavor is generally judged by the subjective method of tasting. However, Dupuy and co-workers (46) have developed a gas chromatographic method to evaluate the flavor quality objectively.

8.4 Manufacture

Because mayonnaise contains more than seven times more oil than water and the emulsion is an oil-in-water type, preparation of a stable mayonnaise emulsion is difficult. Thus major manufacturers of these products have developed proprietary techniques to achieve this goal. Effect of various factors on the stability of mayonnaise emulsions is not well understood. Therefore, production of stable mayonnaise has remained an art to some degree. Although egg solids are the backbone of a stable emulsion, processing conditions whose interactions are only partially understood play an important role.

A process for making mayonnaise in a batch mixer is to mix egg yolks, sugar, salt, spices, and a portion of the vinegar followed by gradually beating in the oil and then thinning out by mixing the remainder of the vinegar. This method of mixing is said to give a product of better consistency than is obtained by adding all the vinegar at the beginning of the operation.

Gray and co-workers (105) studied the effect of the method of mixing on the consistency of the emulsions and recommended that one-third of the vinegar be added initially and the balance be added at the end. On the other hand, Brown (106) recommended that the mixture of eggs and other ingredients be as stiff as possible when the oil is incorporated. The stiffness can be controlled during the addition of the oil by adding small portions of vinegar. This method of preparation has been claimed to produce small oil droplets, thus making the emulsion highly stable.

The temperature of the oil and other materials during mixing also influences the body characteristics of mayonnaise. A thin product results if the operation is carried out with materials that are warm. Gray and Meier (107) suggest a temperature of 15.5–21.1°C (60–70°F). The inconvenience of attempting to operate at temperatures below 15.5°C (60°F) was not worthwhile as the initial superiority of the low temperature processed product is lost with slight aging of the product. Brown (106), however, recommends a temperature of 4.4°C (40°F).

The equipment most popular in the manufacture of mayonnaise and salad dressings are the Dixie mixer and the Charlotte colloid mill connected suitably with valves and pumps. A colloid mill is a mechanical device in which the product passes between a high-speed rotor (3600 rpm) and a fixed stator. Product enters the area at a low velocity and is subjected to a high shear with reduction of the particle size. Clearance between these two parts determines (*1*) the amount of shear imposed and the viscosity of the final product and (*2*) throughput of the mill. Weiss (90) recommended an opening within a range of 25–40 mils. The mixer is used to prepare a coarse emulsion that is then passed through the colloid mill to give the creamy texture.

A second system is the AMF system (108), which consists of a premix feed tank and two mixing stages. In the premix tank, slow agitation is maintained to keep the ingredients well dispersed but not to thicken the emulsion. The mix containing all the ingredients except a portion of the vinegar from the premix tank is converted to a coarse emulsion in the first stage of mixing and is then pumped to the second stage. At this stage, the balance of the vinegar solution is added. This mixer rotates at about 475 rpm, which is much slower than a colloid mill, but because of its teeth design the mixing is very intense. It is claimed that this system produces much smaller and more uniform droplets of oil than the colloid mill does and thus makes a highly stable emulsion.

Other equipment suggested are the Girdler CR mixer (109) and the Sonalator (110), which is an ultrasonic homogenizer. A comparative study of homogenizers and colloid mills for producing dressings was made and it was concluded that homogenizers can be used for certain types of dressing but not for mayonnaise (111). Automation techniques for the salad dressing industry have been discussed (108). Advanced techniques for making mayonnaise and other salad dressings have been reviewed (112). Continuous systems that can improve productivity and have built-in versatility to permit production of different dressings have been described (113–116). A streamlined production line intended to safeguard quality of dressings has been outlined (97).

POURABLE DRESSINGS

Pourable dressings are relative newcomers as a grocery store food category compared with mayonnaise and spoonable dressings. They first appeared on the market in limited flavors, such as French and Thousand Island. Over the last several decades, many more varieties have become available. The only legal definition for pourable dressings pertains to French dressing. To be labeled "French dressing" the product requires a minimum of 35% oil and one or more of the following ingredients: salt, nutritive sweeteners, mustard, paprika or other spice extractives, monosodium glutamate (MSG), any suitable flavoring other than imitations, tomato in any suitable form, sherry wine, color additives, vegetable gums or thickeners, sequestrants, such as EDTA, or crystallization inhibitors. This definition gives the dressing manufacturer flexibility to make any flavored product using almost any formula and call it a "dressing," with the exception of French dressing.

Table 5.3 contains production volumes for pourable dressings (regular, reduced calorie, and fat free). Unlike salad dressings and mayonnaise, pourable dressings have shown a steady increase in production from 1988 through 1993. Reduced-calorie pourable dressings peaked in volume in 1989, which is also the year fat-free pourable dressings were introduced. The fat-free pourables category has continued to grow through 1993 at the expense of the reduced-calorie category. Partially contributing to these trends are the healthier lifestyles and dietary awareness of the buying public during these years. Salads have become a staple food in the American meal, and production of total pourable dressings reflects this trend.

Most pourable dressings are distributed and sold at ambient temperature. Since many of the ingredients used support microbiological growth, spoilage can occur. Most pourable dressings are formulated within acid-salt-preservative tolerances to avoid spoilage on the grocery store shelves. Refrigerated dressings (discussed below) are marketed in the grocery store produce section.

Dressings may be formulated in two different finished forms. These are broadly described as two phase and one phase. This distinction generally represents the presence or absence of homogenization. Two-phase dressings are characterized by the presence of a layer of free oil above the aqueous layer. Two-phase dressings usually contain seasonings and flavors in the aqueous portion and the oil is added at the time of filling. Examples of two-phase dressings include oil and vinegar, Italian dressing, and some French-style dressings. These products are shaken before use, and they return to separated phases after setting.

One phase dressings can be homogenized or blended. If homogenized, they are passed through a homogenizing device that reduces oil droplet size to produce a smooth, creamy dressing. Examples of homogenized dressings are Ranch, French, and Creamy Cucumber varieties. Blended dressings are characterized by the presence of both oil and aqueous portions in a loose blended matrix that is stabilized and thickened with gums. Typical blended examples are certain sweet French-style dressings and Italian varieties.

9.1 Composition

Due to the variety of pourable dressings, the number of ingredients used in their manufacture is large. The following is a discussion of some of the major functional and flavor ingredients used in pourable dressing manufacture.

Vegetable Oils. Vegetable oils have traditionally been a major ingredient in many dressing varieties. The first vegetable oil used in pourable dressings was cottonseed oil. As supplies of high quality soybean oil became plentiful in the United States during the 1960s, its use became predominant. More recently, health concerns have opened the way for use of canola and other low saturated vegetable oils. Canola oil is predominately used in Canada; other oils such as sunflower, rapeseed, and soybean are used in European markets. Vegetable oils impart lubricity and a creamy, pleasant mouthfeel to dressings. The action of homogenization on an oil–water dressing mixture causes an increase in viscosity. Other functions of vegetable oil include the solubilization of certain oil-soluble flavors and the contribution of an overall balanced flavor profile.

Water. Water is present in all dressings. Its presence contributes liquidity to the dressing system and serves as a vehicle for delivering water-soluble flavors. Many ingredients must be dissolved to be functional, and water is the major solute for these ingredients. Water content is a critical consideration for developing a microbiologically stable dressing system.

Vinegar. Vinegar, as in the case of spoonable dressings, has a dual role: flavor and microbiological preservation. Its role in flavor is emphasized in dressing varieties such as zesty-style Italian and deemphasized in buttermilk-based dressings. Shelf-stable dressings are formulated to a pH of 4.2 or less to control pathogenic organisms. Other acidifying agents have been used, and several patents are in effect that claim the use of lactic acid, phosphoric acid, malic acid, citric acid, etc. (117–120). Most of these acids impart tart flavors and control microbial growth. However, certain acids (such as phosphoric and lactic) are more efficient in pH reduction due to their relatively low pK_a. They are valuable in controlling pH in dressings with high levels of buffering ingredients, such as buttermilk and blue cheese.

Dairy Ingredients. Dairy products are used in pourable dressings to give characteristic flavors and textures. Buttermilk is a major ingredient in Ranch and, of course, buttermilk dressings. Sour cream and yogurt have been used in some creamy-style dressings. Blue Cheese is a popular dressing variety; if its process of manufacture includes cooking it yields a smooth creamy dressing. Blue Cheese can also be added after emulsion formation, which yields a chunky Blue Cheese dressing. Buttermilk and cheese also supply protein, which aids in emulsification of the oil. This protein also contributes to development of a pleasing mouthfeel. Protein from milk and cheese buffers pH upward, and formulation adjustment may be needed to meet pH criteria.

Salt. Salt has been a known preservation agent for centuries. Its use in dressings complements vinegar and other preservatives in producing shelf-stable dressing products. Salt also serves as an important source of zesty flavor as well as a flavor enhancer for a number of other ingredients.

Sugar. Sugar is included to add flavor to dressings as well as to counteract the harsher flavor effects of vinegar and salt. Sugar may be used in any of several different forms, which include sucrose, fructose, corn syrup, and maltodextrins. Sugar also increases the solids level in dressings, which contributes to smoothness and a creamy mouthfeel. In addition to its own flavor contribution, which may differ with different sugars, sugar is also a flavor enhancer for other ingredients.

Spices. The essence of pouring a dressing on a salad is to add flavor to otherwise relatively bland vegetables. Spices deliver characteristic flavors to variety dressings. Thus clove is added to Thousand Island dressing, mustard is used in French dressings, and thyme and oregano are associated with Italian dressings. Certain spices also contribute antimicrobial and antioxidant properties.

Emulsifiers. Emulsifiers help form one continuous emulsion phase of water and oil. Two-phase dressings such as oil and vinegar have no emulsifier or a very low level of emulsifier. A small amount of emulsifier may be added to stabilize the loose emulsion of two-phase dressings momentarily, after shaking but before pouring on the salad. Several different ingredients function as emulsifiers and are used in pourable dressings. Eggs (yolks or whole eggs) are used as emulsifiers in Thousand Island dressing and also contribute some flavor. Polysorbate 60 is a commercial emulsifier that is often used in creamy-style dressings. It produces small oil droplets and yields a very white emulsion. Emulsifiers also help solubilize flavors and improve overall flavor perception.

Gums. Gums are added to dressings to produce creaminess, increase or modify viscosity, and control pourability. Commonly used gums include xanthan gum, propylene glycol alginate (PGA), carrageenan, and cellulose gums. Xanthan gum is the most commonly used gum in pourable dressings. It resists hydrolysis by the acidic dressing medium and maintains the product viscosity over shelf life. Each of these materials contributes characteristic rheological properties to the dressing; they are often used in combination. In addition to viscosity modification, certain gums, such as PGA, also aid in emulsion formation and stabilization.

Stabilizers. In dressing manufacture, two different types of ingredients are often referred to as stabilizers: chelating agents (EDTA) and hydrocolloids (gums). Gums have already been discussed.

Chelating agents are used in dressings to stabilize the oil component against oxidation. EDTA is the most commonly used chelating agent, and the usual form of this material in dressings is the calcium disodium salt. Vegetable oils are prone to oxidation, which causes off-flavors in dressings. Oil oxidation is accelerated by the presence of iron and copper ions, but chelating agents inactivate, or chelate, these metal ions so they are not available to promote oil oxidation. EDTA also helps stabilize flavor and color in finished dressings.

Colors. Several different colors used in dressings include β-carotene, β-APO carotenal, FD&C colors, tumeric, and titanium dioxide. They are added to make pourable dressings eye appealing and to augment naturally occurring colors.

Flavors. Flavors are used to extend the range of dressings available to the consumer. They can be either natural or artificial. Flavors are added to supplement the use of more expensive ingredients or when certain flavors cannot be obtained from the ingredients.

Preservatives. Although vinegar is often the preservative of choice, the level of vinegar needed to ensure microbiological stability in mild-tasting varieties such as Ranch produces a dressing product that has too tart and too harsh a flavor. In these products, preservatives can be added to give the additional microbiological stability needed while still maintaining a desirable mildly acidic-tasting dressing. The most commonly used preservative is sorbic acid (sometimes added as potassium sorbate). Sodium benzoate is another preservative; it is not commonly used.

9.2 Quality Control and Stability Measurements

Quality control considerations for pourable dressings are similar to those discussed for mayonnaise. The use of high quality ingredients is an important consideration for producing a high quality dressing that will maintain its flavor for 8–12 months of shelf life. Certain pourable dressings contain a relatively high content of spices and seasonings. Spices in particular can contribute a significant load of microorganisms. Appropriate microbiological specifications must be adhered to, especially ingredients such as root crops. Spices and other ingredients used in pourable dressing manufacture can also contribute a high level of iron and copper ions, which catalyze the oxidation of fats and contribute to poor flavor.

Ranch dressings containing fermented dairy products such as buttermilk are often a source of wild cultures. These cultures, often heterofermentative *Lactobacillus,* can cause gassiness and product spoilage. Processing often includes pasteurization of the dairy component.

Quality measurements most often employed in pourable dressing manufacture include percent fat, moisture, and salt. Acidity and pH are critical control points to ensure a microbiologically stable dressing. Viscosity depends on gum hydration and mixing conditions. Viscosity is most often determined by a Brookfield viscometer, using standard spindles.

9.3 Manufacture

Pourable dressing manufacture is a relatively simple unit process. There are some order of addition considerations that should be incorporated when designing manufacturing procedures. Most gums tend to dissolve rather slowly with simple mixing into water. Their hygroscopic nature causes the outer layer of a gum particle to hydrate rapidly but does not allow water to hydrate the inner layers of the gum particles. This results in partially hydrated gum particles, commonly called "fish eyes." There are three methods used to incorporate gums into water. First, high shear mixing can be used, but this often incorporates a large amount of air. Second, preblending gums with other dry substances (sugar, salt) will keep the individual dry gum particles separated. When this mixture is added to water, the gum will disperse effectively. This method is used in dry mix dressings, sauces, and gravies. Another, more common, method is to disperse the gums in vegetable oil before addition to the water. Vegetable oil effectively separates the individual gum particles, and hydration is effected on addition to water.

Mustard flour should be added to water without vinegar or salt. The hydration of mustard flour initiates an enzymatic process that develops the proper flavor. In the presence of acid, this enzyme process results in less desirable flavor components.

Most pourable dressing manufacturing process are batch designed. Due to the large number of varieties usually produced, variety changeover is more efficient in a batch process than in a continuous or semicontinuous process.

Gum Slurry. The gum slurry tank is usually adjacent to the main mix tank and contains about 0.1 the volume of the main mix tank. Gums are dispersed in a portion of the formula vegetable oil with a lightening or similar type agitator.

Primary Mix Tank. Water is added to the mix tank, and the gum slurry mixture is added. Between 2 and 3 min of mix time is required for the gum to dissolve and hydrate. Additional dry ingredients and water-soluble liquid ingredients are then added. Certain ingredients such as mustard flour, PGA, garlic powder, and microcrystalline cellulose should be added before salt and acids. These ingredients are sensitive to salt and acid and may lose functionality if added in their presence. Addition of the acid follows most other ingredients, and vegetable oil is usually the last ingredient to be added.

Homogenization. Homogenization produces a uniform, stable emulsion. There are several pieces of equipment that can be used for homogenization, including piston type, static shear, and colloid mills.

Particle Addition. Following homogenization, additional ingredients that cannot be homogenized are added in the secondary mix tank. These items are often large spice particles (typical of Italian dressings) and pickle relish (a common ingredient in Thousand Island dressing).

REDUCED-CALORIE DRESSINGS

Before 1993, dressings (both spoonable and pourable) could be manufactured and claimed to be reduced calorie if the caloric contribution was at least 30% reduced from a similar full-calorie product. The Nutrition Labeling Education Act (NLEA) of 1993 (121) changed the requirements for reduced-calorie claims. Reduction in calories is largely obtained by reduction of oil. This causes quality problems in terms of lower viscosity, shelf-life stability, and flavor–mouthfeel perceptions.

Use of proper stabilizers to improve the quality of reduced-calorie dressings has been discussed (122). The advantages of using apocarotenal, a naturally occurring carotenoid that can be synthesized, as a coloring agent in nonstandardized dressings and spreads has been reviewed (123). Freeze–thaw stable salad dressings have been prepared using cooked starch mixtures made from freeze-resistant starch and nonwinterized oil (124). Use of highly esterified sugar esters to reduce the caloric content of dressings has been described (125). A stable low-calorie, cream-type salad dressing containing only 10–12% of fat and 5–20 Kcal per tablespoon has been claimed (126). A patent has claimed the use of agar-agar and methocel mixture to produce a low-calorie dressing (127). An emulsion base prepared by homogenizing fats and oils—like sunflower, coconut, lard, and cottonseed—with a solution containing proteins from whey, whole milk, sunflower seed globulins, or rapeseed has been used for preparing low-calorie dressings (128). Combination use of carrageenan and carob bean or guar gum to prepare low-calorie dressings has been patented (129). Weidemann and Reinicke (130) prepared a dietetic salad dressing with only 15–35% oil. In addition to protection provided by the low pH, pasteurization of the product at 95°C for 40 min gives the product longer life.

10.1 Composition

In 1993, the NLEA (121) was implemented. This law set a new standard for reduced-calorie dressings. Briefly, with reference to pourable dressings, the act states that to be declared "reduced calorie," a dressing must contain 25% fewer calories than a reference dressing. The reader is referred to the text of the act for more information (121). In addition, the act declared the serving size for pourable dressings at 2 tablespoons. (Previous regulations were based on a 1 tablespoon serving size.)

This law set a new direction for dressings with lower calories. Many products on the market were reformulated; some were discontinued. Fat-free dressings were coming on the market at about the same time, and these new products offered another option for the consumer.

Reduced-calorie pourable dressings generally follow the flavor trends of full-calorie dressings. Similar ingredients are used. However, adjustments for microbiological stability must be made. *Reduced calorie* is usually synonymous with *reduced fat,* because carbohydrates are present in low amounts and contribute only a small portion of calories in most products. To produce a reduced-calorie dressing, the fat is often reduced and replaced with water and additional gums. The higher level of water increases the susceptibility to microbial spoilage. Acetic acid (vinegar) and salt are concentrated in the aqueous phase and control microbial growth. As fat is reduced, the percentage of water must be increased. More acid and salt are needed to maintain the concentrations necessary to control spoilage. In practice, salt is usually held constant, and additional vinegar is added to maintain microbiological stability. The additional acidity contributes to a harsher acidic flavor in the reformulated dressing. The challenge the product developer faces is to use the correct amount of additional acid needed to ensure microbiological stability while choosing other ingredients that will offset the higher acidic taste. Reduced-calorie dressings, both spoonable and pourable, are processed similarly to regular (full-fat) dressings.

FAT-FREE DRESSINGS

Health trends in the 1980s initiated a relentless spiral of food product development activity centered around products with very low fat or no fat. Oil-free (fat-free) dressings had appeared on the market before 1980, but these products gained little consumer attention and effectively were niche products. Most if not all could be described as poor quality; they exhibited watery textures and harsh acidic flavors due to the relatively high amount of acid needed to stabilize them against microbial spoilage. The definitions surrounding what constituted fat free abounded, but with little uniformity either in composition or product labeling. In the late 1980s, there was some agreement in the pourable dressing category that less than 0.5 g of fat per serving (1 tablespoon = 16 g) was fat free.

11.1 Composition

In 1993, the FDA unveiled the NLEA (121). This act set the maximum amount of fat permitted for a dressing to be called fat free. It also defined a format for labeling the nutritional composition of most food products and redefined the serving size for pourable dressings. Previously, the serving size was 1 tablespoon (16 g); NLEA defined a serving of pourable dressing as 2 tablespoons. To be labeled fat free, pourable dressings must contain less than 0.5 g of fat per serving. This is equal to 1.56% fat in the product formulation.

The serving size for viscous (spoonable) dressings is 1 tablespoon. Here too, fat free requires less than 0.5 g of fat per serving. Therefore, a fat-free viscous dressing product may contain about 3% oil or fat in the formulation.

11.2 Fat Replacement

Fat or vegetable oil plays a key role in the mouthfeel and flavor release of dressings. Addition of enough oil to produce a reduced-calorie dressing yields a product with considerable mouthfeel and acceptable flavor quality. In essence, small quantities of oil, about 5–15% contribute generously to the eating quality of dressings. When no oil is present, the flavors are perceived as harsh, the mouthfeel is absent and the dressing just does not taste good. Consumers generally are unwilling to trade flavorful eating satisfaction for the benefits associated with a fat-free dressing.

Many types of fat replacers have been developed with the goal of replacing fat in food products. These materials have generally been classified as fat substitutes and fat mimetics.

Fat Substitutes. A fat substitute is a material that has fat properties, but reduced calories or no calories. Several of the latter materials have been developed. These materials have been reviewed but none has received FDA approval for food use (131,132).

Fat Mimetics. A fat mimetic is a material that mimics the role of fat. Many different food ingredients have texture and flavor properties that mimic some of the properties of fat. The challenge for the product developer designing fat-free dressings is to choose existing food ingredients and process options that mimic the eating qualities of fat-containing dressings.

11.3 Flavor Quality

From an empirical view, fat in a dressing causes a slow buildup of flavor in the mouth. The desirable flavor plateaus for a distinct time and then slowly dissipates, with a lingering desirable aftertaste. The same dressing formulation without fat will immediately give a very sharp flavor peak, sometimes two to three times higher in intensity than the fat-containing product, then rapidly decline with minimal or no desirable lingering after taste. The flavor developer is challenged to shape the flavor perception of fat-free products to match more closely the fat-containing dressing through choice of ingredients.

Bulking Agents. A common approach to improving low fat dressings is the addition of bulking agents. A number of different bulking agents are used in dressings; among them microcrystalline cellulose, polydextrose, starches, and dextrins. Patents have been granted for specially processed microcrystalline cellulose (133) and a combination of microcrystalline and other bulking agents (134). Because these materials are soluble or dispersible in water, they help build body and mouthfeel in fat-free dressings. Bulking agents also increase solids, which lowers water activity and aids in preservation. The technology of fat replacement through the use of various bulking agents has its roots in the fat mimetic theory, and many different types of bulking agents have been developed for dressings.

Flavor Technology. Flavor compounds are complex chemicals some of which are fat soluble and some of which are water soluble. In conventional fat-containing products, the partition of flavors seeks a natural equilibrium between the fat and aqueous phases. This natural partitioning provides a desirable release of flavor throughout the eating experience. If fat or oil is removed, the natural partitioning is removed. If the same flavor system used in fat-containing products is attempted in fat-free versions, the flavor will seem unbalanced.

Fat also functions as a coating for the flavor receptors in the mouth, effectively slowing the perception of the flavor. Again, the absence of fat results in a quick perception of flavor that often seems unnatural.

Many attempts have been made to overcome these flavor problems. Flavor suppliers have developed many prototype products that address the flavor problem. Suffice to say the technology of flavor development for oil-free dressings and other foods is in an evolutionary state.

11.4 Microbiological Stability

The issue of microbiological stability, discussed earlier, is more significant with fat-free dressings than with reduced-calorie dressing since the entire formulation is aqueous based. Microbiological stability is commonly addressed through use of various preservatives such as sorbic acid–potassium sorbate and sodium benzoate in addition to adjusting the level of vinegar. Acid harshness can be a problem and can be addressed through the selection of combinations of fillers and bulking agents that mask the harsher acidic sensation. Commonly, microbiological stability is determined by inoculation of product with cultured organisms isolated from spoiled dressings.

11.5 Processing

Conventional dressing manufacture involves the mixing of liquid and dry ingredients together with the development of viscosity during the process. Fat-free dressing manufacture involves the hydration of various bulking agents. This unit operation often requires high shear mixing to disperse and hydrate the fat mimetics adequately. Since there is no oil in these formulas, gums are dry blended with sugar before dressing manufacture. This dry blending helps prevent the formation of "fish eyes." No homogenization step is necessary, but through intimate mixing is necessary to effect full functionality from the fat replacer system.

REFRIGERATED DRESSINGS

Refrigerated dressings comprise a small portion of the pourable dressing category. They are marketed in the produce section of the supermarket and appeal to the consumer interested in freshness. Although generally thought of as refrigerated and distributed under refrigeration, they are often displayed at room temperature. As a result, they are generally formulated in a similar manner as shelf-stable products to resist spoilage during periods of nonrefrigeration.

HEAT-STABLE DRESSINGS

Improvement in the shelf stability of dressings can be expected if the product can be heated to high temperatures without losing body or color. Furthermore, such heat-stable dressings can be incorporated into canned meat, fish, vegetable salads, and other foods. A certain microcrystalline cellulose imparts this stability (135), even when the product is retorted at 115.5°C (240°F) in the presence of food acids. Similarly, xanthan gum has been shown to produce a product with no visible signs of oil separation and with a viscosity that remains practically unchanged over a wide range of temperatures.

A dry preparation of the ingredients that may be reconstituted to form a salad dressing simply by the addition of water has been patented (136). Several such preparations are being marketed by different companies in the United States.

REFERENCES

1. G.P. Lensack, *Food Eng.* **12,** 98–100 (1969).
2. ICI United States, Inc., *Bulletin*, 222–228.
3. E. Deffense, *J. Am. Oil Chem. Soc.* **70,** 1193–1201 (1993).

4. I. Newton, P. Clough, and D. Mazinger, paper presented at the American Oil Chemists' Society Meeting, Atlanta, 1994.
5. R.K. Chandra, *Health Effects of Fish and Fish Oils,* ARTS Biomedical Publishers and Distributors Ltd., St. John's Newfoundland, Canada, 1989.
6. F.J. Baur and E.S. Lutton (to Procter & Gamble Co.), U.S. Pat. 3,158,490 (Nov. 24, 1964).
7. F.J. Baur (to Procter & Gamble Co.), U.S. Pat. 3,211,558 (Oct. 12, 1965).
8. F.R. Hugenberg and E.S. Lutton (to Procter & Gamble Co.), U.S. Pat. 3,353,966 (Nov. 21, 1967).
9. G. Fuller, G.O. Kohler, and T.H. Applewhite, *J. Am. Oil Chem. Soc.* **43,** 477–478 (1966).
10. G.N. Fick (to Sigco Research Inc.), U.S. Pat. 4,627,192 (Dec. 9, 1986).
11. USDA, *Stat. Bull.,* 867 (1993).
12. R. Purdy, *J. Am. Oil Chem. Soc.* **62,** 523–525 (1985).
13. T.L. Mounts, K. Warner, G.R. List, W.E. Neff, and R.F. Wilson, *J. Am. Oil Chem. Soc.* **71,** 495–499 (1994).
14. S.S. Chang, T.H. Smouse, R.G. Krishnamurthy, B.D. Mookherjee, and B.R. Reddy, *Chem. Ind.* 1926–1927 (1966).
15. G. Hoffmann, *J. Am. Oil Chem. Soc.* **38,** 1–3 (1961).
16. S.S. Lin, T.H. Smouse, and R.R. Allen, paper presented at the American Oil Chemists' Society Spring Meeting, Mexico City, 1974.
17. W.E. Neff, T.L. Mounts, W.M. Rinsch, and H. Konishi, *J. Am. Oil Chem. Soc.* **70,** 163–168 (1993).
18. E.N. Frankel and S.W. Haung, *J. Am. Oil Chem. Soc.* **71,** 255–259 (1993).
19. N.R. Artman, *Adv. Lipid Res.* **7,** 245–330 (1969).
20. H.W. Schultz, E.A. Day, and R.O. Sinnhuber, *Symposium on Foods: Lipids and Their Oxidation,* The AVI Publishing Co., Westport, Conn., 1962.
21. Anon., *BIBRA Bull.* **10,** 4–8 (1971).
22. R. Guillaman, *Rev. FR Corps. Gras* **18,** 445–456 (1971).
23. J.P. Freeman, *Food Process Mark. London* **38,** 303–306 (1969).
24. R.J. Sims and H.D. Stahl, *Baker's Dig.* **44,** 50–52 (1970).
25. U. Shimura, *J. Jpn. Oil Chem. Soc.* **19,** 748–756 (1970).
26. E. Yuki, *J. Jpn. Oil Chem. Soc.* **19,** 644–654 (1970).
27. W.W. Nawar *J. Agr. Food Chem.* **17,** 18–21 (1969).
28. C. Cuesta, F.J. Sanchez-Muniz, C. Gorrido-Polonio, S. Lopez-Varela, and R. Arroyo, *J. Am. Oil Chem. Soc.* **70,** 1069–1073 (1993).
29. B.R. Reddy, K Yasuda, R.G. Krishnamurthy, and S.S. Chang, *J. Am. Oil Chem. Soc.* **45,** 629–631 (1968).
30. K. Yasuda, B.R. Reddy, and S.S. Chang, *J. Am Oil Chem. Soc.* **45,** 635–628 (1968).
31. R.G. Krishnamurthy and S.S. Chang, *J. Am Oil Chem. Soc.* **44,** 136–140 (1967).
32. M.C. Dobarganes, M.C. Perez-Camino and G. Marquez-Ruiz, *Fat Sci. Technol.* **8,** 308–311 (1988).
33. D.M. Lee, *Bull. Brit. Food Manuf. Ind. Res. Assoc.,* 80, (1973).
34. L.M. Wright, *J. Am. Oil Chem. Soc.* **53,** 408–409 (1976).
35. V.C. Mehlenbacher, *Analysis of Fats and Oils,* Gerrard Press, New York, 1960.
36. S.S. Chang and F.A. Kummerow, *J. Am. Oil Chem. Soc.* **32,** 341–344 (1955).
37. M. Loury and L. Garber, *Rev. FR Corps. Gras.* **15,** 301 (1968).
38. U. Holm, K. Ekbom, and G. Wode, *J. Am. Oil Chem. Soc.* **34,** 606–609 (1957).
39. G.R. List and co-workers, *J. Am. Oil Chem. Soc.* **51,** 17–21 (1974).

40. G. Johansson and V. Persmark, *Oil Palm News* (10–11), 3 (1971).
41. R.G. Scholz and L.R. Ptak, *J. Am. Oil Chem. Soc.* **43,** 596–599 (1966).
42. C.D. Evans, G.R. List, R.L. Hoffmann, and H.A. Moser, *J. Am. Oil Chem. Soc.* **46,** 501–504 (1969).
43. G.A. Jacobson, J.A. Kirkpatrick, and H.E. Goff, *J. Am. Oil Chem. Soc.* **41,** 124–128 (1964).
44. J.A. Fioriti, M.J. Kanuk, and R.J. Sims, *J. Am. Oil Chem. Soc.* **51,** 219–223 (1976).
45. B. Tsoukalas and W. Grosch, *J. Am. Oil Chem. Soc.* **54,** 490–493 (1977).
46. H.P. Dupuy, S.P. Fore, and L.A. Goldblatt, *J. Am. Oil Chem. Soc.* **50,** 340–342 (1973).
47. A.D. Waltking and H. Zaminski, *J. Am. Oil Chem. Soc.* **54,** 454–457 (1977).
48. H.W. Jackson and D.J. Giacherio, *J. Am. Oil Chem. Soc.* **54,** 458–460 (1977).
49. J.L. Williams and T.H. Applewhite, *J. Am. Oil Chem. Soc.* **54,** 461–463 (1977).
50. D.B. Min and T.H. Smouse, *Flavor Chemistry of Lipid Foods,* American Oil Chemists' Society, Champaign, Ill., 1989.
51. H.A. Moser, C.D. Evans, J.C. Cowan, and W.F. Kwolek, *J. Am. Oil Chem. Soc.* **42,** 30–33 (1965).
52. R. Radtke, P. Smits, and R. Heiss, *Fette Seifen Anstrichmit* **72,** 497–504 (1970).
53. A. Sattar, J.M. deMan, and J.C. Alexander, *Lebensm. Wiss. Univ. Technol.* **9,** 149–152 (1976).
54. A. Sattar, J.M. deMan, and J.C. Alexander, *J. Can. Inst. Food Sci. Technol.* **9,** 108–113 (1976).
55. H.S. Olcott and E. Einst, *J. Am. Oil Chem. Soc.* **35,** 161 (1958).
56. W.M. Gearhart, B.N. Stuckey, and J.J. Austin, *J. Am. Oil Chem. Soc.* **34,** 427 (1957).
57. B.S. Mistry and D.B. Min, *J. Food Sci.* **52,** 831 (1987).
58. N.W. Labuli and P.A. Bruttel *J. Am. Oil Chem. Soc.* **63,** 79 (1986).
59. J.M. deMan, F. Tie, and L. deMan, *J. Am. Oil Chem. Soc.* **64,** 993 (1987).
60. T.A. Jebe, M.G. Matlock, and R.T. Sleeter, *J. Am. Oil Chem. Soc.* **70,** 1055–1061 (1993).
61. J.C. Allen and R.J. Hamilton, *Rancidity in Foods,* Elsevier Applied Science, Publishers, Ltd., Barking, U.K., 1989.
62. British Food Manufacturing Industries Research Assoc., *Technical Circular,* No. **605,** Leatherhead, U.K., 1989.
63. D. Firestone, R.F. Stier, and M. Blumenthal, *Food Technol.* **45,** 90–94 (1991).
64. M. Ahrens, G. Guhr, J. Waibel, and S. Kroll, *Fette Seifen Anstrichmit.* **79,** 310–314 (1977).
65. U.J. Salzar and J. Wurziger, *Fette Seifen Anstrichmit.* **73,** 705–710 (1974).
66. M.R. Sahasrabudhe and V.R. Bhalerao, *J. Am. Oil Chem. Soc.* **40,** 711–712 (1963).
67. E.G. Perkins, R. Tanbold, and A. Hsieh, *J. Am. Oil Chem. Soc.* **50,** 223–225 (1973).
68. G. Billek and G. Guhr, paper presented at the American Oil Chemists' Society Meeting, New York, 1977.
69. A.E. Waltking and H. Zaminski, *J. Am. Oil Chem. Soc.* **47,** 530–534 (1970).
70. E.R. Sherwin, *J. Am. Oil Chem. Soc.* **53,** 430–436 (1976).
71. Q. Chen, H. Shi, and C. Ho, *J. Am. Oil Chem. Soc.* **69,** 999–1002 (1992).
72. Z. Djarmati, R.M. Jankov, E. Schwirtlich, B. Djulinac, and A. Djordjevic, *J. Am. Oil Chem. Soc.* **68,** 731–734 (1991).
73. Y. Hara, *American Biotechnology Laboratory,* 1994, p. 48.
74. R.G. Cunningham, R.D. Dobson, and L.H. Going (to Procter & Gamble Co.), U.S. Pat. 3,415,658 (Dec. 10, 1968).
75. H. Enci, S. Okumura, and S. Ota (to Aginomoto Co., Inc.), U.S. Pat. 3,585,223 (June 15, 1971).
76. S.S. Chang and P.E. Morne (to Swift and Co.), U.S. Pat. 2,966,413 (Dec. 27, 1960).
77. E.R. Purves, L.H. Going, and R.D. Dobson (to Proctor & Gamble Co.), U.S. Pat. 3,415,660 (Dec. 10, 1968).

78. R.W. Fallat, C.J. Glueck, R. Lutmer, and F.H. Mattson, *Am. J. Clin. Nutr.* **29,** 1204 (1976).
79. S. M. Lee, *Fat Substitutes: A Literature Survey,* British Food Manufacturing Industries Research Association, Leatherhead, U.K., 1989.
80. P. Seiden (to Proctor & Gamble), Eur. Pat. Appl. 322 027 A2 (1989).
81. S.F. O'Keefe, V.A. Wiley, and D.A. Knauft, *J. Am. Oil Chem. Soc.* **70,** 489–492 (1993).
82. A.C. Nielsen 1994.
83. CPC International, Brit. Pat. 1,473,208 (1977).
84. F.J. Baur (to Proctor & Gamble Co.), U.S. Pat. 3,027,260 (Mar. 27, 1962).
85. M.H. Joffe, *Mayonnaise and Salad Dressing Products,* Emulsol Corp., Chicago, 1942.
86. S. Doores in A.L. Branen and P.M. Davidson, eds. *Antimicrobials in Foods,* Marcel Dekker, Inc., New York, 1983.
87. R.H. Forsythe, *Cereal Sci. Today* **2,** 211 (1957).
88. W. Fluckinger, *Fette Seifen Anstrichmit.* **68,** 139–145 (1966).
89. W.J. Stadelman and O.J. Cotterill, *Egg Science and Technology* 3rd ed., Food Product Press, 1990.
90. T.J. Weiss, *Food Oils and Their Uses,* AVI Publishing Co., Westport, Conn., 1970.
91. J.W. Corran, *Food Manuf.,* **9,** 17 (1937).
92. D. Cummings, *Food Technol.,* **18,** 1901–1902 (1964).
93. C.H. Japikse (to Proctor & Gamble Co.), U.S. Pat. 3,425,843 (Feb. 4, 1969).
94. D.M. Gray, *Oil Fat Ind.,* **4,** 410 (1927).
95. D.M. Gray, *Glass Packer,* **2,** 311 (1929).
96. R.B. Smittle, *J. Food Protect.* **40,** 415 (1977).
97. A. McKenzie and J.V. Ziemba, *Food Eng.* **36,** 96–98 (1964).
98. D.R. Beswick, *Food Technol. N.Z.* **4,** 332–339 (1969).
99. G.T. Muys and J.A. Schaap (to Unilever Ltd.), Brit. Pat. 1,130,634 (1968).
100. R.D. McCormick, *Food Prod. Dev.* **1,** 15–18 (Feb.–Mar. 1967).
101. Anon. (to Kyowa Hakko Kogyo Co., Ltd.), Brit. Pat. 1,155,490 (1969).
102. H. Enei, A. Mega, O. Ayako, S. Olumura, and S. Ota (to Ajinomoto Co.), Brit. Pat. 1,152,966 (1969).
103. H.A. Gardner and A.W. VanHeuckeroth, *Ind. Eng. Chem.* **19,** 724–726 (1927).
104. L.B. Kilgore, *Glass Packer* **4,** 65–67, 90 (1930).
105. D.M. Gray, C.E. Maier, and C.A. Southwick, *Glass Packer* **2,** 397–400 (1929).
106. L.C. Brown, *J. Am. Oil Chem. Soc.* **26,** 632–636 (1949).
107. D.M. Gray and C.E. Maier, *Glass Packer* **4,** 23–25, 40 (1931).
108. M.H. Joffe, *Food Eng.* **28**(5), 62–65, 100 (1956).
109. J.P Bolanowski, *Food Eng.* **39**(10), 90–93 (1967).
110. O.C. Samuel, *Food Process. Market.* 81–84 (1966).
111. L.H. Rees, *Food Prod. Develop.* **9,** 48–50 (1975).
112. A.J. Finberg, *Food Eng.* **27**(2), 83–91 (1955).
113. S.E. Potter, *Food Process.* **31,** 43–44 (1970).
114. V.R. Carlson, *Food Eng.* **42,**(12), 54–55 (1970).
115. M. Lipschultz and R.E. Holtgrieve, *Food Eng.,* **40**(11), 86–87 (1969).
116. F. Taubrich, *Fette Seifen Anstrichmit.* **65,** 475–478 (1963).
117. J.M. Antaki and D.T. Layne (to the Clorox Co.), U.S. Pat. 4,927,657 (May 22, 1990).
118. R.W. Wood, J.V. Parnell, and A.C. Hoefler (to General Foods Corp.), U.S. Pat. 4,352,832 (Oct. 5, 1982).

References

119. J.E. Tiberio and M.C. Cirigliano (to Thomas J. Lipton, Inc.), U.S. Pat. 4,477,478 (Oct. 16, 1984).
120. J.G. Oles (to Kraft, Inc.), U.S. Pat. 4,145,451 (Mar. 20, 1979).
121. *Federal Register* January 6, 1993.
122. G. Meer and T. Gerard, *Food Process.* (5), 170–171 (1963).
123. A.J. Finberg, *Food Prod. Develop.* **4,** 46–47 (1971).
124. A. Partyka (to National Dairy Products Corp.), U.S. Pat. 3,093,485 (June 11, 1963).
125. F.H. Mattson and R.A. Volpenhein (to Proctor & Gamble Co.), U.S. Pat. 3,600,186 (Aug. 17, 1971).
126. A.S. Szczesniak and E. Engel (to General Foods Corp.), U.S. Pat. 3,300,318 (Jan. 27, 1967).
127. J.C. Spitzer, L.S. Nasareisch, J.L. Lange, and H.S. Bondi (to Carter Products, Inc.), U.S. Pat. 2,944,906 (July 12, 1960).
128. J. Kroll, G. Mieth, M. Roloff, J. Pohl, and J. Bruecker, Ger. Pat. 106,777 (1974).
129. U. Steckowski (to Carl Kuhne KG), Ger. Pat. 2,311,403 (1974).
130. H. Weidemann and H.P. Reinicke, Ger. Pat. 1,924,465 (1970).
131. B.F. Hausmann, *J. Am. Oil Chem. Soc.* **63,** 278 (1986).
132. R.G. LaBarge, *Food Technol.* **42**(1), 84 (1988).
133. C.C. Baer and co-workers (to Kraft General Foods), U.S. Pat. 5,011,701 (Apr. 30, 1991).
134. R. Bauer and co-workers (to Thomas J. Lipton Co.), U.S. Pat 5,286,510 (Feb. 15, 1994).
135. C.T. Herald, G.E. Raynor, and J.B. Klis, *Food Process. Mark.* **11,** 54–55 (1966).
136. M.H. Kimball, C.G. Harrell, and R.O. Brown (to Pillsbury Mills, Inc.), U.S. Pat. 2,471,435 (May 31, 1949).

6

Marine Oils

INTRODUCTION

Until the past few decades most people thought of marine oil or fish oil as that smelly material called cod liver oil that kids had to take in order to receive their adequate supply of oil-soluble vitamins A and D. Today, the image of these oils, the lipid fraction of fish and other aquatic plants and animals, has a much broader acceptance (1). However, it is the structure of the lipid fraction of all aquatic plants and animals, and not only that of fish from marine waters, that is of interest to the nutrition, medical, and commercial worlds. By using the term fish oil or marine oil, one is normally referring to a class of lipids originating from the marine and freshwater bodies of the world as opposed to those from land plants and animals. Hence, by historical practice, marine oils, fish oils, and aquatic plant and animal oils are used interchangeably throughout the world. Since the majority of the oils currently produced commercially are from fish, the term fish oil will be used throughout this discussion.

Lipids, a general component of plants and animals, are arbitrarily classified as to their physical state at room temperature, fats being solids and oils being liquid. Oils are the principle lipid component of fish and fat is the most prevalent lipid in land animals. Mammals that are associated with the aquatic environment normally have a high percentage of their mass as lipids, both in the form of fat and oil.

The principal reason that aquatic plant and animal oils are thought of as fish oils is that the bulk of all commercial oils available from these sources is a by-product of fish that arc being rendered into fish meal for animal consumption. Some 28–30% of the approximately 100 million metric ton (MMT) annual world production of fish is reduced to fish meal and oil, resulting in approximately 6.5 MMT of fish meal and 1.5 MMT of fish oil (Table 6.1). The principal raw materials are the so-called industrial fish (i.e., herring, menhaden, anchovy, sardine, and mackerel) that contain high fractions of oil. Most of these species

Table 6.1 World supply of fats and oils (millions of metric tons)^a (2)

Source	Production
Soy	15.57
Cottonseed	3.67
Groundnut	3.73
Sunflower	7.45
Rapeseed	7.76
Sesame	0.65
Corn	1.29
Olive	1.83
Palm	9.57
Palm kernel	1.25
Coconut	3.01
Butterfat	6.39
Lard	5.42
Fish	1.50
Linseed	0.71
Castor	0.39
Tallow/grease	6.63
Total	76.82

^{a} Five-year average (1986–1991).

are bony, oily fish that are not popular as human foods. Fish meal and oil is also produced from waste fractions of processed edible fish.

THE UNIQUE PROPERTIES OF FISH OILS

2.1 Composition of Fish Oils

The basic composition of fish oils include triglycerides (or triacyglycerols), diglycerides, and monoglycerides. Saponifiable triglycerides account for the majority of these constituents. Other significant components include unsaponifiable phospholipids, sterols, wax esters, and hydrocarbons, and some species of fish contain diacyl glyceryl ethers (alkoxydiglycerides). Minor fat soluble fractions include vitamins, pigments, and contaminants. Oil that has been exposed to oxygen will have varying amounts of oxidation products.

Triglycerides are composed of three fatty acid chains containing up to 24 carbon atoms (C), the majority containing 12–24 carbon atoms, esterified to a glycerol backbone. A saturated fatty acid has no double bonds while unsaturated fatty acids have one or more double bonds. The melting point of a lipid is dependent on both the degree of unsaturation and the chain length. Fatty acids in the triglyceride portions of a lipid range from highly saturated to highly unsaturated, the degree of unsaturation determining the physical and chemical characteristics of the oil. In general, fats from animal sources (e.g., muscle meats, butter, cheese, eggs) contain a larger portion of saturated fatty acids than oils from plants. However, certain plant oils (e.g., coconut and palm) have a high degree of saturation and most animals associated with the aquatic environment contain highly unsaturated fatty acids in their fats and oils. Oils from some plants and animals such as olive oil, canola oil, peanut oil, and poultry fat contain relatively high levels of monounsaturated fatty acids (one double bond). The resulting classification of fatty acids depends on the number and position of the double bonds.

Oils from plants and animals living in aquatic environments are unique in that the first double bond of a major fraction of the fatty acids is between the third and fourth carbon atom from the end of the chain, designated as the omega-3 (n-3 or ω3) position. The comparable unsaturated fatty acids in land plants and animals have the first double bond between the sixth and seventh carbon atom and are known as omega-6 (n-6 or ω6) fatty acids. Fatty acids having more than one double bond are known as polyunsaturated fatty acids (PUFA). The n-3 fatty acids in fish oils have a higher degree of unsaturation (more double bonds) than n-6 fatty acids. To indicate the PUFA that are important in fish oil, n-3 PUFA that contain five or more double bonds are referred to as Highly Unsaturated Fatty Acids (HUFA). This distinction is becoming more important as research continues to expand on the knowledge that HUFA provide many health benefits that are not obtained from PUFA components of an oil.

2.2 Fish Oils in Human Nutrition and Health

Lipids, and particularly the PUFA component, have long been known to be essential to proper growth and nutrition. In 1929 (3) it was reported that a deficiency disease was associated with complete exclusion of fat in the diet. The essential fatty acid was later identified as linoleic acid, C18:2n-6. It was shown to produce a product of metabolism, an eicosanoid, called prostaglandin that stimulated the contraction of smooth muscles. Today, the knowledge of different prostaglandins and their relationship to fatty acid metabolism is well known. However, there was a lapse of meaningful fatty acid research for about 30 years due to the lack of sufficiently sensitive analytical techniques.

Interest in effects of fish oil on human health was stimulated in 1952 (4) when it was reported that blood cholesterol levels were controlled by the type of lipid (animal or vegetable) rather than by the dietary level of lipid. This was followed by evidence that the total unsaturation of the fatty acids, rather than the source, was the important factor (5,6). Some of these early studies indicated that dietary fish oil was more effective than corn oil in reducing serum cholesterol. However, the scientific community essentially ignored these results for almost ten years before the reporting of work by Bang and Dyerberg (7).

In the early 1960s it was concluded that saturated fat was thrombogenic and that polyunsaturated fat was nonthrombogenic (7,8). By the 1970s many scientific and popular publications reported that premature coronary heart disease was epidemic in the United States and that this could be attributed to the high saturated fat and cholesterol intake (9). In 1952 Dr. Avery Nelson began a 19 year study in which several hundred patients with previous histories of one or more heart attacks were placed on high fish diets vs. a normal diet with no fish. The results of this work were not reported until the experiment was completed in 1972 (10) so it did not receive due credit for pioneering work with human diets containing high levels of fish oil.

Accelerated interest in working with fish oils to reduce heart disease was the result of studies carried out by Bang and Dyerberg in 1972 (11). They reported on the rarity of heart disease among Greenland Eskimos and its relationship with consumption of marine lipids high in n-3 HUFAS. Many research and review papers were published in the 1980s that related the consumption of n-3 fatty acids to reduction of heart and other diseases (12–18).

Lipids are necessary as a source of energy, the transport of fat soluble vitamins, and the source of certain essential fatty acids. However, it has been rather recently that science is beginning to really understand the radically different effects of certain fatty acids, especially HUFA, as compared to others (19). The fraction of particular interest in fish oils involves the omega-3 fatty acids, especially eicosapentaenoic acid ($C20:5n$-3, commonly called EPA) and docosahexaenoic acid ($C22:6n$-3, commonly called DHA).

Cholesterol is an animal sterol which is a precursor to many hormones, vitamin D, and bile acids. Being water-insoluble, cholesterol can only be transported in the blood by combining with water-soluble proteins (myofibrilar proteins) to form lipoproteins. It is the form of these lipoproteins that has been the subject of many investigations involving heart disease. There are several types of lipoproteins that have different densities. These are classified as high-density lipoprotein (HDL), low-density lipoproteins (LDL), and very low-density lipoproteins (VLDL). High levels of LDL and VLDL cholesterol can cause the deposition of cholesterol, accompanying cell debris, and other compounds on the linings of blood vessels. This deposit, called plaque, can increase with time until the cross section of blood vessels is greatly restricted. An aggregation of sticky platelets (a thrombus) can block the artery to the extent that a stroke or heart attack may occur. Although there has been much attention given to the consuming of low-cholesterol foods in order to decrease heart disease, it is the lipoprotein form of the cholesterol carrier that is controlling. Perhaps more important than n-3 HUFA effects on lipoprotein cholesterol levels is their effects on which eicosanoids are produced by the cell; n-6 PUFA metabolites enhancing plate aggregation, n-3 HUFA metabolites causing decreased aggregation and less chance of thrombus formation. In fact, the body synthesizes cholesterol and this synthesis may be stimulated by excluding the essential nutrients from the diet. However, it is mostly saturated fat that stimulates cholesterol production.

Of particular interest is the relationship of HUFA (believed to be EPA) to the cholesterol which is carried in the blood to various body tissues. There have been many studies that indicate that fish oil will increase HDLs in the blood and decrease LDLs and VLDLs. However, the major increases are noted with high doses, above that normally consumed by people on a high fish diet (20–24). Harris, investigating the conflicting results of fish oil, summarized the work to 1989 (25). In fish oil studies with human subjects, the oil intake has varied from as little as 1.6 to over 100 g/day. The n-3 fatty acid intake in these studies ranged from 0.5 to 25 g/day and studies lasted from 2 weeks to over 2 years. In general HDL levels usually rise with fish oil supplementation. The majority of the placebo-controlled, crossover trials showed that fish oils increase HDL cholesterol levels 5–10%, indicating that the n-3 fatty acids were responsible. Although much of this research is somewhat outdated in light of present research which is definitizing the chemistry of lipids in the body, there is certainly no question but that the Greenland Eskimos eat large amounts of fatty acids from fish and marine animal oils, consuming a pound or more of fish or other animals from the sea each day.

Recent work in Norway and elsewhere has indicated that the monounsaturated fatty acids in seal oil (61%) may be as responsible as n-3 fatty acids (23%) in significantly lowering the incidence of coronary heat disease among Greenland Eskimos (26). Minke whale blubber was found to have a similar high monounsaturated fatty acid content to that of seals (27). However, the position distribution of the n-3 long-chain fatty acids may be a more important factor in the effectiveness of fatty acids in seal and whale oil as compared to cod liver oil. In minke whale oil 81% of 20:5n-3 and 94% of 22:6n-3 are in the 1 and 3 positions of triglycerides, while the corresponding amounts for cod liver oil in these position are respectively 58 and 22% (27,28). In addition, other factors, such as the temperature of the fish oil extraction process, can greatly affect the acceptability and quality of fish oil (29).

EXTRACTION AND REFINING OF MARINE OILS

3.1 Sources of Marine Oils

Any aquatic animal living in an environment where there are other aquatic plants and animals is a potential source of fish oil. However, the nearly 1/3 of the industrial fish that are reduced to fish meal and oil are certainly the best source for large or commercial volumes of the product.

Although there are plants in many parts of the world for reducing industrial fish, by-catch fish, and waste from fish processing operations, the majority of fish oil is produced by fish meal plants in South and North America that are processing industrial fish that have high oil content, dark meat, and many bones. Peru and Chile process predominantly anchovy meal and the Eastern and Southern coasts operations in the United States process menhaden. In fact, almost one-half of the total fish landed in the United States are menhaden, used only for this purpose. Menhaden and anchovies have high oil content, off-flavored dark and unattractive meat, and many bones that make them unsuitable for human consumption.

One growing market for fish meal is the so called white fish meal that is made from waste and trimmings of edible fish that are being prepared for the human market. This meal is especially demanded by aquaculture farmers who desire high protein meal with low oil content. This gives the maximum in protein nutrition and the minimum in undesirable properties created by rancid oil.

The development of fish oil has not paralleled that of the fish meal, and, in the past, has been relegated to industrial uses, the exception being that some countries refine and hydrogenate fish oil for margarine production.

3.2 Conventional Fish Meal Process

There are several techniques by which high quality fish oils can be prepared from whole fish or portions of fish. However, the majority of fish oil available for the market is a comparitively low grade by-product from the reduction of fish or waste to fish meal. Much of the degradation is caused by heating and holding the oil at high temperatures during the processing and initial storage. Interest in high quality fish oils for human consumption is increasing as more is being learned about the health value of n-3 fatty acids. The process for converting fish to dry, stable meal for animal feed supplements actually followed the development of various techniques for removing oil from fish. Although oil has been produced for several centuries, until the twentieth century the waste from the process was discarded or used for fertilizer. Prior to the development of fish meal operations in response to a growing demand for animal feeds, oil was the most valuable product recovered from fish. From a steady production of about 2 million gallons per year during 1873–1911, the rather crude product was produced at an increased level of 6.6 million gallons per year from 323,000 MT of fish in 1912 (30). Production remained at that level until the mid 1940s when the demand for animal feed and improved production techniques encouraged significant increase in the processing of industrial fish.

Although there have been improvements and modernization of machinery and equipment, for the past 50 years the conventional meal process has not had major procedural changes. The major industrial plants are located in areas of the world where industrial fish (e.g., anchovy, herring, menhaden, and sardine) are harvested in large quantities. Fish meal plants can process from a few tons up to 50 or more tons per hour. As shown in Figure 6.1, the process

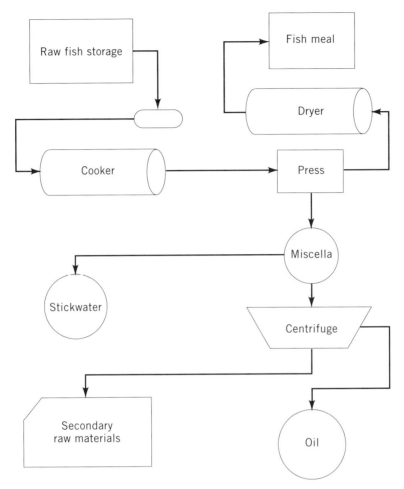

Figure 6.1 Conventional fish meal process.

involves cooking whole fish or waste portions, pressing to separate the liquid phase (miscella) from the solids, and drying the solids (31). Miscella consists of water, oil, extracted water-soluble materials, and insoluble suspended solids that have been carried through the press screens. The oil fraction is removed from the miscella by centrifugation, pumped to storage tanks, and shipped in bulk to secondary processors. The remaining water fraction containing water solubles and some suspended solids is called stickwater.

Cooking as Received Raw Material. Miscella, and particularly the oil fraction, only can be effectively released from the fish flesh if the raw material is cooked to denature and coagulate the proteins. The cookers are of two general types, namely direct steam injection and indirect cookers. Direct steam injection involves the direct introduction of steam into the mass of fish being cooked. Modern indirect cookers are heated by introducing steam into a jacket that surrounds the cylindrical cooker. Within the cooker is a cylinder with a steam heated auger that conveys the fish through the cooker. The dwell time in the cooker is controlled by the temperature of the steam heated surfaces and the speed of the cooker. During the cooking, some of the water is released in the cooker. This portion is passed through screens to remove suspended solids and then pumped to the miscella being pressed from the cooked fish.

Pressing the Cooked Fish to Remove Miscella. As cooked fish leaves the cooker it is conveyed (normally by a screw conveyor) to the press for removing the oil/water that has been released during cooking. Fish solids, called the press cake, are separated from the miscella by a continuous single screw or multiscrew press. Single screw presses are tapered so that the pressure builds up as the material travels through the press. Soft fish sometimes come from the cooker due to their being low quality (e.g., enzyme hydrolyzed or degraded by bacteria so that the proteins will not denature), cooked for the wrong period of time, having excessively high oil content, or having other physical and chemical properties preventing denaturation. A twin screw press is the best for these products since they have heavily supported strainer plates, both screws are on tapered shafts that are pitched opposite to the taper of the shaft, and the screws turn in the opposite direction. This combination minimizes the amount of soft flesh that slips through the press without having the miscella extracted (32). At this point in the process the press cake, containing approximately 50% solids, and the miscella have been separated. The press cake is then dried to fish meal and the miscella is treated to separate the oil and water fractions.

Drying Press Cake to Final Fish Meal Product. Press liquor (miscella) contains a considerable quantity of solids, both soluble and suspended. The soluble solids are small-molecular-weight compounds that are water soluble and water-soluble proteins (myofibrilar proteins) that leach into the water during cooking. Insoluble materials include coagulated proteins, as well as scale and bone fragments that are extruded through the small openings in the press. The suspended solids are removed by course screening followed by centrifugation. Solids recovered during screening are added to the press cake prior to drying. In order to recover solubles and reduce pollution problems from dissolved and suspended solids, the water fraction is sprayed into the fish meal dryer during the drying process. The solids recovered are usually fed back through the presses, pressed separately, or added directly to the press cake before the drying operation.

Press cake is dried to a moisture content that prevents microbial degradation during subsequent storage and use. There are two types of continuous dryers used in large fish meal operations. A direct dryer subjects the drying meal to direct contact with hot flue gases from a furnace. There are several types of indirect dryers that depend on external heating of the heat transfer surfaces. By far the most extensively used indirect dryer is jacket heated by steam.

Over the past decade more gentle drying techniques have been developed that employ indirect hot-air, vacuum, and fluid beds (33). Some of the modern plants use a combination of direct-fired flame dryers for partial drying to 20–25% moisture and final drying in an indirect dryer (34). This allows more rapid drying since the wet meal is not as subject to protein damage as it is during the latter stages of drying.

The effect of over heating fish meal during drying causes a disease called Vometaria Negra (black vomit). Formation and subsequent reaction of the histidine inidazolic–ethyl radical with the lysine amino-γ-radical causes erosion and ulceration in the gizzard of fowls (35). Many buyers require that meal, especially from areas using older direct drying processes, must be certified by bioassays to determine that the level of toxicity will not cause the disease in commercially raised fowl.

Fish meal contains considerable lipid content (2–10% and even higher) that is not removed during pressing of the cooked fish. An antioxidant is added to the final meal to prevent oxidation of the product during storage (34). Oil from species of fish from northern areas (e.g., herring and capelin) with a relatively low degree of unsaturation is less vulnerable to oxidation than species with higher amounts of unsaturated fatty acids.

3.3 Refining Fish Oil

As discussed, fish oils are obtained primarily as by-products of fish meal processing. The screened miscella, or press liquor resulting from the pressing of meal prior to drying, still contains some suspended solids, water (containing water solubles), and oil. In many plants the oil is separated from the suspended solids and water fraction by centrifugation and then stored as a crude oil without further processing. However, it is becoming common in many plants to polish the crude oil by a hot water or steam stripping process to remove many of the contaminants that contribute to degradation during storage. The general refining of fish oils involves removing moisture and chemical impurities, including free fatty acids, various pigments, water-soluble components, pigments, diglycerides, and fatty acid oxidation products. As will be discussed in a later section, with the current interest in utilizing fish oils for human and animal food components, further refining to deodorize and stabilize the product must be carried out. Without further refining, beyond that practiced for oils destined for industrial use, the off-flavors and odors caused by oxidation and other degradation during processing, extracting, and storing make the oil inedible for humans.

MARINE OILS IN INDUSTRIAL PRODUCTS

An extensive survey of the uses of fish oils for industrial (nonedible) processes and products was published in 1993 by Stansby (36). In order of importance, the major industrial uses include:

tanning of leather
protective coatings (paint and varnish)
pharmaceutical uses
lubricating oils
soap
insecticidal sprays
protecting metal surfaces

The use of fish oils for many industrial purposes is dependent on the market price as compared to other suitable oils. This is primarily due to the fact that many industrial uses of oils are not dependent on the properties of a specific oil or source of oil.

4.1 Tanning of Leather

Publications on the use of fish oil for industrial purposes are dominated by those discussing oil, or, more frequently, sulfonated fish oil that is being used in one step of leather tanning. Although the first paper describing this use of fish oil was published in 1919 (37), fish oils have been used for leather manufacture for many years prior. Until 1950, cod liver oil was considered to be the only fish oil that could be used in the leather industry. In fact, many early papers stated that only cod liver oil was suitable for tanning leather (36). The use of fish oils by the leather industry did not expand to include other fish oils until a 1959 study showed menhaden oil to be as good as cod liver oil (38).

A condition called fish oil spew, due to an oxidation product, causes an undesirable appearance of tanned leather (39). The undesirable reaction increases when iron is present. However, it never occurs during chrome tanning operations (40) and the problem was essentially eliminated by the addition of 0.3 g pyrogallol or hydroquinone or of 0.1 g of p-nitrophenol per 100 g of oil (41). Most of the publications on leather tanning have been since 1950. Most of these publications have been involved with the improvement of fish oil properties and in methods for testing or selecting oils for this use (36). Some of this information most likely was important to further development of uses for fish oil.

4.2 Fish Oils Used in Paint, Varnish, and Other Protective Coatings

Although there have been earlier utilization of fish oils for use in protective coatings, the value of this component was not widely recognized until 1928 when a research paper showed that paints and varnishes made from fish oils were of good quality and were resistant to sea-air and heat (42). Since that time, many papers have been published in which fish oils were shown to be better than vegetable oils in imparting durability and nonyellowing properties, to have properties that improve paint films under certain conditions, and to give good rust protection in coating materials used on steel (36). It is interesting to note that in 1939, a paper extolling the virtues of fish oils for improving drying used seal oil in the research (43).

4.3 The Use of Fish Oil for Pharmaceuticals

Fish oil has been of interest for its health benefits for at least the past 200 years. Percival (44) reported that in 1783 patients at Manchester Hospital in England consumed more than 500 lb/yr of cod liver oil as a successful treatment for arthritis. A more recent report of fish oil for pharmaceutical use from 1935 (44) reported that oils from salt and freshwater fishes taken internally reduce problems of leprosy. A 1944 French patent (893,112) (45) discussed the substitution of fish oils for fat in various pharmaceuticals or perfumes. The entire fat substitute consisted of fish oils mixed with methyl or ethyl esters of phthalic acid and spermaceti. No major references to fish oils for pharmaceutical purposes was published until 1988, after which numerous papers described the use or development of products containing fish oils. These uses include fish oil in moisturizing cream, foams, and whips for sore throats (provides long-lasting pain relief with minimum numbing of tongue and mouth), skin cleansing and lubricating products, burn treatment, compounds for treating allergic diseases, and other products based on the EPA and DHA component (36).

In 1990, fish oils were reported to have properties beneficial to patients with diabetes. Dietary fish oils benefit patients having diabetic nephorpathy (46). A more dramatic effect of fish oils used for enhancing renal hypertrophy reported that the "results indicate that dietary fish oil has profound effects on renal eicosanoid metabolism in experimental diabetes and that this may participate in the biological events which regulate diabetic renal hypertrophy" (47).

4.4 The Use of Fish Oils as Lubricants

There have been numerous applications whereby fish oils were used for industrial lubricants. Most of the publications have been over the past two decades. Table 6.2 lists patents from Europe, Japan, and the United States that cover some of these uses. In fact, most of the references and information regarding the use of fish oils in lubricants are patents. Technical papers have been published on using fish oil to reduce viscosity (48), impart a well lubricated feel to leather (49), and improve lubricant oil for metal working (50).

Table 6.2 Patents involving the use of fish oils in lubricants

Country	Patent Number	Date	Subject of Patent
Japan	99,215	1/26/33	Lubricating oils
United States	3,175,972	3/30/65	Lubricant compositions
France	1,465,982	1/13/67	Lubricating pastes for cold forming of metals
United States	3,702,822	11/14/72	Extreme pressure lubricant additives
Poland	89,393	9/17/77	Mold lubricant for concrete structures
Czechoslovakia	171,393	2/15/78	Composite lubricant of collagen fibers
Former Soviet Union	SU 958,481	9/15/82	Lubricant for hot working of metals by pressure
Poland	PL120,723	3/31/82	Lubricants for cold working of metals
Japan	62,240,389	10/21/87	Gear oil compositions containing antioxidants and extreme pressure additives
Japan	87,240,389	10/21/87	Gear oil compositions containing antioxidants and extreme pressure additives
Europe	410,849	1/30/91	Polysulfide composition and their preparation and utilization as lubricant additives

4.5 The Use of Fish Oils in Soap

In the early 1900s, a major use of fish oils was in soapmaking. This has declined over the years until today it is of only minor importance. The early papers involving the use of fish oils in soap included preventing fish odors in soap (51–54), reducing lathering (55), and increasing cold lathering (56).

4.6 The Use of Fish Oils in Insecticidal Sprays

During the late 1920s, there was a considerable amount of research in which fish oils were used as adhesive for lead or arsenic sprays. This followed work showing that lead could be eliminated from these sprays (36). Much of the published work was carried out by Washington State University, Pullman, Washington, as related to the use of fish oil in sprays for apples. It has recently been demonstrated that hydrolyzed fish fertilizer containing emulsified fish oil is effective in reducing the pesticide spray required for some crops (57). This is certainly an area that should be further developed since this may be

Marine Oils in Industrial Products **237**

a means of significantly reducing the use of certain undesirable pesticides. With the use of fish fertilizer increasing as an effective substitute for many petrochemical fertilizers, the growing interest in reducing the run-off of chemicals not kind to the environment should further the interest in the fish fertilizer containing controlled amounts of emulsified fish oil.

4.7 Cutting Oils for Cleaning and Protecting Metal Surfaces

Most of the value in fish oils for lubricating and protecting metal surfaces has been reported during the past decade or so. As in the case of using fish oils for lubricants, most of the information has been disclosed in patents as shown in Table 6.3.

4.8 Other Uses

There have been numerous other industrial uses of fish oils published in the literature (36). However, most of these have been of rather minor importance

Table 6.3 Patents involving the use of fish oils in metal protection

Country	Patent Number	Date	Subject of Patent
Former Soviet Union	189,309	6/17/66	Wetting agent
Former Soviet Union	209,610	1/26/66	Ground coat for rusty metal surfaces
Germany	2,153,672	10/71	Coating and sealing compositions for protecting motor vehicle underside against corrosion and flying rocks
Poland	64,696	5/15/72	Agent for steel surface priming
Former Soviet Union	958,481	9/15/82	Lubricant for hot working of metals by pressure
Poland	120,723	3/31/82	Lubricant for cold working of metals
Japan	61,666,793	3/28/86	Maleated fish oil for metal working lubricating oils
Former Soviet Union	SU 342,917	1/17/86	Cutting fluid for cold working of metals by pressure
Japan	62,257,488	11/4/87	Cutting oil emulsions
Japan	91 17,188	1/25/91	Cold rolling oils for steel plates
Europe	PREP 10,849	1/30/91	Polysulfide compositions from unsaturated fatty acids and their use as lubricant additives

238 Marine Oils

as compared to the uses discussed above. Fish oils have been used as bases for conversion to other chemicals, ore flotation, softening agents for textile finishing, cosmetics, fuel, printing inks, ceramic moldings, and core oils for foundries.

The use of fish oil for fuel has increased somewhat over the past decade in high sea fleets processing fish. This is especially true of the surimi operations where the oil from fish is removed during washing of the flesh. Although the fish, such as pollock, used for surimi are low in oil content, the large volume of fish processed results in considerable oil recovered from the wash water. It has proven more economical under present market conditions to burn the oil with the ship fuel than to transport it to the currently low-priced markets.

FISH OIL IN ANIMAL DIETS

5.1 Land Animals

Although there is a considerable amount of fish oil consumed in the fish meals fed to animals, the specific feeding of fish oil has not been practiced to a large extent. Fish meal containing fish oil is fed in the diets of poultry, pigs, fish, crustaceans, ruminants, fur-bearing animals, and pets. Special efforts are being made to improve the quality of the protein and oil in fish meal so that many commercial animals, currently not fed significant amounts in their diet or needing better quality meals, can utilize this high protein supplement (58,59). These products, produced from menhaden, have proven to be beneficial for early-weaned pigs, high yielding dairy cows, and aquaculture raised fish.

Poultry could be a major source of n-3 fatty acids for humans if their diets include high quality fish oil high in HUFA content. Recent studies have indicated that chickens can be a source of n-3 fatty acids equal albeit different to that in cod fish (60–62). These studies have shown that significant amounts of n-3 HUFA can be incorporated in poultry diets without affecting the meat flavor. However, it should be emphasized that only specially processed fish meal to minimize heat degradation and oil oxidation is utilizable. Many conventional meals are limited as to the amount that can be fed in poultry diets or the fish flavor, caused primarily by rancid oil and other degradation products, will be transferred to the meat.

Since a large part of the world's population consumes poultry, especially chickens, as a significant portion of the meat protein in their diets, considerable effort is being expended to produce low-cholesterol, high n-3 fatty acid eggs from layer hens (63,64). Up to 6% refined menhaden oil can be added to the total diet without affecting the flavor (65). A layer hen diet containing 3% menhaden oil was shown to increase the EPA and decrease the ratio of ω6 to ω3 from 18 to 3 (66). Other experiments have shown that regular menhaden oil, stabilized with antioxidants, could be fed to layer hens at a level of 3% of the diet without causing a fishy flavor (67).

5.2 Aquatic Animals

Oil from terrestrial plants are high in *n*-6 fatty acids while oil from aquatic plants contain significant levels of *n*-3 fatty acids. While many land plants have significant amounts of C:18*n*-3 PUFA (e.g., linolenic acid, C18:3*n*-3), it is the C20:5*n*-3 and C22:6*n*-3 HUFA (eicosapentaenoic acid and docosa hexaenoic acid respectively) in aquatic foods that are the significant fatty acids in reducing certain premature heart disease, inflammatory disorders, and certain other health problems. Wild fish in marine and fresh waters have high HUFA content in their bodies due to the consumption of aquatic plants. Aquaculture-raised fish do not have natural HUFA containing oils unless they are fed *n*-3 fatty acids in their diets. As the consuming public becomes more aware of the beneficial effects of consuming *n*-3 fatty acids present in fishery products, and the proportion of aquaculture-raised fish increases, it is increasingly important that aquaculture fish be fed diets that contain this important ingredient (68). One good source of essential fatty acid is fish meal made from fish; however, an essential fatty acid content alone can be misleading (Table 6.4) since the meal in a low-fat fish (e.g., white fish) is actually much lower in total fat than fat fish, such as herring, anchovy, and menhaden. Table 6.5 shows the comparison of HUFA between wild fish and farmed fish that have no significant HUFA in their diets. Although the wild fish have much higher HUFA content than farmed fish, note that farmed crayfish have about the same HUFA as wild crayfish. This is due to the fact that crayfish are in contact with the bottom of the ponds and tend to eat algae and other naturally growing vegetation. Table 6.6 shows results of feeding wet diets to hatchery rainbow trout (*Salmo gairdneri*). These diets, containing a high percentage of fresh fish portions, are representative of those normally prepared and fed to hatchery salmon and trout. Although these data do not represent that from a precisely controlled test, since there is normally a difference in the moist diet oil content over the entire period of raising fish, the HUFA in the feed does reflect a much higher HUFA content than commercially raised trout fed a dry diet. Table 6.7 also shows the favorable comparison of HUFA in wild and cultured

Table 6.4 Essential fatty acid content of oil in selected fish meals (59)

Essential Fatty Acid	Herring (%)	White Fish (%)	Anchovy (%)	Menhaden (%)
C18:2*n*-6	2	1	1	1
C18:3*n*-3	1	1	1	1
C18:4*n*-3	2	2	2	2
C20:4*n*-6	1	NA	1	1
C20:5*n*-3	6	12	16	12
C22:5*n*-3	1	2	2	3
C22:6*n*-3	13	19	14	9

Table 6.5 Relationship between n-3 and n-6 content of oil from edible portions of wild versus pond-reared shrimp, crayfish, and catfish (69)

Species[a]	Total PUFA (%)	Fatty Acids (%)		Ratios	
		n-6	n-3	n-3/n-6	HUFA/n-3
Marine shrimp	45.15	16.88	28.28	1.67	1.33
Pond-reared prawns	41.64	23.04	18.60	0.81	0.66
Wild crayfish	50.12	16.38	33.74	2.06	1.55
Pond-reared crayfish	47.50	16.64	30.84	1.86	1.49
Wild catfish	39.77	12.13	27.64	2.54	2.00
Pond-reared catfish	26.07	15.85	10.22	0.62	0.48

[a] Scientific names not given.

fish when the cultured fish are fed diets containing significant quantities of ω3 fatty acids.

There is also a large variation between n-3 and n-6 fatty acid content in wild fish as shown in Table 6.8. In addition to the variations between different species, there are also major changes within a given species or group at different periods of growth and development, especially during the spawning cycle. This most likely accounts for many of the differences in research involving the effects of HUFA in human and animal nutrition, especially during the early work when the product was simply cited as fish oil. It is extremely important that a complete analysis of any given fish oil be made prior to feeding in nutritional or clinical tests. Furthermore, over long extended tests, the oil should be periodically analyzed to check on chemical and oxidative changes taking place over time.

Table 6.6 Fatty acid groups (% of total fatty acids) of oil from edible portions of hatchery trout (Salmo gairdner) fed wet diets containing fresh fish portions (68)

Fatty Acids	Fish (Trout)	Diet[a]
Saturated	28.27	18.90
Monounsaturated	53.82	32.27
n-6 PUFA	2.92	6.24
n-3 PUFA	14.86	36.76
C20+C22 HUFA	14.72	36.76
Total PUFA	17.78	43.00
n-3/n-6 Ratio	5.09	5.89
C20+C22 HUFA/n-6	5.04	5.90

[a] Wet diet containing approximately 50% frozen commerical fish scrap and carcasses of wild hatchery return salmon.

Table 6.7 Fatty acid composition of oil in edible portion of carp (Cyprinus carpio) rainbow trout (Salmo gairdner), and eel (Anguilla japonica) (70)

Fatty Acid	Fatty Acid in Oil (%)					
	Carp		Trout		Eel	
	Cultured[a] (%)	Wild (%)	Cultured[a] (%)	Wild (%)	Cultured[a] (%)	Wild (%)
14:0	2.0	1.4	1.4	0.4	3.8	4.3
16:0	22.2	19.5	24.8	22.8	22.9	20.1
16:1n07	7.4	8.1	3.7	9.5	8.1	12.5
18:0	4.7	7.0	5.2	4.3	3.8	3.9
18:1n-9	32.8	22.9	16.6	21.6	46.1	39.8
18:2n-6	15.2	6.0	10.1	4.9	2.0	3.1
18:3n-3	1.1	1.8	0.9	6.9	0.3	1.4
20:1-9	2.6	1.6	1.0	0.8	3.5	1.0
20:4n-6	0.9	7.5	1.5	1.7	0.3	1.8
20:5n-3	2.5	7.0	5.2	5.3	2.3	2.9
22:6n-3	6.0	7.0	25.8	11.7	4.0	2.7

[a] Cultured fish fed diets high in *n*-3 fatty acids.

5.3 Adding Refined Fish Oil to Diets

Due to the economics of production and logistics, most commercial fish diets are sold in the dried or stable forms. The normal source of high protein is fish meal and vegetable meals (e.g., soya, wheat, cottonseed, etc.). However, there is normally insufficient HUFA from fish meal and no HUFA from

Table 6.8 Relationship between n-3 and n-6 fatty acid content in the oil from commercial wild fish[a] (68)

Species[a]	Fatty Acids (%)			Ratios	
	n-3	C20+C22	*n*-6	*n*-3/*n*-6	HUFA/*n*-6
Menhaden[b]	29.20	25.68	1.02	28.63	25.18
Salmon, minced[c]	38.52	15.42	1.12	29.93	13.77
Salmon, pink[c]	29.15	18.84	0.92	31.68	20.48
Salmon, sockeye[c]	28.70	16.17	0.99	29.00	16.33
Hake Pacific[c]	22.29	12.33	0.10	222.90	123.00
Herring[b]	19.25	6.79	0.49	39.29	13.86
Shark[c]	27.39	16.48	0.66	41.50	24.97
Shark liver	17.93	6.36	0.42	42.69	15.14
Cod[c]	49.31	48.01	0.47	104.90	102.20

[a] Scientific names not available; commercial lots taken at random.
[b] Commercial oil from whole fish.
[c] Edible portion.

vegetable and grain meals or flour in these diets. As the quantity of farmed fish increases there will be mounting pressures by wild fish producers to emphasize the differences in nutritional value of the products. With the growing expectation by the consumer that fish are a good source of ω3 HUFA, it is important that fish oil or other foods from aquatic sources be included in fish diets. With the improvements in refining and handling fish oils, resulting in good shelf life and stability, the addition of fish oil to aquaculture fish diets is now a practical means of ensuring that wild and aquaculture fish are comparable sources of HUFA.

FISH OIL IN HUMAN DIETS

It has only been during the past 50 years, and particularly the past decade or so, that lipids contained in aquatic plants and animals have been considered important in human nutrition. With the present knowledge that HUFA fractions of these lipids can be important in preventing or reducing certain premature heart diseases, inflammatory disorders, and many other health problems, there has been a growing interest in seafoods (aquatic foods) with this important lipid component (71). Humans have been consuming many foods from the oceans and freshwaters of the world since recorded times. However, with the world population increasing at approximately 86 million persons per year, and the present world catch having reached the maximum sustainable yield of approximately 100 million metric tons per year, ways are being sought to increase the availability of this valuable source of protein and lipids (72). Improving the utilization of raw materials, improving processing technology and quality control, using currently underutilized species, and improving longer term storage for saving products during oversupply are increasing availability of present harvests. However, there is a definite limit to these measures. The ultimate solution for increasing aquatic food must come from aquaculture.

In the past, major products from aquatic waters, fish and shellfish, have been considered good sources of high-quality proteins without consideration of the lipid fraction. With the growing knowledge about the healthful value of fish oils, interest has concentrated on the source and quality of these oils. As has been discussed, this is emphasizing the need for controlling the diets of aquaculture fish to ensure sufficient HUFA in the oil. As shown in Tables 6.9 and 6.10, even the amount of HUFA being consumed in wild fish can be misleading. The HUFA in lean fish being eaten by a large portion of consumers is much lower than that in fatty fish.

Of particular importance is the insurance that the highly unsaturated oils in the fish being offered to the buyer do not degrade by oxidation prior to consumption. Modern packaging, holding, freezing, and processing techniques are quite satisfactory for maintaining all of the desirable components in fish until it reaches the consuming market. Extracted lipids, currently available

Table 6.9 N-3 fatty acid content in fillets of fatty fish available in U.S. markets (73)

Species	EPA g/100g Fillet	DHA g/100g Fillet
Mackerel, Atlantic	0.65	1.10
Mackerel, Pacific (eviscerated)	1.10	1.30
Salmon, Atlantic	0.18	0.61
Salmon, Chinook	1.00	0.72
Salmon, chum	0.24	0.31
Salmon, coho	0.82	0.94
Salmon, pink	0.64	0.86
Salmon, red	1.30	1.70
Trout, rainbow	0.22	0.62
Tuna, alvacor (white meat)	0.63	1.70
Average	0.63	0.99

from conventional meal operations and certain other extraction processes, must be refined and stabilized before they are acceptable for constituents in human food.

6.1 Stabilizing Fish Oils for Human Consumption

As the desire to increase HUFA consumption continues, considerable scientific and commercial efforts are being made to utilize fish oil in a manner comparable to that of vegetable oils. Hence, major goals are the extraction, refining, stabilization, and packaging of fish oils that are acceptable as components of many types and forms of food products. This is an interesting change of events in that for many years the loss of markets for fish oil having high vitamin content, the low price of edible vegetable oils (e.g., soy, peanut,

Table 6.10 N-3 fatty acid content in fillets of lean fish available in U.S. markets (73)

Species	EPA g/100g Fillet	DHA g/100g Fillet
Bass, striped	0.17	0.47
Cod, Atlantic	0.08	0.15
Cod, Pacific	0.07	0.12
Flounder, yellowtail	0.11	0.11
Haddock	0.05	0.10
Sole, lemon	0.09	0.09
Average	0.68	1.00

sunflower, cottonseed), and the disallowance of fish oil for human consumption by the U.S. Food, Drug, and Cosmetic act of 1938 relegated fish oil to a minor position in the human food market. The exception to this was the large-scale use of fish oil for margarine, in countries other than the United States, that involves the hydrogenation of the oil, significantly reducing most of the HUFA content. The first change in the official status of fish oils in the United States was the 1988 FDA affirmation of generally recognized as safe (GRAS) status for partially hydrogenated and hydrogenated menhaden oil as a direct human food ingredient. This encouraged research to improve the acceptability of fish oils in foods and further FDA applications for using all refined fish oils other than menhaden.

Other than the objectionable off-flavor caused by oxidation, fish oils contain minor amounts of nontriglyceride substances that cause discoloration, foaming, and precipitation of solids (74).

Young (75) classified the nontriglycerides in fish oils according to their effects, as follows:

1. Hydrolytic: moisture, insoluble impurities, free fatty acids, mono- and diglycerides, enzymes, and soap.
2. Oxidative: trace metals, oxidation products, pigments, tocopherols, and phospholipids.
3. Catalyst poison: substances that inhibit the hydrogenation reaction, for example, phosphatides, oxidation products, and compounds containing nitrogen, sulfur, and halogens.
4. Miscellaneous: hydrocarbons, resins, sterols, waxes, trace metals, and sugars whose effect is less well known but can be classified as contaminants and also may have an effect on the final flavor of the oil.

The general processing steps used to purify fats and oils by reducing or removing the impurities are shown in Figure 6.2 (76,77). In recent years, additional steps have been added to various processes to remove other natural and environmental contaminants and to make the taste palatable. These include vacuum steam stripping, steam deodorization, and supercritical fluid extraction. It should be emphasized that the basic process for producing a clear, edible fish oil was developed to prepare oil for subsequent hydrogenation to make margarine. Even today, the majority of fish oil being refined for human consumption is destined for margarine. This product has few of the HUFA advantages since the hydrogenation process saturates and solidifies many of the fatty acids in the triglycerides.

Degumming. Degumming is often not carried out with animal fats and fish oils because they are low in phosphatides. In some refineries, however, an acid pretreatment designed to hydrate gums and remove phosphorous and other trace metals is applied as the oil enters the alkali-refining plant. The

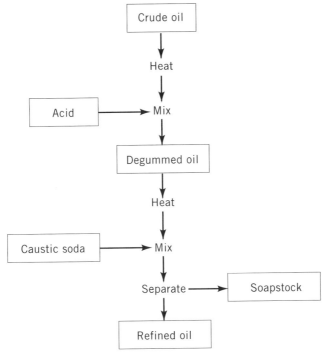

Figure 6.2 Basic fish oil refining process.

pretreatment of fish oils with phosphoric acid ahead of caustic refining is standard practice in Europe. From 0.002 to 1% phosphoric acid (85%) is added at temperature of 70–90°C (158–195°F) and mixed for one minute (78). In the United States, one of the major processors of fish oil prefers to carry out the commercial degumming process for at least 45 min. at 71°C (160°F) in a 40% phosphoric acid soloution (79). The completion of the process results in oil that has a greenish cast and a significant increase in free fatty acids.

Refining Methods. The basic refining process consists of adding caustic soda to the crude oil to remove some of the nontriglycerides previously discussed. Sodium hydroxide is responsible for removing the contaminants as follows:

The gums absorb the alkali and are coagulated by hydration.
The insoluble material is entrained with other materials that have been coagulated.
The alkali combines with any free fatty acids present in the oil and forms water-soluble compounds.
Much of the coloring matter is degraded, made water soluble by the alkali, or absorbed on the gums.

The solvent extraction of vegetable oils has been practiced for many years and is a possible technique that could be used to extract oil from fish. There are many advantages to solvent extraction in that the process can be designed so that oil does not have to be heated to the high temperatures required in conventional fish meal processing. Although a high quality fish oil can be obtained by solvent extraction techniques, the choosing of a solvent that is approved for food-grade products does limit the chosen solvent. This is demonstrated by the research program backed by what is now the U.S. National Marine Fisheries Service, National Oceanographic and Atmospheric Administration. During the 1960s and 1970s, alcohol, a polar solvent that is not efficient for extracting oil, was used to extract oil from fish since it was considered safe. The entire program, costing the taxpayer a tremendous amount of money, was a complete failure due to the poor economics of the process. However, solvent extraction of fish oil could be a major area of commercial interest for initial extraction when final refining techniques are improved to the point where it is commonly accepted in food products.

Another technique of refining has received a considerable amount of attention. Preparation of fish oil for human consumption has been carried out using supercritical carbon dioxide extraction. Carbon dioxide, under high pressure and above its critical temperature of 31°C, has solvent properties for purifying oil. It is selective for the removal of odor compounds, pigments, and products of autoxidation that contribute to off-flavors (80). However, it is possible that carbon dioxide removes such compounds as natural antioxidants (tocopherols and phospholipids), and there are some concerns about the residual trace metals, pesticides, cholesterol, or other nontriglyceride compounds (81). In addition, the current state of the processing technique is not economically practicable.

There have been numerous other methods suggested to improve the efficiency of refining fish oils, including physical methods and gas chromatography. A method that shows good potential for commercial refining of fish oils was developed at the University of Washington (82) and is currently being refined for pilot plant production. This process, which involves the passing of the oil through a cation strong-acid microporous resin, has produced oils that made high quality mayonnaise and oil and vinegar dressings.

Bleaching. Bleaching involves the absorption of color components by activated clay. This not only improves the color, but results in a better flavored oil with better stability to oxidation. Impurities created during the previous refining step, such as soap, are also removed by the activated clay. Although the process is carried out in batch and continuous operations at atmospheric pressure or under vacuum, the vacuum reduces the amount of clay required, and results in less oxidation and improved oil color. The resin process (82) incorporates the refining and bleaching operations in one step, significantly improving both the oil quality and the economics of operation.

Processes to Meet Special Market Requirements. Health food and pharmaceutical markets have been developing over the past decade in response to demands for *n*-3 fatty acids as "heart healthy" food supplements. In addition to the interest in EPA as a major contributor to prevention of heart disease, interest in DHA is also increasing as it is being shown to be an important factor in other health factors, especially those involving the immune system.

With the interest being the specific *n*-3 fatty acid components in fish oil, the amount of these components in a given species take on added importance. Stansby (83) has prepared an excellent summary on the sources and health effects of *n*-3 fatty acids. The oil content in edible portions of fish from different species has been shown to vary from as little as 0.3% to 14%, or greater. Also, the oil content varies over a wide range during different periods of the growth and spawning cycles. There is a major problem with controlling the *n*-3 fatty acid content of a diet since there is tremendous variation between species and within the same species. Many of the values for fatty acid content available in the literature were using only a few fish taken at the same time or in the same location and are extremely erroneous when the oil composition is accepted as a fixed value for a species. Table 6.11 shows the average and range of oil content of 17 common species of fish and Table 6.12 shows the specific range of fatty acids in menhaden oil. The variation of oil in different portions of

Table 6.11 Average and ranges of oil content of 17 common species of fish[a] (83)

Species	Range Oil Content (%)	Average Oil Content (%)
Carp	1.0–12.0	5.0
Chub, lake	4.0–13.0	8.5
Cod, Atlantic	0.2–0.9	0.4
Flounder	0.3–3.4	1.4
Haddock	0.2–0.6	0.35
Mackerel, Atlantic	2.7–25	14.0
Ocean perch, Pacific	3.0–6.0	4.2
Perch, yellow	0.8–1.2	0.9
Pike, yellow	0.8–3.0	1.3
Rockfish (Sebastes)	1.2–4.3	2.5
Salmon, chum	2.2–7.3	4.0
Salmon, coho	3.3–11.2	7.0
Salmon, pink	3.2–11.6	6.5
Salmon, sockeye	7.8–13.7	11.3
Smelt, marine	4.6–8.8	6.3
Smelt, lake	1.5–3.3	2.4
Whitefish, lake	4.7–18.8	9.6

[a] Average value based upon fish taken during the main industrial harvest season, although even then a fairly large range of fat content usually occurs.

Table 6.12 Range of fatty acid content in large batches of commercial menhaden oil from 1977–1988 (83)

Fatty Acid	Range (% of Total FA)	Ratio (Highest/Lowest)
C14:0	7.2–12.1	1.7
C15:0	0.4–2.3	5.8
C16:0	15.3–25.6	1.7
C16:1	9.3–15.8	1.7
C16:2	0.3–2.8	9.4
C16:3	0.9–3.5	3.9
C16:4	0.5–2.8	5.6
C17:0	0.2–3.0	15.0
C18:0	2.5–4.1	1.6
C18:1	8.3–13.8	1.7
C18:2	0.7–2.8	4.0
C18:3	0.8–2.3	2.9
C18:4	1.7–4.0	2.4
C20:0	0.1–0.6	6.0
C20:4	1.5–2.7	1.8
C20:5	11.1–16.3	1.5
C22:1	0.1–1.4	10.0
C22:5	1.3–3.8	2.9
C22:6	4.6–13.8	3.0

edible flesh is shown in Table 6.13. Stansby (83) has also summarized the use of fish oils for many diseases for which PUFA has a proven or possible beneficial effect and has clarified the problems with using published data for calculating the amount of n-3s consumed in a given species. There are many research projects currently being carried out to study the effects of fish oil and n-3 fatty acids on heart disease, cancer, blood pressure, eye and brain disorders, skin diseases, arthritis, diabetes, malaria, and others. The work to date has established the specific value of n-3 fatty acids for health, especially coronary heart disease. However, the jury is still out as to the exact relationship

Table 6.13 Variation in the oil content of different portions of the edible flesh of the same fish (83)

Species	Section Analyzed	Oil
Red King salmon	Near head	20.2
Red King salmon	Near tail	11.1
White King salmon	Near head	15.1
White King salmon	Near tail	8.1
Yellowfin tuna	Middle	3.2
Yellowfin tuna	Near tail	1.4

between the specific n-3 fatty acid and the amount and/or form necessary to prevent or treat many diseases.

The extreme variation in oil content of fish somewhat limits control of the amount of HUFA that is consumed by eating fish. This has created a demand for fish oil containing a known amount of HUFA. Techniques to increase the n-3 content by solvent extraction, cold filtration, enzymatic digestion, and combinations of these and other processes have been developed to satisfy this demand. The resulting concentrated product, protected against oxidation by approved antioxidants, is available in gel capsule form for the health food market.

As previously discussed, the Norwegians have pioneered work in improving the effectiveness of PUFA in fish oils (26–29,84) based on indications that the position of the n-3s on the triglyceride are an important factor in the effectiveness of EPA and DHA on coronary heart and other diseases. This has stimulated research in controlling the position as well as increasing the amount of HUFA in a given triglyceride. Also, improved separation techniques and more gentle processing methods will allow further work with other compounds in fish oils that may be important to the pharmaceutical and health food industries in the future. Such compounds as squalene (C_3OH_5O), found in certain types of shark liver and a good oil for skin creams; alkoxyglycerol in Greenland shark which increases the production of white blood cells and improves the immune system; and other interesting compounds found in fish oil and fish liver oil should be made available in commercial quantities.

Microencapsulation of fish oil is a physical process that holds potential for making fish oils more usable to the food industry (85). This process can convert fish oils to a more convenient powdered form for adding and mixing with other ingredients in a food product. Based on forming microcapsules with a continuous uninterrupted film, the product can be made as a free flowing powder or as particles suspended in water.

6.2 Marine Oils as Food Supplements

Flavors and Odors. A major problem with incorporating fish oils in foods is the susceptibility to oxidative deterioration which affects the flavor and odor of the oil, often making it unacceptable as a food component. Off-flavors and oil stability have been of interest to scientists for some time. Autoxidation of unsaturated fatty acids is the most important oxidation occurring in food systems in absence of light and enzymes. Hydroperoxide, the major initial reaction products, are highly reactive. The three stages of the reaction involve the production of free radicals by thermal dissociation, hydroperoxide decomposition, or exposure to light (initiation); rapid production of peroxyl radicals which slowly react with the substrate (propagation); and interaction of peroxyl radicals (termination). The resulting monohydroperoxides formed in this overall lipid oxidation are unstable and result in the ensuing reactions that account for many of the carbonyls, alcohols, esters, and hydrocarbons responsible for the oxidized flavor. Some of the many compounds found in a fish oil, including the chemical identification and the odor characteristic, are shown in Table 6.14. Experience has shown that consumers can detect extremely low levels of oxidation products in an oil. These products, that affect both flavor and aftertaste, are caused by degradation of phospholipids and other compounds as well as triglycerides.

The problem of oxidative degradation is increased by the large number of double bonds (up to six on a given fatty acid per triglyceride molecule) as compared to 2 or 3 double bonds per molecule in other animal and vegetable lipids. Hence, fish oils will deteriorate in the flesh of fish faster than the oils from other foods, but are especially susceptible to oxidation after they have been extracted. The greater number of double bonds in fish oil also reduces

Table 6.14 Some compounds with their chemical identification odor characteristic, and quantity (crude menhaden oil) (83)

Chemical Identification	Odor Characteristic	Concentration (Parts/Billion)
(E)-Butenal	Painty	130
Hexanal	Cut grass, green	1380
(E)-Pent-2-enal	Greasy, musty	980
Heptan-3-one	Sickly sweet, cooling	8530
Heptanal	Waxy green, grassy	410
(E)-Hex-2-enal	Sharp, green, oily	540
1,3,5-Trimethylbenzene	Pesticide-like	20
1,2,4-Trimethylbenzene	Pesticide-like	40
Octanal	Citrus, fatty, orange	240
(E)-Hept-2-enal	Sharp, green, greasy	140
1,2,3-Trimethylbenzene	Pesticide-like	20
Nonan-2-one	Musty with citrus topnote	620
Nonanal	Fatty floral	500
(E)-Oct-2-enal	Musty, waxy, floral	280
Acetic acid	Irritative, vinegar-like	1380
Hepta-2,4-dienal	Vegetable green	1880
Decanal	Sweet, green fruity, fatty with citrus topnote	640
Benzaldehyde	Cherry almond, sweet fruity	60
Nonenal	Fatty, waxy, musty	200
Propanoic acid	Astringent, acidic	2410
Isobutanoic acid	Sweaty, dirty socks	100
Butanoic acid	Dirty socks	8110
Pentanoic acid	Parmesan cheese, dirty socks	330
Hexanoic acid	Sweaty, dirty socks	810
Decatrienal	Oxidized fish oil	70
Phenol	Medicinal disinfectant	110

Table 6.15 Tocopherol (vitamin E) content in the oil of flesh from several species (83)

Species	Tocopherol Content (Micrograms/gm Oil)
Sardine	40
Menhaden	70
Tuna	160
Herring	140
Whale	220
Sablefish	630

the effects of both natural (e.g., tocopherol) and added antioxidants. However, there is a wide variation of tocopherol in different fish species as shown in Table 6.15. The effect of the amount of natural tocopherol is emphasized in that sardine oil is one of the most unstable (40 micrograms per gram of oil) while sablefish oil (630 micrograms per gram of oil) is one of the most stable.

Not all off-odors and flavors in fish oil are caused by oxidative degradation of the oil. Fish or fish portions used for fish meal are often partially spoiled prior to processing. This is caused by not holding the raw material at adequate low temperatures prior to processing. This most often occurs in warm climates or when fish must be hauled for long distances prior to processing. Although immediate icing of the fish to reduce temperature can alleviate these problems, the cost of ice is often prohibitive in warm areas where large ratios of fish-to-ice is required. In these cases, spoilage of the fish results in protein degradation releasing trimethylamine and/or ammonia or, in advanced stages of decomposition, sulfur compounds (e.g., hydrogen sulfide) that impart medium to extremely disagreeable taste and odor to the oil.

Forms of HUFA for Human Consumption. As the information about the benefits of HUFA continues to be publicized, many people are interested in increasing their HUFA intake. The Danish government has taken a lead in recommending HUFA intake in that the National Food Agency recommends a daily intake of 30 grams of fish. Among other things, this recommendation was the result of an extensive epidemiological investigation carried out over 20 years on 852 healthy men in their forties (86). The results were that there was more than 50% reduction in mortality from vascular disease by eating as little as 30 grams of fish (mixed species) per day. They point out that the 30 grams of mixed specie fish corresponds to 350 mg/day of EPA/DHA (86).

Although, like the Greenland Eskimos, high levels of HUFA can be ingested by eating large amounts of fish, many people do not eat fish. This can be due to not having learned to combine fish in an overall diet pattern or simply that, as is the case of many foods, the people do not like fish. Taking fish oil is certainly not the solution for this group to ensure an intake of HUFA since the flavor would be quite revolting to them. In fact, many of the older generation remember the trying days when they took cod liver oil for vitamin D. Today, fish oil capsules are available for people who do not eat fish or who want to ingest higher levels of HUFA. Fish oil in gel capsules are widely available and can be taken without detecting a fishy odor or flavor.

Probably the best means of incorporating fish oil in foods is by the addition of a high-quality refined oil in the microencapsulated form. The state of microencapsulation is developed to the extent that highly sophisticated powder can be prepared that does not affect the taste, smell, or shelf-life of any food to which it is added (87). The technology of preparing high-quality oils and introducing them into food is now well on its way to becoming a regular practice over the ensuing years.

REFERENCES

1. G.M. Pigott and B.W. Tucker, *Seafood: Effects of Technology on Nutrition,* Marcel Dekker, Inc., New York, 1990.
2. Anon., *Oil World* **34**(3), 1 (1991).
3. G.O. Burr and M.M. Burr, *J. Biol. Chem.* **82,** 345–367 (1929).
4. L.W. Kinsell, J. Partridge, L. Boling, S. Margren, and F.P. Michaels, *J. Chem. Endocrinol.* **12,** 909–913 (1952).
5. E.H. Ahrens, D.H. Blanhenhorn, and T.T. Tsaltas, *Proc. Exp. Biol. Med.* **86,** 872–878 (1954).
6. E.H. Ahrens, J. Hirsch, and W. Insul, *Lancet* **1,** 943–953 (1957).
7. W.E. Conner, *J. Clin. Invest.* **41,** 1199–1205 (1962).
8. W.E. Conner, J.C. Hook, and E.D. Warner, *J. Clin. Invest.* **42,** 860–866 (1963).
9. F.J. Stare, *Atherosclerosis* MEDCOM, Inc., New York, 1974.
10. A.M. Nelson, *Geriatrics* **27,** 103–116 (1972).
11. H.O. Bang, J. Dyerberg, and N. Hjorne, *Acta. Med. Scand.* **192,** 85–94 (1972).
12. J.E. Kinsella, *Food Tech.* **40**(2), 89–97 (1986).
13. J.E. Kinsella, *Nutr. Today* **21,** 7–14 (1986).
14. J. Dyerberg, *Nutr. Rev.* **44,** 125–134 (1986).
15. W.E. Lands, *Fish and Human Health,* Academic Press, Inc., Orlando, Fla., 1986.
16. W.S. Harris, *Contemp. Nutr.* **10**(8) (1985).
17. J. Nettleton, *Seafood Leader* (July/Aug. 1990).
18. P.J. Nestel, *Am. J. Clin. Nutr.* **43**(5), 752–757 (1986).
19. G.M. Pigott and B.W. Tucker, *Fd. Rev. Intl.* **3**(1&2), 105–138 (1987).
20. W.S. Harris, W.E. Connor, S.B. Inlelesm, and D.R. Illingwoth, *J. Metabolism* **33,** 1016–1019 (1984).
21. W.S. Harris, W.E. Connor, N. Alam, and D.R. Illingwoth, *J. Lipid Res.* **29,** 1451–1460 (1988).
22. D.R. Illingworth, W.S. Harris, and W.E. Connor, *Arteriosclerosis* **3,** 270–275 (1984).
23. P.J. Nestel and co-workers, *J. Clin. Invest.* **74,** 72–89 (1094).
24. B.E. Phillipson, D.R. Rothrock, E.E. Connorm, E.S. Harris, and D.R. Ilingworth, *New Engl. J. Med.* **312,** 1210–1216 (1985).

25. W.S. Harris, *J. Lipid Res.* **30,** 785–807 (1989).
26. E.O. Elvevoll, P. Moen, R.L. Olsen, and J. Brox, *Atherosclerosis* **81,** 71–74 (1990).
27. E.O. Elvevoll, H. Barstad, and R.L. Olsen, *Lipid Forum,* Sandefjord, Norway, Mar. 10–11, 1994.
28. B. Osterud, personal communication, Institute of Medical Biology, University of Tromsa, Norway, July 1994.
29. S. Jansson, T. Stram, R.L. Olsen, and E.O. Elvevoll in Ref. 27.
30. D.S. Fitzgibbon, *Historical Statistics—Fish Meal, Oil and Solubles,* U.S. Fish and Wildlife Service, Bureau of Commercial Fisheries, Current Fisheries Statistics no. 5105, 1969.
31. G.M. Pigott and B.W. Tucker, *Encyclopaedia of Food Science, Food Technology and Nutrition, Fish Oils,* Harcourt Brace Jovanovich, London, 1993.
32. R. Onarhein and A.O. Utvik, *IAFMM New Summary* **46,** 105–131 (1979).
33. G. Sand and J. Burt, *IAFMM Processing Bulletin* **2,** 11 (1987).
34. A.P. Bimbo, *Production of Fish Oil,* Van Nostrand Rienhold, New York, 1990, pp. 141–180.
35. T. Masumara and M. Sugahara, Poultry Sci. **64,** 356–361 (1985).
36. M.E. Stansby, *Industrial Uses of Fish Oils Particularly with Reference to Practices in the United States,* U.S. Department of Commerce, National Marine Fisheries Service, Northwest Fisheries Science Center, Seattle, Wash. 1993.
37. H. Schlosstein, *J. Am. Leather Chemists Assoc.* **14,** 41–49 (1919).
38. V. Mattei, *J. Am. Leather Chemists Assoc.* **54,** 12–27 (1959).
39. F. Strather, *Collegium,* 74–83 (1932).
40. M.P. Balfe, *I. Intern. Soc. Leather,* 436–441 (1939).
41. M.P. Balfe, *I. Intern. Soc. Leather,* 442–451 (1939).
42. A. Julia-Souri, *Quim. Ind.* **5,** 252–261 (1928).
43. B. Kulikov, *Khim. Referal Zhur* **2**(2), 118 (1939).
44. O. Calcagno, *Semanca Med. (Buenos Aires)* **2,** 557–562 (1935).
45. J. de Granville, Fr. Pat. 893,112, (May 1944).
46. U.O. Barcelli, *Am. J. Kidney Dis.* **16**(3), 244–251 (1990).
47. J.L. Logan, *Diabetes Res. Clin. Pract.* **10**(2), 145 (1990).
48. Y. Tanacka, *J. Soc. Chem. Ind. Japan* **39,** 235–236 (1936).
49. K.J. Kedlalya, *Leather Sci. (Madras)* **20**(7), 239–244 (1973).
50. Anon., *Infofish,* (Mar./Apr. 1992).
51. Anon., *Industrie Saponier* **3,** 1873 (1910).
52. F.R. Tappert, *Z. Deut Oel-Eu Ind.* **45,** 446–448 (1925).
53. M. Hirase, *J. Soc. Chem. Ind. (Japan)* **32,** 233–236 (1929).
54. M. Hirase, *J. Soc. Chem. Ind. (Japan)* **32,** 381–384 (1929).
55. M. Tsujimoto, *J. Chem Ind. Japan* **18,** 1062–1072 (1915).
56. C. Ellis, U.S. Pat. 1,178,142, (Apr. 4, 1916)
57. G.M. Pigott, *New Developments in Seafood Science and Technology,* Proceedings of conference, Canadian Institute of Food Science and Technology, May 11–14, 1994, in press.
58. A.P. Bimbo and J.B. Crowther, *JAOCS* **69**(3) (1992).
59. J. Opstvedt, *Fish Lipids in Animal Nutrition,* International Association of Fish Meal Manufacturers, technical bulletin no. 22, 1985.
60. R.R. Lamothe, H.R. Hulan, and F. Proudfoot, *N-3 News* **3**(1) (1988).
61. H.R. Hulan, W. Ackman, W. Ratnayake, and F. Proudfood, *Can. J. Anim. Sci.* **68,** 533 (1988).
62. H.R. Hulan, W. Ackman, W. Ratnayake, and F. Proudfood, *Poultry Sci.* **68,** 163 (1989).

63. W. Stadelman, paper presented at the *30th Annual Fisheries Symposium,* National Fish Meal and Oil Association, Baltimore, Md., Mar. 15, 1989.
64. M. Yu and J. Sim, *Poultry Sci.* **66,** 195 (1987).
65. S. Oh, T. Hsieh, J. Ryuie, and D. Bell, *FASEB Meeting,* Las Vegas, Nev., May 1988.
66. P.S. Hargis, M.E. Van Elswyk, and B.M. Hargis, *Poultry Sci.* **70,** 874 (1991).
67. Z.B. Huang, H. Leibovitz, C.M. Lee, and R. Miller, *J. Agric. Food Chem.* **38,** 743 (1990).
68. G.M. Pigott, *World Aquaculture* **20**(1) (1989).
69. P. Channugam, M. Boudres, and D.H. Hwang, *J. Food Sci.* **51,** 1556–1557 (1986).
70. H. Suzuki, K. Odazaki, S. Hayakawa, S. Wasa, And S. Tamura, *J. Agri. Fd. Chem.* **34,** 58–60 (1986).
71. G.M. Pigott and B.W. Tucker, *Fd. Rev. Intl.* **3**(1&2), 105–138 (1987).
72. J.W. Avault, Jr., *Aquaculture Mag.* **20**(4) 80–84 (1994).
73. J. Exler and J.L. Weihrauch, *J. Am. Diet. Assoc.* **69,** 23–48 (1976).
74. F.A. Norris in D. Swern, ed., *Bailey's Industrial Oil and Fat Products,* Vol 2, 4th ed., Wiley John & Sons, Inc., New York, 1982, pp. 253–314.
75. F.V.J. Young, *Fish Oil Bulletin 17,* International Association of Fish Meal Manufacturers, 1985, 27 pp.
76. F.V.J. Young, *Chem. Ind.,* 692–703 (Sept. 16, 1978).
77. FAO, *Dietary Fats and Oils in Human Nutrition,* Food and Nutrition, paper no. 3 Food and Agriculture Organization of the United Nations, Rome, 1977, 94 pp.
78. A.P. Bimbo in M.E. Stansby, ed., *Fish Oils in Nutrition,* Van Nostrand Reinold, New York, 1990, pp. 141–225.
79. B.H. Booth, Zapata Haney Corp., personal communication, 1994.
80. J. Spinelli, V.F. Stout, and W.B. Nilsson, U.S. Pat. 4,692,280, (1987).
81. V.J. Krukinis, *J. Am. Oil Chem. Soc.* **66,** 818–821 (1989).
82. C. Fernandez, Ph.D. Thesis, University of Washington, Seattle, 1986, 350 pp.
83. M.E. Stansby, *Fish and Fish Oil in the Diet and its Effects on Certain Medical Conditions,* special report published by the U.S. National Oceanic and Atmospheric Administartion, National Marine Fisheries Service, Northwest Fisheries Science Center, Seattle, Wash., July 1991.
84. B. Hjaltason, S.J.C. Somogyi, and H.R. Muller, eds., *Bibl. Nutr. Dieta. Basel, Karger* **1989**(12), 96–106 (1989).
85. S. Hegenbart, *Fd. Prod. Design,* (Apr. 1993).
86. D.B. Kromhout, E.B. Bosschieter, and C. de L. Coulander, *New Engl. Med.* **312**(1), 39–44 (1985).
87. D. Lauritzen, *IFI NR 1/2* (1994).

7
Fats in Feedstuffs and Pet Foods

Fats and oils are used for many purposes in animal feeds and pet foods, including the following:

Increasing caloric density of feeds (by about 2.25 times that of similar dry weights of proteins or carbohydrates).
Improving feed palatability and appearance.
Reducing total feed intake, increasing feed efficiency, and minimizing feed costs.
Increasing blood glycogen levels and endurance in working animals like horses and sled dogs.
Lowering the heat of reaction during digestion and metabolism—important for comfort and productivity of large animals in hot weather.
Delaying digestion of feedstuffs beyond the rumen by use of inert forms of fats and coatings.
Providing needed molecular structures through dietary essential fatty acids (EFA) and phospholipids.
Improving appearance of skin and hair, and prevention of dermatitis.
Modifying fatty acid profiles in "designer food" animal products.
Carrying fat-soluble vitamins and color compounds.
Binding heat-sensitive flavorings, vitamins, medications, and "instant gravy" mixes to pet foods and feeds after extrusion or drying.
Improving dispersion of dry mixes, e.g., lecithin in calf milk replacers.
Preventing segregation of mixed feeds.
Reducing dustiness of feeds and feeding operations and of grain elevator dust.
Lubricating feed processing machinery.

Nutritional values of fats and oils differ with fatty acid composition, with the species, age, physiological stage, environment of the animal, and adequacy of the overall diet. Care should be exercised in regard to the following:

Avoid overtaxing the fat-handling capabilities of digestion systems of the respective species, especially young animals and ruminants.

Avoid contamination of fats, oils, and oil-bearing materials with toxic or unwholesome constituents.

Handle oil-bearing feedstuffs with active antinutritional or toxic constituents, including mycotoxins, properly and legally.

Avoid excessive levels of fat that interfere with making cohesive pellets (more than 4%) or restrict the desired puffing of starches and development of lamellar soy protein textures (more than 6%) in extruded feeds.

Edible animal fat in the United States can be rendered only in food grade plants under inspection of the U.S. Department of Agriculture (USDA) (1). The majority of tallows and greases used domestically in animal feeds and pet foods are feed grade. The National Renderer's Association (NRA) describes rendering as a process that

> heats raw animal by-products to release fat (71.1–82.2°C; 160–180°F) and remove moisture (115.6–126.7°C; 240–280°F). Ninety percent of the fat is removed from the protein by presses, leaving approximately 10% in the protein meal. Fat quality is determined by hardness, color, moisture, impurities, stability, and free fatty acids (FFA) content.

In the United States, inedible fats are called *tallows* if, after saponification in the American Oil Chemists' Society's titer test (AOCS Cc 12-59), solidification occurs above 40°C (104°F), and *greases* if solidification occurs below this temperature (1). The demarcation temperature varies among countries and 38°C (100.4°F) is used by some. Throughout this chapter, the term *tallow* refers to both tallows and greases, which also may contain vegetable oils, recycled by renderers.

Fat and *oil* usually refer to semisolid and liquid forms of triacylglycerols (triglycerides), respectively, although the use of these terms is not consistent throughout industry. In this chapter, the term *fats* means the group of commercial lipids (mono-, di- and triacylglycerols, fatty acids, and phospholipids), whether in solid or liquid state. In endorsement of the movement to improve the value of by-products of various agribusiness operations, the term *coproduct* is used throughout this chapter, except where established regulations are being quoted.

Individuals who are not fats' and oils' chemists may benefit from the following orientation. Polyunsaturated fatty acids is abbreviated as PUFA. Essential fatty acids (EFA) are required by the host animal but are not synthesized and are available only through the diet. The notation C18 indicates a fatty acid chain, 18 carbons long. C18:0 and C18:3 identify the number of unsaturated ("double") bonds. C18:0 represents a completely saturated fatty acid, 18 carbons long, in this case, stearic acid. C18:3 represents a fatty acid, 18 carbons long and with 3 unsaturated sites, but does not indicate the location of the double bonds.

Linolenic acid (9,12,15-octadecatrienoic acid; IUPAC convention) is also 9,12,15-C18:3, meaning an 18-carbon fatty acid with double bonds occurring after the 9, 12, and 15 carbons counted from the carboxyl end of the chain; this is the major linolenic acid isomer and also is called α-linolenic acid. Another C18:3 fatty acid that has received considerable interest from nutritionists in recent years is 6,9,12-octadecatrienoic acid (γ-linolenic acid), actually a reduced member of the linoleic family. The symbol Δ has been retained from the Geneva naming convention for limited use here. For example, a $\Delta 5$ desaturase enzyme installs a double bond after the fifth carbon, counting the carboxyl carbon as number one. When working with polyunsaturated fatty acids, sometimes it is more convenient to count from the methyl end. The terms omega (ω) and n are sometimes used for this purpose. The notation C18:2n-6 signifies linolenic acid, an 18-carbon chain fatty acid with two unsaturated positions, the first one occurring after the sixth carbon atom counting from the methyl end. (Linolenic acid is represented by C18:2n-3.) Throughout this chapter, the all-cis forms of fatty acids are intended unless identified otherwise.

HISTORY

1.1 Evolution of Uses

Occasional feeding trials of extracted fats and oils to animals have been reported in the United States since about the 1890s when the first state agricultural experiment stations were established. However, significant quantities did not become available at prices affordable for volume feeding until the late 1940s. Feeding applications in the 1950s and 1960s focused on high energy broiler rations and on improving palatability of dry pet foods. Research on feeding fats and oils to swine and ruminants grew rapidly in the 1970s and 1980s, and was extended to aquaculture species in recent years.

1.2 Demands for Better Feeds and Foods

Improvements in genetics and biotechnology manipulations of metabolic systems, including feeding of growth-stimulating hormones (bovine somatotropin, BST, and porcine somatotropin, PST), have developed animals with increased capabilities for producing milk and meat. Increasing the caloric and amino acid intakes to realize these abilities are difficult with traditional feedstuffs, and calorie-dense and nutritionally balanced feedstuffs are increasingly sought.

Concerns by the public about the amounts and types of fats in the human diet have led to increased demands for poultry, fish, and leaner red meats. The latter need is being met by hand and mechanical trimming until animal genetics and feeding practices are developed to produce lean animals directly. Various snack and convenience food manufacturers, and fast food shops, have changed from frying in tallow to frying in vegetable oils. Currently, rendered animal fats and inedible recycled tallows are in surplus in the affluent countries and are sold at prices that place them among the lower cost sources of concentrated metabolizable energy for feed uses.

The types of fat fed affect the fatty acid composition of animal tissues (meat) and products (milk and eggs) as well as the fat metabolism in consumers of these products. This has led to the "designer foods" concept—animal products with fat compositions intentionally altered by feeding for human health benefits (2). Although nutritionists are not fully in agreement on optimum fatty acids profiles, designer eggs with increased contents of linolenic and other n-3 PUFA fatty acids are being marketed. The question of which fats or fractions are suitable for achieving the desired designer foods is materializing and may dominate feeds applications research for the next decade or longer.

1.3 Feeding of Whole Oilseeds

Oilseed producers have recognized that it may be more profitable to process and feed whole oilseeds on the farm, rather than sell them into the elevator–oil mill–refinery infrastructure and purchase extracted protein meals, tallows, and spent restaurant yellow greases (mixtures of animal fats and vegetable oils). Technologies have been developed to minimize the effects of natural toxic and growth-inhibitor components in raw cottonseed, soybeans, and canola seed. Feed processors and livestock feeders can now switch between using in-seed and extracted fats and oils, depending on price. Both options must be considered in a comprehensive review of feeding fats.

INFORMATION SOURCES, AUTHORITIES, AND OBLIGATIONS

2.1 Early Technology Transfer

The modern feeding of commercial animals and pets has become a quantifiable, information-based science with alternative choices and accompanying responsibilities. The exploration and codifying of animal nutritional requirements and feedstuff nutrient compositions have long been ongoing processes.

Major contributions to transferring knowledge have been made by the 22 editions of *Feeds and Feeding,* published from 1898 to 1956 (by W.A. Henry and later by F.B. Morrison). This bible was succeeded in 1978 by *Feeds and Nutrition—Complete* (3). Updated information is published in the more compact *Feeds and Nutrition Digest* (4). Other feeding and feedstuffs reference books also have been published during the years (5,6). The newer editions of books on feedstuffs and feeding management are more guarded in generalized statements about nutrients metabolism than publications of a decade ago. This is probably because different metabolic pathways are being found between species as well as alternative pathways within the same species.

2.2 National Research Council Nutrient Requirements

The developing knowledge about nutrient requirements of different species in various life stages has been monitored, interpreted, and summarized by expert subcommittees of the Committee on Animal Nutrition of the Board on Agriculture of the National Research Council (NCR). Various editions of *Nutrient Requirements* are currently available for beef cattle (7), cats (8), dairy cattle (9), dogs (10), fish (11), goats (12), horses (13), mink and foxes (14), nonhuman primates (15), poultry (16), rabbits (17), sheep (18), and swine (19). Each publication contains recommendations for various life stages of the respective species, a literature review and bibliography of feeding trials, and tables of ingredients commonly fed to that species. Respective nutrients composition, digestibility, and metabolizable energy values are available in the more recent editions. Gradually, the older general nutrients composition references, including the *United States–Canadian Tables of Feed Composition* (20), are being replaced by species-specific information.

Some of the recent *Nutrient Requirements* publications, notably for dairy cattle (8) and horses (13), include computer discs with interactive programs for calculating nutrients requirements based on age, weight, sex, physiological stage, and work or performance of the animal. Various compendiums of feedstuff compositions and animal requirements, based on NRC publications, are published periodically, including the annual *Feedstuffs Reference Issue.*

2.3 Association of American Feed Control Officials Regulations

NRC nutrient requirements are advisory and have no regulatory status. The Association of American Feed Control Officers (AAFCO) establishes the official definitions and names of feed ingredients, nutritional and labeling requirements for dog and cat foods, regulations on medicated feeds, labeling practices, and other requirements for selling feeds. AAFCO official ingredient names must be shown in listings on the product package, tag, or delivery invoice, in order of diminishing content for fixed ("closed") formula products and percentage-wise for open formula mixed feeds. Uniform regulations have been developed to expedite interstate commerce in feedstuffs, but each state also has reserved the right to establish independent regulations and interpretations. All commercial feedstuffs must be registered directly with the feed control official in each state where sold. AAFCO definitions and regulations are updated and published annually in the *Official Publication.* Copies can be obtained by contacting local feed control officials.

Ingredients and regulations continuously evolve. For example, the 1994 *Official Publication* includes the first listing and temporary status of feed grade "hydrolyzed sucrose polyesters"; these are nutrient-containing coproducts from the manufacture of low calorie fat substitutes. The establishment of an ingredient definition in the *Official Publication* is not an endorsement of its efficacy for specific nutritional objectives, but rather affirmation of its general safety under the conditions of intended use within the scope of current expert knowledge.

2.4 International Feed, AAFCO Ingredients, and FDA Numbers

Ingredients in NRC publications are identified by their international feed numbers (IFN) and names, currently developed and maintained by the Feed Composition Data Bank (FCDB) at the National Agricultural Library in Beltsville, Maryland. Ingredient definitions in the *Official Publication* use AAFCO numbers and also include the IFN. Several different IFN specifications and commercial quality trading grades for a similar material may qualify under the same AAFCO name. Some AAFCO ingredients (essentially isolated chemicals) also carry FDA identity numbers under the Food Additives Amendment of the Code of Federal Regulations (21 CFR).

2.5 Formulating Feeds for Minimum Ingredients Costs

The published tables of animal nutrient requirements and feedstuffs composition are crude approximations across the industry at best. A formulator usually will be more effective if he or she applies local experience in feeding animals of similar genetic origin under the environmental conditions at hand and relies on actual analyses of the ingredients received or the historical performance of the specific supplier—if such information is available.

Modern techniques for measuring metabolism and instrumental analysis have enabled rapid estimations of animal needs and nutrient availability in specific feedstuffs. Uniformity in feedstuffs, despite origin, is being advanced by global trade, harmonization of product definitions, and trading standards and by total quality management (TQM) programs like the ISO 9000 movement in Europe. The availability and decreasing costs of desktop and notebook computers in the last decade has made the calculation of least-cost feed formulas by linear programming available to many feed formulators. These factors, coupled with instantaneous knowledge of costs and availability of alternative feedstuffs, regionally and globally, through on-line communication networks are rapidly exhausting the undiscovered bargains in feedstuffs. In the future, profits in animal feeding are increasingly likely to result from private knowledge about (*1*) responses of animals with known genetic abilities, (*2*) ensuring animal health and comfort in extreme climates and weather conditions, and (*3*) improved effectiveness in feeding management practices. Claims of improved feed conversion abilities and improved health status already are being made for genetically controlled broilers and swine. Computer programs are now able to formulate feeds at least cost on the basis of digestible, metabolizable, or net energy, protein, crude fiber, ash, essential amino acids, and specific vitamins and minerals for different growth stages of selected species. Formulas also may be flexed for specific EFA and triacylglycerol structures in the future.

2.6 Consequences of Feedstuffs Contamination

The consolidation of feeds processing and rendering into fewer but bigger installations has also set up the conditions for large and widespread economic losses when mistakes or accidents occur. One example is the chick edema and death problems in the early to mid 1970s that were traced to feeding fats and fatty acids contaminated with polychlorodibenzo-*p*-dioxins (21,22) and polychlorophenols (23) possibly of herbicide origin. This led to purchase specifications that tallows be guaranteed free of chick edema factors by renderers or distributors.

Another example is the 3-year outbreak of polybrominated biphenyls (PBB) poisoning that occurred in Michigan during 1973–1976 from a one-time accidental mislabeling of a toxic fire retardant and its inclusion in an animal feed concentrate mix. An unknown quantity of meat, dairy products, and rendered materials entered the food and recycling chains before the problem was recognized and the involved farms quarantined. By 1978, an estimated 8 million of Michigan's 9.1 million residents had detectable levels of PBB in their bodies (24). Later studies continued to find PBB in samples of serum, body fat, and breast milks of most Michigan residents (25).

Finally, dairy cows fed whole cottonseed that contains aflatoxins, at levels higher than the 20 ppb permitted by FDA, can transfer them to milk in amounts greater than the 0.5 mμg/kg FDA action level (26). State and national programs for monitoring aflatoxin content in milk have been implemented as the feeding of whole cottonseed to cattle has increased and expanded to states not producing cotton. Even nontoxic substances like polyethylene packaging materials, which melt at rendering temperatures, can cause nuisance problems by solidifying at 80°C and forming lumps that clog fat application spray nozzles (27).

AVAILABILITY, CHARACTERISTICS, AND COMPOSITION

3.1 Supply and Uses of Feeding Fats

Stability of Obligatory Coproducts Supplies. Many fats and oils are generated as obligatory coproducts of other endeavors, with the supply relatively inelastic to price change. Examples include (*1*) slaughtering and dressing of animals and poultry, with hand trimming of the fatty tissues at the packing plant or in-store meats department; (*2*) disposal of spent frying oils from restaurants, skimmings from grease traps, and dead animals from pastures and feedlots—all necessary for sanitation, public health, and environmental interests; (*3*) production of high protein content soybean and fish protein meals; (*4*) growing of cotton fiber for domestic and world markets, with cottonseed as a coproduct; and (*5*) development of the domestic corn starch and sweeteners industry, which has generated enough corn germ to make corn oil the second major oil currently produced in the United States.

Because of these situations, the production of the major portion of feed grade fats and some oil-bearing feedstuffs is likely to continue regardless of their market prices. However, future prices will reflect (*1*) domestic competition between feed, oleochemicals, and detergent industries needs; (*2*) opportunities to ship into global markets; and (*3*) abilities to import lower cost fats like palm oil and stearin.

Sources of Fats, Oils, and Tallows. The total world production of fats and oils is estimated at 76.2 million MT. It consists of 59.2 million MT of edible vegetable fats and oils (soybean oil, 16.9 million; palm oil 11.5 million; rapeseed and canola oil, 9.1 million; sunflower seed oil, 7.6 million; cottonseed oil, 4.2 million; peanut oil, 3.4 million; coconut oil, 2.9 million; olive oil, 2.1 million; and palm kernel oil, 1.5 million), butter fat, 5.3 million; total marine oils, 1.1 million; and total tallows and greases, 7.0 million (28).

Approximately 4.1 million MT of inedible animal fats are rendered in the United States annually (Table 7.1). The major sources, in order of decreasing tonnage, are beef packing, pork packing, spent restaurant fats, and broiler and turkey processing. Only about 5% of the total supply of inedible fat is recovered from dead stock (1).

Use of Domestically Produced Tallows. Currently, approximately 35% of domestically produced inedible tallow is exported, leaving about 2.7 million MT available for domestic use (1). Rouse (29) reported that domestic use of inedible tallow increased by 63%, from 0.81 million MT in 1950 to 1.3 million MT in 1991. In 1950, about 72% of the available domestic inedible tallow (0.58 million MT) was used in making soap and hardly any in animal feeds. With the development of synthetic detergents, the use of edible tallows in soap making dropped to 0.15 million MT or 12% of the total in 1991, and animal feeds rose to using about 62% of the domestic supply.

The largest use for inedible fats worldwide is in animal feeds. Domestic

Availability, Characteristics, and Composition   263

Table 7.1 U.S. sources of rendered animal fats and oil (1)

		Fat Produced	
Source	Number Slaughtered (or Kilogram of Product)	Percent of Total	Metric Tons
Steers and heifers	28,000,000	37.3	1,527,000
Cull cows and bulls	7,000,000	3.9	159,000
Market pigs	83,000,000	21.3	874,000
Cull sows and boars	5,000,000	2.2	91,000
Broiler	5,200,000,000	8.7	355,000
Turkey	242,000,000	1.4	57,000
Dead stock	(1,636,000,000)	5.2	213,000
Restaurant grease	(1,023,000,000)	16.7	682,000
Miscellaneous	—	3.3	136,000
Total domestic inedible	—	100.0	4,125,000

use in various feeds is shown in Table 7.2. Approximately 56.2% is used in broiler and turkey feeds, and another 2.7% in feeding poultry layers (29). It has been estimated that inclusion of 3–4% fat in all feeds, a level favored by some nutritionists and feeders, would use about 2.5 million MT of fats in domestic animal feeds (1).

3.2 Definitions of Fats and Related Products Used in Feeding

Industrywide definitions for feed fats products are still being developed. USDA and NRA definitions of rendering and tallows and greases were already discussed.

Table 7.2 Estimated use of fats and oils in domestic animal feeds in 1993 (29)

	Fat Used	
Type of Feed	Percent of Total	Million Metric Tons
Broilers	34.8	591
Turkeys	21.4	364
Pet foods	16.0	273
Swine	10.7	182
Beef cattle	5.4	91
Dairy cattle	4.8	82
Layers	2.7	45
Veal	2.1	36
Fish	2.0	34
Total	100.00	1,698

Table 7.3 *American Fats and Oils Association grading standards for tallow and grease (1)*

Grades	Titer (Min. °C)	FFA (Max. %)	FAC Color (Max.)	Specifications (R&B, Max.)	MIU[a] (%)
Edible tallow	41.0	0.75	3	None	[b]
Lard (edible)	38.0	0.50	[c]	None	[b]
Top white tallow	41.0	2	5	0.5	1
All-beef packer tallow	42.0	2	None	0.5	1
Extra fancy tallow	41.0	3	5	None	1
Fancy tallow	40.5	4	7	None	1
Bleachable fancy tallow	40.5	4	None	1.5	1
Prime tallow	40.5	6	13-11B	None	1
Special tallow	40.0	10	21	None	1
No. 2 tallow	40.0	35	None	None	2
"A" tallow	39.0	15	39	None	2
Choice white grease	36.0	4	13-11B	None	1
Yellow grease	[d]	15	39	None	2

[a] Moisture, insolubles, unspecifiables.
[b] Moisture, maximum 0.20%; insoluable impurities, maximum 0.05%.
[c] Lovibond color 5.25-inch cell, maximum; 1.5 REd. Lard Peroxide Value 4.0 ME/K maximum.
[d] Titer minimum, when required, to be negotiated between buyer and seller on a contract by contract basis.

NRA Recommended Standards and Definitions. AFOA grading standards for tallows and greases for industrial uses are shown in Table 7.3 (1). Typical analyses of feed-grade fats of different origins are shown in Table 7.4 (30). Quality specifications suggested by the NRA for commonly traded feeding fats are shown in Table 7.5 (1). The NRA additionally suggests the following.

Table 7.4 *Typical analyses of feed grade fats (30)*

Fat Source	Titer (°C)	FAC Color (Maximum)	MIU[a] (%)	Iodine Value	FFA[b] (Maximum %)	Percent Fatty Acids		
						Saturated	Unsaturated	Linoleic
FGF (for all feeds)	34–38	37	2	55	15	44	56	10
FGF (for milk replacers)	38–41	9	1	45	5	50	50	4
All-beef tallow	38–42	7	1	40	5	56	44	2
All-pork fat	32–38	37	2	58	15	36	64	12
All-poultry fat	28–35	19	2	65	15	28	72	20
Butter fat	28–35	—	—	32	—	63	37	2
Vegetable fat (palm oil)	28–36	—	2	53	—	42	58	10

[a] Moisture, insolubles, and unsaponifiables.
[b] Free fatty acids.

Table 7.5 National Renderer's Association suggested fat quality standards for feeds (1)

Standard	Tallow	Choice White Grease	Yellow Grease	Hydrolized A/V Blend
Total fatty acids (%)	90	90	90	90
Free fatty acids (%)	4–6	4	15	40–50
FAC color	19	11A	37–39	45
Moisture (%)	0.5	0.5	1.0	1.5
Impurities (%)	0.5	0.5	0.5	0.5
Unsaponifiables (%)	0.5	0.5	1.0	2.5
Total MIU (%)	1.0	1.0	2.0	4.0
Iodine value	48–58	58–68	58–79	85
AOM (h)	20	20	20	20

Feeding fats pass a minimum of 20-h active oxygen method (AOM) test.

Blended feeding fats contain only tallow, grease, poultry fat, and soapstocks; any other products should be included only with the knowledge and approval of the purchaser.

All fat products must be below tolerances for toxic chemicals and pesticide residues; certification is available from most renderers; fats used in poultry rations must be free of the chick edema factor; and all fats should be devoid of contaminants such as heavy metals.

Fats for poultry rations should not contain cottonseed soapstock or other coproducts.

Changes to new sources of fats, especially in ruminant and swine feeds and pet foods, should be gradual due to potential differences in palatability from previous sources (1).

No standards exist for polyethylene in fats; the practical way to remove it is by filtering tallow at low temperatures using special filter aids; most customers can use tallow containing up to 30 ppm polyethylene, and a few up to 150 ppm (31).

Maximum pesticide residues tolerances are 0.5 ppm for DDT, DDD, and DDE; 0.3 ppm for Dieldrin; and 2.0 ppm for PCB (31).

Some buyers also have a rate of filtration (ROF) specification, which is defined as the milliliters of tallow at 110°C (230°F) that will pass through a filter paper in 5 min under specified test conditions; a commonly sought value is 35–40 ROF; the test identifies fats that may give processing difficulties such as slow filtration, emulsion, and foaming (31).

Poultry fat is rendered mainly from poultry offal collected in packing plants but may include renderings from mortalities, hatchery rejects, and condemned or unmarketable parts of birds. Much of the rendered poultry fat and meal is produced and recycled in integrated growing–processing plant operations, and only limited amounts are available in the open market.

AAFCO Definitions—Fats and Oils. Many feed ingredients are coproducts of agribusinesses and are of secondary interest. With some exceptions, little is done currently to process them further, and they are sold as made. The AAFCO defines feed fats as follows.

Animal fat (AAFCO number 33.1) is obtained from the tissues of mammals and/or poultry in the commercial processes of rendering or extracting. It consists predominantly of triacylglycerol esters of fatty acids and contains no additions of free fatty acids or other materials obtained from fats. It must contain, and be guaranteed for, not less than 90% total fatty acids, not more than 2.5% unsaponifiable matter, and not more than 1% insoluble impurities. Maximum free fatty acids and moisture must also be guaranteed. If the product bears a name descriptive of its kind or origin (e.g., beef, pork, or poultry), it must correspond thereto. If an antioxidant is used, the common name or names must be indicated, followed by the words *used as a preservative.* Includes IFN 4-00-409 (animal poultry fat).

Vegetable fat, or oil (33.2) is the product of vegetable oil origin obtained by extracting the oil from seeds or fruits that are commonly processed for edible purposes. It consists predominantly of glyceride esters of fatty acids and contains no additions of free fatty acids or other materials obtained from fats. It must contain, and be guaranteed for, not less than 90% total fatty acids, not more than 2% unsaponifiable matter, and not more than 1% insoluble impurities. Maximum free fatty acids and moisture must also be guaranteed. If the product bears a name descriptive of its kind or origin (e.g., soybean oil or cottonseed oil) it must correspond thereto. If an antioxidant is used, the common name (or names) must be indicated, followed by the words *used as a preservative.* Includes IFN 4-05-077 (vegetable oil).

Hydrolyzed fat, or oil, feed grade (33.3), is obtained by the fat-processing procedures commonly used in edible fat processing or soap making. It consists predominantly of fatty acids and must contain, and be guaranteed for, not less than 85% total fatty acids, not more than 6% unsaponifiable matter, and not more than 1% insoluble impurities. Maximum moisture must also be guaranteed. Its source must be stated in the product name, e.g., hydrolyzed animal fat or hydrolyzed animal and vegetable fat. If an antioxidant is used, the common name (or names) must be indicated, followed by the words *used as a preservative.* Includes IFN 4-00-376 (animal fat hydrolyzed) and IFN 4-05-076 (vegetable oil hydrolyzed).

Ester, feed grade (33.4), is the product consisting of methyl, ethyl, and other nonglyceride esters of fatty acids derived from animal and/or vegetable fats. It consists predominantly of the ester and must contain not less than 85% total fatty acids, not more than 10% free fatty acids, not more than 6% unsaponifiable matter (2% for methyl esters), and not more than 1% insoluble matter. Its source must be stated in the product name, e.g., methyl ester of animal fatty acids or ethyl ester of vegetable oil fatty acids. Methyl esters must contain not more than 150 ppm (0.015%) free methyl alcohol. If an antioxidant is used, the common name or names must be indicated, followed by the words *used as a preservative.* This feed fat includes FDA regulation 573.640:

IFN 4-00-377 (animal fatty acids of ethyl ester).
IFN 4-00-378 (animal fatty acids of methyl ester).
IFN 4-00-379 (animal fats of nonglyceride ester).
IFN 4-12-240 (vegetable fatty acids of ethyl ester).
IFN 4-05-075 (vegetable fatty acids of nonglyceride ester).
IFN 4-05-074 (vegetable fatty acids of methyl ester).

Fat product, feed grade (33.5), is any fat product that does not meet the definitions for animal fat, vegetable fat or oil, hydrolyzed fat, or fat ester. It must be sold on its individual specification, which will include the minimum percentage of total fatty acids, the maximum percentage of unsaponifiable matter, the maximum percentage of insoluble impurities, the maximum percentage of free fatty acids, and moisture. The above listed specifications must be guaranteed on the label. If an antioxidant is used, the common name or names must be indicated, followed by the words *used as a preservative.* Includes IFN 4-00-414 (animal vegetable fat product).

Corn endosperm oil (33.6) is obtained by the extraction of oil from corn gluten. It consists predominantly of fatty acids and glycerides and must contain not less than 85% total fatty acids, not more than 14% unsaponifiable matter, and not more than 1% insoluble matter. If an antioxidant is used, the common name or names must be indicated, followed by the word *preservative.* Includes IFN 4-02-852 (maize endosperm oil and FDA Reg. 8.322).

Vegetable oil refinery lipid, feed grade (33.7), is obtained in the alkaline refining of a vegetable oil for edible use. It consists predominantly of the salts of fatty acids, glycerides, and phosphates. It may contain water and not more than 22% ash on a water-free basis. It is to be neutralized with acid before use in commercial feed. Includes IFN 4-05-078 (vegetable oil refinery lipid).

Calcium salts of long-chain fatty acids (33.14) are the reaction products between calcium and long-chain fatty acids of vegetable and/or animal origin. They shall contain a maximum of 20% lipid not bound in the calcium salt form and the percent total fat shall be indicated. The unsaponifiable matter (exclusive of calcium salts) shall not exceed 4% and moisture shall not exceed 5%. If an antioxidant is used, its common name must be indicated on the label. Before conducting an assay for total fats, hydrolysis of the calcium salts should be performed to liberate the lipid fraction.

Hydrolyzed sucrose polyesters, feed grade (T33.15) is the product resulting from acid hydrolysis of sucrose polyesters, such as olestra, to make them digestible. It shall consist predominantly of fatty acids and contain, and be guaranteed for, not less than 85% total fatty acids, not more than 2% sucrose polyesters (hex ester and above), not more than 2% unsaponifiable matter, and not more than 2% insoluble impurities. Maximum moisture must also be guaranteed. Its source must be stated in the product name (e.g., hydrolyzed animal sucrose polyesters, hydrolyzed vegetable sucrose polyesters, or hydrolyzed animal and vegetable sucrose polyesters). If an antioxidant is used, the common name must be indicated, followed by the words *used as a preservative*. This definition was proposed in 1993 and is tentative at this time.

Fish oil (51.8) is the oil from rendering whole fish or cannery waste. It includes IFN 7-01-965 (fish oil).

Salts of volatile fatty acids (60.73) is a blend containing the ammonium or calcium salt of isobutyric acid and the ammonium or calcium salts of a mixture of 5-carbon acids/isovaleric, 2-methyl-butyric, and *n*-valeric. The contained ammonium or calcium salts of volatile fatty acids shall conform to the specifications in 21 CFR 573.914. It is used as a source of energy in dairy cattle feed. The label of the product shall bear adequate directions for use, including statements expressing maximum use levels. For ammonium salts of volatile fatty acids, not to exceed 120 g per head per day thoroughly mixed in dairy cattle feed as a source of energy, and for calcium salts of volatile fatty acids, not to exceed 135 g per head per day thoroughly mixed in dairy cattle feed as source of energy. Includes FDA Reg. 21 CFR 573.914.

Soy phosphate or soy lecithin (84.10) is the mixed phosphatide product obtained from soybean oil by a degumming process. It contains lecithin, cephalin, and inositol phosphatides, together with glycerides of soybean oil and traces of tocopherols, glycosides, and pigments. It must be designated and sold according to conventional descriptive grades with respect to consistency and bleaching. Includes IFN 4-04-562 (soybean lecithin).

AAFCO Definitions—Fat-associated Vitamins and Essential Compounds. Fat-associated feed ingredients include sources of natural or synthetic fat-soluble vitamins. AAFCO defined vitamin A sources include

90.3, vitamin A oil (IFN 7-054-141).
90.14, vitamin A supplement (IFN 7-05-1244).
90.25, vitamin A acetate [IFN 7-05-142, FDA Reg. 582.5933 (GRAS)].
90.25, vitamin A palmitate [IFN 7-04-143, FDA Reg. 582.5936 (GRAS)].
90.25, vitamin A propionate (IFN 7-26-311).

Defined vitamin D sources include

90.4, vitamin D_2 supplement (IFN 7-05-149).
90.7, cholecalciferol (D-activated animal sterol) (IFN 7-00-408).
90.8, ergocalciferol (D-activated plant sterol) (IFN 7-03-728).
90.15, vitamin D_3 supplement (IFN 7-05-699).
96.3, irradiated dried yeast, irradiated dried yeast (IFN 7-05-524).

Defined vitamin A and D sources include

90.1, cod liver oil (IFN 7-01-993).
90.2, cod liver oil with added vitamins A and D (IFN-7-08-047).

90.25, herring oil (IFN 7-08-048).
90.25, menhaden oil (IFN 7-08-049).
90.25, salmon oil (IFN 7-08-050).
90.25, salmon liver oil (IFN 7-02-013).
90.25, sardine oil (IFN 7-02-016).
90.25, shark liver oil (IFN 7-02-019).
90.24, tuna oil (IFN 7-02-024).

Vitamin E is a biological antioxidant necessary to prevent cell membrane damage. It is essential for proper growth, hormone functions, and proper muscle and nervous system activity. Vitamin E also retards fats and oils oxidation. Defined sources are as follows.

90.12, vitamin E supplement (IFN 7-05-150).
90.25, tocopherol [IFN 7-00-001, FDA Reg. 582.5890 (GRAS)].
90.25, α-tocopherol acetate [IFN 7-18-777, FDA Reg. 582.5892 (GRAS)].
90.25, wheat germ oil (IFN 7-05-207).

Choline (bilineurine) is a lipotropic factor in the metabolism of fatty acids in the liver of fish, poultry, and swine. It is a nonspecific source of biologically active methyl groups and is essential in the synthesis of acetylcholine (the main chemical in transmission of nerve impulses), for building and maintaining cell structure, and for prevention of perosis (slipped tendon) in poultry (3). Choline also is a precursor for phosphatidylcholine, which is active in absorption of lipids in the small intestine and for their transport. AAFCO defined sources include the following.

90.25, choline chloride [IFN 7-01-228, FDA Reg. 582.5252 (GRAS)].
90.25, choline pantothenate (IFN 7-01-229).
90.25, choline xanthate (IFN 7-01-230, FDA Reg. 573.300).
90.17, betaine (hydrochloride or anhydrous), IFN 7-00-722; this is sometimes included in diets as a methyl group donor and partial replacement for choline.
90.25, inositol (myo-inositol, meso-inositol, *i*-inositol) [IFN 7-09-354, FDA Reg. 582.5370 (GRAS)]; this is a lipotropic growth factor for some animals.
6.12, taurine (IFN 5-09-821, FDA Reg. 573.980), this is a sulfonic amino acid; the bile salts of most mammals are a mixture of taurine and glycine conjugates but are exclusively taurine conjugates in the cat; deficiencies have led to central retinal degeneration and blindness in cats; it is normally found in muscle flesh; supplementation is recommended in kitten and cat diets that might not contain sufficient proteins of animal origin; it is listed as a required nutrient for cats by the NRC (8) and in the AAFCO's *Official Publication*.

AAFCO Definitions—Chemical Preservatives (Antioxidants). Antioxidants are used to retard oxidation of unsaturated fatty acids in fats, oils, and fish and meat meals, and to slow the destruction of fat-soluble vitamins, especially vitamin E. Their presence must be identified on the label when used. Use of the following compounds is permitted alone or in combination at levels not to exceed 0.02% of the fat present.

- 18.1, butylated hydroxyanisole (BHA) (IFN 8-01-044, FDA Reg. 582.3169); either the full name or BHA may be used in the ingredients listing.
- 18.1, butylated hydroxytoluene (BHT) (IFN 8-01-045, FDA Reg. 582.3173); either the full name or BHT may be used in the ingredients listing.
- 18.1, tertiary butyl hydroxyquinone (TBHQ) (IFN 8-04-829, FDA Reg.); in the informal review process.
- 18.1, thiodipropionic acid (IFN 8-04-830, FDA Reg. 582.3109).
- 18.1, dilauryl thiodipropionate (IFN 8-04-789, FDA Reg. 582.3280).
- 18.1, distearyl thiodipropionate (IFN 8-01-792, FDA Reg. 582.3280).
- 18.1, propyl gallate (IFN 8-03-308, FDA Reg. 582.3660).
- 18.1, resin guaiac (IFN 8-03-909, FDA Reg. 582.3336); same as guaiac gum, may be used at 0.1%, or equivalent preservative 0.01%, in fats and oils.
- 18.1, ethoxyquin (FN 8-01-841, FDA Reg. 573.380); may be used not to exceed 0.015% in or on feed.
- 18.1, tocopherols (IFN 7-05-0348, FDA Reg. 582.3890); no statutory maximum exists for tocopherols, but their use should conform to good manufacturing practices.

The formulator needs to know the properties of each antioxidant. For example, most of the antioxidants are easily distilled by heat or steam. Propyl gallate will form strong purple complexes with iron. The preservative effects of antioxidants also can be enhanced by addition of metal chelating agents like citric acid. Furthermore, many crude feedstuffs of plant origin, including soybean meal, crude soybean oil, and lecithin, contain a variety of natural quinone-type compounds with beneficial antioxidant properties that do not require labeling.

AAFCO Definitions—Special-Purpose Products. Fat-based or associated special-purpose products include the following.

- 85.5, diacetyl tartaric acid esters of mono and diglycerides of edible fats or oils, or edible fat-forming fatty acids (IFN 8-07-248, FDA Reg. 582.4101); used as emulsifying agents.
- 85.5, ethoxylated mono and diglycerides, (FDA Reg. 172.834); used as emulsifiers.
- 85.5, lecithin (IFN 8-08-041, FDA Reg. 582.1400); used as stabilizer.

85.5, methyl glucoside coconut oil ester (IFN 8-09-346, FDA Reg. 573.660); used as surfactant in molasses.

85.5, mono and diglycerides of edible fats or oils, or edible fat-forming acids (IFN 8-07-251, FDA Reg. 582.4505); used as emulsifying agents.

85.5, monosodium phosphate derivatives of mono and diglycerides of edible fats or oils, or edible fat-forming fatty acids (IFN 8-07-252, FDA Reg. 582.4521); used as emulsifying agents.

AFIA Definitions. The American Feed Industry Association definitions include the following.

Animal and vegetable fat blends, feed grade, are blends of rendered animal fats, cooking fats, crude vegetable fats, by-products of fat splitting and hydrolyzed fats in any combination that have been processed under good manufacturing practices. They shall be of a quality suitable for use as an animal feed ingredient. Typical analyses include total fatty acids, 93%; free fatty acids, 40%; moisture, 1.0%; impurities, 0.4%; and unsaponifiables, 3.5%. Fatty acid profiles and iodine values should be consistent from one load to another. Metabolizable energy claims should be substantiated by research data. Fat should be stabilized with an acceptable feed- or food-grade antioxidant and at levels recommended by the antioxidant manufacturer. Fish oil and fish oil by-products should be avoided in other fowl, unless agreed on between the purchaser and the supplier. Fats must be within the pesticide and industrial chemical tolerances set by federal and state agencies. Blends for poultry should pass the color test for fat acceptance (a modification of the Liebermann-Burchard test), an indicator of the presence of the chick edema factor. This is not a specific test for the chick edema factor, and results must be carefully evaluated. Physical properties are color, brown to black; odor, typical and not rancid; density, 7.5 lb/gal. animal and vegetable fat blends are primarily used as sources for energy in livestock and poultry feeds, especially for broilers, turkeys, and cattle. Blended fats are also added to dairy and swine feeds. These are not defined by the AAFCO, but tentative definition has been requested. Must be declared on mixed feed labels as animal fat and vegetable fat (31).

Acidulated cottonseed soapstock is the product obtained from the complete acidulation and thorough setting of soapstock, which itself is the by-product obtained from the alkali refining of cottonseed oil. It is sold on a basis of 95% total fatty acid content. If it falls below 85% total fatty acid content, it may be rejected. It should not have more than a total of 6% unsaponifiable plus insoluble matter. Typical analyses are total fatty acids (TFAs), 90% and moisture, 2.5%. The amount of gossypol present in acidulated cottonseed soapstock is variable and difficult to determine with existing methods of analysis. Physical properties include dark brown to greenish black color, appearing as an oily mixture. It has a slightly sour odor with a pH of 4. It is soluble when cold and is a pumpable liquid at 49°C (120°F). It is used as a feed-grade fat for ruminants. Caution must be exercised when used with other species due to the gossypol content. It is not to be used with layer rations. It qualifies under AAFCO 33.3, IFN 4-17-942 or IFN 4-05-076 (31).

Acidulated soybean soapstock is the product obtained from the complete acidulation and thorough setting of soapstock, which itself is the by-product obtained from the alkali refining of soybean oil. It is sold on a basis of 95% total fatty acid content. If it falls below 85% total fatty acid content, it may be rejected. Typical analyses are TFAs, 90%; moisture, 1%; and iodine value 125. In practice, soybean soapstock may be found in combinations with other vegetable oil soapstocks. The buyer should determine if cottonseed soapstock is present, as it may contain gossypol, which is detrimental in nonruminant feeds. Physical properties are medium brown color; odor somewhat typical of soybeans, slightly nutty; solid when cool; and liquid and pumpable at 38–44°C (100–110°F). It qualifies under AAFCO 33.3, IFN 4-17-893 (31).

3.3 Definitions of Extracted and Whole Seed Products

Mechanically Extracted Meals. Solvent extracted oilseed meals typically contain less than 1.5% residual fat unless the gums (hydrated phosphatides) or soapstock have been added back to the meal before the desolventizer–toaster or meal dryer. Mechanically extracted (expeller or screw-pressed) meals can contain 4–9% oil, which can be a significant calorie source in animal feeds. Fat contents of extracted meals are not part of the definition, although typical analyses are shown below.

AAFCO Definitions—Extracted Products. The words *mechanical extracted* are not required when listed as an ingredient in a manufactured feed.

T24.10, cottonseed meal, mechanical extracted; contains anticaking agent; will replace definition 24.10 if adopted (IFN 5-02-045; cottonseeds meal mechanical extracted 36% protein); a typical NRC composition (9) is 4.6% oil ether extract (EE), and dry matter basis (dmb).

71.1, linseed meal, mechanical extracted (IFN 5-16-280; flax seeds meal mechanical extracted); NRC: 6.0% oil EE, dmb.

71.9, peanut meal, mechanical extracted and solvent extracted (IFN 5-03-649: peanut seeds without coats meal mechanical extracted); NRC: 6.3% oil, EE, dmb.

71.25, rapeseed meal, mechanical extracted (IFN 5-03-870: rape seeds meal mechanical extracted); NRC: 7.9% oil EE, dmb.

71.130, safflower meal, mechanical extracted (IFN 5-04-109: safflower seeds meal mechanical extracted); NRC: 6.7% oil EE, dmb.

Availability, Characteristics, and Composition   **273**

- 71.210, sunflower meal, dehulled, mechanical extracted (IFN 5-04-738: sunflower seeds meal without hulls meal mechanical extracted); NRC: 8.7% oil EE, dmb.
- 71.220, sunflower meal, mechanical extracted (IFN 5-27-477: sunflower seeds meal mechanical extracted.)
- 84.60, soybean meal, mechanical extracted (IFN 5-04-600: soybean seeds meal mechanical extracted); NRC: 5.3% oil EE, dmb.

Full-fat Soybeans.   AAFCO definitions do not exist for whole oilseeds per se. Considerable variations in composition occur between species, location and weather during crop maturation. Soybeans typically contain 19% oil at 92% total solids and 21% oil at 100% total solids (dmb). Definitions for full-fat soybean products are as follows.

- 84.1, ground soybeans are obtained by grinding whole soybeans without cooking or removal of any of the oil; includes IFN 5-04-596 (soybean seeds ground).
- 84.11, heat-processed soybeans are the result of heating whole soybeans without removing any of the component parts; they may be ground, pelleted, flaked, or powdered; the maximum pH rise using the standard urease testing procedure should not exceed 0.10 pH units; they must be sold according to their crude protein, crude fat, and crude fiber content; includes IFN 5-04-597 (soybean seeds heat processed).
- 84.15, ground extruded whole soybeans are the result of extrusion by friction heat and/or steam of whole soybeans without removing any of the component parts; the meal must be sold according to its crude protein, fat, and fiber content; includes IFN 5-14-005 (soybean seeds extruded ground).

The AFIA (31) has further elaborated on quality factors and feed applications of whole (full-fat) soybeans. Whole soybeans must be properly heated to provide optimum protein nutrition for critical animals, especially poultry, swine, lambs, and calves as well as pets and fur-bearing animals.

Underheating of whole soybean may fail to destroy the trypsin inhibitor and reduce urease and lipase activity, resulting in low protein efficiency for critical feeds. An underheated soybean meal greatly increases the need for vitamin D to prevent rickets in turkey poults. Overheating of whole soybeans tends to inactivate or destroy the essential amino acids lysine, cystine, and methionine and possibly others (31).

Laboratory tests such as urease activity, protein dispersibility index (PDI), nitrogen solubility index (NSI), thiamine, and water absorption have been found valuable in monitoring daily production for protein quality. But biological chick and/or rat assays are the only reliable means currently available for predetermining the nutritional value of whole soybean protein; they must be conducted periodically to verify results of chemical tests (31). If whole soybeans are to be used in a mixture containing 20% or more soybean meal, 5% or more urea, and 20% or more molasses, or an equivalent mixture, and exposed to hot, humid storage conditions, it is advisable that the urease activity of the whole soybeans not exceed 0.12 increase in pH (31). Extruded or roasted soybeans properly treated for cattle to increase bypass protein should have urease values of less than 0.05 pH rise. A urease rise of 0.05–0.20 is an indication of proper treatment for swine and poultry.

Whole soybean may be fed to gestating swine (ground form) and to mature ruminants (whole or rolled form). However, feeding whole soybeans to all other livestock will result in reduced performance and is not recommended. Extruded soybeans may be fed to all livestock, especially swine and poultry. Extruded soybeans fed to lactating dairy cattle increase by-pass protein but most often will reduce butterfat. Roasted soybeans may be fed to all livestock, especially mature ruminants. Roasted soybeans fed to lactating dairy cattle will increase by-pass protein and slightly increase butterfat.

Poultry and swine will perform as well on pelleted rations containing extruded or roasted soybeans as on isocaloric, isonitrogenous ration containing soybean meal and soybean oil. The improved performance from pelleting with rations containing extruded soybeans is primarily because of increased ration nutrient density. The improvement with rations containing roasting soybeans is primarily because of increased fat digestibility (the oil vesicles are ruptured to allow the oil to be more available for digestion). Bulk densities are whole soybeans, 737–769 kg/m^3 (46–48 lb/ft^3); whole soybeans, ground, 384–545 kg/m^3 (24–34 lb/ft^3); extruded soybeans, 384–497 kg/m^3 (27–31 lb/ft^3); and roasted soybeans, 720–753 kg/m^3 (45–47 lb/ft^3) (31).

Cottonseed and Cottonseed Products. Whole cottonseed and its oil-bearing (screw press/expeller) meals can be economically attractive sources of oil and protein but must be used properly. Currently, approximately 35–40% (1.5 million tons) of the domestic cottonseed crop is fed whole to adult ruminants (primarily dairy cattle) annually. Whole cottonseed is not defined by AAFCO but is identified as IFN 5-01-614; it is called feed-grade cottonseed by the National Cottonseed Products Association's trading rules (32). It consists of the entire seed of the cotton plant after the cotton fibers have been removed by ginning at a fiber:seed ratio of approximately 1:1.65. Naked (Pima-type) cottonseed exists, and operations have been installed between the gin and the oil mill to intercept the seed and remove the linters for other uses. Whole cottonseed is sold in three grades: prime feed-grade cottonseed (moisture, 13% maximum; free fatty acid in oil, 3% maximum; crude protein and crude fat, 34% minimum, dmb); delinted prime feed-grade cottonseed (lint, 5% maximum; moisture, 13% maximum; free fatty acid in oil, 3% maximum; crude protein and crude fat, 34% minimum, dmb; and feed-grade cottonseed, off quality. Fuzzy ("white," "with lint") whole cottonseed contains approximately 21% fat at 92% total solids and naked seed contains about 23% fat at the same moisture level. Rolling (flaking) naked seed before feeding reduces whole seed passage through dairy cattle (33). Bulk density for undelinted cottonseed is 288–401 kg/m^3 (18–25 lb/ft^3) and 401–561 kg/m^3 (25–35 lb/ft^3) for delinted seed.

Cottonseed is added to feeds for ruminant animals as a source of highly concentrated energy and rumen undegradable protein and fiber. It typically is fed whole—not ground or pelleted—although a process for extruding whole seed and soybeans was patented in 1993 (34). Whole cottonseed contains 0.5–1.2% gossypol, a yellow-green pigment that is toxic to monogastric animals and young calves, lambs, and kids whose rumens are not yet functional. It can be fed to cattle, sheep, and other ruminant animals due to bacterial detoxification in the rumen. Care should be exercised in feeding so ruminal detoxification abilities are not overstressed, resulting in gossypol toxicity. The typical feeding rate is 2.3–3.6 kg (5–8 lb) whole cottonseed per day per cow. "Free" gossypol in cottonseed products can be deactivated ("bound,") either during processing or by use in conjunction with iron (ferrous) salts. Levels of gossypol typically present in common cottonseed products, and tolerances in feeds by selected species, are presented in Tables 7.6 and 7.7, respectively (35).

The AFIA (31) advises that (1) 200 ppm (0.02%) dietary free gossypol does not affect egg production, (2) 50 ppm (or 150 ppm with addition of FeSO$_4$ at 4 parts iron to 1 part gossypol) avoids (green) egg yolk discoloration during refrigerated storage, and (3) up to 150 ppm free gossypol (or 400 ppm with addition of FeSO$_4$ at 1 part iron to 1 part free gossypol) can be used for growing broilers. Levels of 100 ppm free gossypol (or higher levels, but no more than 400 ppm FeSO$_4$ added at a weight ratio of 1 part iron to 1 part free gossypol) can be used for growing and fattening pigs.

Table 7.6 Analyzed gossypol levels in common cottonseed products (35)

	Gossypol	
Product	Percent Total	Percent Free
Cottonseed kernel	0.39–1.7[a]	0.39–1.4[a]
Whole cottonseed	—	0.47–0.63[b]
Delinted whole cottonseed	—	0.47–0.53[b]
Cottonseed meal		
Screw press	1.02[a]	0.02–0.05[a]
Prepress solvent	1.13[a]	0.02–0.07[b]
Direct solvent	1.04[a]	0.1–0.5[a]
Solvent (expander process)	—	0.06–0.1[b]
Cottonseed hulls	—	0.06[a]
Glandless whole cottonseed	0.01[a]	—

[a] Dry matter basis.
[b] As-fed basis.

Table 7.7   Reported effect and no effect levels of free gossypol (35)

Class of Livestock	Free Gossypol Intake (ppm)	
	Effect	No Effect
Ruminants		
Preruminant calves	—	100^a
Young lambs	824^b	—
Mature dairy cows	1076^b	—
Nonruminants		
Yearling horses	—	115^a
Weanling horses	—	348
Catfish	—	900^b
Tilapia	—	1800
Rainbow trout	1000^{a,c}	250^{a,c}
Shrimp	—	170^b

^{a} As-fed basis not reported.
^{b} Dry matter basis.
^{c} Fed-as gossypol acetic acid.

Jones (36) reviewed the natural antinutrients of cottonseed protein products—gossypol and the cyclopropenoic fatty acids (CPFA; malvalic and sterculic acids). The CPFAs participate in forming the pink color complex in the Halphen reaction, a test specific for the admixture of cottonseed oil with other oils and fats. They also inhibit Δ9 desaturase, an enzyme that converts stearic acid into oleic acid, and thus increase hardness of fats from animals (e.g., pig backfat and lard) raised or finished on feedstuffs containing high levels of polyunsaturated oils like corn. Feed industry practice is to limit cottonseed lipids to no more than 0.1–0.2% in the diet of laying hens to avoid pink discoloration of egg whites and alterations of the vitelline membrane that cause pasty yolks.

Other Whole Oilseeds. Various oilseeds have been fed whole, or dehulled, when available for feed at competitive prices or in grades substandard for extraction but still wholesome for feeding. Examples include safflower (*Carthamus tinctorius*), fat content 35%, dmb; and *oil-type sunflower seed* fat content 44%, dmb. Currently, interest is high in feeding whole canola seed (41–46% oil) in Canada and northern European countries. "Double-zero" strains of *Brassica napus* (rapeseed, oilseed rape, swede rape, and Argentine rape) and *Brassica campestris* (turnip rape, oil turnip, and Polish rape), which contain greatly reduced levels of erucic acid and thioglucosinolates, are used (37). Canola is the name coined for low-erucic acid rape (LEAR), compared with the traditional high-erucic acid rape (HEAR), in Canada where the crop was developed.

Factors to be considered before feeding whole oilseed include fiber content, natural toxic components, and fungal contaminations. The high hull/fiber content of many oilseeds often restricts their use as ruminant feeds. Almost every oilseed has one or more recognized toxic or antigrowth components. Fortunately, most are labile to processing by dry or moist heat. The only exception is sunflower seed, but animal growth still is improved by heating before feeding (38).

3.4 Compositions of Fat Sources Used in Feeding

Fatty Acid Compositions of Tallows and Fish and Oilseeds Oils. Fatty acid compositions of tallows and fish and vegetable oils used as feed ingredients are shown in Table 7.8 (11). Profiles for palm oils are not shown; they typically are chill crystallized and fractionated into oleins and stearins and can vary greatly in fatty acid compositions. The principal fatty acids of major marine oils are shown in Table 7.9 (39).

Table 7.8 Fatty acid composition of common feed animal fats, fish oils, and vegetable oils[a]

Lipid Source	IFN[b]	14:0	16:0	16:1	18:0	18:1	18:2 n-6	18:3 n-3	18:4 n-3	20:1	20:4 n-6	20:5 n-3	22:1	22:5 n-3	22:6 n-3	Σn-6	Σn-3	n-3:n-6
Animal fat																		
Beef tallow	4-08-127	3.7	24.9	4.2	18.9	36.0	3.1	0.6	—	0.3	—	—	—	—	—	3.1	0.6	0.19
Pork fat	4-04-790	1.3	23.8	2.7	13.5	41.2	10.2	1.0	—	1.0	—	—	—	—	—	10.2	1.0	0.10
Poultry fat	4-09-319	0.9	21.6	5.7	6.0	37.3	19.5	1.0	1.1	0.1	—	—	—	—	—	19.6	1.0	0.05
Fish oils																		
Anchovy		7.4	17.4	10.5	4.0	11.6	1.2	0.8	3.0	1.6	0.1	17.0	1.2	1.6	8.8	1.3	31.2	24.0
Cod liver	7-01-994	3.2	13.5	9.8	2.7	23.7	1.4	0.6	0.9	7.4	1.6	11.2	5.1	1.7	12.6	3.0	27.0	9.0
Capelin	7-16-709	7.9	11.1	11.1	1.0	17.0	1.7	0.4	2.1	18.9	0.1	4.6	14.7	0.3	3.0	1.8	12.2	6.78
Channel catfish, cultured		1.4	17.4	2.9	6.1	49.1	10.5	1.0	0.2	1.4	0.3	0.4	—	0.3	1.3	12.7	3.2	0.25
Herring, Atlantic	7-08-048	6.4	12.7	8.8	0.9	12.7	1.1	0.6	1.7	14.1	0.3	8.4	20.8	0.8	4.9	1.4	17.8	12.71
Herring, Pacific		5.7	16.6	7.6	1.8	22.7	0.6	0.4	1.6	10.7	0.4	8.1	12.0	0.8	4.8	1.0	15.7	15.7
Menhaden	7-08-049	7.3	19.0	9.0	4.2	13.2	1.3	0.3	2.8	2.0	0.2	11.0	0.6	1.9	9.1	1.5	25.1	16.73
Redfish		4.9	13.2	13.2	2.2	13.3	0.9	0.5	1.1	17.2	0.3	8.0	18.9	0.6	8.9	1.2	19.1	15.92
Salmon, sea caught		3.7	10.2	8.7	4.7	18.6	1.2	0.6	2.1	8.4	0.9	12.0	5.5	2.9	13.8	2.1	31.4	15.00
Vegetable oil																		
Canola	4-06-144	—	3.1	—	1.5	60.0	20.2	12.0	—	1.3	—	—	1.0	—	—	20.2	12.0	5.94
Coconut	4-09-320	16.8	8.2	—	2.8	5.8	1.8	—	—	—	—	—	—	—	—	1.8	0.0	0.0
Corn	4-07-882	—	10.9	—	1.8	24.2	58.0	0.7	—	—	—	—	—	—	—	58.0	0.7	0.01
Cottonseed	4-20-836	0.8	22.7	0.8	2.3	17.0	51.5	0.2	—	—	—	—	—	—	—	51.5	0.2	0.0
Linseed	4-14-502	—	5.3	—	4.1	20.2	12.7	53.3	—	—	—	—	—	—	—	12.7	53.3	4.2
Palm		1.0	43.5	0.3	4.3	36.6	9.1	0.2	—	0.1	—	—	—	—	—	9.1	0.2	0.02
Peanut	4-03-658	0.1	9.5	0.1	2.2	44.8	32.0	—	—	1.3	—	—	—	—	—	32.0	0.0	0.0
Safflower	4-20-526	0.1	6.2	0.4	2.2	11.7	74.1	0.4	—	—	—	—	—	—	—	74.1	0.4	0.0
Soybean	4-07-983	0.1	10.3	0.2	3.8	22.8	51.0	6.8	—	0.2	—	—	—	—	—	51.0	6.8	0.13
Sunflower	4-20-833	—	5.9	—	4.5	19.5	65.7	—	—	—	—	—	—	—	—	65.7	0.0	0.0

[a] Adapted with permission from Ref. 11. A dash indicates that the measurement was taken but no values were detected.
[b] International feed number.

Table 7.9 Principal fatty acids of major marine oils of commerce (g/100 g)[a,b]

FA	M	SM	P	C	H	A	CL	MA	HM	NP	S	SA
C14:0	9	7	8	7	7	9	3	8	8	6	1	7
C16:0	20	15	18	10	16	19	13	14	18	13	16	15
C16:1	12	10	10	10	6	9	10	7	8	5	7	8
C18:1	11	15	13	14	13	13	23	13	11	14	16	9
C20:1	1	3	4	17	13	5	0	12	5	11	10	15
C22:1	0.2	2	3	14	20	2	6	15	8	12	14	16
C20:5	14	17	18	8	5	17	11	7	13	8	6	9
C22:6	8	10	9	6	6	9	12	8	10	13	9	9

[a] Reprinted with permission from Ref. 39.
[b] M, menhaden; SM, specially processed marine oil (menhaden); P, pilchard; C, capelin; H, herring; A, anchovy; CL, cod liver; MA, mackerel; HM, horse mackerel; NP, Norway pout; S, sprat; SA, sand eel.

Lecithins. Soybean lecithin is identified as AAFCO 84.10; IFN-4-04-562, and lecithin as AAFCO 87.5, IFN 8-08-041, 21 CFR 582.1400 GRAS in the AAFCO's *Official Publication.* Lecithin is also known as "gums" or "phospholipids" when hydrated and recovered wet from oil. Commercial crude, dried, soybean lecithins are standardized before trading to meet the NOPA specifications for fluid natural lecithin, containing 62% minimum acetone insolubles, 1% maximum moisture, 0.3% maximum hexane insolubles, 32 maximum acid value (AV), 10 maximum Gardner color, and 150 poises maximum viscosity at 25°C (77°F). This is usually accomplished by adding fatty acids and oil to ensure fluidization. The crude lecithin may then be deoiled, fractionated by alcohols and other solvents, and granulated for specific applications (40–42).

The major lecithin sold domestically in commercial quantities is extracted from soybeans. Corn and sunflower seed lecithins are available in limited amounts. Canola is being reviewed as a lecithin source in countries that do not grow significant quantities of soybeans. Lecithins may be added to feeds in crude or refined forms, remain as residuals in solvent- or mechanical-

Table 7.10 Composition of polar lipids in deoiled lecithins of various species[a]

Compound	Soybean (%)	Cottonseed (%)	Corn (%)	Sunflower (%)	Rapeseed (%)	Peanut (%)	Rice Bran (%)
Phosphatidylcholine (PC)	29–39	34–36	30	13–27	16–24	49	20–23
Phosphatidylethanolamine (PE)	20–26	14–20	3	15–18	15–22	16	17–20
Phosphatidylinositol (PI)	13–18	—	16	7–8	8–18	22	5–7
Phosphatidylserine (PS)	5–6	7–26	1	—	—	—	—
Phosphatidic acid (PA)	5–9	—	9	—	—	—	—
Phytoglycolipids (PGL)	14–15	—	30	—	—	—	—
Other phospholipids	12	—	8	—	—	—	—

[a] Reprinted with permission from Ref. 39.

Table 7.11 Composition of fatty acids in deoiled lecithins of various species[a]

Fatty Acid	Soybean (%)	Cottonseed (%)	Corn (%)	Sunflower (%)	Rapeseed (%)	Peanut (%)	Rice Bran (%)
Myristic (C14:0)	0–2	0.4	—	—	—	—	—
Palmitic (C16:0)	21–27	31.9	17.7	11–32	18–22	12–34	18
Palmitoleic (C16:1)	7–9	0.5	—	—	—	—	—
Stearic (C18:0)	9–12	2.7	1.8	3–8	0–1	2–3	4
Oleic (C18:1)	17–25	13.6	25.3	13–17	22–23	30–47	43
Linoleic (C18:2)	37–40	50.0	54.2	42–69	38–48	27–36	34
Linolenic (C18:3)	4–6	—	1.0	—	7–9	—	2
Arachidic (C20:0)	0–2	—	—	—	—	—	—
Total gossypol	—	9.1	—	—	—	—	—
Free gossypol	—	0.02	—	—	—	—	—

[a] Reprinted with permission from Ref. 39.

extracted oilseed meals, be returned to oilseed meals as extracted gums or soapstocks at combined solvent extraction-oil refinery operations, or simply be native to an oilseed fed whole.

As the shortest carbon chain members, the polar lipids in lecithins are traditionally depicted in the *sn*-3 triacylglycerol position. The major polar lipids and fatty acids in deoiled lecithins from various oilseed species are presented in Tables 7.10 and 7.11, respectively (44,45). In the trade, the term *lecithin* sometimes is used interchangeably with *phosphatidylcholine,* and *cephalin* is interchanged with *phosphatidylethanolamine.*

DIGESTION METABOLISM AND FATS FEEDING REQUIREMENTS

4.1 Comparative Fat Digestion Systems

An orderly system does not exist for discussing the use of fats over the breadth of domesticated animals; thus a survey approach is used here. Metabolism is the sum of processes by which nutrients are handled in the living organism. Digestion is the process of reducing foods into smaller particles and finally to compounds that are absorbed for physiological processes. The relative nutritional value of a component is the product of its concentration in the feedstuff times its digestibility coefficient. Current research on fat metabolism is intensive, and new information is continuously reported; the reader is referred to technical journals for the latest details and concepts.

Within a period of several hundred years, people have learned how to breed selectively for broad-breasted turkeys and low-profile dachshund dogs, to get eggs from chickens and milk from cows the year around, and to feed cooked oilseeds and cereals to carnivores like dogs and cats. Although meat, milk, and egg production have been increased, the animals remain limited by their primordial digestive and metabolic systems, which must be respected in feeding management. Currently, recommended levels of fat supplementation often result in doubling the level already consumed by the respective species, with the existing digestive and metabolic system expected to handle about equal parts of indigenous and supplemented fats.

Short-chain fatty acids consist of 6 or less carbons; medium-chain fatty acids, 6–10 carbons; and long-chain fatty acids, 12–24 carbons (45). Considerable interest exists in the medium-chain fatty acids and triacylglycerols (MCTs), especially in young animal nutrition and human medical applications (46). These compounds are quickly catabolized or elongated in everyday nutrition, but will be mentioned only briefly here. Most of the storage triacylglycerols found in nature consist of long-chain fatty acids, called long-chain triacylglycerols (LCTs).

Familiarity with the comparative differences between digestive systems helps identify the opportunities and limitations in feeding fats to various species. Animals have been classified as (1) *carnivores,* flesh eaters who derive their nutrients and energy primarily from proteins and fats (dogs, cats, and mink among the mammals); (2) *herbivores,* vegetarians who depend entirely on plant materials (cattle, goats, sheep, and horses); and (3) *omnivores,* flesh and plant consumers (humans and swine). Carnivores have short digestive tracts and require concentrated food sources. Herbivores have relatively long digestive tracts, with additional structures like rumens and enlarged cecum-colons for microbial fermentation of fiber. The digestive tracts of omnivores are intermediate in length, complexity, and efficiency in using plant matter (3–6).

Birds, a major nonmammalian group without teeth for mastication, but with adaptations for reducing the size of their food, typically are described as *avians;* however, carnivores, herbivores, and omnivores exist among birds. Fish, who do not chew their food, must rely on strong digestive enzymes.

Lipids occur in cell walls, cellular cytoplasm, and in fat storage cells in animals. In contrast, specialized fat storage cells per se do not exist in plants or seeds. Rather, fatty acids are found in surface waxes that reduce the loss of moisture from leaves, stems, fruits, and seeds of plants; in seed cell walls and cytoplasm; and as stored triacylglycerols in dispersed spherical organelles in cells of seed embryos (48).

Fat digestion is a two-step process. The first step is gaining access to fats through proteinaceous matrixes of flesh, matrixes consisting mainly of nonstructural carbohydrates and proteins in seeds and fruits, or more complex matrixes, including structural carbohydrates (fiber). The second step is splitting the triacylglycerols into fatty acids and absorbable components and bringing these into the bloodstream by various pathways.

Overview of Digestion Systems. The basic digestive system can be described as *monogastric* (one stomach) consisting, sequentially, of the following:

A *mouth* for biting off and chewing food and for admixing saliva containing enzymes to initiate hydrolysis of carbohydrates and fats.

An *esophagus* for conveying the digesta.

A *stomach* (pH 1.5–3.5), where many proteins are brought below their isoelectric point (e.g., the curdling of casein), proteolytic enzymes are introduced, and protein-based matrixes are further reduced by churning; gastric lipases may also be introduced into the digesta.

A *small intestine*, where the fat is emulsified by bile salts; the triacylglycerols are hydrolyzed at the *sn*-1 and *sn*-3 positions by pancreatic lipases at near-neutral pH to produce fatty acids and *sn*-2-monoacylglycerols; and these products are absorbed by several alternative processes.

A *colon* (large intestine), where water is absorbed, and fermentation of remaining nutrients may occur.

A *rectum* for holding the extracted digesta.

An *anus* or vent for discharge of feces.

In all species, the small intestine is the main site for simultaneous hydrolysis of fats, proteins, and carbohydrates by selective enzymes and absorption of the resulting nutrients. It consists of three sequential sections: duodenum, jejunum and ileum, each with villi and mucosal linings. During the process, the pH of the digesta is raised from that of the stomach, to near neutrality over the length of the small intestine. In swine, the pH profile is as follows: stomach, 2.4; proximal duodenum, 6.1; distal duodenum, 6.8; proximal jejunum, 7.4; distal jejunum, 7.4; and ileum, 7.5. In sheep, the profile is abomasum, 2.0; proximal duodenum, 2.5; distal duodenum, 3.5; proximal jejunum, 3.6; distal jejunum, 4.7; and ileum, 8.0 (48). Several types of contractive and peristaltic actions mix and move the digesta down the intestine. The lower pH at the proximal duodenum of ruminants plays a critical part in fatty acid reabsorption. Hydrolysis of triacylglycerols by pancreatic lipase starts at the duodenum, but different absorption sites in the three sections are reported for the various species.

In ruminants (cattle, sheep, goats), the digesta is intercepted after the esophagus and processed in a four-section stomach before being returned to the basic digestive system at the small intestine. Fermentation is conducted by bacteria, protozoa, and fungi at near neutrality in the rumen. Fiber is digested to produce two- to 4-carbon acetic, propionic and butyl volatile fatty acids (VFA), which are absorbed through the rumen wall. Maximization of acetic VFA is preferred. Breaking down the fiber matrixes makes the nonstructural carbohydrates, proteins, and fats accessible to microbial enzymes. Within the capacity of the system, the triacylglycerols are hydrolyzed to fatty acids, the glycerol is metabolized, the short-chain fatty acids are absorbed through the rumen wall, and the polyunsaturated long-chain fatty acids are biohydrogenated and allowed to proceed to the small intestine for absorption (3–6).

But the overfeeding of high-grain (carbohydrate) rations can result in a buildup of lactic acid and a reduction of rumen pH to 4–5, resulting in conditions uninhabitable by rumen bacteria. This arrests normal fermentation, and the animal goes off feed with acute digestive problems, at the minimum. Part of this problem can be offset by feeding sodium bicarbonate or other buffers. Ammonia also is lost from proteins digested in the rumen, and the amino acid profiles of microbial protein typically are lower in quality than those of the original animal or oilseed protein supplements. Excessive free oil coats the fiber and interferes with its digestion and with the fermentation process. The need to get more nutrients into dairy cattle to increase milk production, despite the natural physiological limitations of the rumen, has led to the development of various commercial bypass, or escape, proteins and fats. These products are not accessible to rumen bacteria and are digested later in the monogastric part of the digestive system.

The nonruminant herbivores (horses, rabbits, guinea pigs) have a functional cecum and enlarged colon, where microbial fermentations are conducted on the digesta after the small intestine. Capabilities for absorption of selected microbe-digested products, primarily VFA and simple nitrogenous compounds, are extended to the large intestine for these animals. Most mammals have at least rudimentary cecums, but these usually are not efficient sites for fiber digestion due to limited size. Although the fiber matrix is not attacked until after initial digestion and absorption of accessible nonstructural carbohydrates, proteins, and lipids is completed, considerable metabolic energy is still generated and absorbed as VFA. Rabbits and guinea pigs practice coprophagy (the eating of their own feces) to increase the absorption of essential amino acids, vitamin K and the B complex vitamins produced in the cecum (3,4). Horses also will coprophage if fed protein-deficient diets.

Fatty Acids Synthesis, Elongation, and Desaturation. The main objective of feeding fats to animals is to provide a concentrated energy source, not to have the fat stored in the tissues. Recognized EFA requirements are no more than several percent of dry matter at the most, but the critical roles they play in maintaining the metabolic machinery has attracted the majority of current research on dietary fat utilization.

New (*de novo*) fatty acids are synthesized from two-carbon acetyl units produced during metabolism. Two enzyme complexes, acetyl-coenzyme A carboxylase and fatty acid synthetase, work in concert to build up fatty acid chains, two carbons at a time, until released by the complex. The primer in plants and animals is essentially a two-carbon acetyl group and the fatty acid chains have even numbers of carbons. If the primer is a three-carbon propionate group, odd-number carbon chains result. Odd-number fatty acids are common in microbial lipids and also are synthesized *de novo* from propionic VFA by rumen bacteria and deposited in adipose tissue. The length of the fatty acid synthesized depends on the tissue. Palmitic acid is produced in the liver and adipose tissue, and shorter-chain fatty acids are also produced in the mammary glands (49).

Palmitic acid produced by fatty acid synthetase is lengthened (or shortened), two carbons at a time, by (*1*) the mitochondria, which use acetyl-CoA and NADH, or NADPH for reduction, or (*2*) the microsomes, which use malonyl-CoA and NADPH, but in a different pathway from the cystolic enzyme fatty acid synthetase. Palmitic acid is elongated to stearic acid and desaturated to oleic acid (*cis* 9-C18:1) by the acyl-CoA Δ9 desaturase complex. All saturated long-chain fatty acids melt above body temperature range (lauric acid, 44°C; myristic acid, 58.5°C; palmitic acid, 63.6°C; and stearic acid, 69.5°C) and must be combined with polyunsaturated fatty acids in the triacylglycerols to ensure fluidity of body lipids. Desaturation is necessary to maintain fluidity of adipose tissue and of phospholipids in membranes and to enable elongation and conversion of the major dietary essential fatty acids (linoleic and linolenic) into even longer and more unsaturated fatty acids (50).

PUFA are required for synthesis of prostaglandins, cell wall lipids, and various other chemical structures. But, unlike plants, the vertebrate animals do not have enzymes able to insert double bonds between the 9-carbon and the terminal methyl group in C18 fatty acids. It has been known for many years that linoleic acid is the precursor for arachidonic acid synthesis and must be obtained through the diet. Unfortunately, early rat studies led to the mistaken belief that animals are able to interconvert linoleic acid to all other EFA. Gradually, it became apparent that there are at least four families of polyunsaturated fatty acids (*n*-7, palmitoleic; *n*-9, oleic; *n*-6 linoleic; and *n*-3 linolenic). Given the shortest member of each polyunsaturated C18 family, fatty acids can be elongated and desaturated to as many as 22 carbons and six double-bond sites, with various essential intermediate compounds also produced. A schematic diagram of pathways for synthesis of highly unsaturated fatty acids (HUFA) in fish is shown in Figure 7.1 (41). Elongation and desaturation is done by the same enzyme complex, and contemporary pathway diagrams show products of two alternative routes, desaturation first or elongation first. However, only products of the elongation first pathway are reported in typical fish oil analyses. The names of selected HUFA are shown in Table 7.12.

Since animals are able to desaturate up to, but not including, the C12 position, products of the *n*-7 (palmitoleic acid) and *n*-9 (oleic acid) families and the lesser-known *n*-5 myristoleic (9-C14:1*n*-5) family are interesting but not dietary essential. However, *n*-6 and *n*-3 fatty acids must be provided in food or feed to enable synthesis of their other family members. Linoleic and linolenic acids are plentiful in oilseeds and typically the least-cost sources, respectively.

Complete metabolic pathways may not be operational in certain animal species and in human fat metabolism disorders. Some animals require linoleic and arachidonic acid supplementation, although both are members of the *n*-6 family. Some carnivorous fish require eicosapentaenoic acid (EPA; 5,8,11,14,17-20:5*n*-3) and docosahexaenoic acid (DHA; 4,7,10,13,16,19-22:6*n*-3) rather than linolenic acid (9,12,15-18:3*n*-3) alone.

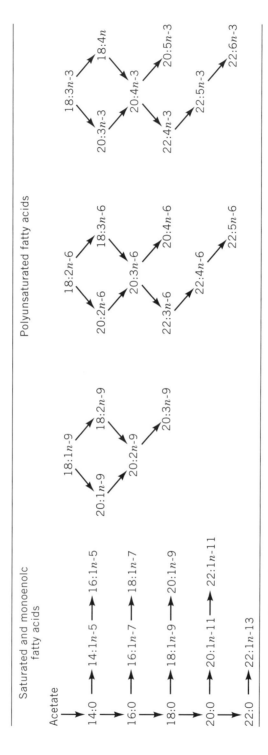

Figure 7.1 Pathways for synthesis of saturated and polyunsaturated acids. Reprinted with permission from Ref. 42.

Table 7.12 Members of polyenoic fatty acid families[a]

Series	Unsaturated	Common Name	IUPAC Nomenclature
Palmitoleic (family)			
C16:1 *n*-7	9–16:1	Palmitoleic acid	9-Hexadecenoic acid
C18:1 *n*-7	11–18:1	*trans*-Vaccenic acid	11-Octadecenoic acid
C18:2 *n*-7	8, 11–18:2	—	8,11-Octadecadienoic acid
Oleic acid (family)			
C18:1 *n*-9	9–18:1	Oleic acid	9-Octadecenoic acid
C18:2 *n*-9	6, 9–18:2	Octadecadienoic acid	6,9-Octadecadienoic acid
C20:2 *n*-9	8, 11–20:2	Eicosadienoic acid	8,11-Eicosadienoic acid
C20:3 *n*-9	5, 8, 11–20:3	Eicosatrienoic acid	5,8,11-Eicosatrienoic acid
Linoleic acid (family)			
C18:2 *n*-6	9, 12–18:2	Linoleic acid	9,12-Octadecadienoic acid
C18:3 *n*-6	6, 9, 12–18:3	γ-Linolenic acid	6,9,12-Octadecatrienoic acid
C20:3 *n*-6	8, 11, 14–20:3	Dihomo-γ-linolenic acid	8,11-14-Eicosatrienoic acid
C20:4 *n*-6	5, 8, 11, 14–20:4	Arachidonic acid	5,8,11,14-Eicosatetraenoic acid
C22:4 *n*-6	7, 10, 13, 16–22:4	Adrenic acid	7,10,13,16-Dicosatetraenoic acid
C22:5 *n*-6	4, 7, 10, 13, 16–22:5	—	4,7,10,13,6-Dicosapentaenoic acid
Linolenic acid (family)			
C18:3 *n*-3	9, 12, 15–18:3	α-Linolenic acid	9,12,15-Octadecatrienoic acid
C18:4 *n*-3	6, 9, 12, 15–18:4	—	6,9,12,15-Octadecatetraenoic acid
C20:4 *n*-3	8, 11, 14, 17–20:4	—	8,11,14,17-Eicosatetraenoic acid
C20:5 *n*-3	5, 8, 11, 14, 17–20:5	EPA	5,8,11,14,17-Eicosapentaenoic acid
C22:5 *n*-3	7, 10, 13, 16, 19–22:5	Culpadonic acid; DPA	7,10,13,16,19-Docasapentaenoic acid
C22:6 *n*-3	4, 7, 10, 13, 16, 19–22:6	DHA	4,7,10,13,16,19-Docosahexaenoic acid

[a] All are cis isomers, unless noted.

Nonruminant Mammalian Systems. Fat digestion is described in greater detail in the following sections. The digestive systems of monogastric animals (swine, mink, and fish), a ruminant (bovine), a nonruminant herbivore (horse), and an avian (hen) are shown in Figure 7.2 (4).

In neonate, suckling mammals, short- and medium-chain fatty acids are preferentially split at the *sn*-3 triacylglycerol position by oral and gastric lipases and are absorbed in the stomach, while the long-chain fatty acids are hydrolyzed at the *sn*-1 and *sn*-2 positions and by pancreatic lipases and are absorbed in the small intestine (50,51). With growth, the neonate fat digestion system becomes less active, and is replaced by the small intestine–pancreatic lipase pathway. But residual oral and gastric lipase activities and direct absorption of short-chain fatty acids in the stomach continue at reduced levels in some species and individuals. These factors help explain why young animals use long-chain fatty acids less efficiently and have inspired the use of medium-chain coconut oil fatty acids in human and calf milk replacers.

The digestion of triacylglycerols in adult nonruminant mammals has been described as initiated in the mouth by lingual lipase released in the saliva at the base of the tongue (52). Up to 6% of the fatty acids are hydrolyzed and initiate emulsion formation in the stomach. The digesta (called chyme at this location) is released from the stomach slowly into the duodenum to ensure complete mixing with the bile salts and emulsification. Lipolysis occurs by association of pancreatic lipase and co-lipase at the surface of the bile salt-stabilized emulsion. Amphipathic molecules (fatty acids, *sn*-2 monoacylglycer-

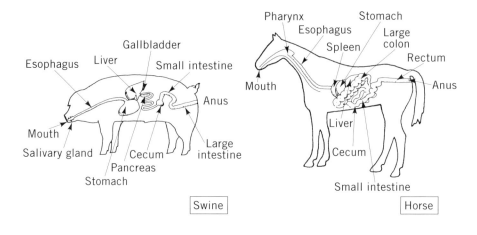
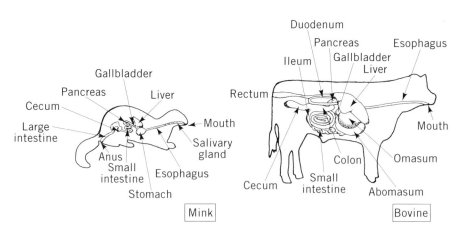
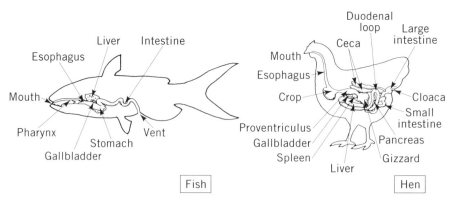

Figure 7.2 Digestive systems of representative animals. (*1*) Monogastrics with nonfunctional cecums: swine (omnivore), mink (carnivore), and catfish (omnivore). (*2*) Nonruminant herbivore with functional cecum and colon: horse. (*3*) Ruminant: bovine. (*4*) Avian: hen. Reprinted by permission from Ref. 4.

ols, and lysolecithins) are produced and associate with the bile salts to form water-soluble micelles from which absorption occurs.

Short-chain fatty acids and some medium-chain fatty acids are absorbed directly through the mucosa in the small intestine and are transported to the portal circulation and liver. In contrast, the long-chain fatty acids and 2-monoacylglycerols are first incorporated into bile salt micelles and transferred across the microvilli membranes. Then the fatty acids are activated into acyl-CoA esters and resynthesized into triacylglycerols. Resynthesis is favored if the *sn*-2-monoacylglycerol is unsaturated. The resulting hydrophobic triacylglycerols are coated with amphiphilic compounds (phospholipids—mainly phosphatidylcholine, cholesterol, and apoproteins) to form chyclomicrons. Chyclomicrons also can be produced by other pathways, including exocytosis (51), also called pinocytosis (4). The chyclomicrons enter the lymphatic system as chyle, pass through the thoracic duct, and enter the blood circulation in the jugular vein (52). Digestibility of fats depends on the unsaturated:saturated (U:S) ratio of fatty acids in the total diet. Digestibility is high (85–92%) at ratios above 1.5, but decreases linearly below U:S of 1.5 (52). This also helps explain the often seen improvements in digestible energy in tallow–vegetable oil blends. Differences in digestibility of various fats between species have been reported.

Ruminants. Ruminants (cattle, sheep) offer many exceptions to the basic one-compartment stomach digestive system model. The first is a four-compartment stomach (located ahead of the small intestine): the rumen, reticulum, omasum, and abomasum. Food in the mouth is mixed with saliva, which provides enzymes, bicarbonate and phosphate buffers, and lubrication, and passes through the esophagus into the atrium ventriculi, a convex area formed by the rumen and reticulum. The reticulum has a honeycomb-like interior that serves as a collection point for foreign objects and as an organ for digestion. The rumen is essentially a fermentation vat, operating at pH 6.2–6.8, and is inhabited by bacteria and protozoa that are later swept down the digestive system as food. Fats deposited in their cells become available to the animal through nonrumen digestion. The omasum contains numerous tissuelike leaves that help grind the food and is also believed to reabsorb water from the partially digested food. The abomasum, with a pH of 1.6–3.2, functions essentially like a true stomach (3,4).

Young ruminants initially function as young monogastric mammals. In calves, nursing stimulates a reflex closing of the esophageal grove, formed by muscular folds of the rumen and reticulum, and results in conveying milk from the esophagus directly to the abomasum. There, it is curdled by acid and rennin action, and the fat is digested by saliva and gastric lipases. The rumen starts developing as the young animal consumes increasing amounts of solid food; it is often considered to be operating at 2 months of age. Mature rumen size in lambs and goats is reached at 8 weeks, and at 6–9 months in cattle (3,4). Rumen microorganisms are able to detoxify limited amounts of compounds, such as free gossypol in whole cottonseed and cottonseed meal, and reduce the effects of trypsin inhibitor in raw soybeans. Despite possible economic pressures for early feeding of cottonseed products to young animals, it is critical that the rumen first be operating.

Not having upper incisor teeth, cattle break off plant stems between their tongue and upper dental plate when grazing or eating dry hay, and swallow the mouthful without chewing. Bulk feed in feedlots is also swallowed by similar action. The feed is later regurgitated as soft lumps called boluses; the "cud" is chewed and again swallowed. This crushing action, after initial soaking, enhances breaking down the typical fibrous matrix. Microbes in the rumen rapidly digest carbohydrates and cellulose, to produce VFAs, which are absorbed through the rumen wall and can provide 60–80% of the animal's energy needs from roughage alone (3).

The ether-extractable material in typical forages is 5–8%, of which about half is fatty acids occurring in monogalatosyldiacylglycerols and digalatosyldiacylglycerols. The fat content of most feeds is less than 5%, unless whole oilseeds or supplemental fats are added. In the rumen, fats are hydrolyzed to fatty acids and glycerol, which is metabolized immediately. Once in the unesterified form, 65–90% of the fatty acids are biohydrogenated, with the major products being stearic acid (C18:0) and vaccenic acid (*trans* 11-C18:1), a member of the palmitoleic (*n*-7) family. The presence of large amounts of free fatty acids and free oils and reduction of rumen pH by high grain feeding, increases the proportion of trans isomers formed. The saturated fatty acids are adsorbed to hydrophobic feed particles and are carried to the duodenum. The only esterified fatty acids reaching the duodenum are in the phospholipids of microbial cells. The fatty acids are dissociated from the feed particles by the detergent action of bile salts at the low pH of the proximal duodenum. In the absence of monoacylglycerols, lysolecithin (formed from the action of a pancreatic lipase on bile and microbial phospholipids) and oleic acid serve as amphipaths to form soluble micelles (52).

Factors favoring inhibition of microorganisms that digest cellulose and produce VFA and other compounds include the following (52):

Feeding supplement fats at concentrations greater than 2–3% of feed dry matter.
Degree of fatty acid solubility, with medium-chain (12- to 14-carbon) and unsaturated (vegetable oils, fish oils) being most inhibitory.
Unesterified fatty acids are more inhibitory than triacylglycerols.
Free oils are more inhibitory than those fed in whole seeds.

Factors minimizing inhibitory effects include:

Increasing roughage in the diet to induce more salivation results in increased buffer supply and rise in rumen pH; there is more surface area for fat absorption.
Removing the fatty acid or source from solution.

Ruminants are equal to or superior to nonruminants in their ability to digest saturated fatty acids, but unsaturated fatty acids fed conventionally have lower digestibility due to rumen biohydrogenation. Unsaturated fatty acids, fed in protected form, are digested equally in ruminants or nonruminants. Moderate levels of added fat (up to 3% of feed dry matter) are 85% truly digestible (52).

Although polyunsaturated fatty acids are biohydrogenated to stearic acid (C18:0) in the rumen, they are desaturated to oleic acid in the small intestine, mucosa adipose tissue, and mammary gland. Thus the ratio of C18:1 to C18:0 is greater in the fatty tissue and the milk lipids than in plasma triacylglycerols. Ruminant milk also contains 40–50% C4:0–C14:0 fatty acids by weight synthesized in the mammary glands from acetate (52).

Acetate is simultaneously produced and used (oxidized) by the gut, possibly accounting for up to 50% of the gut's energy requirement in ruminants. Propionate and butyrate produced in the rumen and/or hind gut fermentations (nonruminant herbivores) are extensively metabolized by the visceral tissues, so that only insignificant amounts reach the portal circulation. Plasma fatty acids, rather than acetate, are the major energy source for skeletal muscle energy in ruminants and other mammals (53).

Avians. Considerable specialization in digestive systems exists among avians. The focus here is on chickens, turkeys, and ducks (essentially omnivores) and ratites (herbivores). The flesh eaters are not discussed. The avian digestive system begins with the mouth and esophagus, which empty into the crop where feed is stored and soaked (see Figure 7.2). The feed then passes to the glandular stomach (proventriculus), where it is mixed with digestive juices, and then to a muscular organ (gizzard), which contains stones or grit to help crush and grind the feed. The digesta then moves through the small intestine where fat is digested and absorbed; to the ceca; the large intestine; and finally, the cloaca (3). Digestion, from the mouth to the cloaca, is rapid and takes about 2.5 h in the laying hen and 8–12 h in the nonlaying hen (4).

Following uptake, triacylglycerol resynthesis from long-chain fatty acids occurs in the small intestine mucosa of fowl by both the monoacylglycerol and glycerol-3-phosphate pathways. In contrast to mammals, the lymphatic system is poorly developed in poultry, and all nutrients are absorbed by the mesenteric portal system. The absorbed lipids are principally transported as triacylglycerols of the very low density lipoprotein fraction (VLDL). Lipoprotein synthesis in the liver is active in the laying hen. VLDL are present in the serum at high concentrations and appear to be the major lipid precursor for egg yolk synthesis (51,54). Thus poultry and poultry products seem to be an almost direct way for transferring selected fatty acids from feed into foods with little modification. Fats high in unsaturated fatty acids content are more readily absorbed by poultry. This partially explains the "extra caloric value" observed when vegetable oils are mixed with tallows in feeding broilers (55).

Ratites are flightless birds with heavily muscled legs for running and defense against enemies. The group includes the ostrich (*Sutruhio camelus*), emu (*Dromaius novaehollandiae*), common rhea and Darwin's rhea (*Rhea americana* and *Pterocnemia pennata,* respectively), Australian cassowary (*Casuarius*), and the kiwi (*Apteryx australis*). Although they are generally considered as zoo animals in the United States, there has been considerable interest recently in raising ostriches and emus for their hides, plumage, meat, and oil.

Ratites are equipped to subsist primarily on roughage in nature and have ceca of various activities. Emus have been reported to digest 35–45% of neutral-detergent fiber (NDF) in their diets (containing 26–36% NDF). This can contribute up to 63% of the standard metabolism and 50% of maintenance requirements for emus. Passage of feed though the gastrointestinal tract is rapid and 4.1 ± 0.2 h has been observed (56,57). Relatively little has been published on the use of fat in raising ratites, with about 6% total being the norm.

Fish. It is estimated that at least 130 aquaculture species are being cultivated. Even so, today's emerging industry already deals with far more species than the economic land animals and pets combined. Each species offers a potential challenge in unravelling a unique combination of metabolic pathways and nutrient needs. Fish are cold blooded, and are at one with their water environment in temperature and disposal of metabolic wastes.

Fish have been grouped into four classes: (*1*) *herbivores,* who eat green plants (carp, giant gourami, milkfish, perch, rabbit fish, tilapia, and Siamese gourami); (*2*) *detritus feeders,* who eat dead organic matter at the bottom of the pond, (mud carp); (*3*) *omnivores* (channel cat fish, common carp, gray mullet); and (*4*) *carnivores* (black carp, catfish, grouper, marble goby, salmon, sea bass, and trout). The relative intestine length of fish is shorter than that of land animals, but as with the land animals, carnivorous fish have shorter digestive tracts than the herbivores. The approximate ratios of intestine to body lengths of fish and land animals are trout, 1.0–1.5; carp, 2.0–2.5; dog and cat, 5; horse, 12; swine, 15; cattle, 20; and sheep, 30 (41).

Approximately 97% of the total fatty acids in fish oils consist of even-number carbon chains, but odd-number carbon chains and branched-chain odd-carbon acids are also present (58). Digestibility of fish fatty acids in poultry and swine feeds decreases with increasing chain length and increases with increasing unsaturation (59).

Green plants are able to synthesize the most complex lipids. As with land animals, fish function as gatherers and concentrators—in this case, of fatty acids that originated in plankton or may first have been cycled through lesser aquatic animals. Some carnivorous fish, in particular, have lost the ability to synthesize certain fatty acids and require supplementation. The marine food chain above 30°N latitudes is particularly rich in *n*-3 family fatty acids. In North Atlantic and Pacific fish, eicosapentaenoic acid ($C20:5n$-3) and docosahexaenoic acid ($C22:6n$-3) account for about 90% of total PUFA, and linoleic acid ($C18:2n$-6) plus arachidonic acid ($C20:4n$-6) account for less than 2%. As a result, the *n*-6:*n*-3 ratio is low (0.15 ± 0.1) compared with 0.38–0.93 in fish from Australian waters, where arachidonic acid is the major fatty acid in the southern latitude food chain. As the habitat moves farther south from 10°S Australia to 70°S Antarctica, the content of $C20:4n$-6 decreases in fish lipids, but $C20:5n$-6 increases. Among other documented observations is the fact that all desaturase enzymes are more active at lower (5°C) than higher (10°C) temperatures. A shift from $C20:4n$-6 to $C22:6n$-3 occurs in oil composition as fish adapt from freshwater to saltwater. The total lipids content is higher in cultivated fish than in wild fish and eels, but the total contents of $C20:4n$-6 and $C18:3n$-3 are higher in wild fish (60).

The feeding of soybean lecithin to cold-water fish and cold-water crustaceans—as a source of choline, inositol, ethanolamine, and PUFA—has been promoted in recent years. Levels as high as 7–8% of diet dry weight have been recommended (40,41).

4.2 Energy Requirements and Availability

Feedstuffs provide water, energy, protein, mineral elements, and also organic chemical structures that the animal cannot make itself, and therefore are termed *dietary essential*. This group includes vitamins, certain amino acids, and certain fatty acids. All organic materials, including the essential structures, also have value as energy sources for muscular movement, body heat, synthesis of tissue, and storage nutrients, and metabolic reactions. The buildup of complex compounds requires energy, and their hydrolysis or breakdown gives up energy.

NRC Definitions. The majority of terms, used for describing energy assessments of feedstuffs have been defined by the NRC (61). A calorie (cal) is the heat required to raise the temperature of 1 g of water from 16.5 to 17.5°C, and is equivalent to 4.184 international joules (a term used for relating mechanical, chemical, and electrical energies as well as heat). A kilocalorie (kcal) consists of 1,000 calories, and a megacalorie (Mcal) equals 1 million cal, 1,000 kcal or 1 therm. The British thermal unit (BTU) is 252 calories, the amount of heat energy required to raise 1 lb of water 1°F, but is rarely used in nutrition. The joule equals 10^7 ergs (1 erg is the amount of energy expended to accelerated mass of 1 g by 1 cm/s) and was selected by Le Système International d'Unités (International System of Units) and the U.S. National Bureau of Standards as the preferred term for expressing all forms of energy (16). The kilocalorie is used here because it is the standard terminology used by nutritionists and industry in the United States.

Somewhat different terms and techniques are used between species for estimating or calculating different types of metabolic and production energies. The general relationships used in poultry are shown in Figure 7.3 (16).

Gross energy (E, GE) is the energy released as heat when a substance is completely oxidized to carbon dioxide and water. It is generally measured at 25–30 atmospheres of oxygen in a bomb calorimeter and is called the heat of combustion.

292 Fats in Feedstuffs and Pet Foods

Figure 7.3 Disposition of dietary energy ingested by a laying hen. Reprinted with permission from Ref. 16.

Apparent digestible energy (DE) is the gross energy of feed consumed minus the gross weight of the feces. Birds excrete feces and urine together at the cloaca, and it is difficult to separate the feces and measure digestibility. DE values are used for many animals, but generally not with poultry.

Apparent metabolizable energy (ME) is the gross energy of the feed consumed, minus the gross energy contained in the excreta (feces, urine, and gaseous products of digestion). A correction term for nitrogen retained in the body is usually applied to obtain ME_n, the most common measure used in formulation of poultry feeds.

True metabolizable energy (TME) for poultry is the gross energy of the feed consumed minus the gross energy of the excreta of feed origin. A correction for retention of nitrogen is applied to give TME_n. Most ME_n values are determined by assays in which the test material is substituted for some ingredient of known ME value. When birds are allowed to consume the feed on an *ad libitum* basis, the ME_n values obtained approximate the TME_n values for most feedstuffs.

Net energy (NE) is metabolizable energy minus the energy lost to the heat of increment. NE may include the energy used for maintenance only (NE_m) or for maintenance and production (NE_{m+p}). Because NE is used at different levels of efficiency for maintenance or for production, there is no absolute NE value for each feedstuff. Productive energy, once a popular measure of the energy available to poultry from feedstuffs and an estimate of NE is seldom used (16).

Energy requirements can be partitioned in various ways, according to the needs for tending the species. For example, the DE requirement for the growing pig is the sum of its requirements for maintenance (DE_m), protein retention (DE_{pr}), fat retention (DE_{fr}) and cold thermogenesis (keeping the body warm) (DEH_c) (19) and is expressed as:

$$DE = DE_m + DE_{pr} + DE_f + DEH_c$$

Net energy of lactation (NEL, NE_L) is used for assessing energy requirements of dairy cattle (9). The relationships between different energy partitions for fish (11) are shown by a balance sheet in Figure 7.4. By this scheme, NE for fish is conceptually similar to NE_{m+p} for poultry, and recovered energy (RE) is equivalent to NE_p.

Gross (combustion) energies for carbohydrate, fat, and protein are 4.15, 9.40, and 5.65 kcal/g, respectively. Digestible and metabolizable energies have not been determined for all feedstuffs, and it is common nutrition practice to use 4 kcal/g for carbohydrates and proteins, and 9 kcal/g for estimating metabolizable energy (ME) values. However, calculated ME values for feedstuffs generally overestimate actual feeding trial results.

Analytical Techniques. Lipids in feedstuffs typically are determined by Soxhlet extraction, using diethyl ether as solvent. But this technique also extracts waxes, chlorophyll, and other pigments as well as galactosides and other lipid-soluble materials, thus overestimating the quantity of lipid energy available to the animal. Fatty acid contents of ether extracts are 90% in triglycerides, 70–80% in cereal grains, and as low as 50% in forages (52).

Intake energy (IE)
    Minus feces excretions (FE)

Digestible energy (DE)
    Minus gill excretions (ZE)
    Minus urine excretions (UE)

Metabolizable energy (ME)
    Minus heat increment (HiE)
        Waste formation and excretion (HwE)
        Product formation (HrE)
        Digestion and absorption (HdE)

Net Energy (NE)
    Minus maintenance (HeE)
        Basal metabolism (HeE)
        Voluntary activity (HjE)
        Thermal regulation (HcE)

Recovered energy (RE)
    Available for
        Growth
        Fat
        Reproduction

Figure 7.4 Schematic presentation of fate of dietary energy for fish. RE is available for new tissue.

Because some insoluble salts of fatty acids are formed during digestion, it is important to use acidified solvents in determining fatty acid content of feces. The differences between conventional ether extract and an acidified ether extract of feces is often reported as "fecal soap."

4.3 NRC Species Fat Feeding Considerations

The NRC subcommittees prepare extensive critical reviews when updating nutrient requirements for different species. Their reports probably have the broadest consensus of any available sources of nutritional information.

Pet Foods. Pet foods are a special consideration among the feedstuffs. Although formulated essentially from feed-grade ingredients, they are expected to have an anthropomorphic, or people foodlike, appearance and odor when used for house pets. Substantial quantities of pet foods are sold in discount chains and farm supply stores. "Professional" diets are also produced for dog breeders and trainers, guard services, police, and the military.

Dogs and cats are carnivores that have adapted to also eating cooked cereal carbohydrates and oilseed proteins. Cats are the more obligate carnivores, typically requiring higher protein content foods on a dry matter basis. Currently, they are the only domesticated animal for which a dietary essential level of taurine, an amino acid found mainly in meat, has been established.

Unlike the production-oriented economic animals (poultry, swine, cattle, fish, and shrimp) pets typically are allowed to live their full life cycles, and thus experience problems of aging, including kidney, vision, backbone, and hip failures, and obesity if food intake is not adjusted to the animal's activity level. Special puppy, kitten, and life-cycle series pet foods have been marketed.

Pet foods are the main products, regulated by the Association of American Feed Control Officials, that reach urban and suburban households. AAFCO has chosen to establish its own nutrient requirements, rather than enforce the NRC nutrient requirements (8,10). The following information was taken from the NRC recommendations (8,10) and perhaps are still the most broadly examined and published consensuses available on dog and cat nutrition.

Dogs. Far greater extremes in breed adult size and expected performance are seen among dogs than cats. Mature body weights range from 1 kg for the Chihuahua to 90 kg for the St. Bernard. Dogs are exposed to extreme temperatures and physical activities, like guard duty, pulling sleds, and hunting, not required of cats.

The breeds differ in the effectiveness of their coats to keep them comfortable over the range of temperatures encountered and the DE required to stay warm. Commercial dry dog foods usually contain 5.0–12.5% fat dmb. Nutritional deficiencies may be encountered in feeding high fat diets formulated without consideration for the higher energy value of fat. Dogs generally eat less of a high fat diet to maintain near normal energy intake, but in doing so, the daily intake of protein, minerals, or vitamins may be inadequate.

The apparent digestibility of fat by dogs varies from 80 to 95% when mixtures of plant and animal triacylglycerols are fed. Fatty acids are the primary source of energy for skeletal muscle during exhaustive exercise. Compared with a high carbohydrate diet, a high fat diet has been shown to lengthen the time to exhaustion of beagles on a treadmill by approximately 30% and to cause a greater elevation in plasma-free fatty acid concentration during exercise of sled dogs. Sedentary adult dogs have a greater tendency for obesity when fed high fat diets *ad libitum* rather than high carbohydrate diets.

Linoleic acid is the only essential fatty acid recommended, and a minimum linolenic acid requirement has not been established. Signs of linoleic acid deficiency in puppies progress from coarse, dry hair to scaly skin, to lesions. The NRC recommends that a dog food contain at least 5% fat on a dry basis, including 2.7 g linolenic acid in the diet per 1,000 kcal ME, or 1% on the dry basis for diets containing 3.67 kcal ME/g) (10).

Cats. Mature body weights of domestic cats (*Felis domesticus*) range from 2 to 6 kg. Dry commercial cat foods usually contain 8–12% fat, and purified diets, containing 25–30% fat and 30–40% protein have been commonly fed. Fat adds to the palatability of diets, with cats having shown preference for beef tallow over butter and chicken fat, but no preference among beef tallow, lard, or partially hydrogenated vegetable oil. A diet containing 25% hydrogenated coconut oil has been reported to be unpalatable. Diets containing medium-chain triacylglycerols have been poorly accepted, and as little as 0.1% caprylic acid (C8:0) has caused diets to become unpalatable.

Cats have the capacity to tolerate and utilize high levels of dietary fat. Diets containing 25% fat have been selected over those containing 10–50%. Apparent digestibilities of 90% of fat (when fed at 10% of dry matter) and 97–99% (when fed at 25–50% of dry matter) have been reported. Fat in experimental diets has been raised to 64% of dry matter, without an increase in the proportion of fecal fat, indications of ketonuria, or significant pathological changes in the cardiovascular system.

Experiences in cat nutrition underscore the fallacy of assuming that metabolic pathways found in one species are automatically present in others. Early studies on metabolism of PUFA were conducted on rats, which have high $\Delta 6$ and $\Delta 5$ desaturase abilities to convert linoleic acid (18:2n-6) to the prostaglandin precursors dihomo-γ-linolenic acid (20:3n-6) and arachidonic acid (20:4n-6), respectively. This led to the assumption that other species can desaturate polyunsaturated fatty acids equally well. Over a period of time, it was shown that cats are not able to convert 18:2n-6 to 20:3n-6 or 20:4n-6. The NRC currently recommends the inclusion of 5 g linoleic acid and 0.2 g arachidonic acid/kg diet dry matter.

Symptoms of EFA deficiency in cats include listlessness, dry hair coat with dandruff, poor growth, and increased susceptibility to infection. The effects of prolonged deficiencies include reduced feed efficiency without change in body weight, water loss through the skin, fatty livers, fat infiltration, and mild mineralization of kidneys as well as changes in blood platelet aggregation. Steatitis (yellow fat disease, a presumed vitamin E deficiency) has been often associated with consumption of high fish based diets, particularly canned tuna, red meat, and tuna oil. Supplemental vitamin E (D-α-tocopherol acetate) has been commonly included in diets containing tuna oils to prevent steatitis. A minimum of 30 mg α-tocopherol/kg diet has been set for cats by the NRC (8). Compositions and metabolizable energies for fats from various sources, including animal organ meats, that might be used in cat foods are shown in Table 7.13 (8).

Beef Cattle. Nearly 50% of the tallow and grease produced domestically is of bovine origin, but only about 12% of the amount used in domestic feeds is consumed by the combined beef, dairy and veal industries. The principles of feeding fats to beef cattle, sheep, and meat goats are similar to those for dairy cattle. In recent years, dairying has essentially abandoned pastures and switched to intensive feedlot nutrition and animal health care practices. In contrast, beef cattle continue to be grazed on dispersed pastures and rangelands as the most profitable use of the land and might be finish fed in feedlots before marketing. Opportunities to observe individual animals and keep daily records on performance have not existed, and economics have not favored use of the concentrated feedstuffs common in dairying.

Although knowledge of bypass proteins have been applied in beef operations, there has been relatively little use of bypass fats or direct feeding of whole cottonseed, again perhaps because of economic considerations. However, beef cattle nutrition and feeding management currently are two of the more active research areas rejuvenated by the challenge of designer foods. The beef cattle industry has recognized that the ability to deliver leaner meats with a more favorable nutritional image, at a cost competitive to other meat, poultry, and fish alternatives, is critical to its future (7).

Dairy Cattle. Perhaps the greatest strides in modifying feedstuffs for use by any species has occurred with dairy cattle. The challenge is shown in Figure 7.5 (9). When a high yielding dairy cow calves, dry matter intake greatly declines and remains depressed on an average of 15% during the first 3 weeks of lactation. Body energy reserves are quickly mobilized, the animal falls into negative energy balance, and weight is lost. Peak milk production is reached in 4–8 weeks, but the animal does not start gaining weight until about week 10 when milk production has started to decline. In simple terms, full advantage is not taken of the animal's ability to increase and extend milk production because not enough nutrients are provided during the critical period. Various feeding strategies, including the use of protected fats as a high energy source, have been improvised in attempts to restore energy levels and maximize milk production.

Table 7.13 Fat, fatty acid, and estimated metabolizable energy contents of oils for potential cat feeding[a]

Common Name	International Feed Number	Dry Matter (%)	Ether[b] Extract (%)	Saturated[b] Fat (%)	Unsaturated[c] Fat (%)	Linoleic[b] Acid (%)	Linolenic[b] Acid (%)	Arachidonic[b] Acid (%)	ME (kcal/kg)
Bran oil	4-14-504	100.0	100.0	18.5	81.1	36.5	—	0	8047
Fat									
Swine (lard)	4-04-790	100.0	100.0	35.9	64.1	18.30	—	0.3–1.0	7850
Bacon	4-15-582	100.0	100.0	42.3	56.7	6.8	0.6	—	—
Beef	4-25-306	100.0	100.0	44.9	55.1	1.9	1.2	1.0	—
Lamb	4-24-921	100.0	100.0	52.1	47.9	2.4	2.4	—	—
Rabbit	4-24-923	100.0	100.0	43.3	56.7	19.9	9.4	1.8	—
Turkey	4-24-924	100.0	100.0	36.5	63.5	19.0	1.0	4.8	—
Brain, lamb	4-15-583	100.0	100.0	41.4	58.6	0.2	0.8	2.4	—
Kidney, lamb	4-15-584	100.0	100.0	45.4	54.6	6.1	3.0	5.3	—
Kidney, beef	4-15-585	100.0	100.0	56.3	42.7	3.6	0.4	1.9	—
Kidney, swine	4-15-586	100.0	100.0	43.5	56.5	8.7	0.4	5.0	—
Liver, beef	4-15-587	100.0	100.0	49.6	50.4	5.5	1.9	4.8	—
Liver, swine	4-15-588	100.0	100.0	41.6	58.4	11.0	0.4	10.7	—
Margarine									
Hard animal and vegetable oils	4-15-589	84.0	96.4	37.5	62.5	4.2	0.1	6.6	—
Hard vegetable oily only	4-15-590	84.0	96.4	38.2	61.8	9.1	0.5	0	—
Soft animal and vegetable oils	4-15-591	84.0	96.4	30.7	69.3	8.1	0.4	6.1	—
Soft vegetable oils only	4-15-592	84.0	96.4	33.1	66.9	19.3	1.8	0	—
Soft, polyunsaturated vegetable oils only	4-15-593	84.0	96.4	24.7	75.3	49.3	0.6	0	—
Offal fat, poultry	4-09-319	100.0	100.0	39.1	60.9	22.30	—	0.5–1.0	—
Oil									
Coconut	4-09-320	100.0	100.0	90.3	9.7	1.10	—	0	8047
Corn	4-07-882	100.0	100.0	12.3	87.7	55.40	1.6	0	8047
Fish, menhaden	7-08-049	100.0	100.0	40.0	60.0	2.70	—	20.0–25.0	—
Flax, common (linseed oil)	4-14-502	100.0	100.0	8.2	91.8	13.90	—	0	8047
Safflower	4-20-526	100.0	100.0	10.5	89.5	72.70	0.5	0	8047
Evening primrose	4-15-591	100.0	100.0	8.5–13.5	86.5–91.5	73.0	10.4	0	8047
Soybean	4-07-983	100.0	100.0	14.7	85.3	51.9	7.4	0	7283
Cotton seed	4-20-836	100.0	100.0	26.8	73.2	53.0	1.4	0	8047
Rapeseed low erucic acid	4-20-834	100.0	100.0	6.9	93.1	23.0	10.0	0	8047
Sunflower	4-20-833	100.0	100.0	10.4	89.6	65.7	—	0	8047
Tallow, animal	4-08-127	100.0	100.0	47.6	52.4	4.3	—	0.0–0.2	8343
White grease	4-20-959	100.0	100.0	—	—	1.1	—	—	—

[a] Reprinted with permission from Ref. 8
[b] Expressed as percent (by weight) of the ingredient on a dry basis (100% matter).
[c] Expressed as percent (by weight) of the total fatty acids in the ingredient as fed; fatty acids make up about 95% of the weight of triglycerides, assuming the average triglyceride contains one glycerol, one 16-carbon fatty acid, and two 18-carbon fatty acids; conversion factors for fat in brain, kidney, liver, and eggs were as recommended by Ref. 62.

The newborn dairy calf requires fat in its diet until the rumen becomes functional. About 10% fat in milk replacers is sufficient to supply the EFA, carry fat-soluble vitamins, and provide adequate energy for normal growth. Higher fat (15–20%) content milk replacers are recommended for colder climates and for veal production. Forages and grains usually contain less than

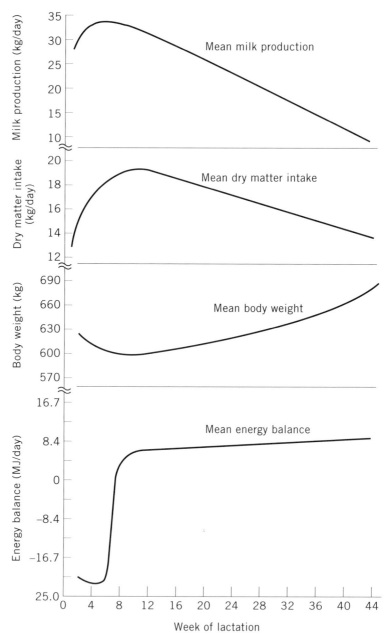

Figure 7.5 Relationships between milk production, dry matter intake, body weight, and mean energy balance of dairy cows during a typical lactation cycle. Reprinted with permission from Ref. 9.

3–4% fat. Dairy cows can use approximately 0.45 kg fat/day, or 2–3% added fat in their diet.

Maximizing forage (fiber) intake is an important consideration in successful feeding of supplementary fat. Also, partial substitution of fat for starch and grain can correct the low milk fat syndrome caused by inadequate fiber and excessive grain diets. As much as 20–30% supplemental calcium and magnesium above recommended levels should be included to provide for calcium and magnesium soaps formed in the rumen when dairy cattle are fed unprotected long-chain fatty acids. Feeding unsaturated fats to dairy cattle is less desirable because of their inhibitory effects on rumen fermentation and digestion if the biohydrogenation ability of rumen microorganisms is exceeded. Animal fats, and blended animal–vegetable fat mixtures, have given the best results. The principles of feeding a maximum of about 0.5 kg fat/day, or to limit cottonseed to 2.5–3.0 kg/day/cow, have generally been successful.

An increase in long-chain fatty acids in the diet increases their secretion in the milk and inhibits synthesis of short- and medium-chain fatty acids in mammary tissue. Added dietary fat and whole cottonseed decrease the protein content of milk by about 0.1%. "Dry fats"—calcium salts of fatty acids that are 82% fat content—and prilled (beadlet-form) fats have been introduced in recent years. These products are free-flowing and easy to mix with the ration, do not coat the fiber, and are not biohydrogenated in the rumen. It has been suggested that cows can be fed up to 15% of their requirements as fat, the equivalent of 6–7% of total dry matter (9).

Horses. Horses are among the most visible nonruminant herbivores and have a highly active cecum–colon that enables microbial fermentation of fiber, but after opportunities for digestion and absorption of fat, proteins and nonstructural carbohydrates in the small intestine have been passed. Apparent prececal digestibilities of 71% for nonstructural carbohydrates have been reported for a high corn diet and 46% for a high alfalfa diet. Fermentation produces the VFA (acetic, propionic, isobutyric, isovaleric, and valeric acids). Approximately 7% of total glucose production (gluconeogenesis) from cecal propionate has been observed in ponies.

Body lipids are mobilized during exercise and are oxidized readily during strenuous exercise like galloping. Physically conditioned horses oxidize fat more efficiently than nonconditioned horses. Hyperlipidemia is an important clinical problem in small pony breeds. It is most common in mares in late gestation and lactation and occurs when the animal is in negative energy balance.

Horses accept added fat in the diet readily if it is not rancid. Corn oil was found to be preferred among different fats and mixtures evaluated at 15% of the diet. Increasing fat levels in the diet, in the form of rendered fat, has resulted in decreased feed consumption, early growth advantage for foals, and increased milk fat concentration. Horses fed a diet containing 12% fat (9% added corn oil) performed better and had higher blood glucose levels at the end of the ride than horses fed 3% fat. Rendered animal fat has been shown to spare muscle glycogen reserves. EFA requirements of horses are not known, and inclusion of 0.5% linoleic acid in the diet is recommended until more data are available (13).

Swine. Cold thermogenesis, the amount of energy required for the animal to stay warm, is important in swine nutrition. A total of 25 g (80 kcal of ME) of feed/day to compensate for each 1°C below the critical temperature has been recommended for 25–60-kg pigs during their growth period, and 39 g/ (125 kcal) for 100-kg pigs during their finishing period. It is further estimated that DE is reduced by 0.017% for each degree Celsius that the effective ambient temperature (EAT) exceeds the upper critical temperature of the animal (19).

The value of adding fat to the diet of weaning pigs is uncertain. Pigs cannot synthesize linoleic and arachidonic acids, which must be supplied in the diet. The linoleic acid requirement of 0.1%/kg diet is more than adequately met in typical corn–soy diets, and requirements have not been set for arachidonic acid. The nutritional value of fat as an energy source for pigs is influenced by its digestibility, quantity of fat consumed and its ME, and the environmental temperature in which the pigs are housed. Substitution of fat for carbohydrate calories in a diet for pigs maintained in a thermoneutral environment increases growth rate and decreases the ME required/unit of live weight gain. For pigs housed in a warm environment, voluntary ME intake increases by 0.2–0.6% for each additional 1% of fat added to the diet. This increase is thought to occur because of the reduced heat increment of fat compared to that of carbohydrate.

The age of the pig, chain lengths of the fatty acids, and the unsaturated : saturated (U : S) fatty acid ratio influence the apparent digestibility of short- or medium-chain fatty acids (14 carbons or less). The apparent digestibility of fatty acids of 14 carbons or less is 80–95%; and that of fats from diets containing a ratio of U : S fatty acids greater than 1.5 : 1 is 85–92%. Apparent fat digestibility decreases by 1.3–1.5% for each additional 1% of crude fiber in the diet. Digestible energy (DE) and metabolizable energy (ME) of different fats for pigs is estimated as lard, 7,860 and 7,750, respectively; poultry fat, 8,635 and 7,975; tallow, 8,200 and 7,895; corn oil, 7,620 and 7,350; and soybean oil, 7,560 and 7,280 (19).

Poultry. The science of nutrition economics is perhaps more advanced for chickens than for any other species. The most appropriate energy level in selecting diets is the one that results in the lowest feed cost per unit of product (weight gain or eggs). Depending on the relative cost of high energy grains and feed-grade fats, this may vary between low energy and high energy diets throughout the world. Fat is usually added to the feed for meat-type poultry to increase overall energy concentration and improve productivity and feed efficiency. The moisture, insoluble, and unsponifiable compounds (MIU components) of rendered fats usually are of no value and act as diluents. Fatty acid chain length, extent of unsaturation, and nature of esterification all influence intestinal absorption. The percentage MIU and percentage digestibility combine to influence the ME_n value. All feed fats should be stabilized by an antioxidant to preserve unsaturated fatty acids.

Improved utilization of dietary fats has been shown to occur after 2–6 weeks of life for chickens, and is particularly evident with long-chain saturated fatty acids. When animal tallow is added to feed at a low level, it may be advantageous to blend it with a small amount of vegetable oil (55). The resulting ME_n values of blends are higher than can be accounted for on an arithmetic basis. Use of saturated fatty acids is improved by the presence of unsaturated fatty acids in the blend. In trials, fat utilization increased rapidly in the U:S ratio of 0–2.5, nearly reaching an asymptotic maximum at U:S of 4. For broilers, nearly 75% of the variation in fat utilization and ME_n was due to differences in the chemical composition of the fat fraction. High level fat feeding apparently increases intestinal retention time of feed and allows for more complete digestion and absorption of the nonlipid constituents. Adding fat to feed as an isogenic substitute for carbohydrate usually improves productive energy at the same level of ME_n. Characteristics and metabolizable energy contents of fats that have been fed to poultry are shown in Table 7.14. Equations for estimating metabolizable energy content of fats for feeding poultry are presented in Table 7.15 (16).

Fatty acid synthesis in fowl occurs primarily in the liver. The rate of fatty acid synthesis and deposit of body fat increase rapidly just preceding sexual maturity. Fat absorbed from the intestine is transported to the liver; the

Table 7.14 *Characteristics and metabolizable energy of various fats used for feeding poultry*[a]

Sample	MIU[b] (%)	Free Fatty Acids (%)	Selected Fatty Acids, Percentage of Total Fatty Acids						Energy Content "As Fed"	
			16:0	16:1	18:0	18:1	18:2	18:3	kcal ME/kg	Methodology[b]
Animal tallows										
Beef	0.3	4.3	26.1	5.1	25.2	37.4	1.9	—	6,683–6,916	—
	—	—	35.4	2.7	36.5	24.5	0.9	—	7,268–7,780	Me_n poults 10%
Commercial	3.5	1.6	22.0	2.7	15.8	47.6	8.7	1.9	7,628	—
	0.5	2.4	25.8	3.7	18.1	42.1	4.6	—	6,808–8,551	Me_n poults 2–8 weeks
	2.2	4.8	26.9	3.3	17.4	41.5	7.5	0.1	6,020–7,690	ME_n chicks 10–20%
	4.1	6.0	19.9	1.5	14.0	47.2	12.7	1.7	6,060	—
	3.0	10.2	21.2	5.9	15.5	45.4	9.6	1.2	7,148	—
	4.0	15.5	22.0	3.6	13.1	49.6	8.4	1.7	6,258	Me_n chicks 9%
	3.6	16.5	22.5	3.0	16.0	47.9	7.0	1.6	6,709	—
	2.9	19.1	25.5	4.0	19.3	40.0	4.9	<0.1	6,633–9,353	ME_n chicks 2–6%
Soap stocks	5.9	65.1	36.2	0.9	9.6	44.1	8.2	—	4,900	—
Animal–vegetable blends										
Commercial	3.6	61.0	21.0	1.4	6.0	25.4	38.6	4.2	7,114–8,924	ME_n poults 2–8 weeks

Table 7.14 (Continued)

Sample	MIU[b] (%)	Free Fatty Acids (%)	16:0	16:1	18:0	18:1	18:2	18:3	kcal ME/kg	Methodology[b]
Commercial edible	—	—	21.1	2.1	16.2	41.3	10.3	0.6	9,360	—
Tallow, crude canola	—	—	16.8	2.2	10.3	47.6	12.1	4.6	8,710	—
Tallow, crude soy	0.8	13.6	19.8	1.6	10.3	34.4	29.9	6.3	7,660	ME$_n$ chicks 10%
	0.9	2.6	19.0	1.7	10.7	34.3	27.8	3.8	8,110–8,820	ME$_n$ chicks 10%
Tallow, refined corn	—	—	20.9	2.1	10.4	32.2	30.5	0.4	9,570	—
Tallow, refined soy	0.7	13.8	19.4	1.5	10.3	34.8	29.5	6.4	7,830	—
Tallow, vegetable oil soap stocks	1.5	49.2	24.7	2.3	9.6	34.6	21.9	0.5	8,490	—
Animal soap stock, soy soap-stock	1.7	68.7	23.9	0.5	6.9	34.1	32.6		5,834	—
Beef, crude soy	0.9	36.3	17.7	1.0	12.5	34.5	31.2	3.9	7,571	Me$_n$ chicks 9%
	0.8	36.2	16.0	3.1	12.2	32.4	31.0	3.9	7,788	ME$_n$ chicks 9%
Lard, crude canola	—	—	17.2	1.3	9.5	51.1	13.7	3.2	10,000	—
Canola oil										
Crude oil	—	—	4.9	0.4	1.9	61.0	18.8	7.7	9,210	TME 15%
Soap stock	—	—	9.9	0.4	4.8	52.4	22.4	7.5	7,780–8,930	ME$_n$ TME regression
Coconut oil										
24 oils, MCFA = 57%	—	—	8.2	0.4	3.0	5.7	1.8	—	—	—
Undefined, MCFA[d] = 34%	—	—	12.8	—	2.9	13.7	23.1	—	8,812	ME$_n$ chicks 9%
Corn oil										
Refined	—	—	12.2	0.5	0.7	24.7	60.5	1.4	9,639–10,811	ME$_n$ poults 10%
	—	—	12.4	0.1	1.9	26.9	57.0	0.7	9,870	TME 15%
Fish oil										
Menhaden	—	—	—	—	—	—	—	—	8,450	ME$_n$ chicks 4–12%
Hydrogenated	—	—	18.6	5.8	4.8	18.5	24.1	1.3	6,800	ME$_n$ chicks 9%
Lard										
Edible	—	—	28.7	2.1	19.6	40.9	8.7	—	9,114–9,854	ME$_n$ poults 10%
	—	—	28.9	2.2	16.9	38.0	9.7	0.2	9,390	TME 15%
	0.2	0.1	26.6	3.1	15.8	42.4	9.1	<0.1	9,926–10,236	ME$_n$ chicks 2–6%
	1.1	0.2	22.4	2.1	17.7	46.1	8.0	2.1	7,337	ME$_n$ chicks 9%
Palm oil										
Fatty acid composite	—	100	46.4	0.2	5.0	38.7	6.9	0.1	7,710	TME 15%
Refined oil	1.8	0.2	40.7	0.3	5.2	41.6	11.4	—	5,800	ME$_n$ chicks 9%
Used in cooking	1.8	1.0	38.0	1.5	5.5	44.3	9.0	—	5,302	—
Poultry fat										
Commercial	0.7	0.7	21.6	4.8	7.2	42.3	23.0	—	8,625–8,916	ME$_n$ TME chick 7%
	3.9	0.5	18.1	5.9	4.6	46.2	23.3	1.1	9,360	TME 7%
Soybean oil										
Crude	1.4	0.6	11.3	0.3	3.9	27.2	49.8	7.5	8,650–8,020	ME$_n$ chicks 10–20%
Dried gums	1.3	12.2	21.0	0.3	4.5	17.1	45.9	1.8	6,440	—
Crude	—	—	12.2	0.1	3.2	26.0	51.6	6.3	9,510	TME 15%
Refined	2.0	1.3	10.6	<0.1	3.9	25.1	52.1	7.0	9,687–10,212	ME$_n$ chicks 2–6%
	1.8	0.1	11.6	—	3.9	19.8	57.9	6.8	8,375	ME$_n$ chick 9%
Soap stocks	4.2	72.3	7.9	—	4.1	24.0	56.9	7.1	6,111	—
Used in cooking	4.0	1.1	28.5	—	5.0	35.8	28.0	2.7	6,309	—
Sunflower oil										
Refined	—	—	6.7	0.1	4.3	27.4	57.1	3.7	9,659	ME$_n$ chick 2–8%

[a] Reprinted with permission from Ref. 16.
[b] Moisture, ether insolubles, and unsaponifiable matter contents as a percentage of the fat.
[c] ME$_n$, apparent metabolizable energy corrected for nitrogen retention; TME, true metabolizable energy using the rooster unless otherwise stated, and level(s) of fat used in the test diet. Some ME values are not corrected for nitrogen retention, particularly those before 1970.
[d] Medium-chain fatty acid contributions (C8:0 + C10:0 + C12:0).

Table 7.15 Equations for estimating energy value (kcal/kg dry matter) of feed ingredients from proximate composition for feeding poultry[a,b]

Ingredient	Prediction Equation
All fats and oils	$ME_n = 8{,}227 - 10{,}318^{(-1.1685)[\text{unsaturated:saturated ratio}]}$ $ME_n = 28{,}119 - 235.8\,(C18{:}1 + C18{:}2) - 6.4\,(C16{:}0) - 310.9\,(C18{:}0) + 0.726\,(IV \times FR_1) - 0.0000379\,[IV(FR_1 + FFA)]^2$
Vegetable oils (free fatty acid <50%)	$ME_n = -10{,}147.94 + 188.28\,IV + 155.09\,FR_1 - 1.6709\,(IV \times FR_1)$
Vegetable oils (free fatty acid >50%)	$ME_n = 1{,}804 + 29.7084\,IV + 29.302\,FR_1$
Animal fats (free fatty acid <40%)	$ME_n = 126{,}694 + 1645\,IV + 838.4\,C16{:}0 - 215.3\,C18{:}0 + 746.61\,FR_1 + 356.12\,(FR_1 + FFA) - 14.83\,(IV \times FR_1)$
Animal fats (free fatty acid >40%)	$ME_n = -9{,}865 + 194.1\,IV + 300.1\,C18{:}0$

[a] Reprinted with permission from Ref. 16.
[b] ME_n, nitrogen-corrected metabolizable energy; IV, iodine value; C16:0, percent palmitic acid; C18:0, percent stearic acid; C18:1, percent oleic acid; C18:2, percent linolenic acid; FFA, percent fatty acid, calculated as oleic acid equivalents; FR_1, first fraction from a column chromatography separation that contains the practically unaltered triglycerides plus other apolar components.

unsaturated fatty acids are unchanged for the most part, but the saturated fatty acids, especially stearic acid may undergo desaturation to be converted to oleic acid. Elongation and further desaturation of 18:2n-6 and 18:3n-3 may occur in the liver. Most lipid in egg yolk is formed in the liver using fatty acid obtained from the diet or formed by *de novo* synthesis. Providing fats avoids the cost of synthesis and is more energy efficient than synthesis from carbohydrate. Directly employing dietary fat in the assembly of either body or egg lipids results in a fatty acid composition similar to that of the diet. Depot fat is most affected by the source of dietary fat and is more influenced by the vegetable oils having high proportions of polyunsaturated fatty acids than by saturated animal fats (16). The feeding of fish oil to chicks significantly improves growth and antibody production (63).

Linoleic acid (18:3n-6 and α-linolenic acid (18:3n-3) are metabolically essential fatty acids, but linoleic acid is the only essential fatty acid for which a dietary requirement has been established. Characteristic EFA deficiency symptoms observed in poultry include an increased need for water and decreased resistance to disease. A dietary requirement for linoleic acid has been set at 1% of diet dry matter. No major special considerations are mentioned for turkeys, ducks, ring-necked pheasants, Japanese quail, and bobwhite quail (16).

Fish. Energy requirements in fish are lower than in animals that maintain constant body temperatures, thus leaving more of the net energy in production diets available for growth and maintenance. Proteins and lipids are highly available energy sources for fish, while the value of carbohydrates is variable among species. Nile tilapia and channel catfish (warm-water omnivorous species) digest more than 70% of the gross energy in noncooked starch; rainbow trout (cold-water carnivore) digests less than 50%. Extrusion processing and other forms of cooking increase digestibility of starch for fish. Studies have shown extrusion-processed corn to have 38% higher DE for channel catfish than compression-pelleted corn; gelatinized starch has a 75% higher DE for rainbow trout than raw starch. The challenge in the coming years will be to learn how to make low cost carbohydrate feedstuffs usable in feeding fish and other aquaculture species.

Maintaining life takes priority over growth and other functions, and energy concentration is the first consideration in developing diets for fish. Protein usually is given first attention because it is more expensive than other energy-yielding diet components. Fish convert dietary protein to tissue as efficiently as warm-blooded animals. A dietary balance is then established between protein and energy, which may be expressed as carbohydrates—or as lipids in the case of the salmonoids.

In common with other vertebrates, fish cannot synthesize linoleic or linolenic acids *de novo* and vary considerably in abilities to convert 18-carbon unsaturated fatty acids to longer-chain, more highly unsaturated fatty acids of the same series. The EFA requirements for selected species of fish are presented in Table 7.16 (11). In general, freshwater fish require either linoleic acid ($18:2n$-6) or linolenic acid ($18:3n$-3) or both, whereas stenohaline marine fish (those unable to withstand a wide variation in water salinity) require dietary eicosapentaenoic acid (EPA, $20:5n$-3) and/or docosahexaenoic acid (DHA, $22:6n$-3).

Among the freshwater species, the ayu, channel catfish, coho salmon, and rainbow trout require $18:3n$-3 or EPA and/or DHA. Chum salmon, common carp, and Japanese eel require an equal mixture of $18:2n$-6 and $18:3n$-6. Nile tilapia and Zillii's tilapia require only $18:2n$-6 for maximum growth and feed efficiency. Stripped bass cannot elongate $18:3n$-3 and require $22:5n$-3 and $22:6n$-3. The EFA function as components of phospholipids in all biomembranes and as precursors of eicosanoids required for a variety of metabolic functions. Principal gross signs of EFA deficiency reported for various fish include fin rot, a shock syndrome, myocarditis, reduced growth rate, reduced feed efficiency, and increased mortality.

Lipids serve as an important source of dietary energy for all fish and to a greater extent for cold-water and marine fish, who have a limited ability to use dietary carbohydrates for energy. Increasing the lipid content of some diets has also been effective in decreasing protein requirements in some applications (11).

Table 7.16 Essential fatty acid requirements of fish[a,b]

Species	Fatty Acid Requirement
Freshwater fish	
Ayu	1% linolenic acid or 1% EPA
Channel catfish	1–2% linolenic acid or 0.5–0.75% EPA and DHA
Chum salmon	% linoleic acid; 1% linolenic acid
Coho salmon	1–2.5% linolenic acid
Common carp	1% linoleic acid; 1% linolenic acid
Japanese eel	0.5% linoleic acid; 0.5% linolenic acid
Rainbow trout	1% linolenic acid; 0.8% linolenic acid; 20% of lipid as linolenic acid or 10% of lipid as EPA and DHA
Nile tilapia	0.5% linoleic acid
Zillii's tilapia	1% linoleic acid or 1% arachidonic acid
Striped bass	0.5% of EPA and DHA
Marine fish	
Red sea bream	0.5% EPA and DHA or 0.5% EPA
Giant sea perch	1% EPA and DHA
Striped jack	1.7% EPA and DHA or 1.7% DHA
Turbot	0.8% EPA and DHA
Yellowtail	2% EPA and DHA

[a] Reprinted with permission from Ref. 11.
[b] Linolenic acid, 18:3(n-3); EPA (eicosapentaenoic acid), 20:5(n-3); DHA (docosahexaenoic acid), 22:6(n-3); linoleic acid, 18:2(n-6); and arachidonic acid, 20:4(n-6).

FAT UTILIZATION PRACTICES

5.1 Fat Selection Principles

Extensive reviews on feeding of fats to animals have been prepared (64,65). It has been reported that feeding fats at high levels to ruminants decreases digestibility of protein and fat by overwhelming the fat absorption capacity (66). Fatty acid intake affected fatty acid digestibility quadratically.

Sources and compositions of fatty materials, and characteristics of animals that will use them, were reviewed above. Approximately 3000–4000 citations on research in feeding fats to various species are available in computer-searchable databases, now approaching their twenty fifth year. The reader is referred to these for the role of fat in additional feedstuffs and animal species. Several principles stand out, which are discussed in the following paragraphs.

Fats are best fed back to the same species. Tallows are best used by cattle. Fish oils should only be used in levels that do not cause off-flavors in eggs, broilers, pork, or bacon. They are not recommended for cats and mink because of potential steatitis problems (67), unless the diet is adequately supplemented with vitamin E. Fish oils are valuable in growing fish, and are more effective in feeds if obtained from species whose metabolism focuses on the same HUFA family (*n*-6 or *n*-9) as the species cultivated.

In nonruminant species (poultry and swine), digestibility increases as chain lengths of fatty acids shorten and as polyunsaturation increases. Corrected digestibility and energy equivalent (the product of digestibility and heat of combustion) of beef tallow have been improved appreciably by adding up to 20% soybean oil (Figure 7.6) (55).

Fats not protected for "escape" or "bypass" will be altered if fed to ruminants. Within limits, triacylglycerols are hydrolyzed, and the polyunsaturated fatty acids are biohydrogenated by rumen bacteria enzymes.

Consistent performance in today's feed conversion–animal production industries requires consistent feedstuffs. In addition to providing products free of chick edema factors, pesticides, herbicides, polychlorinated biphenyls (PCBs), and other hazardous contaminants, the tallow and grease supplier is expected to deliver products with increasingly consistent fatty acid profiles. This requires appropriate raw materials purchasing, classification, segregation, and blending abilities. Fat suppliers likely have users with different needs. Modern quality-control procedures for fats already have gone beyond iodine value (IV)—a measurement of unsaturation in the product)—to gas liquid chromatography (GLC) for monitoring fatty acids profiles as well as potential contaminants.

Increasing use is being made of the peroxide value (PV, POV) as a means of screening feeding fats for potential toxic fat oxidation products. Fish larvae and fingerlings are the most sensitive to these compounds, but PV is also used

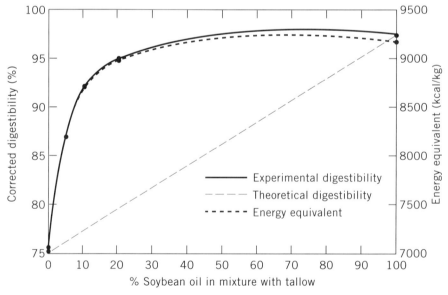

Figure 7.6 Synergistic effect of soybean oil on corrected digestibility and energy equivalent of beef tallow. Data from Ref. 55.

for monitoring fat quality in feeding poultry and other animals. Peroxides are unstable compounds formed during oxidative deterioration of fats and oils and may not themselves be the toxic constituents. Thus a fat that has undergone oxidation may have reduced peroxide values but still contain significant quantities of toxic components. Despite these limitations, PV currently is preferred in practice over 2-thiobarbituric acid (TBA), diene value (DV), triene aldehyde value, p-anisidine value, and totox (two times the peroxide value plus the anisidine value).

As background, the AOCS active oxygen method (AOM) test uses 100 meq peroxide/1000 g sample as the end point for predicting hours of stability of a fat sample heated at 97.8 ± 0.2°C, under controlled aeration. But, in food-grade applications, major buyers of frying oils typically specify 1 or 2 as the maximum allowable PV at receipt. In soy oil, 1.0–5.0 PV is considered low oxidation, 5.0–10.0 PV moderate oxidation, and 10+ high oxidation. The flavor of oil with a PV above 20 usually is so unacceptable that the product is of little interest.

Currently, PV generally is not included in industry product definitions or trading specifications, but increasingly appears in buyer purchase specifications. Oils with wide ranges of PV values have been reported fed to different species with varying toxic effects. The AOCS will table the AOM method as means for predicting the oxidative stability of fats and oils in 1996, with endorsement of oil stability index (OSI) and other methods that estimate the time for an oil to reach the state of oxidation induction, instead of a specified PV value. Interest in the healthfulness of polar compounds formed in oils during deep fat frying of foods is also increasing and is likely to lead to reviews on use of spent restaurant frying oils in animal feeds. Anything that is written on this topic today is likely to become outdated rather quickly. It is recommended that readers check with the prevailing industry practices regarding measurement and use restrictions of oxidized and heated oils in feeding animals.

Processing fats and oils for feeds use includes:

Mixing liquid tallows, greases, and oils into dry feeds.
Coating ("enrobing") extruded and pelleted feeds.
Extrusion of high fat content feeds.
Preparation of free-flowing granular fats for mixing into dry feeds.
Preparation of free-flowing rumen-protected fats.
Feeding whole cottonseed.
Extrusion of full-fat oilseeds.
Dry roasting of soybeans.

Not all decisions regarding the use of fats are technical or engineering-based. For example, liquid fats wick through kraft paper feed bags. In using a multiwalled bag, the fat barrier should be inside, next to the product. Some feed manufacturers prefer to use "harder" (higher melting point) fats to minimize this problem and possibly save on packaging costs. Undesirable fatty odors, and tendencies of products to oxidize (become "rancid") during storage, also can be reduced using higher melting point fats.

5.2 Fat Storage and Application

Storage Tanks. There is no major reason for not making full use of permitted antioxidants and metal scavengers in feed fats. These ingredients should be added to fats, with thorough dispersion, as soon as they are made. Nitrogen blanketing, to form a cover of inert gas, is often used in storage and handling of edible fats. Costs of maintaining a nitrogen blanket may not be justified for storage of inedible fats, but at least every effort should be made to avoid mixing in air, including tight seals on the suction side of truck unloading pumps and bringing fat into tanks through internal down pipes with tips submerged or floating to avoid splattering of fat.

Fats must be liquid to be pumpable, whether for addition to feeds or for further processing. Heat accelerates all deterioration processes. Quality of the final product is improved by storing fats at temperatures slightly above the melting point (50°C; 121°F). They should then be heated as needed, by heat exchangers to 80°C (175°F), to improve mixing and absorption in feeds. In cold climates, or in wintertime, it may be necessary to heat to higher temperatures to avoid formation of "fat balls" when added to cold ingredients at the mixer.

Tanks typically are made of mild steel, but stainless steel may be required when continuously handling fatty acids, acidified soapstocks, or fats containing high levels of free fatty acids. Approved epoxide-type internal coatings are sometimes used. Moisture in the oil should be kept low; an increase in moisture content from 1 to 2% is reported to double the corrosion rate. The tank should be built and leveled so the bottom drains to the discharge port. An external thermostatically controlled jacket steam heating ring or electrical element should surround the tank near the bottom. Internal steam heating coils can develop unnoticed leaks and should be avoided. Some operators prefer to take oil from the tank slightly off the bottom, allowing water and precipitate to collect in the low spot. The installation of a side-mounted mixer may be desirable if the fat is of a quality that separates or forms precipitates on standing. The propeller should be of a size, speed, and placing that does not incorporate air into the fat. A side manhole is preferred for cleaning and inspecting the tank. The top of the tank should have a vent or vacuum release valve to avoid collapsing the tank when unloading, and equipped to avoid intake and condensing of moist air. Insulation of the tank is a good investment in almost any location. The tank should be enclosed by a containment dyke large enough to hold its contents in case of leakage.

If the installation is large enough, welded piping is preferred, and an expansion loop may be warranted if long horizontal runs are required. Galvanic corrosion always occurs between dissimilar metals and can be reduced by the use of bolted flanges and gaskets. Copper and brass valves should be avoided. All lines should be chased with steam-heated copper tubing or electrical heating wire, with thermostat controls to prevent hardening of tallows. Condensate traps should be included to drain the tubing. Unless the pipes are designed to drain completely, they should be blown down by compressed air to prevent solidification of fat when the installation is not operating. The fat should be filtered at the time of receipt, after storage, and before application to the product. Two parallel filters with crossover valves are preferred at each location to enable continued flow of fat while one is being cleaned. Characteristics of typical feed fats include specific gravities of 0.893 and 0.866 at 49°C (120°F) and 93°C (200°F), respectively, weights of 7.44 and 7.22 lb/gal, and viscosities of 24 and 8 centipoise, respectively (68,69).

Most deliveries of feeding fats are made by insulated tank truck, which allows users without railroad sidings to be serviced, just-in-time deliveries, reduced numbers of local storage tanks, and rapid delivery while the fat remains hot. The industry and its contract haulers take pride in their equipment, and an inedible tallow tank truck looks little different from other milk or liquid ingredients trucks if kept clean.

Liquid Fat Application to Mixed Feeds. The typical operation consists of a horizontal batch dry feed mixer, equipped with fat addition ports. Hot fat is added to the batch after the dry ingredients have been mixed to minimize segregation of ingredients into fat balls. Some operators spray the fat, and others prefer to let it run into the mixer through small distributor lines or horizontal pipes with diagonal cuts. Much can be said in favor of the distributor line approach. Spray nozzles clog, even if equipped with screens. A spray fan pattern, that doesn't deposit fat on the sides of the mixer, must be chosen and the installation examined periodically to ensure the nozzles stay in alignment. Also, the production of spray mists cool the fat, reducing its temperature and wicking into the dry feedstuff.

In recent years, the addition of tallow has moved directly to dairy and beef feed lots. Small, heated, insulated tanks have been developed to receive and hold fat shipments (Figure 7.7). As described earlier, beef tallow is the most compatible fat for ruminants. It may be sprayed or flowed, at 2–3% of dry matter onto rations of chopped hay, other roughages, grains, or concentrates while in the mixer, or top-dressed on hay or other forages. Feed mixing operations like these require a positive displacement pump, like a Moyno or gear pump with variable-speed drive set to deliver a known amount of fat within a specific time period. As an alternative, mixers in feed mills can be mounted on load cells and equipped with controllers to stop the flow of fat when the desired weight has been added.

Hydrolyzed animal and vegetable fats, containing more than 90% fatty acids but no cottonseed oil, are available and used in feeding poultry. Their estimated metabolizable energy is 8360 kcal ME/kg.

310 Fats in Feedstuffs and Pet Foods

Figure 7.7 Insulated tanks for holding liquid fat at feeder lots. Note thermostatic controls for electrical heater (bottom center of each tank) and waterproof vinyl-coated covers. Courtesy of Griffin Industries, Cold Spring, Kentucky.

Fat Coating. Historically, about 2–3% fat at the maximum could be added to feed ingredients before pelletizing and still obtain pellet integrity. About 6% fat, total, could be present in extruded feeds and retain expansion. (Levels of 10–12% fat can be included in extrusion premixes for "semimoist" pet foods, preserved by controlled-water activity, but these products typically are not expanded.) The upper limits of internal fat content have been raised in recent years by moist heat conditioning and holding ("ripening") of starch-containing feedstuffs before pelleting and by more thorough precooking of starch in extrusion.

Fat coating of the pellet machine or extruded pellets improves animal acceptance of the feed because of surface fat aroma; binds heat-sensitive vitamin mixtures, flavorings, and rapidly hydrating "instant gravies" to dry pet foods; and enables addition of antioxidants, which steam distill easily and would be lost if included before extrusion. If emulsion-type coatings are used, they may be added to extruded feeds before the drying oven. Otherwise, the typical practice is to apply coating fats to warm pellets after the dryer, at 60–70°C (140–160°F) to enable the hot fat to soak partially into the product. The maximum amount of fat used for coating is about 5%. Cooling before packaging may be necessary to minimize wicking of liquid fat into the feed bag liner or container wall.

A slowly turning horizontal batch mixer, equipped with fat spray nozzles, can be used for coating small amounts of pellets or extruded collets. The mixing action should be gentle to avoid breakage of pellets. A simple continuous fat application system can be made by using a flat belt conveyor, straddled by a hopper and adjustable height gate to deliver a constant height (volume) of pellets. These are allowed to cascade off the end through a spraying chamber and then to continue through a slightly inclined barrel tumbler to finish distributing the fat.

Figure 7.8 shows a commercial fat applicator. The pellets enter through a hopper, are metered by a loss-in-weight belt feeder to cascade through a spray box, and then are mixed while being conveyed by mixing screw. A less expensive fat applicator, which uses a volumetric belt rather than a weight belt feeder, is also available from the same manufacturer.

A rotary pellet coating system is shown in Figure 7.9. Pellets and fat both are spun off separate rotating disks and mixed in a conveyor mixer. Elimination of spray nozzle clogging is claimed as the major advantage of this design.

Nozzles have the limitation of spraying only fat and fat-soluble materials alone. Dry materials must then be dusted onto the particles while sufficient liquid fat is at the surface to bind them. Another option is to add slurries of

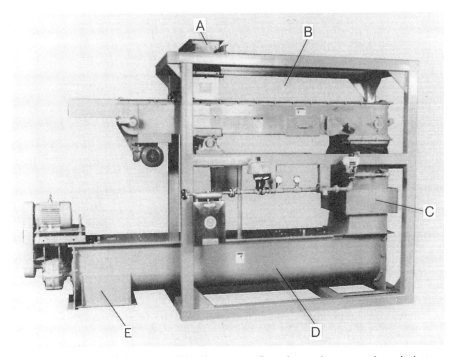

Figure 7.8 Gravimetric fat coating–blending system. Granular product enters through chute or hopper (A). A constant feed rate is maintained by a gravimetric weigh belt feeder (B). Liquid fat falls through a spray chamber (C). Coating is completed by tumbling in continuous open screw conveyor (D). Product is discharged (E). A lower cost volumetric coating–blending system, with the gravimetric feeder replaced by an adjustable-flow drag feeder, is available. Courtesy of Hayes & Stolz, Ft. Worth, Texas.

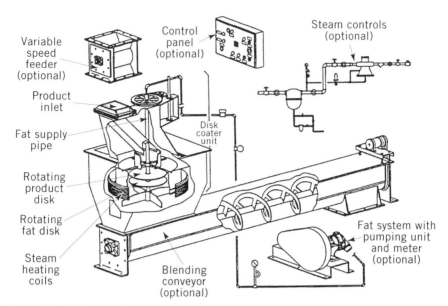

Figure 7.9 Disk-Coater fat coating system. Product is metered into a steam-coil heated chamber through a rotary lock feeder and is thrown off the upper rotating disk by centrifugal force. The particles fall through a thin spray of fat simultaneously thrown off the lower rotating disk. Mixing and absorption are continued in the takeaway conveyor. Courtesy of ASIMA Corp., Independence, Kansas.

all coating ingredients to pellets passing through a continuous screw conveyor mixer. Fine particles often occur in the pellet stream, either from breakage, polishing of coarse edges, or disintegration of pellets. As shown in Figure 7.10, fine particles can be removed (scalped) from the stream either before or after fat coating.

Extrusion of High-Fat Content Feeds. Aeration (expansion) is required to produce floating fish feeds. In turn, this limits the amount of fat used in the extrusion formula. Calorie-dense feeds can be made if the ability to float in water is not a necessity. Dry pet foods containing 9–12% fat have been produced for many years. Trout and salmon have evolved as carnivores that obtain their caloric needs from either fat or protein. Special techniques have been developed to produce 20–25% fat content feeds on single-screw extruders and 25–30% fat content feeds on twin-screw extruders. (The twin-screw extruder has corotating, intermeshing, self-wiping double screws that essentially enable it to operate like a positive displacement pump. In contrast, fat-caused slippage between the screw and barrel limits the ability of single-screw extruders ability to push product against the resistance of steam locks and the die plate.)

Techniques used in making high fat content feeds include (*1*) using nonex-

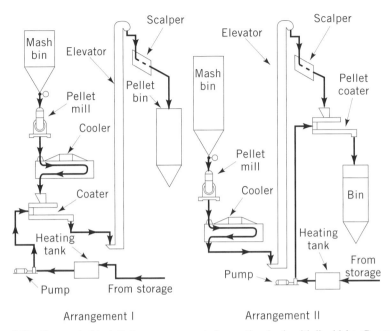

Figure 7.10 Two typical installation arrangements for coating feeds with liquid fat. Courtesy of Hayes & Stolz, Ft. Worth, Texas.

tracted cereals and oilseeds with full content of indigenous fats, (2) using starch-containing cereal fractions, (3) operating the extruder at moisture levels above 30% (well within starch gelatinization requirements, rather than the lower-moisture dextrinization range), (4) long-time (4–5 min) moistening and heating of cereals and other feed ingredients in the preconditioner, (5) use of special-design screw worm and barrel segments in single-screw extruders, (6) injection of steam into the extruder barrel to enhance starch cooking and formation of a starch-fat gel, and (7) pressure injection of fat into the screw barrel. After shaping, cutting, and drying, an additional 5% fat may be added by coating to obtain single-screw extruded feeds containing about 30% fat and twin-screw extruded feeds containing 35% fat (70).

A major advantage in extrusion is the ability to make various size feeds simply by changing the die plate. Particle sizes can range between 1.5 mm diameter for shrimp and eel larvae feeds, up to 2.5 cm (1 in.) for cattle feeds. One domestic manufacturer of commercial trout/salmon feeds offers products with the following particle diameters:

Meal. <0.600 mm (<U.S. mesh 30).
Crumbles. No. 1, 0.850–0.600 mm (U.S. mesh 20/30); no. 2, 1.18–0.850 mm (U.S. mesh 16/20); no. 3, 2.00–1.18 mm (U.S. mesh 10/16); and no. 4, 3.35–2.00 mm (U.S. mesh 6/10).

Pellets. 2.4 mm (3/32 in.), 3.2 mm (1/8 in.), 4.0 mm (5/32 in.), 4.8 mm (3/16 in.) (pigment available on request).

Brood Pellets. 6.4 mm (1/4 in.).

A large twin-screw extruder that is capable of making 6.3–15 MT of aquatic feeds and pet foods per hour and requires 450 kW (600 hp) is shown Figure 7.11. A process flow chart for making dry-expanded or semimoist pet foods and aquatic feeds is shown in Figure 7.12.

Free-flowing Fats. The objective in making free-flowing fats is to produce "dry" ingredients that can be readily mixed into feeds when rumen bypass is not required (primarily swine feed and milk replacers). Various mixtures of fats and cereal fractions, peanut and other hulls, and vermiculite were tried before the current spray-dried or chill-granulated products were developed. Choice white grease (including lard with its higher levels of unsaturated fatty acids) and coconut oil (containing medium-chain fatty acids) are often included. Milk or whole or delactosed whey solids, soybean protein, or corn syrup solids are included to help in processing the products. Typically, hardness of purchased fat (partially estimated by iodine value) increases with the

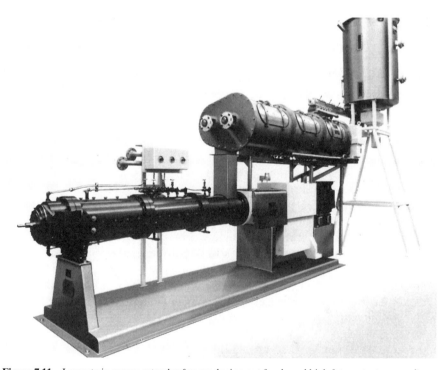

Figure 7.11 Large twin-screw extruder for producing pet foods and high fat content aquaculture feeds. Notice the differential diameter preconditioner (DDC). Machines available in sizes to 450 kw/600 hp, 15 MT/h. Courtesy of Wenger Manufacturing, Inc., Sabetha, Kansas.

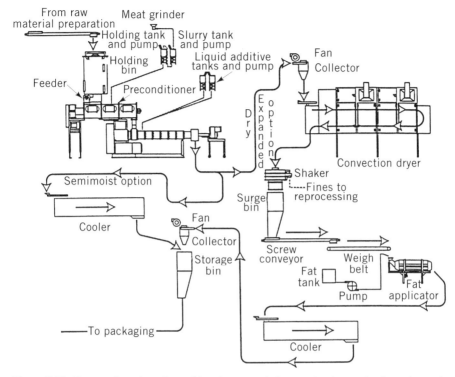

Figure 7.12 Process flow sheet for making dry expanded or semimoist pet foods, and aquatic feeds. Courtesy of Wenger Manufacturing, Inc., Sabetha, Kansas.

percent fat content of the product. Products sold domestically currently include (*1*) 40% choice white grease and 7% protein from nondelactosed whey; (*2*) 15% coconut oil, 45% edible lard, and 7% dairy protein; (*3*) 60% choice white grease and 7% dairy protein; (*4*) 80% edible beef tallow and 4% soy protein; (*5*) 80% coconut oil and 4% soy protein; (*6*) 80% choice white grease and 4% soy protein; (*7*) 80% animal fat, 4% dairy protein, and corn syrup solids; and (*8*) 90% animal fat. Lower fat content products (30% fat or less), spray-dried with skim milk, whey, or sodium caseinate, are available in addition to dry complete calf, lamb, and pig milk replacers.

Rumen-protected Fats. Rumen-protected fats are designed to be unavailable to rumen microorganisms, and pass to the stomach (abomasum) and small intestine for hydrolysis by pancreatic lipase and absorption. Free-flowing products are also desired. There is little advantage over the tallows, greases, and vegetable oils in using protected fats for nonruminants. Inertness in the rumen can be achieved by encapsulating the fat with a nondigestible protein, making the fat insoluble as a fatty acid–calcium soap, or by hydrogenation

316   Fats in Feedstuffs and Pet Foods

to raise the melting point of tallows (triacylglycerides) or fatty acids above the temperature of the rumen.

Scott and co-workers pioneered the development of rumen inert fats by demonstrating that the polyunsaturated fatty acids content of milk can be increased by feeding polyunsaturated oils encapsulated in formaldehyde-treated casein (71–74). Various types of domestically sold rumen inert fats are show in Table 7.17. Palmquist and co-workers pioneered the feeding of calcium soaps to dairy cattle (75–77). Their work became the basis for Megalac, made from palm oil and stearin.

Calcium soaps dissociate as they move from the rumen into the acidic abomasum. pKa values have been estimated for the following calcium soaps: soy oil, 5.6; palm fatty acid distillate, 4.6; tallow, 4.5; and stearic acid, 4.5. Unsaturated fatty acids soaps are less satisfactory for maintaining normal rumen function because their pH of dissociation is higher (78). The feeding of calcium salts of long-chain fatty acids to goats significantly increased milk fat content from 3.4 ± 0.1% to 3.7 ± 0.2%, but the content of short- and medium-chain fatty acids decreased (79). Supplemental feeding of methionine, protected by coating with C14:0–C18:0 fatty acids and calcium carbonate, has been reported to increase the protein content of milk (80,81).

Preformed calcium soap is highly available to laying hens, with about a 99.2% availability. True metabolizable energy of calcium soap has been reported as 7,200 kcal ME/kg, estimated by regression and 8,140 kcal ME/kg calculated from retention (82).

Table 7.17   Domestic rumen inert fat products

Product	Company	Ingredient Composition	Fat (%)
Alifet	Alifet U.S.A	Hydrogenated tallow mixed with wheat starch and crystallized	95
Biopass	Bioproducts, Inc.	Hydrogenated long-chain fatty acids	98
Booster fat	Balanced Energy Co.	Tallow plus soybean meal treated with sodium alginate	95
Carolac	Carolina Byproducts	Hydrogenated tallow—prilled	98
Dairy 80	Morgan Mfg.	Hydrogenated tallow—prilled; contains some phospholipid, flavor, and coloring agents	92
Energy booster	Milk Specialities Co.	Relatively saturated free long-chain fatty acids—prilled fat	98–99
Megalac	Chuch & Dwight Co.	Calcium salts of palm oil	82

Feeding of Whole Cottonseed. Whole cottonseed was one of the earliest forms of protected fats fed to ruminants, although its mechanism did not become apparent for many years. Feeding research on this crop was reported as early as 1890 in Mississippi (83) and 1894 in Texas (84). Because of its ready availability, cottonseed became the main source of edible oil domestically, from the 1890s until the late 1930s when soybean acreage increased rapidly. Competition and improvement in soybean and corn oil quality narrowed the price premium once enjoyed by cottonseed oil. Interest in feeding whole cottonseed was rekindled by a feasibility study by Stanford Research Institute in 1972. Currently, an estimated 35–40% of the cottonseed produced domestically is used for feeding dairy cattle. This practice has essentially spread throughout the country. Relative to other nutrient sources, cottonseed is worth enough to the dairy industry to outbid oil milling for the first 1.5–2 million tons produced in the United States even in low harvest years.

A typical response from feeding 3.6 kg (8 lb) cottonseed daily has been reported (85). Milk production rose by 4.9%, milk fat content by 15.3%, and total milk solids secreted by 1%, but milk protein content decreased by 6.4% compared with a no-cottonseed diet. Other studies have had similar results (86). It also has been reported that feeding cottonseed decreases the proportion of medium-chain fatty acids (C6-16) in milk and increases stearic and oleic acids (87). Since substantial premiums were paid for butterfat at that time, the return through the increased value of fat in the milk made the feeding of whole cottonseed profitable.

In contrast, milk production and fat contents typically decreased in feeding studies of other oilseeds and liquid oils (88–90). Various studies were conducted to explain the different response in feeding whole cottonseed. While this question has not been answered, various researchers have noted that protein in cottonseed has a higher bypass property than soy protein. Furthermore, the kernel in fuzzy cottonseed is enmeshed in a porous cocoon and shell, which the cow crushes but does not disintegrate while chewing its cud. One study that used nylon bags and rumen fistula showed only 16% digestibility of linters during 24 h in the rumen, 0.0% protein digestibility in whole (non-chewed) cottonseed, and 64% protein digestibility in finely ground cottonseed. The linters cocoon and seed protein may act together to provide a physical bypass effect in the rumen, which shifts the major portion of the protein and fat digestion to the small intestine.

Extrusion of Full-fat Oilseeds. Because oil in oilseeds exists in a proteinaceous matrix, its usability by animals depends on the treatment given to protein. Whole soybeans contain about 18% fat, and 38% protein, with the best balanced essential amino acid profile available in significant quantities from one plant source. Ruminants are able to tolerate trypsin inhibitor (an antigrowth factor) to a limited degree; however, most of the trypsin inhibitor should be inactivated before the soybeans are fed to monogastric species, especially poultry and swine. Extruders are used to shred soybeans and inactivate trypsin inhibitor simultaneously. During this process, other toxic or antinutritional factors (hemagglutinins, goitrogens, and others) are also inactivated.

A farm-size dry extruder, with a capacity of 273 kg (600 lb) per hour, is shown in Figure 7.13. This model also can be run by the power take-off of a farm tractor. The manufacturer also makes larger units. A diagram of a processing line for making full-fat soybean meal by wet extrusion is shown in Figure 7.14. The manufacturer recommends that the machine be operated, using soybeans at 8–10% moisture content and a product exit temperature of 145°C (295–300°F). Metabolizable energies (kcal ME/kg) for growing poultry (3850), adult poultry (3960), swine (4180), and ruminants (3335) have been reported for full-fat soybean meal made by this machine (91).

Almost a parallel relationship exists between deactivation of trypsin inhibi-

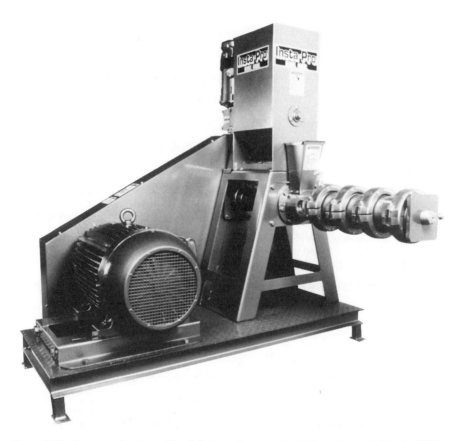

Figure 7.13 Dry extruder for making full-fat soybean meal and shaped extruded feeds; 272 kg (600 lb)/h. This machine also can be run by a farm tractor power take-off (PTO). Courtesy of InstaPro International, Inc., Des Moines, Iowa.

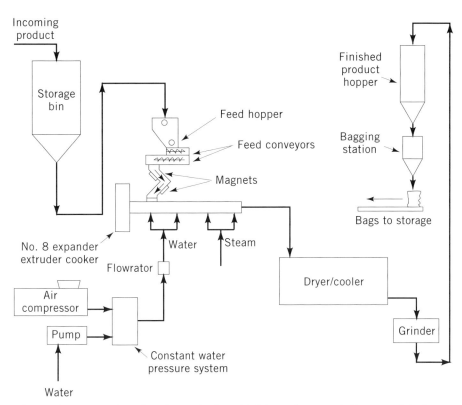

Figure 7.14 Process flow sheet for production of full-fat soybean meal. Courtesy of Anderson International Co., Cleveland, Ohio.

tor and urease activity by heat. The latter component is often tracked because it is easier to analyze. Moisture plays a significant role in the rate of enzyme deactivation. For example, at 15% moisture content and an extrusion temperature of 135°C (275°F), approximately 12% trypsin inhibitor is inactivated with a corresponding urease activity of 1.0 pH units. But at 20% moisture content, 89% of the trypsin inhibitor is inactivated (urease activity 0.1 pH units). The protein efficiency ratio (PER) is 1.82 and 2.15 at these moisture levels, respectively (92).

Extruders are also used for extrusion of secondary resources (coproducts rendering). Poultry mortalities, eggshells, feathers, shrimp heads, and various other meat and fisheries coproducts have been mixed with solvent-extracted soybean meal and extruded at sterilization temperatures for use as animal feeds. Significant amounts of fats can be recycled in this manner, while fresh and near the site of production (92).

Extrusion of whole double-O rapeseed, with 30–70% peas or 47–53% wheat meals, at 150°C (302°F) decreased trypsin inhibitors by 20–40%, total glucosinolates by 20–40%, and progoitrin by 46–60%. Rapeseed lipids had apparent digestibilities of 70.1% and 80.5% in pigs 4 and 7 weeks of age, respectively, and were similar to corn (maize) oil (93).

Dry Roasting. Roasters are simple to operate and maintain and do not have parts that wear as rapidly as those of extruders. A diagram of a commercial unit is shown in Figure 7.15. Roasting of soybeans is done for two effects: deactivation of trypsin inhibitor, and increasing bypass properties of the protein. The manufacturer recommends a roaster exit temperature of 115–127°C (240–260°F) for feeding pigs and a final temperature of 149°C (300°F) plus 30 min hot holding for increased bypass. Roasted soybeans with a protein dispersibility index (PDI) of 9–11 have been deemed as optimally heated; PDI 11–14, as marginally underheated; and PDI above 14, as underheated (94).

In one study, lactating cows were fed 8% soybean oil, either by direct addition of oil or as 50% coarsely ground raw soybeans (95). Both rations decreased content and yield of $C10:0$, $C12:0$, $C14:0$, $C14:1$, $C16:0$, and $C16:1$ fatty acids in milk fats but increased content and yield of $C18:0$, $C18:1$, and $C18:2$ fatty acids. Direct addition of soybean oil to the diet also reduced the percentage and yield of $C6:0$ and $C8:0$, while inclusion of raw soybeans increased the percentage of $C4:0$ and yields of $C4:0$ and $C6:0$ in the milk fat.

The feeding of roasted soybeans increases milk yield over the feeding of raw soybeans. But particle size has no apparent effect, indicating that whole or cracked roasted soybeans can be used as feed (96,97).

It has been shown that heating soybeans reduces their protein degradability in the rumen and postpones digestion to the small intestine (98). However, the production of unavailable protein is also increased (Figure 7.16). The relative effects of feeding soybean meal, roasted soybeans, extruded soybeans, and raw soybeans on milk yield, protein, and fat content are shown in Table

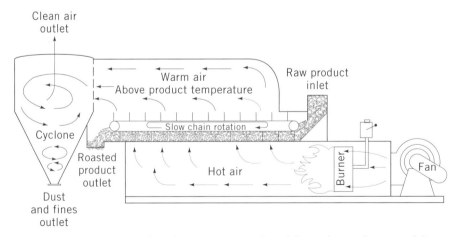

Figure 7.15 Schematic drawing of a crop roaster adapted for soybeans. Courtesy of Sweet Manufacturing Co., Springfield, Ohio.

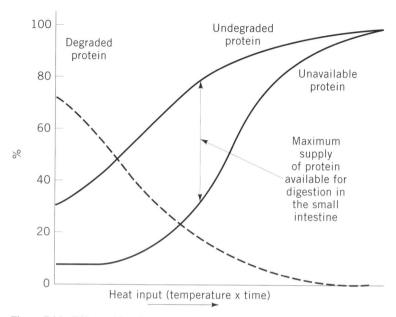

Figure 7.16 Effects of heating soybean on protein (and fat) bypass in cattle (98).

7.18. Extruded soybeans gave the highest yield of 3.5% fat corrected meal (FCM) and was second to soybean meal in total protein produced. However, fat and protein contents of milk from extruded soybeans were lower than for the other treatments.

The effects of two commercial rumen inert fats and whole roasted soybeans on yield of milk and its fatty acid profile were evaluated (99). Fatty acid profiles of the three fat sources are shown in Table 7.19. Whole roasted soybeans produced more total and 3.5% FCM milk, total protein, and total fat than the two rumen inert fat sources (Table 7.20). However, Megalac produced the milk with the highest fat content. Protein content in the milk decreased by about 0.1% for all fat sources. Fats of milks produced with Megalac (of palm oil origin) and Alifet (a hydrogenated tallow) contained more C16:0 and less C18:0, C18:2, and C18:3 than milk produced with roasted soybeans (Table 7.21). Feeding roasted soybeans whole apparently achieves fat bypass and milk production equivalent to rumen inert fats costing $0.40–0.50/lb.

Dust Control. The use of soybean oil (100,101) and inedible animal fats (102) to reduce dustiness in swine feeding operations, especially during months of low humidity, have been reported. Particle counts and total dust on a mass basis have been reduced by 50–99% at levels of 2% oil. Fewer bacterial spores and respiratory problems, improved pig health and survival, and improved feed efficiency and daily gain have been claimed.

Table 7.18 Effect of supplemental protein and fat on feed intake, body weight change, milk yield, and milk composition (98) (^{A,B,C,D}Means in the same row with different superscripts differ at given p value)

Item	Soybean Meal	Roasted Soybeans	Extruded Soybeans	Raw Soybeans
Intake				
DM, kg/d^a	24.7^A	22.5^B	25.1^A	22.7^B
DM, % of BW^a	4.01^{A,B}	3.83^{A,B}	4.10^A	3.74^B
CP, kg/d^b	4.32	3.87	4.29	3.88
UIP^c, kg/d	1.49^C	1.61^B	1.79^A	1.20^D
NEL^d, Mcal/d	38.3	35.6	39.7	35.9
Yield, kg/d				
Milk^{a,e}	36.0^B	37.5^{A,B}	39.0^A	35.9^B
FCM, 3.5%^{a}	35.7^{A,B}	36.6^{A,B}	37.6^A	35.2^B
Protein^a	1.16^A	1.06^B	1.14^A	1.06^A
Milk Composition, %				
Fat^{a,f}	3.48^A	3.35^{A,B}	3.25^B	3.45^A
Protein^{a,g}	3.10^A	2.95^{B,C}	2.89^C	3.00^B

^{a} Means were covariately adjusted.
^{b} No statistical analysis.
^{c} Undegraded intake protein.
^{d} Net energy lactation.
^{e} Significant week × treatment interaction ($p < .10$).
^{f} Significant week × treatment interaction ($p < .01$).
^{g} Significant week × treatment interaction ($p < .001$).

It is estimated that, each time grain is handled, about 0.15% weight is changed into dust, which potentially could reach explosive levels. The loss is equivalent to 1.5 kg/MT (3 lb/short ton). Use of 200 ppm (0.02%) of soybean oil sprayed onto the grain at the point of entry into commercial elevator and trading channels has been claimed to suppress 99% of the dust that typically becomes airborne. The FDA and the U.S. Federal Grain Inspection Service

Table 7.19 Fatty acid profile of fat sources (99)^a

Fat Source	C14:0 (%)	C16:0 (%)	C18:0 (%)	C18:1 (%)	C18:2 (%)	C18:3 (%)	Saturated^b (%)	Unsaturated^c (%)
Soybeans	—	12.5	4.5	22.0	52.9	8.13	17.0	83.0
Megalac^d	1.8	50.9	4.1	35.1	8.1	—	56.8	43.2
Alifet^e	3.6	26.7	37.2	31.7	0.8	—	67.5	32.5

^{a} Percent by weight of fatty acids reported.
^{b} Saturated fatty acids (C14:0, C16:0, C18:0).
^{c} Unsaturated fatty acids (C18:1, C18:2, C18:3).
^{d} Calcium salt of palm oil fatty acids.
^{e} Hydrogenated tallow.

Table 7.20 Effects of fat source on milk yield, milk composition, and blood (99)

Parameter	Treatment[a]			
	CTL	RSB	MG	AL
Milk yield (kg/d)	32.3	34.0	32.1	33.8
3.5% FCM[b] (kg/d)	31.6	33.8	32.6	32.5
Fat (%)	3.41[b]	3.45[b]	3.62[a]	3.25[c]
Protein (%)	3.01[a]	2.95[a,b]	2.92[b]	2.91[b]
Fat yield (kg/d)	1.10	1.18	1.16	1.11
Protein yield (kg/d)	0.96	1.00	0.94	0.99
Gross feed efficiency[c]	1.27[b]	1.31[b]	1.45[a]	1.31[b]
Blood plasma glucose (mg/100 mL)	70.7	69.8	70.3	69.9

[a] CTL, no fat supplement; RSB, roasted soybeans as fat supplement; MG, Megalac as fat supplement; AL, Alifet as fat supplement.
[b] 3.5% FCM, kg/d = 0.432 kg milk + 16.2 kg fat.
[c] Gross feed efficiency values were calculated on an individual cow basis (kg 3.5% FCM/kg DMI).

Table 7.21 Effect of dietary fat supplementation on milk fatty acid composition (99)[a] ([A,B,C,D]Means in the same row with different superscripts differ ($p < .001$))

Fatty Acid	Treatment[b]			
	CTL	RSB	MG	AL
C4:0	2.5	2.5	2.5	2.6
C6:0	2.1[A]	2.0[B]	1.8[C]	2.0[B]
C8:0	1.5[A]	1.4[B]	1.2[C]	1.3[B]
C10:0	3.7[A]	2.9[B]	2.4[C]	2.8[B]
C12:0	4.5[A]	3.2[B]	2.9[C]	3.3[B]
C14:0	13.6[A]	10.8[C]	10.5[D]	11.9[B]
C16:0	37.1[A]	27.2[C]	37.4[A]	32.8[B]
C18:0	10.2[C]	14.0[A]	9.8[C]	11.8[B]
C18:1	20.2[B]	27.3[A]	26.2[A]	26.7[A]
C18:2	3.3[C]	6.1[A]	3.5[B]	2.9[D]
C18:3	1.5[C]	2.2[A]	1.5[C]	1.9[B]

[a] Fatty acid percent by weight.
[b] CTL, not fat supplemented; RSB, roasted soybeans as fat supplement; MG, Megalac (calcium salt of palm oil) as fat supplement; AL, Alifet (hydrogenated tallow) as fat supplement.

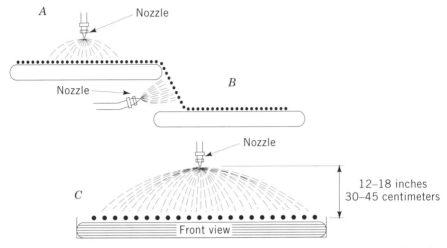

Figure 7.17 Recommended spraying procedures for applying soybean oil to reduce dust in grain. Oil may be sprayed on top of grain on conveyor (A) and/or on the underside of a cascade (B). The spray pattern (C) should be broad enough to cover the grain but not the sides of the conveyor. Courtesy of American Soybean Association, St. Louis, Missouri.

have approved the application of refined soybean oil at this level. Suggested techniques for applying food-grade (refined) soybean oil by spraying onto flowing grain are shown in Figure 7.17. Application should be made at temperatures (10.0°C, 50°F) where the oil flows freely (105).

5.3 Modification of Fatty Acid Profiles—Designer Foods

A recent survey of consumer attitudes and viewpoints found that the majority of U.S. consumers believe that natural substances in food can play a role in disease prevention (104). The designer foods movement to modify food components is broad and promotes increased uses of natural source phytochemicals; bioflavonoids; fiber, calcium, and/or vitamin-enriched milks and cereal products; probiotic yogurt, and isotonic beverages in addition to meat and eggs (105).

Fatty acid profiles of animals are automatically altered somewhat by the feed. Designer foods are modified for specific objectives. Chickens and swine readily translate fatty acids from feeds into meat and other products. Cattle also have this ability, provided the fat is rumen inert.

It has been shown that the fatty acid profile of milk can be changed through feeding (106). Table 7.22 shows the fatty acids profiles of fat sources fed, and Table 7.23 shows the fatty acids profiles of the resulting milks. The pattern as seen by earlier researchers is repeated. Supplemental dietary fats reduce the level of medium-chain fatty acids but tend to increase the content of long-chain $C18:0$ and $C18:1$ fatty acids. Many of the researchers have also

Table 7.22 Fatty acid content in feed and composition of test fats in a feeding trial on fat digestibility and milk fat composition[a]

Fat Source	Fatty Acid (% of DM)	Fatty Acid (g/100 g)								
		C14:0	C16:0	C16:1	C18:0	Trans C18:1	Cis C18:1	C18:2	C18:3	IV
Basal diet	1.1	5.2	19.7	—	2.9	—	20.2	37.3	3.1	94
Animal–vegetable blend	88.7	0.7	21.7	5.1	5.1	—	34.9	28.6	1.5	92
Calcium soap	83.8	1.5	50.1	—	4.2	—	34.1	7.8	0.3	45
Hydrogenated animal fat	83.6	3.0	24.8	1.2	34.5	12.8	15.3	1.1	0.4	30
Saturated fatty acids	100	2.2	44.1	0.9	34.7	2.9	9.4	0.9	—	14
Tallow	81.0	2.8	23.7	2.8	20.9	4.0	37.7	2.0	0.4	45

[a] Reprinted by permission from Ref. 106.

found that blood cholesterol levels increase in the presence of long-chain fatty acids.

As mentioned earlier, the lymphatic system is poorly developed in poultry. Absorbed lipids pass into the portal circulation and then are principally transported as triacylglycerols of the VLDL fraction. VLDLs are present in the serum at high concentrations and appear to be the major lipid precursor in the liver for egg yolk synthesis. Furthermore, fats high in unsaturated fatty acids are absorbed more readily than fats low in unsaturated fatty acids (51,54).

Designer eggs with higher n-3 fatty acids content and lower saturated fatty

Table 7.23 Effect of dietary fat source and level on milk fatty acid composition in a feeding trial[a,b] *(*[A,B,C]*Values with different superscripts differ ($p < .05$);* [X,Y,Z]*values with different superscripts differ ($p < .01$))*

Fatty Acid	Basal	Animal–Vegetable Blend	Calcium Soap	Hydrogenated Animal Fat	Saturated Fatty Acids	Tallow
C4:0	3.34	3.66	3.81	3.79	3.62	3.49
C6:0	2.70[A]	2.40[AB]	2.48[AB]	2.53[AB]	2.46[AB]	2.34[B]
C8:0	1.75[X]	1.34[Y]	1.35[Y]	1.39[Y]	1.41[Y]	1.34[Y]
C10:0	3.97[X]	2.51[Y]	2.57[Y]	2.63[Y]	2.72[Y]	2.60[Y]
C12:0	4.64[X]	2.75[Y]	2.84[Y]	2.88[Y]	3.03[Y]	2.89[Y]
C14:0	13.01[X]	9.33[Y]	9.54[Y]	10.28[Y]	10.10[Y]	10.30[Y]
C14:1	1.46[A]	1.08[B]	1.07[B]	1.26[AB]	1.26[AB]	1.31[AB]
C15:0	1.28[X]	0.87[Z]	0.84[Z]	1.07[Y]	1.06[Y]	1.04[Y]
C16:0	29.87[B,Y]	26.45[C,Z]	34.15[A,X]	28.42	32.67[A,XY]	28.41[BC,Z]
C16:1	1.68[Y]	1.64[Y]	1.64[Y]	1.72[Y]	1.99[Y]	1.80[XY]
C17:0	0.60[Y]	0.52[Y]	0.39[Z]	0.88[X]	0.78[X]	0.82[X]
C18:0	9.05[B,YZ]	11.50[A,XY]	7.71[C,Z]	11.68[A,X]	9.86[B,YZ]	10.43[AB,XY]
C18:1	17.22[C,Z]	25.74[A,X]	22.80[AB,XY]	22.89[AB,XY]	20.30[B,YZ]	23.26[AB,XY]
C18:2	2.24[B,XY]	2.00[BC,YZ]	2.58[A,X]	1.67[C,Z]	1.74[C,Z]	1.59[C,Z]
C18:3	0.55[C,Z]	1.16[A,X]	0.63[C,YZ]	0.72[BC,YZ]	0.62[C,YZ]	0.91

[a] Reprinted with permission from Ref. 106.
[b] n = 12; numbers are in g/100 g of methyl esters.

acids and total fat contents have been in U.S. and Canadian supermarkets since 1992. Typically, they are produced by feeding poultry diets containing whole flax seed or flax seed oil or by eliminating animal coproducts and feeding poultry diets containing vitamin E, kelp, and canola oil (107).

General experience in increasing the n-3 lipids contents of eggs and poultry meat includes the following. The feeding of linseed oil, menhaden oil, and soybean oil to chickens has resulted in similar blood plasmas with VLDL and LDL lower than those fed chicken fat. The levels of $C20:5n$-3 in tissue lipids of chickens fed linseed oil approached those of chickens fed menhaden oil. Chickens fed linseed oil had the highest levels of polyunsaturates in tissues while those fed chicken fat had the lowest. Chickens fed soybean oil maintained the highest levels of linoleic ($C18:2n$-6) and arachidonic ($C20:4n$-6) acids in tissue lipids, but linseed oil and menhaden oil resulted in reduced $C20:4n$-6 content (108). Other researchers have also reported similar decreases in arachidonic acid, and increased levels of eicosapentaenoic ($C20:5n$-3) and docosahexaenoic ($C22:6n$-3) in the fatty acids of egg yolks from chickens fed menhaden oil (109–112).

Human trials of four eggs/day have shown that control eggs significantly increased plasma cholesterol levels but were unchanged by eggs with high n-3 fatty acids contents. Mean plasma triglyceride concentration was decreased by n-3 eggs but increased by control eggs. Both systolic and diastolic blood pressures were significantly decreased in one group of participants consuming n-3 eggs, and systolic pressure was significantly decreased in a second trial. Blood pressure did not change when participants consumed control eggs (113).

REFERENCES

1. NRA, *Rendered Animal Fats and Oils,* National Renderer's Association, Inc., Washington, D.C., July 1992.
2. National Research Council, *Designing Foods: Animal Product Options in the Marketplace,* National Academy Press, Washington, D.C., 1988.
3. M.E. Ensminger and C.G. Olentine Jr., *Feeds and Nutrition—Complete,* Ensminger Publishing Co., Clovis, Calif., 1978.
4. M.E. Ensminger, J.E. Oldfield, and W.E. Heinemann, *Feeds Nutrition Digest,* 2nd ed., Ensminger Publishing Co., Clovis, Calif., 1990.
5. D.C. Church, ed., *Livestock Feeds and Feeding,* 3rd ed., Prentice-Hall, Inc., Englewood Cliffs, N. J., 1989.
6. D.C. Church and W.G. Pond, *Basic Animal Nutrition and Feeding,* O&B Books, Inc., Corvallis, Oreg. 1978.
7. National Research Council, *Nutrient Requirements of Beef Cattle,* 6th rev., National Academy Press, Washington, D.C., 1984.
8. National Research Council, *Nutrient Requirements of Cats,* rev. ed., National Academy Press, Washington, D.C., 1986.
9. National Research Council, *Nutrient Requirements of Dairy Cattle,* 6th rev. ed., National Academy Press, Washington, D.C., 1989.

10. National Research Council, *Nutrient Requirements of Dogs,* rev. National Academy Press, Washington, D.C., 1985.
11. National Research Council, *Nutrient Requirements of Fish,* National Academy Press, Washington, D.C., 1993.
12. National Research Council, *Nutrient Requirements of Goats,* National Academy of Sciences, Washington, D.C., 1981.
13. National Research Council, *Nutrient Requirements of Horses,* 5th rev. ed., National Academy Press, Washington, D.C., 1989.
14. National Research Council, *Nutrient Requirements of Mink and Foxes.* rev. ed., National Academy Press, Washington, D.C., 1989.
15. National Research Council, *Nutrient Requirements of Nonhuman Primates,* National Academy Press, Washington, D.C., 1989.
16. National Research Council, *Nutrient Requirements of Poultry,* 9th rev. ed., National Academy Press, Washington, D.C., 1994.
17. National Research Council, *Nutrient Requirements of Rabbits,* 5th rev. ed., 2nd ed., National Academy Press, Washington, D.C., 1977.
18. National Research Council, *Nutrient Requirements of Sheep,* 6th rev. ed., National Academy Press, Washington, D.C., 1985.
19. National Research Council, *Nutrient Requirements of Swine,* 9th rev. ed., National Academy Press, Washington, D.C., 1988.
20. National Research Council, *United States–Canadian Tables of Feed Composition,* 3rd. rev., National Academy Press, Washington, D.C., 1982.
21. D.F. Fick, D. Firestone, J. Ress, and R.J. Allen, *Poultry Sci.* **52,** 1637 (1973).
22. C.Y. Pang, G.D. Phillips, and L.D. Campbell, *Can. Vet. J.* **21,** 12 (1980).
23. G.R. Higginbottom, J. Ress and A. Rocke, *J. Assoc. Off. Anal. Chem.* **53,** 673 (1970).
24. L.B. Brilliant and co-workers, and H. Price, *Lancet* **2** (8091), 643 (1978).
25. M. Barr Jr., *Environ. Res.* **12,** 255 (1980).
26. R.A. Frobish, B.D. Bradley, D.D. Wagner, P.E. Long-Bradley, and H. Hairston. *J. Food Protection* **49,** 781 (1986).
27. A.J. Howard in J. Wiseman, ed., *Fats in Animal Nutrition,* Butterworth & Co., London, 1984, pp. 483–494.
28. USDA, *World Oilseed Situation and Outlook,* USDA FAS Circular Series **FOP 6-93,** Foreign Agriculture Service, Washington, D.C., 1993.
29. R.H. Rouse, *Fat Quality—The Confusing World of Feed Fats,* Rouse Marketing, Inc., Cincinnati, Ohio, 1994.
30. NRA, *Pocket Manual,* National Renderers Association, Washington, D.C., 1993.
31. AFIA, *Feed Ingredient Guide II,* American Feed Industry Assoc., Inc., Arlington, Va., 1992.
32. NCPA, *Rules of the National Cottonseed Products Association, 1993–1994,* National Cottonseed Products Association, Memphis, Tenn. 1993.
33. C.E. Coppock and co-workers, *J. Dairy Sci.* **68,** 1198 (1985).
34. G.D. Buchs, U.S. Pat. 5,270,062 (Apr. 1, 1993).
35. S.D. Martin, *Feedstuffs* **62**(33), 14 (Aug. 6, 1990).
36. L.A. Jones in R.L. Ory, ed., *Antinutrients and Natural Toxicants in Foods,* Food & Nutrition Press, Inc., Westport, Conn., 1981, pp. 77–98.
37. F. Shahidi, ed., *Canola and Rapeseed,* Avi-van Nostrand Reinhold Co., Inc., New York, 1990.
38. H.E. Amos, *J. Dairy Sci.* **40,** 90 (1975).
39. A.P. Bimbo, *J. Am. Oil Chem. Soc.* **64,** 706 (1987).

40. G.R. List in B.F. Szuhaj, ed., *Lecithins, Sources, Manufacture and Uses*, American Oil Chemists' Society, Champaign, Ill., 1989, pp. 145–161.
41. J.W. Hertrampf, *Feeding Aquatic Animals with Phospholipids. I. Crustaceans*, Lucas Meyer, Co., Hamburg, Germany, 1991.
42. J.W. Hertrampf, *Feeding Aquatic Animals with Phospholipids. II. Fishes.* Lucas Meyer, Co., Hamburg, Germany, 1992.
43. *Lecithin—Properties and Applications*, Lucas Meyer Co., Hamburg, Germany, 1993.
44. J.P. Cherry and W.H. Kramer in Ref. 41, pp. 16–31.
45. National Research Council, *Recommended Dietary Allowances*, 10th ed., National Academy Press, Washington, D.C., 1989.
46. V.K. Babayan in J. Beare-Rogers, ed., *Dietary Fat Requirements in Health and Development*, American Oil Chemists' Society, Champaign, Ill., 1988, pp. 73–86.
47. M.I. Gurr in Ref. 28, pp. 3–22.
48. J.H. Moore and W.W. Christie in Ref. 28, pp. 123–149.
49. M. Enser in Ref. 28, pp. 23–52.
50. D.N. Brindley in Ref. 28, pp. 85–103.
51. C.P. Freemen in Ref. 28, pp. 105–122.
52. D.L. Palmquist in E.R. Ørskov, ed., *Feed Science*, Elsevier Science Publishing Co., Inc., Amsterdam, the Netherlands, 1988, pp. 293–311.
53. D.E. Pethick, A.W. Bell and E.F. Annison in Ref. 28, pp. 225–248.
54. C.C. Whitehead in Ref. 28, pp. 153–166.
55. D. Lewis C.G. Payne, *Br. Poultry Sci.* **7,** 209 (1966).
56. M.E. Fowler, *J. Zoo Wildlife Med.* **22,** 204 (1991).
57. R.M. Herd and T.J. Brown, *Physiol. Zool.* **57,** 70 (1984).
58. M.E. Stansby, H. Schlenk and E.H. Gruger Jr., in M.E. Stansby, ed., *Fish Oils in Nutrition*, Avi-van Nostrand Reinhold, New York, 1990, pp. 6–39.
59. J. Opstvedt in Ref. 28, pp. 53–82.
60. D.H. Greene in Ref. 59, pp. 226–246.
61. National Research Council, *Nutritional Energetics of Domestic Animals and Glossary of Energy Terms*, 2nd rev. ed., National Academy Press, Washington, D.C., 1981.
62. Ref. 8.
63. K.L. Fritsche, N.A. Cassity and S-C. Huang, *Poultry Sci.* **70,** 611 (1991).
64. D.L. Palmquist and T.C. Jenkins, *J. Dairy Sci.* **63,** 1 (1980).
65. D.L. Palmquist in Ref. 28, pp. 357–382.
66. D.L. Palmquist, *J. Dairy Sci.* **74,** 1354 (1991).
67. N.L. Karrick in Ref. 59, pp. 247–267.
68. R.E. Atkinson in Ref. 28, pp. 495–503.
69. D.E. Sayre in R.R. McElhiney ed., *Feed Manufacturing Technology III*, American Feed Industry Association, Inc., Arlington, Va., 1985, pp. 99–103.
70. *Process Description: Pet and Aquatic Food Production*, Wenger Manufacturing Co., Sabetha, Kans., 1993.
71. L.J. Cook, T.W. Scott, K.A. Ferguson, and I.W. McDonald, *Nature* **228,** 178 (1970).
72. T.W. Scott, L.J. Cook, and S.C. Mills, *J. Am. Oil Chem. Soc.* **48,** 358 (1971).
73. T.W. Scott and co-workers, *Aust. J. Sci.* **32,** 291 (1970).
74. T.W. Scott, P.J. Bready, A.J. Royal, and L.J. Cook, *Search* **3,** 170 (1972).
75. T.C. Jenkins and D.L. Palmquist, *J. Animal Sci.* **55,** 957 (1983).
76. T.C. Jenkins and D.L. Palmquist, *J. Dairy Sci.* **67,** 978 (1984).
77. D.L. Palmquist, T.C. Jenkins, and A.E. Joyner Jr., *J. Dairy Sci.* **69,** 1020 (1984).

78. P.S. Sukhija and D.L. Palmquist, *J. Dairy Sci.* **73,** 1784 (1990).
79. A. Baldi, F. Cheli, C. Coirino, V. Dell'Orto, and F. Polidori, *Small Ruminant Res.* **6,** 303 (1992).
80. D.J. Schingoethe and co-workers, *J. Dairy Sci.* **69**(Suppl. 1), 111 (1986).
81. D.P. Casper and D.J. Schingoethe, *J. Dairy Sci.* **69**(Suppl. 1), 111 (1986).
82. R. Rising, P.M. Maiorino, R. Mitchell, and B.L. Reid, *Poultry Sci.* **69,** 768 (1990).
83. E.R. Lloyd, *Feeding for Milk and Butter,* Bulletin **13,** Mississippi Agricultural Experiment Station, Mississippi State, 1890.
84. J.H. Connell and J. Clayton, *Feeding Milk Cows,* Bulletin **33,** Texas Agricultural Experiment Station, College Station, 1894.
85. J.E. Tomlinson and co-workers, *J. Dairy Sci.* **64**(Suppl. 1), 141 (1981).
86. M.J. Anderson, D.C. Adams, R.C. Lamb, and J.L. Walters, *J. Dairy Sci* **62,** 1098 (1979).
87. E.J. DePeters, S.J. Taylor, A.A. Frank, and A. Aguirre, *J. Dairy Sci.* **68,** 897 (1985).
88. H.E. Amos in *Proceedings of the 1984 Georgia Nutrition Conference for the Feed Industry,* Atlanta, Georgia Agricultural Experiment Station, Athens, Ga., 1984, pp. 119–129.
89. H.E. Amos, *Feed Management* **35**(12), 26, 28, 32, 34 (1984).
90. D.L. Palmquist, *J. Dairy Sci.* **67**(Suppl. 1), 127 (1985).
91. M-J Kiang, *Dry Extrusion of Whole Soybeans,* Insta-Pro International, Des Moines, Iowa, 1993.
92. *Fullfat Soya Handbook,* American Soybean Association, Brussels, Belgium, 1983.
93. T.R. de Souza and co-workers, *J. Rech. Porcine France* **22,** 151 (1990).
94. L.D. Satter, J-T Hsu, and T.R. Dhiman paper presented at the Advanced Dairy Nutrition Seminar for Feed Professionals, Wisconsin Dells, Wisc., Aug. 18, 1993.
95. W. Steele, R.C. Noble, and J.H. Moore, *J. Dairy Res.* **38,** 43 (1971).
96. E.M. Tice, M.L. Eastridge, and J.L. Firkins, *J. Dairy Sci.* **76,** 224 (1993).
97. M.A. Faldet and L.D. Satter, *J. Dairy Sci.* **74,** 3047 (1991).
98. L.D. Satter, M.A. Faldet and M. Socha in *Symposium Proceedings: Alternative Feeds for Dairy and Beef Cattle,* National Invitation Symposium USDA Extension Service, in Cooperation with University Extension Conference Office, University of Missouri, Columbia, 1991, pp. 22–24.
99. T.R. Dhiman, K. Van Zanten, and L.D. Satter, personal communication, November 25, 1994.
100. R.M. Gast and D.S. Bundy, "Control of Feed Dusts by Adding Oils," ASAE Technical Paper No. 86-4039, Am. Soc. Ag. Engineers, St. Joseph, Mich., 1986.
101. A.J. Heber and C.R. Martin, *Trans. ASAE* **31**(2), 558 (1988).
102. E.R. Peon, Jr. and L.I. Chiba, *Tech. Newslett.* (141984) (1984).
103. *The Soybean Oil Dust Suppression System,* American Soybean Association, St. Louis, Mo., 1986.
104. K.L. Wick, L.J. Friedman, J.K. Brewda, and J.J. Carroll, *Food Technol.* **47**(3), 94 (1993).
105. M.E. van Elswyk, *Nutr. Today* **28**(2), 21 (1993).
106. D.L. Palmquist, D. Beaulieu, and D.M. Barbano, *J. Dairy Sci.* **76,** 1753 (1993).
107. B. Fitch-Haumann, *INFORM* **4**(41), 371 (1993).
108. H.W. Phetteplace and B.A. Watkins, *J. Food Composition Anal.* **2,** 104 (1989).
109. Z-B Huang, H. Leibovitz, C.M. Lee, and R. Millar, *J. Agr. Food Chem.* **38,** 743 (1990).
110. H.W. Phettplace and B.A. Watkins, *J. Agr. Food Chem.* **38,** 1848 (1990).
111. P.S. Hagris, M.E. van Elswyk, and B.M. Hargis, *Poultry Sci.* **70,** 874 (1991).
112. M.E. Van Elswyk, A.R. Sams, and P.S. Hagris, *J. Food Sci.* **57,** 342 (1992).
113. S.Y. Oh, J. Rye, C-H Hsieh, and D.E. Bell, *Am. J. Clin. Nutr.* **54,** 689 (1991).

8
Oils and Fats in Bakery Products

INTRODUCTION

1.1 Shortening in Bakery Usage

Fats and oils have been important bakery ingredients for centuries. Indeed, "shortening" is a baker's term; fat in a bakery item "shortens" (tenderizes) the texture of the finished product. In bakery foods, shortenings impart tenderness, give a moister mouthfeel, contribute structure, lubricate, incorporate air, and transfer heat.

Properties of a fat or oil that determine its ability to carry out these functions are:

The ratio of solid to liquid phase.
The plasticity of a solid shortening.
The oxidative stability of the fat or oil.

Oil (or the oil fraction of a plastic shortening) in a baked food gives a tender bite, moist mouthfeel, and lubricity (the product clears more readily from the surfaces in the mouth). The solid portion of a shortening contributes to the structure of the dough and the final product, and entraps air bubbles during mixing. These two functions are the key to selecting the proper shortening for a given application. Also, the fatty acids in the oil fraction are generally more unsaturated than those in the solid part, and polyunsaturated fatty acids are subject to the development of oxidative rancidity. Oxidative stability is especially important in shortenings that are exposed to air (cracker spray oils) and high temperatures (frying fats).

1.2 Other Uses: Icing, Filler, Coating Fats

In addition to inclusion in the dough, fats are also applied to the baked piece, providing additional enhancement of consumer appeal. Examples are: creme icings on cakes, aerated fillings in cornettos, fillings in sandwich cookies and sugar wafers, and chocolate and compound coatings on snack cakes.

These enhancements are primarily carriers of flavors and sweetness (they are typically high in sugar content). The solid portion of the fat is a major structural element in the material, while the oil part also enhances mouthfeel, particularly lubricity.

BREAD AND ROLLS

In yeast-leavened baked foods the volume and texture (fine-grained crumb with small, evenly distributed air cells) depends upon the strength of the gluten matrix. This is a mixture of hydrated storage proteins (glutenin and gliadin, found in wheat flour), surface active lipids (both those native to flour and added surfactants), and nonpolar fats. The structure of this matrix is still not clearly defined, but it is known that shortening is an integral part of it. During baking the protein is denatured, forming a glass that is in part responsible for the elasticity of the cooled bread. Shortening modifies both the baking process and the characteristics of the final product.

2.1 Effect on Texture

Up to 5% fat (flour weight basis) may be used in ordinary white pan bread, although the usual level is 3 to 4% of a plastic fat such as all-purpose shortening or lard. These amounts produce the optimum effect in the bread. In soft rolls, such as hamburger buns, 6 to 8% may be used, to give a softer bun. This tenderizing effect also slows down the staling process, so bread made with shortening remains softer after storage for several days, as compared to bread made with the same formula but without fat in the dough.

The shortening is simply added to the mixer at the beginning of dough mixing. The statement has been made that the presence of fat slows down development of gluten (lengthens the required mixing time), but in experiments both in the laboratory and in the plant this thesis has not been confirmed.

2.2 Effect on Volume

Loaf volume of bread increases as the amount of plastic shortening increases, up to about 5% (flour basis), then remains roughly constant (Figure 8.1). This is because the dough expands in the oven for a longer time when shortening is present, as compared to a dough made without added fat (1). Fat (or oil)

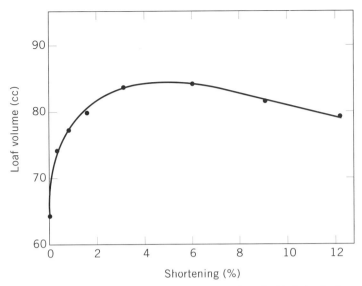

Figure 8.1 Final loaf volume of white bread, as a function of added plastic shortening.

seems to interact with dough components (starch and gluten) and delay the reactions that end loaf expansion during baking. In other words, in bakery terminology, the addition of shortening increases ovenspring of the bread.

2.3 Use of Oil Versus Plastic Fat

The tenderizing effect of shortening is due to the liquid phase. Since all-purpose shortening contains about 25% solid fat at room temperature, 3 kg of vegetable oil is equivalent to 4 kg of plastic fat in softening bread crumb. The solid fat portion of traditional bread shortening does play a structural role. It provides dough strength during proofing, contributing resistance to shocks during transfer from the proofer to the oven, and also appears to strengthen sidewalls of baked bread, minimizing "keyholing" of the finished loaf. In the continuous bread process hard flake [fat hydrogenated to an iodine value of about 5, and a melting point of about 60°C (140°F)] is often melted into the warm oil being pumped to the developer. When a switch is made from plastic fat to oil, it is necessary to use a dough strengthener such as sodium stearoyl lactylate (SSL) or diacetyl tartrate esters of monoglyceride (DATEM) to get good volume.

2.4 Frozen Doughs

Extremely lean bread doughs deteriorate more rapidly in frozen storage than doughs that contain sugar and shortening. For a lean French-type bread the

334   Oils and Fats in Bakery Products

inclusion of 0.5 to 1% vegetable oil in the formula will extend the shelf life by several weeks. In regular doughs (i.e., for baking pan bread or hamburger buns) the usual amount of shortening is used: 3 to 4% in bread, 6 to 8% in buns. The function in these cases is the same as in standard baking, that is, the shortening improves ovenspring and tenderizes the finished product.

LAYERED DOUGHS

In many bakery products, fat is layered between sheets and dough, and this is manipulated to make a dough sheet consisting of up to 100 alternating layers of dough and fat. Such roll-in doughs include Danish pastry, puff pastry, and croissants. The dough is mixed, divided into pieces of about ten kilograms each, then cooled to 5–10°C in a retarder. The cooled dough is rolled out, two to three kilograms of fat is spread over part of the sheet, and the dough is folded over to cover the fat. The "sandwich" is then rolled out and folded; this is repeated several times, often with a retarding step included to keep the dough/fat mass cool.

The primary goal during the roll-in process is to preserve the structure of alternate layers of dough and fat. There are several important factors to consider in selecting the correct shortening or margarine for such a process. Some of them are: solid fat index (SFI) and plasticity of the fat; complete melting point of the fat; consistency (softness) of the dough; retarder temperature; number of folds given the dough before returning it to the retarder; and proofing temperature. Many of these factors are unique to a given product in a particular bakery, and they influence the specification for the roll-in fat which gives the best final product in that bakery.

3.1   Danish Pastry

The dough for Danish pastry contains higher levels of sugar and water than a bread dough, making it quite soft at the usual mixing temperature of around 20°C. After it is mixed (the gluten matrix is fully developed) it is divided and cooled as described above. The cooling and rest time (during which some fermentation occurs) makes the dough cohesive enough to handle, but if it is over-worked or allowed to warm up too much during the roll-in step, it again becomes soft, sticky, and easily torn. The plastic range of the roll-in fat must be rather broad. The consistency of the fat should nearly match that of the dough across the temperature range, which usually includes the temperature of the retarder (on the cool end) to room temperature or above. If the fat is significantly harder than the dough at the cool temperature, then when a retarded dough is rolled out, the fat does not spread into a uniform layer between the dough layers, but is likely to tear holes in the dough sheet. If the fat is softer than the dough at room temperature, then as the dough mass warms up (during rolling and folding) the shortening soaks into the dough, the adjacent dough layers knit together, and the layering effect is lost. The proper plasticity of roll-in fat requires a relatively shallow SFI profile, and stabilization in the β' crystal phase. The latter is particularly important. If the shortening has begun to transform to the more solid β phase, the additional hardness due to β crystals will cause excessive tearing during roll-in.

The melting point of the fat must be higher than the proofing temperature for the Danish. If the proof box temperature is above the fat melting point, the fat layers turn to oil. This allows the dough layers to knit together to some extent during proofing, and the final product is less flaky than desired. As a general rule, the complete melting point should be at least 5°C higher than proof box temperatures.

A standard all-purpose shortening having a complete melting point of about 45°C usually works well in most Danish pastry production lines. This gives a flaky finished product. Some manufacturers use an emulsified shortening containing 3% monoglyceride, but with similar SFI and melting point specifications. This produces a Danish having a somewhat gummier mouthfeel, which is preferred in some parts of the country.

3.2 Croissants

A croissant dough is similar to Danish dough, although it generally contains somewhat less sugar and water, so it is somewhat firmer at the optimum mixed dough temperature of 20°C. As with Danish, the dough pieces are usually retarded before roll-in is begun. The best quality croissants are produced using unsalted butter for roll-in. The amount used varies from 20 to 35% of dough weight; higher levels give a product that has less volume and flakiness, and is often perceived as greasy in the mouth. The optimum is usually about 25% roll-in fat.

The factors involved in successful processing are similar to those discussed above for Danish pastry. Because butter has a steeper SFI curve than all-purpose shortening (it is much harder at retarder temperatures) more care is required to prevent tearing the dough as it is being rolled. The melting point is lower than that of shortening, so proofing temperatures are lower than for Danish. Puff pastry margarine is an acceptable substitute, although it does not contribute as much flavor as butter. Since puff pastry margarine has a higher melting point, the proofing may be done at a higher temperature (and for shorter times) if time is a factor.

3.3 Puff Pastry

Rolled-in doughs that contain no yeast (puff pastry) depend upon steam generation in the oven for their leavening. Usually margarine (which contains about 17–20% water) is used for the roll-in fat for these doughs. The water is trapped and held in the fat layers in the dough. It evaporates and expands in the oven, giving an expanded structure to the final product. If the fat portion of the margarine is too soft, the water migrates into the dough during the roll-in step, and the leavening action in the oven is decreased. Puff pastry margarine has a higher SFI curve than all-purpose shortening, and it is somewhat more brittle during roll-in. In this case a smooth, continuous layer of fat between dough layers is not particularly desirable. The roll-in process is adjusted somewhat to achieve numerous discrete particles of margarine between the dough layers. These produce somewhat larger voids in the finished product, desirable in a puff pastry.

CAKES

4.1 Emulsified Plastic Fats

In layer cakes and related items the closeness of the internal grain, and to some extent the final volume, are strongly influenced by the characteristics of the shortening used. In the finished cake a high percentage of the total volume is open space, present as finally divided cells. These spaces are created by carbon dioxide (from the leavening system) and steam, formed during baking. When these gases are generated, by heat, they migrate to the nearest air bubbles that have been incorporated into the cake batter during mixing. If there are many small air bubbles in the batter the leavening gases are distributed widely. Each of the bubbles is small, and does not rise rapidly to the surface of the cake. The leavening gases are retained in the cake and contribute to final volume. If the air incorporated during mixing is present as relatively few, larger, bubbles, then during baking the bubbles are expanded by the leavening gases, and many of them are large enough that they rise to the surface, escaping and yielding a lower final cake volume.

If the original air bubbles are fewer and larger, the final cake will have less volume and a coarse (open) grain. If the original batter contains many small air cells, the final cake will have larger volume and a fine (close) grain. These two situations are exemplified in Figure 8.2. The shortening used plays a large role in determining the degree of subdivision of the air.

In the usual two-stage process of cake production, the shortening and sugar are combined and mixed. During this step the air is dispersed into the solid phase. Then the eggs are incorporated, followed by the flour, liquids, and other ingredients. During the first creaming step, the plastic shortening entraps air bubbles. In the presence of an emulsifier such as 4% monoglyceride, these bubbles are divided into numerous small air cells by the action of the beater. The shortening must be solid (so the bubbles don't escape), but also plastic so it can fold around each air pocket. This is best accomplished by a plastic shortening, crystallized in the β' phase. If the shortening has transformed into the β phase, the large plates of solid fat are much less effective in entrapping

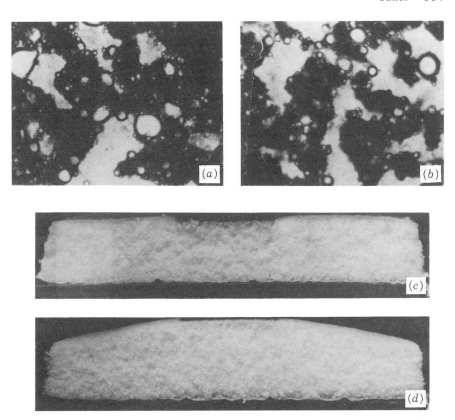

Figure 8.2 Emulsifiers help subdivide air particles in shortening. (*a*), Microphotograph of batter made with unemulsified shortening, showing large air bubbles; (*c*), cake baked from this batter. (*b*), Microphotograph of batter made with emulsified shortening, showing smaller air bubbles; (*d*), cake baked from this batter. (From Reference 2 by permission.)

the fat. A good shortening for this type of cake batter production has the SFI profile of all-purpose shortening, containing added monoglyceride. Plastic shortening containing 5 to 8% of an α-tending emulsifier such as propylene glycol monoester plus 2 to 4% monoglyceride is also sold. These products are successfully used in one-stage cake production.

4.2 Oil Shortening with α-Tending Emulsifiers

It is also possible to make cakes in a one-stage production process, in which all the ingredients are added at the beginning and the batter is mixed. In this case the air is entrapped in the water phase rather than in the shortening. In order to form the air-in-water foam, which is stabilized by proteins contributed from flour and eggs (2), it is necessary to prevent the defoaming action usually associated with fats and oils. This is accomplished by including an α-tending

338 Oils and Fats in Bakery Products

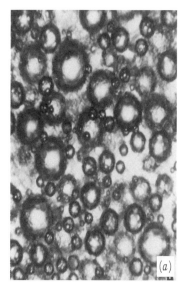 

Figure 8.3 (*a*) Microphotograph of batter of oil cake, with PGME in the oil, showing good air incorporation. (*b*) A water droplet was suspended in oil containing PGME, then some water was withdrawn through the syringe tip. The crumpled material is solidified emulsifier at the interface. (From Reference 3 by permission.)

emulsifier in the shortening; typical ones used are propylene glycol monoesters (PGME) or acetylated monoglyceride (AcMG). At high enough concentrations these emulsifiers form a solid film at the oil/water interface (Figure 8.3). This solid interfacial film segregates the fat from the water phase, so it cannot destabilize the protein foam.

The film-forming tendencies of α-emulsifiers enable the use of liquid oil as the shortening in a cake or muffin batter. Cakes made with oil shortening are more tender than when they are made with a plastic shortening. The cake gives an impression of moistness when eaten, even after storage for a week or longer. Packaged dry cake mixes, for use in the home, are usually made with oil containing 10 to 15% α-tending emulsifier dissolved in the oil. Other emulsifiers such as Polysorbate 65 are also used in oil cakes. Because of the tenderness of the cake layer, this type of formulation is usually not suitable for commercial cake production.

4.3 Muffins, Cupcakes

Muffins and cupcakes are quite similar to layer cakes in the batter characteristics and reactions during baking (timing of release of leavening gases, setting of the structure). They are less sweet than most cakes, and generally have a somewhat more open crumb structure (a close, fine grain is not desired). The shortening used is an all-purpose (nonemulsified) type. In some instances a vegetable oil shortening is used, when an extremely open texture accompanied by a moist mouthfeel is desired, for example, in a high fiber muffin. Shortening levels vary widely, from 18 to 35% based upon the amount of flour; tenderness of the finished product varies accordingly.

4.4 Creme Icings

Layer cakes are generally covered with an icing, giving increased sweetness and flavor. Creme icings are made by creaming together the fat and sugar (in a ratio of about 3 to 4), then adding flavor and egg whites and perhaps a small amount of water, and mixing at high speed until the specific density is in the range of 0.4 to 0.6 g/cc. The fat used varies widely. Traditionally butter was used, giving a flavorful icing. Due to economics, some or all of the butter is frequently replaced with an emulsified shortening having an SFI curve similar to that of butter. It is important that the melting point of the shortening be about 37°C (body temperature), otherwise the residual hard fat leaves an undesirable greasy or waxy impression in the mouth.

Monoglyceride at 2 to 3% is commonly used as the emulsifier. At higher levels (e.g., the 4% level found in emulsified cake shortening) the aeration of the icing is less efficient. Polysorbate 65 at a concentration of about 0.5% is also included in some icing shortenings. It has a higher HLB, and allows the incorporation of somewhat more water in the icing.

CAKE DONUTS

Donuts are fried in hot (180–190°C) fat rather than being baked. Yeast-raised donuts are a sweet dough, cut into rings and proofed before being fried. Cake donuts are chemically leavened and similar to cake or cupcake batter in composition. In a cake donut batter the degree of subdivision of air entrapped in the shortening phase has much to do with the openness of grain in the final donut. Thus an emulsified cake shortening is generally used. The batter is extruded in a ring shape into the hot fat, where it is fried. Initially the ring sinks in the fat, but the leavening soon causes it to rise to the surface; it is flipped halfway through the frying cycle so both sides are cooked. A significant amount of the batter water is evaporated during the process, and an equal volume of frying fat is absorbed into the donut, accounting for about half the fat in the finished donut. The nature of the frying fat has much to do with the eating characteristics of the finished donut. If it has a rather high SFI curve and a high melting point, the cooled donut will have a solid texture and a dry mouthfeel. On the other hand, if a high stability frying *oil* is used (with a melting point at or below room temperature) the donut will be greasy and

Table 8.1 **Typical physical properties of bakery shortenings**[a]

Shortening Type	Solid Fat Index					AOM (min. h)
	10°C (50°F)	21°C (70°F)	27°C (80°F)	33°C (92°F)	40°C (104°F)	
RBD oil[b]	0	—	—	—	—	10
Lightly hydrogenated oil	<5	<1.5	—	—	—	25
Deep frying fat	47 ± 3	32 ± 3	25 ± 2	12 ± 1	<2	200
All-purpose	28 ± 3	20 ± 2	17 ± 1	13 ± 1	7 ± 1	75
Cake, icing[c]	32 ± 3	25 ± 2	22 ± 2	16 ± 1	11 ± 1	75
Pie crust	26 ± 3	17 ± 1	14 ± 1	10 ± 1	6 ± 1	75
Wafer filler fat	55 ± 4	39 ± 3	29 ± 3	4 ± 1	<1	100
Sandwich cookie filler	38 ± 3	24 ± 2	18 ± 1	9 ± 1	<2	100
Coating fat	64 ± 5	52 ± 4	44 ± 4	20 ± 2	0	200
Puff pastry margarine	34 ± 3	30 ± 3	27 ± 2	22 ± 2	16 ± 1	200
General use margarine	28 ± 2	21 ± 2	18 ± 1	15 ± 1	10 ± 1	200
Butter[d]	32	12	9	3	0	—
Cocoa butter[d]	62	48	8	0	0	—

[a] General chemical characteristics: peroxide value, 1 meq/kg max.; free fatty acid (as oleic acid), 0.05% max.; phosphorous content, 1 ppm max.
[b] RBD = Refined, bleached, and deodorized oil.
[c] Contains 3–5% α-monoglyceride for cake, 2–3% for icing.
[d] Typical values. As natural products, these can vary somewhat.

unappetizing. A good donut frying fat has an SFI somewhat higher than an all-purpose shortening, but with a lower melting point.

If the donut is to be coated with powdered sugar, the consistency of the absorbed fat is important to obtaining proper coverage. If the fat in the cooled donut is too hard, sugar pickup will be too small. On the other hand, if the fat is too soft the sugar will "melt" when the packaged donut is stored. For best results, in warm weather the SFI curve of the frying fat should be on the upper side of the specifications given in Table 8.1, while in cold weather it should be on the low side. The proper selection of frying fat is of major importance to the quality of the finished cake donut.

COOKIES

5.1 Effect on Texture, Spread

Entrapped air bubbles in the shortening phase of a cookie dough serve as nuclei for leavening gas during baking, as in cakes. Thus if a fine-grained cookie, for example, a soft sugar cookie, is desired an emulsified shortening should be used. For a thin, crisp cookie a lightly hydrogenated oil may give the best product characteristics. For most wirecut cookies an all-purpose shortening works well. As the amount of shortening in the formulation increases, the tenderness of the baked cookie also increases, as would be expected.

The diameter of the baked cookie is important in commercial bakeries, where the finished product must fit a previously designed container. The height is also sometimes a factor, depending upon the type of packaging equipment in use. The amounts of both sugar and shortening (relative to flour) in the dough affect diameter and height, but in a rather complex fashion (4,5). Briefly, in a 50% sugar (flour basis) dough, increasing shortening increases diameter and decreases height. In a 90% sugar formula, increasing shortening decreases diameter and has little effect on height.

5.2 Processing Considerations

Dough. For wirecut cookies, where the dough is extruded through a die and portions are cut and dropped onto the baking band, the main dough function is incorporation of finely divided air bubbles. As the cookies bake, leavening gases (steam, carbon dioxide) collect in the air bubble nuclei. If the air cells are few and large, the grain of the resulting cookie will be open, while if the air nuclei are many and small, the cookie will have a finer grain. The shortening also lubricates the dough as it is extruded. If a low-fat cookie is being developed, difficulty is often seen at the wirecut depositor; the dough does not extrude readily. Using oil as the added allowable lipid may help this problem.

In rotary-molded cookies, air incorporation is important, but shortening also makes a structural contribution to cookie processing and quality. The solid phase of the shortening influences the consistency of the dough; if the fat is too soft (the SFI profile is too low) the dough will be soft, and not machine properly at the molder. Also, oil will tend to leak from the shaped pieces, and saturate the cloth takeaway belt, causing problems. If the SFI profile is too high, the dough will be stiff, it will not fill the mold properly, and it will not release cleanly from the die. In the former case the shortening is making an inadequate contribution to dough structure, while in the latter case lubrication by the shortening is lacking. It is necessary to maintain the proper balance between the solid and liquid phase in the shortening to have good machinability. To obtain the precise balance required it may be necessary to use a combination of lightly hydrogenated oil plus all-purpose shortening in the dough. The exact ratio for best results is determined by experiment: if bits of dough stick in the mold, use a little more oil, while if the pattern tends to smear out, use less oil (or a plastic shortening with a higher SFI profile).

For sheeted cookies, such as Marias biscuits, dough structure is primarily due to a modest amount of gluten development, and the grain of the final product depends upon proper extrusion and sheeting. The main contribution of shortening in these items is tender eating quality in the finished product. This is an attribute of the oil phase, and a liquid oil shortening should give good results, when used at about 75% of the amount of plastic fat in the formula. Cookies are a long shelf life item, and they must be acceptable to the consumer for up to six months or more. If oil is used in the dough, it should have good oxidative stability, so rancidity does not develop in the product during storage.

Wafer batter does not ordinarily contain fat or oil. However, on occasion a small amount of oil is included in the batter, to facilitate release of the baked wafer from the griddle. The amount used is just enough to effect release (perhaps 0.5% of total batter weight). An oil with high oxidative stability would give the best lubrication in this application.

Filling Fat. Fillings for sugar wafers or sandwich cookies consist of fat and sugar, with flavor and color added as desired. The consistency of the filling is determined to a large extent by the SFI profile of the fat used. This profile must meet three requirements:

1. The blend must have a soft consistency, so it can be extruded onto the basecake or wafer
2. The extruded filling must be firm at room temperature and below, so that it does not slide when the cookie is eaten
3. The fat must melt almost completely at mouth temperature, so it does not have a waxy mouthfeel

To achieve these goals, the SFI profile of a filler fat is rather steep, higher than that for all-purpose shortening at low temperatures, and lower at high temperatures. The plastic range is much narrower, and close control of temperature at the mixer and extruder is necessary.

The solid fraction of filler fat for wafers and sandwich cookies gives body to the filling. When the filled wafer or sandwich cookie is cooled to room temperature the cream sets up to the firm consistency needed. In the production of filled wafers it is crucial that the crystal structure be β'. If the shortening has started to go beta, the resulting oiliness causes the wafer sheets to slide during the transport and cutting operation. Also, β crystals set up slower than β' crystals, causing delays between the extruder and the cutting operations. In sandwich cookies this is not such a factor, because they are not cut after filling. Wafer filler fat has a higher SFI profile than sandwich cookie filler fat, because it must prevent slippage between the top and bottom wafer sheets during cutting. Filler fats with a wide range of SFIs are available and individual plants may want to develop their own specification to fit their equipment, processing conditions, and geographical locations.

Coating Fat. Cookies and other snack items are frequently coated with chocolate. The SFI profile of cocoa butter is unique among natural fats, being very high at room temperature and below, but melting rather sharply and completely at about 32–35°C. This characteristic is accepted as the norm for coating fats. A number of substitutes for cocoa butter have been sold. These are based upon shea butter, fractionated palm kernel oil, or soy and similar vegetable oils that have been hydrogenated in a special fashion. Such a fat is called a "hard butter."

Hard butters do not have such a sharp SFI profile as cocoa butter and their melting point is generally slightly higher, around 38–42°C (the melting point is adjustable, by making slight changes in process parameters). When used to make a coating for a cookie or wafer, they are blended with cocoa powder, sugar, and milk solids. While the fat does not melt completely at mouth temperature, this is not a problem because it is chewed along with the other, nonmelting parts of the cookie and the slight residual solid fat is not noticed. The confectionery coating has an advantage over chocolate; it does not melt as readily when held in the fingers or when exposed to warm summer temperatures.

The crystallization behavior of cocoa butter is complex. Careful tempering of the chocolate is necessary, to obtain a covering that is smooth, glossy, and stable. If the cocoa butter in the covering undergoes crystal transformation, because of temperature fluctuations or a variety of other reasons, it takes on a dusty look, referred to as chocolate "bloom." Hard butters are generally less complex in their crystal habit, tempering is easier, and bloom formation less likely to be a problem. The two kinds of fat generally are not compatible, and a coating made with hard butter should not contain any more cocoa butter than the few percent that is present in the cocoa powder, used for flavor and color. Likewise, trying to extend chocolate by adding hard butter enhances the tendency for bloom formation.

Crackers. Two main types of crackers are produced commercially: soda and snack. In both types about 10 to 12% of a plastic shortening is incorporated in the dough, providing tenderness and a desirable crisp "bite" in the final product. In addition, snack crackers are sprayed with oil as the hot product exits the oven to improve mouthfeel and help seasonings adhere to the piece. Traditionally lightly hydrogenated coconut oil with a complete melting point of 33°C has been used, but in recent years selectively hydrogenated soy oil has been successfully used. The polyunsaturated fatty acids are reduced to almost zero, but the m.p. is around 35°C. If a shiny appearance is desired an oil with an SFI around 10 at 20°C is used, while one with an SFI of about 20 at 20°C will give a drier appearance. Because the oil is on and near the surface of the cracker, and because the cracker should have a shelf life of up to six months, the spray oil must have good oxidative stability; AOM values should be at least 100 hours.

PIE CRUST, BISCUITS

6.1 Texture

Shortening functions in pie crust and American-style biscuits analagous to its action in layered doughs; layers of fat create regions of low tensile strength within the dough, giving it a flaky texture. The means of achieving this effect, however, is different, necessitating adjustment in the properties of the plastic shortening used. The shortening is mixed into the blended dry ingredients in such a way as to form small (pea-sized) pieces. Then liquid is added, the dough is gently mixed until it is just cohesive (the flour gluten is not developed), then it is sheeted out, the piece (crust or biscuit) is cut, and baked. The dispersion of shortening is more like that in puff pastry than in croissants. Intimate contact between all the shortening and the flour is not desired. The shortening should be slightly harder than the all-purpose shortening used, for example, for Danish roll-in. Alternatively, the crust or biscuit dough should be kept cool (even refrigerated) during processing to maintain the integrity of the fat pieces.

On the other hand, shortening used in biscuits should not have a melting point markedly higher than body temperature, or a waxy mouthfeel results. A shortening having an SFI profile similar to that suggested for sandwich cookie filler fat has been successfully used to accomplish all these aims.

SPECIFICATIONS FOR BAKERY SHORTENINGS

Certain properties of shortening are of particular importance to bakers. The solid fat index, plasticity, and oxidative stability of shortening are determined by the supplier's production process. The source of the starting oils, the conditions and extent of hydrogenation, the blending and crystallization of various basestocks, the storage conditions after packaging, these production variables determine the factors which influence shortening functionality. Some understanding of the nature of the three factors mentioned above clarifies their role in the bakery production process, and contributes to improved selection of shortenings for different bakery products.

7.1 Solid Fat Index/Content

The solid fat index (SFI) relates to the percent of shortening which is solid at various temperatures. This curve can have a variety of shapes, being rather humped like cocoa butter, or almost straight over most of the range, with a steeper or shallower slope. The whole curve cannot be predicted from a determination made at just one temperature. Curves for different fats may cross; the whole SFI curve is required in order to understand the properties of the shortening at different temperatures.

Solid fat index is measured by placing a sample of the fat in a dilatometer and measuring the volume at various temperatures. When a solid fat melts it expands. A solid triglyceride has a coefficient of expansion of about 0.00040 mL/g/°C, while a liquid triglyceride has a coefficient of expansion of about 0.00084 mL/g/°C. A shortening is a mixture of triglycerides that melt over a range of temperatures, and the actual volume change resembles the solid line shown in Figure 8.4. In this hypothetical example, the actual percent of solids in the fat at 20°C is S/T, or about 47%. While the specific volume line for oil is easy to determine, the corresponding line for fully solid fat is difficult to define. The standard SFI method (AOCS Method Cd 10-57) circumvents this difficulty by adopting a convention: the lower reference line is given the same slope as the line for the liquid, and is located 0.100 units below it (the dotted line in Figure 8.4). For the example the SFI value is S/T', or about 40%.

The dilatometric method is time-consuming and subject to the bias introduced by the convention described. More recently pulsed Nuclear Magnetic Resonance (pNMR) has been used to measure the relative amounts of liquid and solid fat in a sample, based upon the difference in rates of relaxation of protons in the two phases after the sample has been pulsed (AOCS Method Cd 16-81). With proper calibration this gives a direct determination of the percentage of solid fat, and the results are termed solid fat content (SFC). The analysis takes less time than dilatometry, but the equipment is more expensive.

The relationship between SFI and SFC is a complex function of both temperature and the level of SFI. A comprehensive study of 46 plastic shorten-

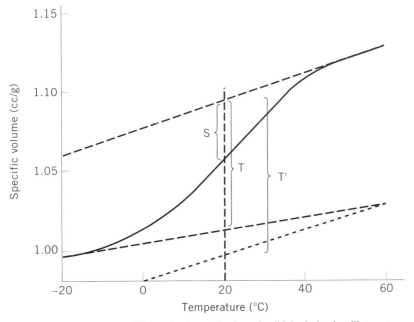

Figure 8.4 A curve exemplifying the determination of solid fat index by dilatometry.

ings across the temperature range of 10–45°C provided data for deriving equations relating the two values (6). A study of 14 hard butters (7) showed that the relationship was not as complex as that for plastic shortenings.

Analytical measurements of SFI and SFC are relatively precise; duplicate determinations should agree within ±1 unit. The suggested values given in Table 8.1 for SFI and SFC have rather large tolerances specified, because it is difficult to control the steps in shortening production to closer tolerances than those given.

Functionality of a plastic fat in the bakery depends not only on solids content, but also on the slope of the SFI curve. Meeting an SFI specification implies that the values for a particular batch lie within the specified ranges, and also that deviations from the target values are all on the same side, either higher or lower. This is particularly important in producing basestocks for blending for margarine and shortening production.

7.2 Plasticity

The plasticity of a fat is defined operationally; the shortening is smooth, not grainy, deforming readily when squeezed but holding its shape when set on a flat surface. No precise method of measuring these characteristics objectively has been developed to date. Various sorts of penetration tests give approximate results which are useful, although they must be used with some caution. One such test is the cone penetrometer method (AOCS Method Cc 16-60). A metal cone is set on the top surface of the fat, and the depth of penetration after a fixed time period is determined. The extent of penetration is larger for a soft fat than for a hard fat. A second method which is of value on the shortening production line employs a thick needle. A tube is held vertically on the surface of the fat, and the needle is dropped from the top of the tube. The depth of penetration is read from markings on the needle.

The plastic range refers to the range of temperatures over which a shortening will have the properties listed above. Plasticity is a function of two factors: SFI and crystal structure. Assuming the shortening has the proper β' crystal structure then it will be plastic over a range of about 10 to 25 SFI units (Figure 8.5). The choice of upper and lower limits of plasticity depend upon the experience of the individual choosing them, and the application; they are broader for bakery shortening than for margarine for use in the home.

Triglyceride fats crystallize in three different crystal forms. Rapid cooling of melted fat forms a waxy solid called the alpha (α) form (Figure 8.6a). This is a rather unstable crystal which quickly changes into long needle-like clusters of beta prime (β') crystals (Figure 8.6b). This is the preferred crystal form for plastic shortenings. The long, thin crystals join together into a "brush heap" which immobilizes several times its own weight in liquid oil. The thin needles are readily broken when squeezed (and reform when deformation ceases) so the overall feeling is one of a very smooth, creamy solid.

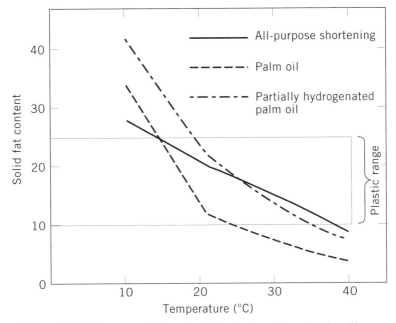

Figure 8.5 Plastic range of three shortenings having different SFI profiles.

If this crystal phase isn't stabilized by proper tempering at the time of manufacture, or if the shortening is stored at too warm a temperature, the solid phase reorganizes into the most stable structure, beta (β) crystals (Figure 8.6c). These are platelike, firm structures. Because there is less surface area per gram than in the case of β', the β crystals immobilize less liquid. A fat that has converted to the β form feels grainy or sandy, and also oily. A shortening that has "gone beta" has lower plasticity than the same shortening stabilized in the β' phase.

Shortening made from 100% hydrogenated soy or sunflower oil converts to β crystals rather readily, but the addition of 5 to 7% hydrogenated palm or cottonseed oil stabilizes the β' phase. Most plastic shortening used in the United States today is made from partially hydrogenated soy oil plus a small amount of palm or cottonseed hard flakes (iodine value of 5). There are a few shortenings for which some β crystals are preferred, mainly in fluid shortenings and in puff pastry margarine.

The function of plasticity in various bakery products was discussed previously. In ordinary cookie production the plastic shortening is combined with sugar and then mixed, or creamed, to incorporate air bubbles. These air bubbles are the nuclei for gas expansion during baking. The air is physically trapped, and the brush heap structure of β' shortening is better able to do this than the plates of β crystals. The air is actually in the liquid oil, so if SFI is too high there is not enough oil volume for adequate aeration. On the other

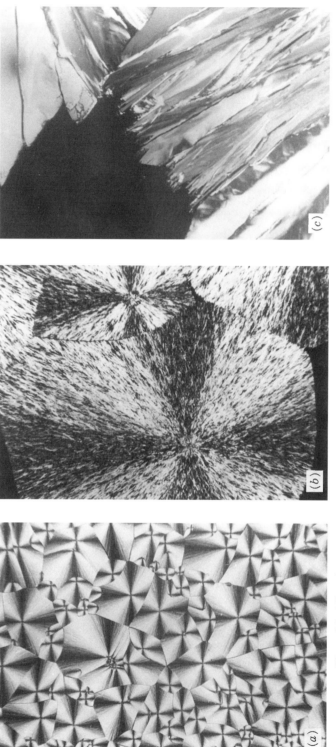

Figure 8.6 Microphotographs of fat crystals, taken in polarized light. From left to right: α crystals; β' crystals; β crystals.

hand, if SFI is too low the air is not trapped, and it escapes before dough mixing is complete. There is a range of SFI values giving optimum aeration of the creamed shortening, corresponding to the plastic range.

The machining of doughs, preparatory to baking, often causes some warming. If the SFI gets too low oiling out may be observed. The plasticity is suitable for the temperatures experienced in mixer operation, but because of warming during machining the shortening causes problems at the front of the oven. The plastic range of the shortening is too short to accommodate both the mixing and the forming operations. This is most often seen when a simple shortening (e.g., palm oil, as depicted in Figure 8.5) is used. Switching to a shortening with a broader plastic range will usually solve the problem.

Filling fat for wafers and sandwich cookies must meet a different set of requirements. If the solid fat index is too low at room temperature the wafers will tend to slide. The plastic range of filler fat is narrower, with less temperature tolerance at the mixer. When the wafer is cooled to room temperature the cream sets up to the firm consistency needed. In the production of filled wafers it is crucial that the crystal structure be β'. If the shortening has started to go beta, the resulting oiliness makes the wafer sheets slide during the transport and cutting operation. Also, β crystals set up slower than β' crystals, causing delays between the extruder and the cutting operations.

7.3 Oxidative Stability

The importance of oxidative stability of a fat or oil depends upon the intended use of the oil, in particular the use temperature and the finished product storage time. Two extremes are bread and a deep-fat fried snack. In bread the maximum temperature to which fat is exposed is 95°C (the final internal temperature during baking) and the storage time is at most 7 days; the least stable shortening, RBD oil, can be used for this application without danger of rancidity. The oil for deep-fat frying is heated to about 180°C, exposed to air during the frying operation, and has an expected shelf life up to 1 year. A stable fat is needed, so that rancidity does not make the snack prematurely unacceptable.

Autoxidation of fats occurs with unsaturated fatty acid chains. The relative rates of oxidation of oleic, linoleic, linolenic, and arachidonic acids (one, two, three, and four double bonds, respectively) are 1, 12, 25, and 50. Autoxidation is a free radical reaction, initiated and propagated by free radicals reacting with methylene ($-CH_2-$) groups adjacent to double bonds (Figure 8.7). A hydrogen radical is extracted and one of the double bonds migrates into a conjugated position, moving the radical site to the outer carbon. Dissolved oxygen adds to this site generating a peroxyl radical; this abstracts a hydrogen from a donor, perhaps another methylene group, making a hydroperoxide. The hydroperoxide splits to generate two free radicals, a hydroxyl and an alkoxyl radical. This cleavage is catalyzed by traces of metal ion such as copper

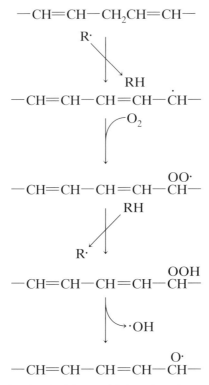

Figure 8.7 Reactions involved in autoxidation of a polyunsaturated fatty acid.

or iron. The net result is three free radicals, each of which can initiate another chain of reactions. The rate of reaction is self-enhancing, i.e., it is an autocatalytic reaction.

The reactions detailed above can occur in the dark, as long as molecular oxygen and an initiating free radical species is present. If the oil is exposed to light, dissolved oxygen may be photoactivated to singlet oxygen, which can initiate the reaction chain at the second stage shown in Figure 8.7.

The extent of autoxidation of a fat is often measured as the Peroxide Value (PV), which is the amount of chemically reactive peroxide in the sample (AOCS Method Cd 8-53).

Four main factors contribute to oxidative rancidity:

1. Chain initiation by trace free radicals
2. Chain propagation by molecular oxygen
3. Hydroperoxide cleavage catalyzed by metal ions
4. Chain initiation by photoactivated oxygen

Proper refining and deodorization removes peroxides which are the source of trace free radicals. Oil should be processed, transported, and stored under a nitrogen atmosphere. Metal ions in the oil can be inactivated by chelation with citric acid. Finally, the exposure of oil to light should be minimal. With these precautions oxidative stability can be increased several-fold.

Antioxidants also increase oxidative stability. These bind active free radicals, thus preventing initiation of new reaction chains. If free radicals continue to form, due to the presence of oxygen and trace metals, eventually all the antioxidant will react, and the autocatalytic sequence will develop without hindrance.

Fat oxidative stability is measured by the active oxygen method (AOM, AOCS Method Cd 12-57). Oil or fat is held at 97.8°C while air is bubbled through it. The time required to develop a peroxide concentration of 100 meq/kg is the AOM stability of the sample. A closely related method, the oil stability index (OSI, AOCS Method Cd 12b-92), also bubbles air through hot oil. One of the breakdown products is formic acid, which is trapped in a water cell. The machine continuously monitors conductivity of the water, and records the time when it rises sharply. Rancimat times obtained at 110°C are 40–45% of the AOM times, so an OSI stability of 4 h is equal to an AOM stability of 10 h.

Oxidative stability is important for obvious reasons. The baker does not want the shortening or oil to develop rancidity in storage before use, and wants to avoid rancidity in the finished product as long as possible. The fat or oil as delivered to the bakery should have a peroxide value below 1 meq/kg. If the amounts of oil used are large enough, the baker may consider blanketing his storage tanks with nitrogen to extend storage stability.

7.4 Typical Specifications

Complete specifications for any bakery shortening include many factors beyond those discussed above and those listed in Table 8.1. These may include such items as the source material desired, packaging, storage, and numerous other factors related to Good Manufacturing Practices and HACCP guidelines. The items presented in Table 8.1 are related to the direct functionality of shortening in baked goods, as discussed in this Chapter. The conscientious Quality Assurance person writing the actual specification will want to include many more items.

The maximum peroxide value given relates to the potential resistance of the shortening against oxidative rancidity; a higher peroxide value means that the shortening will develop rancidity sooner. The value for AOM speaks directly to this characteristic. The free fatty acid and phosphorous limits given should be met by any supplier doing a good job of refining the initial raw materials (e.g., soybean oil) used to produce the shortening.

No melting points are given, because in general these have little relevance to functionality in baking. For reference, it is noted that different methods are used to determine m.p., and they relate to the SFI curve roughly as follows:

Wiley m.p. corresponds to an SFI of 3; Mettler dropping point m.p. corresponds to an SFI of about 1.5; and the complete m.p. is at an SFI of 0.

Finally, it is emphasized that Table 8.1 gives typical physical specifications for a number of bakery shortenings; adjustments can and should be made to meet requirements for a particular product or equipment line. The values given are meant as guidelines, and a complete ingredient specification should be developed in consultation with shortening suppliers.

REFERENCES

1. W.R. Moore and R.C. Hoseney, *Cereal Chem.* **63,** 172–174 (1986).
2. N.B. Howard, *Bakers' Dig.* **46**(5), 28–30, 32, 34, 36–37, 64 (1972).
3. J.C. Wootton, N.B. Howard, J.B. Martin, D.E. McOsker, and J. Holme, *Cereal Chem.* **44,** 333–343 (1967).
4. J.L. Vetter, D. Blockcolsky, M. Utt, and H. Bright, *Tech. Bull. Am. Inst. Baking* **6**(10), 1–5 (1984).
5. J.L. Vetter, J. Zeak, and H. Bright, *Tech. Bull. Am. Inst. Baking* **10**(9), 1–6 (1988).
6. J.C. van den Enden, A.J. Haighton, K. van Putte, L.F. Vermaas, and D. Waddington, *Fette Seifen Anstrichm.* **80,** 180–186 (1978).
7. J.C. van den Enden, J.B. Rossell, L.F. Vermaas, and D. Waddington, *J. Am. Oil Chem. Soc.* **59,** 433–439 (1982).

9
Oils and Fats in Confections

Fats and oils are found in a variety of confectionery products. Applications and levels of usage cover a broad range. For example, they might be employed as a vehicle for flavoring agents (processing aid) and thus make a negligible contribution to the total fat content. At the opposite end of the spectrum, fat levels exceeding 50% for meltaways and truffles and 60% for frozen novelty coatings are common. Fats and oils contribute to flavor, texture, eye appeal, aroma, mouthfeel, and generally significantly dictate the overall quality of the eating experience. They are found in centers, coatings, solid and hollow molded pieces, and also serve as roasting media for frying nutmeats, release agents/lubricants for equipment and wrappers, and vehicles for flavorants, colorants, and polishing agents and glazes. Given the myriad roles satisfied by fats and oils in confectionery products, it is easy to appreciate that this ingredient is often paramount to the quality delivered in the finished confection. Demands placed on the fat and oil might include (*1*) the ability to deliver a unique characteristic flavor (e.g. cocoa butter, butterfat); (*2*) conversely, not contribute to the flavor at all but rather provide structure, shortness, or lubricity or perhaps to suspend other ingredients selected for their ability to deliver flavor and aroma in an acceptable form between factory and consumer; (*3*) provide gloss, surface finish, and eye appeal; (*4*) serve as a heat transfer medium during roasting processes; (*5*) provide a moisture barrier; and (*6*) postpone oil migration. For many confectionery applications, where fat is more than a processing aid, lipids constitute the continuous phase and are present at significant levels.

CHOCOLATE

Typically, chocolate comes to mind when one thinks of confections. The cocoa bean is key to the production of this tasty treat. Seeds of the *Theobroma cacao* tree are the cocoa beans of commerce. The species *T. cacao* is divided into two main groups, *T. criollo* and *T. forastero*. A third group, *T. trinitario*, is believed to be a naturally occurring cross between *T. criollo* and *T. forastero*. The cocoa tree is a native of the equatorial Americas and today thrives in West Africa, Asia, and the West Indies. The ideal warm humid climate required by the cocoa tree lies within 10° or 15° of the equator. *Theobroma criollo* types are grown primarily in Venezuela, Indonesia, and Ecuador; *T. forastero* in West Africa, Brazil and Malaysia; and *T. trinitario* in Trinidad, Jamaica, and Papua New Guinea. Depending on the variety, trees require 3–5 years to bear the first fruit and may not reach full yield until the tenth year. The productive life may be 40 years or more. The fruit, or pod, is unique in that it grows from the main trunk and large branches. It is similar in appearance to a small furrowed melon and typically measures 4 inches in diameter and 8 inches in length. Twenty to 40 pulp-covered beans are contained within each pod. The beans constitute about 40% of the total pod weight. Water represents some 65% of bean weight, skin 10–15%, and the remainder is cotyledon. Cotyledon (Table 9.1) is about 8–10% of total pod weight and represents the yield of interest (1–3). Subsequent to harvesting, the pods are split open and the beans and white mucilaginous pulp removed.

Bacterial action upon the pulp breaks it down, and the fermented liquid drains away from the beans. This process is critical to the development of desirable color and flavor characteristics in the beans. Cocoa beans are fermented over a period of 3–8 days typically in covered heaps or boxes during which time sugar is converted to alcohol and carbon dioxide; the alcohol is

Table 9.1 Typical composition of pulp and cotyledon (4)

Pulp		Cotyledon	
Component	%	Component	%
Water	84.5	Water	32.5
Glucose, fructose	10.0	Fat	31.0
Pentosans	2.7	Proteins	9.0
Sucrose	0.7	Polyphenols	5.5
Proteins	0.6	Pentosans	5.0
Acids	0.7	Starch	5.0
Salts	0.8	Cellulose	2.5
		Sucrose	2.5
		Theobromine	2.5
		Salts	2.5
		Acids	1.0
		Caffeine	1.0

subsequently oxidized to acetic acid by bacterial action. This process (*1*) raises the temperature of the beans high enough to kill them thus preventing germination, (*2*) browns the beans, (*3*) reduces bitter, sour, astringent, floral, and sweet notes, (*4*) increases cacao and nutty notes, and (*5*) hardens the skin to a shell. The beans are then dried, either naturally or mechanically, to about 6% moisture and bagged. This step is required to preclude mold growth and concomitant development of objectionable flavors that cannot be removed by subsequent processing steps (1,2).

1.1 Products from the Cocoa Bean: Chocolate Liquor, Cocoa Butter, and Cocoa Powder

The cocoa bean yields three products fundamental to the manufacture of chocolate and chocolate products: chocolate liquor, cocoa butter, and cocoa powder. Beans are cleaned, roasted, shelled, and the broken pieces of kernel, or nibs, recovered for further processing. Roasting develops both flavor and aroma and generally is conducted over a range of about 100–140°C; during the initial phase, the moisture content is reduced to around 1%. Historically, whole beans were roasted and then cracked. The shell inhibits moisture evaporation from the bean interior, and consequently higher roasting temperatures were applied to overcome its insulating effect. The net result was the risk of overroasted burnt notes as well as entrapment of some of the volatiles detrimental to the development of desired flavor profiles. Thus, today beans are typically predried, broken and shelled, and then the nibs roasted. Flavor development during the roasting process is very complex and certainly paramount to the quality of finished chocolate. Well over 400 compounds have been isolated from roasted beans. *T. criollo* and *T. trinitario* types are known as fine or flavor cacaos—the former, when processed, is very light brown and has a sweet, milky flavor frequently accompanied by a nutty character and delicate aroma; the latter exhibits a wide variation in color and flavor notes that range from mild and nutty to overpowering with underlying tones of licorice, raisin, and wine. *T. forastero* dominates the market today and is often referred to as bulk cacao; it is darker and more strongly chocolate flavored than *T. criollo* (3,5).

The fat content of the nibs, cocoa butter, averages about 55%. The cocoa butter is locked within the cellular structure of the nib and therefore either a grinding or expeller process is required to free the fat. A variety of grinding devices are available and include hammer, pin, ball, and roller mills. Typically, a particle size distribution averaging between 15 and 70 μm is desired, depending on the intended use. Four groups of components are reduced in size during grinding: shell, germ, cell walls, and cellular contents. Shell thickness typically ranges between 200 and 250 μm and cell walls about 2 μm; the cell diameter is 20–30 μm. Cellular contents (predominately starches and proteins) are substantially smaller than 20 μm and really do not need to be ground (6).

Chocolate liquor, then, is simply finely ground nibs—particles of cocoa solids suspended within cocoa butter (also often referred to as cocoa mass, unsweetened chocolate, or baking chocolate). Chocolate liquor can be separated into cocoa butter and cocoa powder under hydraulic pressure in a pot press. Residual fat content of the press cake is typically controlled to either 22% (breakfast cocoa, high-fat cocoa) or 10–12% (cocoa, medium-fat cocoa); recently, powders with less than 1% residual cocoa butter (low-fat cocoa) have become commercially available. The cake is crushed and pulverized to the desired fineness for cocoa powder. Alternatively, roasted nibs may be softened by a steam treatment and then fed to an expeller press where the fat content can be reduced to below 10%. The shear forces imposed upon the nib by this process tears rather than grinds them and the resultant press cake is thick flakes. The flakes are subsequently ground to the desired fineness. Table 9.2 summarizes the composition, physical/chemical properties, and bacteriological standards for some typical commercial 10–12% cocoa powders (7).

The beans, nib, liquor, powder, or flake are often treated with an alkali solution, potassium or sodium carbonate, for example, to improve color and flavor. Product so treated is referred to as "dutched" or "alkalized" while nonalkalized cocoa products are often spoken of as "natural." The color strength and hue can be altered over a wide range by careful control of reaction conditions. Colors can range from very light brown to red tints to almost black. Minimizing alkaline off-flavors is a challenge with heavily alkalized darker colors because the acids are neutralized, consequently reducing the original astringency. Furthermore, it is often difficult to predict the buffering capacity of a given cocoa product lot and over neutralization is certainly a risk associated with this treatment. However, it is possible to readjust the pH with small additions of approved food acids; the Codex Alimentarius permits phosphoric acid, citric acid, and L-tartaric acid for this purpose. The ability to neutralize the alkali after dutching gives the possibility of an even greater range of color as well as some flavor improvement to highly alkalized black powders. Control over pH is especially important for cocoa powders used in the baking industry because it interacts with leavening systems.

The poor wettability of cocoa powders represents an issue for some applications and can be overcome via the addition of lecithin. A classic example of this problem is illustrated by the steps necessary to prepare an acceptable chocolate-flavored beverage. It is extremely difficult to thoroughly disperse cocoa powder into an aqueous media like milk or water. The traditional approach involves either (*1*) making a paste by blending the cocoa with a small portion of milk or water and then adding boiling milk or water to the paste and whisking to promote dispersion or (*2*) boiling all ingredients together. Wettability of cocoa powders can be substantially improved by adding between 1.5–5% lecithin (8,9). The aim when lecithinating powders is to coat each cocoa particle to derive the maximum surface activity from the lecithin. Given the substantial quantity of lecithin involved as well as the tremendous surface area it presents, only very low flavor, high-quality stable lecithins are

Table 9.2 Typical properties and composition of commercial 10–12% cocoa powders

Item	Amount
Composition	
Fat	9–12%
Water	3–5%
Protein	19–25%
Theobromine	2.2–2.8%
Caffeine	0.3–0.6%
Polyhydroxyphenols	7–15%
Starch	9.5–14.5%
Sugars	1–4%
Cell wall constituents	21–25%
Organic acids	2.5–3.5%
Ash	4–13%
Shell content	1.75% max. calculated on alkali free nib
Calories	210 kcal/100 g
Potassium	1600–5400 mg/100 g
Sodium	24–56 mg/100 g
Calcium	100–200 mg/100 g
Magnesium	400–600 mg/100 g
Iron	10–30 mg/100 g
Phosphorus	500–700 mg/100 g
Chlorine	24–56 mg/100 g
Sulfate (SO_3)	100–200 mg/100 g
Physical/chemical properties	
Bulk density	0.35–0.38 cm^3/g
Specific heat	1717 J/kg °C
Ignition temperature	165°C
pH	5.9, nonalkalized (light color)
	7.2, alkalized (medium color)
	8.1, alkalized (dark color)
Fineness	99.5–99.9%, wet, through 200 mesh sieve (0.075 mm)
Bacteriological standards	
Standard plate count	5000 max., median 300
Molds per gram	50 max., median 5
Yeasts per gram	50 max., median 5
E. coli per gram	Negative
Enterobacteriaceae per gram	Negative
Salmonellae	Negative
Lipase activity	Negative

selected. This improvement makes possible the formulation of cold-wettable as well as instant products. The instantized drink mixes are typically an agglomerated mixture of about 70% sugar and 30% cocoa powder with perhaps small additions of flavors and salt. Over time, gravitational sedimentation of cocoa solids from the medium within which they are suspended will occur. This does not represent an issue for beverages prepared for immediate consumption; however, prepackaged drinks are affected. This problem is generally allayed via the addition of a suspending agent like carrageenan, which forms a loose network with the casein molecules (10).

1.2 Refining

The manufacture of chocolate entails four basic steps: (*1*) mixing the ingredients, (*2*) particle size reduction or refining, (*3*) conching, and (*4*) crystallization. A typical basic sweet dark chocolate formula is represented in Table 9.3.

The chocolate liquor, granulated sugar (or 50–60 μm pulverized sugar might be used instead), and perhaps a portion of the cocoa butter (enough to wet the dry ingredients and form a paste) are blended. The resultant paste is fed into a roll refiner. The objective is to grind the particles to an acceptable range; generally, particles larger than about 30 μm tend to be perceived as gritty while those smaller than 6 μm are not detected by the tongue. Therefore, the particle size distribution target normally lies somewhere between these upper and lower limits. The refiner is typically a bank of five very closely spaced rotating steel rollers. Paste is fed into the gap (about 100 μm) between the bottom two rollers, which form a very thin film of paste. The film is picked up by the next roller, which rotates in the opposite direction to and slightly faster than that immediately below. The slit between adjacent rollers becomes progressively narrower moving up the refiner. At the second refining slit, the film is reduced by a factor of 1.5 or more (60 μm). This roll-to-roll transfer continues up the refiner (particle size is reduced to about 40 μm at the third slit and to 20 μm at the last gap), and the layer of film becomes thinner and thinner as it moves from roll to roll because of the progressive increase in rotation speed. The pressure between adjacent rolls is set hydraulically to

Table 9.3 Sweet dark chocolate

Component	%
Chocolate liquor	42.0
(23.1% cocoa butter	
18.9% cocoa solids)	
Cocoa butter	8.4
Sugar	49.6
Lecithin	0.4

maintain a regular and continuous film of chocolate mass (11). The rollers are water cooled to maintain control over the heat of friction. Another approach involves blending nibs, sugar, and enough cocoa butter to form a coarse paste. This mass is passed through breaker rolls to produce a paste that can be efficiently handled by the refiners. A third option brings nib, sugar, cocoa butter, and sometimes a portion of the lecithin together in a kneader. This is typically a heavy-gauge cylinder fitted with a reciprocating toothed shaft that comminutes the ingredients to produce a paste that is fed to the refiners. Regardless of the process employed, the dry, refined paste is transferred off the top refiner roll with the aid of a doctor blade and moved to the conche.

During the milling process, whether it be impact or roll refining, some of the sucrose crystals are cracked. They can also be flattened to very thin transparent plates when passed quickly through the roll gaps. In either case, amorphous sugar is formed. When a sugar crystal fractures, so much energy is converted that temperatures in the fracture area may reach 2300°C for less than a microsecond (sucrose has a melting point of about 170°C). Only a few layers of molecules into the broken surface are affected, and heat conducted into the broken piece nearly instantaneously cools the molten surface. The high fracture surface temperature produces objectionable burned flavor notes and is of particular concern when sugar is milled separately from other ingredients. However, it seems that when milled with aromatics, like cocoa solids or milk powder, the negative taste is largely masked by absorption of the associated aromatic compounds. These phenomena are important because amorphous sugar is extremely unstable and hygroscopic; it can take up the necessary water for crystallization within a few seconds of milling. As the sugar crystallizes, the water is released—and the quantity is substantial, in the neighborhood of 3 kg per ton of sugar. In an enclosed storage environment, the water is largely absorbed on the surface of the sugar, causing the crystals to agglomerate (12,13).

1.3 Conching

The refining step produces a dry friable flake that is fed to the conche, or "taste changer." This change in paste appearance is primarily a function of the reduction in particle size accomplished by the refiner. The greatly enhanced surface area presented by the refined solids exhausts the free cocoa butter (and any other fat present in the formula) that was available in more than sufficient quantity to coat the surface of each solid particle present in the paste fed to the refiner. The conche serves to intimately disperse the cocoa butter and to coat each sugar and cocoa particle with fat (the continuous phase) as well as to break up the agglomerates formed during refining. Kneading, shearing, and thermal control are key to a successful conche. Physical changes that occur include dispersion, dehumidification, homogeneity, improvement of flow properties, and removal of volatiles. Dark chocolates are generally conched at 70°C and as high as 82°C while milk chocolates are conched at lower temperatures because of their protein content. Crumb milk chocolate is typically conched at 49–52°C, full-fat milk powder chocolate up to 60°C, and nonfat milk powder chocolate up to 70°C (14).

Traditional longitudinal conches consist of a trough with a flat granite bed over which large heavy granite rollers move back and forth. Eventually, the conched chocolate must fluidize and develop the required rheological properties. In the case of the longitudinal conche, the only means by which this can be accomplished is with the immediate addition of more cocoa butter and lecithin in order to induce a consistency that can be handled by the conche. Longitudinal conches require highly fatty masses rather than the powdery flake produced by the refiners because the powdery material compacts and could damage the conche. The process normally requires about 3–4 days to develop the desired flavor and smoothness because the shearing stress is so low; the development of more powerful conches as well as new techniques have shortened this time considerably and today conche times of between 6 and 12 h are common (15).

Such mechanical improvements have led many processors to dry conching. Refined paste is subjected to high friction and high shear stress resulting in rapid and intensive evaporation of water together with many water-soluble acids. Water evaporation is most efficient when the refiner paste is still dry and powdery because the solids have not been covered with fat. The moisture leaves the mass from both sugar solution on the surface of sucrose (and lactose if milk solids are present in the recipe) as well as cocoa solids. This is an especially important process because many of the undesirable volatiles are associated with the cocoa solids; and, in general, water-soluble volatiles are undesirable for flavor while the fat-soluble flavors are preferred (16). The heat of friction also encourages the Maillard reaction and concomitant organoleptic changes. As water is released, the mass becomes progressively softer; and the dull appearance brightens and changes to a glisten. Trapped fat is released as well, adding to the fluidity of the mass. Application of extreme physical stress to the semiplastic mass encourages the chemical processes so critical to the development of desirable flavor notes. Near the end of the process, the balance of the cocoa butter and lecithin are added to achieve the required viscosity and fat content at a final moisture level of perhaps 0.3–0.5% (15,17). Much effort has been devoted over the years toward understanding the chemistry and mechanics of this process. Clearly, this step is critical to the production of acceptable chocolate—it serves to round out the flavor profile and reduce the harshness. Furthermore, conching is time consuming and requires a great deal of energy; and of course each of these elements has an associated cost.

The predominant odor in the vicinity of an operating conche is that of acetic acid. Hoskin and Dimick report that over 95% of the volatile fraction in chocolate is acetic acid; propionic, isobutyric, isovaleric, and butyric (in milk chocolate) acids make up most of the balance (18,19). Acetic acid is among the many volatile fatty acids produced during cocoa bean fermentation;

however, many of these are removed during the roasting process. It is generally believed that conching further releases additional volatile acids; however, the magnitude of the change in volatiles is not clear. For example, Hoskin and Dimick (18) found a significant decrease in volatile fatty acid content after conching for some, but not all, treatments. They used four types of conche for their work: pug mill, longitudinal, vertical, and horizontal rotary. Total volatile fatty acids in their starting chocolate were reported as 823 μg/g. Both the vertical and horizontal rotary conches reduced total volatile fatty acids to 634 and 614 μg/g, respectively, while neither the pug mill nor the longitudinal conche altered the volatile fatty acid content significantly (909 and 721 μg/g, respectively). These researchers point out that the boiling point for the two carbon volatile fatty acid, acetic, is 118°C, well in excess of temperatures reached during the conching process. The three-carbon propionic acid and four-carbon isobutyric acid have even higher boiling points (141 and 154°C, respectively). Because the associated aldehydes of these fatty acids have boiling points almost 100°C lower, it is proposed that they may in fact represent a more significant class of compounds vis-à-vis flavor development. Therefore, it could be expected that during conching the aldehydes would decrease.

However, Maniere and Dimick report that total carbonyls and monocarbonyls were not significantly affected during conching (19). Additionally, Rohan and Stewart (20) reported that a 48-h longitudinal conche at 71°C had no measurable impact upon the quantity of either reducing sugars or amino acids. Ley (21), on the other hand, supports Strecker's theory of free amino acid development after roasting. One-third to half of the amount of free amino acids are developed during conching as are during roasting, but roasting destroys roughly half of the free amino acids that are formed. Amino acids form more slowly in the conche because of the low temperature involved. Free amino acids and reducing sugars are important precursors for the flavors developed during heating via Maillard reactions.

Analysis of headspace volatiles provides a possible explanation for this apparent discrepancy. Manier and Dimick (19) measured and plotted the relative intensity of headspace volatiles generated during the conche of a dark semisweet chocolate from low-roast beans. During the first 16 h of conching, the relative intensity of headspace volatiles score fell from 7 to about 1.5 (nearly an 80% decrease) and remained relatively constant thereafter. This finding is consistent with what one might typically expect as a result of prolonged mixing at elevated temperature. And, of course, as these researchers point out, changes in flavor characteristics are not universally associated with major changes in volatile levels. Hoskin and Dimick (18) pose a provocative theory concerning the nature of the flavor changes that occur during conching. They suggest the possibility that the complete coating of each cocoa particle and sugar particle with cocoa butter as a result of the temperature and shear encountered in the conche might mute the bitter notes of cocoa solids and also dampen the clean sweetness of sugar.

1.4 Milk Products

The sweet dark chocolate formula in Table 9.3 can be altered to include up to 10% milk solids to offset some of the sweetness; the sweet milk chocolate described in Table 9.4 is another formulation option that incorporates even higher milk solids and lower cocoa solids.

Liquid milk is rarely used in the confectionery industry; rather cream, evaporated milk, sweetened condensed milk, and milk powders are the typical forms. Dried milk products are most generally used in the production of chocolate. They can be either drum (roller) dried or spray dried, and each process produces a characteristically different product. Most chocolate manufacturers tend to prefer roller-dried milk. The higher temperatures required for this process initiate the Maillard reaction, giving these powders a somewhat spicy and salty taste, whereas the spray dried products have a distinct milky flavor. Roller-dried milk has a substantially higher surface fat content than does spray dried (>95 vs. <10%). The available fat improves rheological properties of the chocolate during mixing and conching and thus reduces energy requirements (22). Whey products, derived from sweet whey, are sometimes used as lower-cost alternatives to dried milk or condensed milk products in nonstandard identity confections. Following pasteurization, the mineral salts are removed by electrodialysis. This step is extremely important because if not removed, the mineral salts deliver an unsatisfactory astringent note to many confectionery formulations. The native protein is largely preserved (present at 11–16% in dried whey products) because of the low heat treatment and the dialysis process.

The crumb process makes possible the production of milk chocolate having a very characteristic, caramelized taste. The process is as follows. Fresh milk is preheated to 75°C and then evaporated to 30–40% total solids. Sugar is added and the mixture condensed under vacuum to a concentration of 90% solids. From here, the condensed sugar/milk is pumped slowly into a melangeur or other type of robust mixer that has been previously charged with a specified quantity of chocolate liquor. Crystallization continues with mixing over a period of about 20 min. The paste is discharged and dried to under 1%

Table 9.4 Sweet chocolate and sweet milk chocolate

Component	Sweet Chocolate (%)	Sweet Milk Chocolate (%)
Chocolate liquor	42.0	15.0
Cocoa butter	8.4	20.0
Sugar	41.6	42.0
Skim milk powder	8.0	8.0
Whole milk powder	—	15.0
Lecithin	0.4	0.4

moisture, broken into lumps or further refined to a coarse powder, and packaged. A white crumb is made by deleting the chocolate liquor addition. Crumb is combined with cocoa butter, mixed, refined, and conched to make milk chocolate (23).

1.5 Rheology

Molten chocolate does not behave as a true liquid but rather exhibits non-Newtonian properties (Figure 9.1). This property is largely a function of the nonfat solids (primarily carbohydrates) present because the cocoa butter is a Newtonian material (Figure 9.2).

The introduction of particles into a Newtonian material, for example, cocoa solids and sugar into cocoa butter, greatly complicates the flow properties. The hydrophilic nature of both cocoa solids and sucrose leads to a high degree of interaction between the two, and consequently a high viscosity in the matrix. Furthermore, a minimum quantity of energy must be applied to initiate flow—a yield stress must be overcome. Application of continually increasing rates of shear, beyond the yield point, results in a progressive decrease in viscosity; thus, chocolate is shear thinning. This is a function of the solid particles [cocoa solids, sugar, and milk solids (if present)] aligning themselves in the field of flow. The particle size distribution, quantity, and/or shape can be altered to influence the resultant flow properties for a given chocolate. Viscosity determinations are typically made with a rotational viscometer and both plastic viscosity as well as yield value are reported to fully characterize the rheological properties of the product under evaluation.

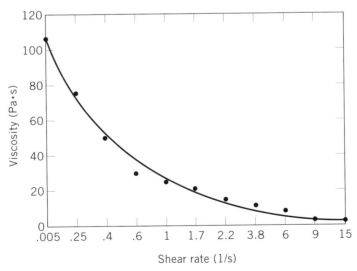

Figure 9.1 Shear thinning flow behavior typical of chocolate.

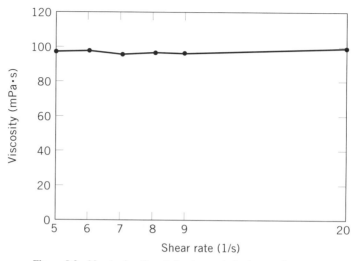

Figure 9.2 Newtonian flow behavior typical of cocoa butter.

To reduce chocolate viscosity to a manageable range, each nonfat particle must be covered with cocoa butter, thereby reducing interaction between hydrophilic particles and permitting them to slide over one another. As more fat is added, the viscosity becomes progressively lower. Particle size significantly affects viscosity since large particles require substantially less fat for complete surface coating than do smaller particles. Hence, the fat content will be higher for a more finely ground chocolate than a coarser grind at any given viscosity. Surface-active agents will further improve fluidity, and consequently it is common practice to add a few tenths of one percent soya lecithin to chocolate. Minifie (24) says that the viscosity-reducing effect of lecithin is primarily a function of its interaction with sugar particles rather than cocoa solids. He demonstrated this by measuring the viscosities of (*1*) chocolate, (*2*) a sugar–fat mixture, and (*3*) a cocoa–fat mixture to which increasing portions (0.0–0.7%) of lecithin were added. Viscosities for both the chocolate and sugar–fat mixture decreased markedly and steadily with increasing levels of lecithin addition while the cocoa–fat mixture showed only a minimal decrease in viscosity across the range of additions evaluated. The viscosity of molten chocolate is quite high compared to molten fat, in the neighborhood of 150- to 200-fold. Generally, about one-third fat is required in the recipe to achieve the desired flow properties. The addition of lecithin permits achievement of a given viscosity at much lower cocoa butter contents than if it were not included and at typical levels of addition reduces the fat requirement by about 5% (a 13% reduction on total fat basis). Table 9.5 summarizes Minifie's evaluation of the cocoa butter sparing effect of lecithin (24).

Lecithin replaces over 10 times its weight in cocoa butter, and because it is significantly less costly than the butter, it provides an obvious economic

Table 9.5 *Cocoa butter contents required to replace lecithin in dark chocolate (24)*

Percent Lecithin	Percent Cocoa Butter		
	In Formula	Reduction Formula Basis	Reduction Total Fat Basis
0	38.0	—	—
0.1	36.2	1.8	4.7
0.2	35.1	2.9	7.6
0.3	34.3	3.7	9.7
0.4	33.6	4.4	11.6
0.5	33.3	4.7	12.4
0.6	32.9	5.1	13.4
0.7	32.8	5.2	13.7

advantage. However, there is a practical limit, about 0.5%, beyond which additional lecithin no longer reduces viscosity and in fact increases it. In addition, higher levels adversely affect flavor and crystallization properties of chocolate. Researchers at Cadbury Bros. Ltd. patented a synthetic lecithin, YN, that is more efficient than soya lecithin and has a bland flavor (25). YN is produced by glycerolysis of hydrogenated (70–80 iodine value) rapeseed oil, followed by phosphorylation, neutralization with dry ammonia gas, and finally screening. Minifie gives the composition of the final product, YN, as

Triglyceride	40%
Neutral phospholipid	15%
Mixed phosphatidic acids as ammonia salts	40%
Ammonia salts of phosphoric acids	5%

YN more effectively reduces viscosity compared to soya lecithin over the standard range of 0.1–0.5% (the viscosity-reducing power is about 5/3 that of soya lecithin); the thinning effect continues up to 0.9% addition (24).

Other surface-active agents useful for rheological modification include polyglycerol polyricinoleate (PGPR) and sucrose dipalmitate. Polyglycerol polyricinoleate is especially interesting because it has a major impact upon yield value (with the possibility of reducing it to essentially zero) and simultaneously minimal impact upon plastic viscosity. Harris (Table 9.6) has reported on the influence of soya lecithin, YN, sucrose dipalmitate, and PGPR (each at 0.3% addition) on the flow characteristics of chocolate (26), and Seguine (Figure 9.3) has assessed the impact of incremental additions of PGPR in milk chocolate (27).

Viscosity is also profoundly affected by moisture. Chocolate typically contains from 0.5 to 1.5% moisture; the addition of extremely small increments of free moisture into finished chocolate substantially and rapidly increases

Table 9.6 *Influence of selected surfactants upon flow properties of chocolate (26)*

Surfactant at 0.3%	Casson Plastic Viscosity (Poise)	Casson Yield Value (dyn/cm²)
Soya lecithin	6.1	92
YN	10.3	30
Sucrose dipalmitate	8.6	166
PGPR	32.5	25

viscosity. Harris (28) suggested that syrup layers form on the surface of sugar particles subsequently impeding their mobility thereby increasing viscosity. This friction effect is reduced to some extent by the addition of phospholipids, hence chocolate containing a surfactant tolerates higher moisture levels. Additionally, lower-fat-content chocolates (35%) suffered a more severe viscosity increase than higher-fat-content chocolates (39%) when water was added according to Rasper (29).

1.6 Cocoa Butter—Physical and Chemical Properties

Fat is the key to successful confections. It is typically a major component of most recipes and generally represents the highest ingredient cost. The fat phase, cocoa butter, is largely responsible for chocolate's characteristic brittle

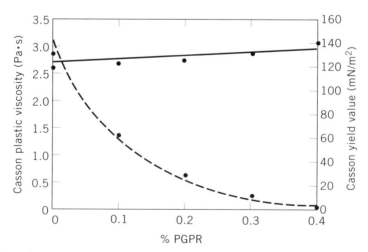

Figure 9.3 Influence of PGPR upon the plastic viscosity and yield value of milk chocolate (27). ——, Plastic viscosity; ----, yield value.

nongreasy texture at and below room temperature, excellent keeping qualities, and characteristically rapid melting near body temperature. Cocoa butter consists of 98% triglycerides, about 1% free fatty acids, 0.3–0.5% diglycerides, 0.1% monoglycerides, 0.2% sterols (mainly sito- and stigmasterol), 150–250 ppm tocopherols, and 0.05–0.13% phospholipid (30). The triglycerides are responsible for the unique melting and crystallization properties of cocoa butter. Three fatty acids predominate: palmitic (25%), stearic (36%), and oleic (34%); and the unsaturated oleic moiety is almost always esterified at the β or 2-position while saturates (palmitic and stearic) occupy the α or 1,3-positions (Table 9.7). Consequently, three symmetrical triglycerides represent about 80% of the triglycerides found in cocoa butter:

Palmitic–oleic–stearic (POSt), 36–42%
Stearic–oleic–stearic (StOSt), 23–29%
Palmitic–oleic–palmitic (POP), 13–19%

Cocoa butter is hard and brittle at temperatures up to about 27°C; most melting occurs over the narrow range of 27–33°C and is essentially complete at 35°C. This melting profile (Figure 9.4) is key to the performance and suitability of cocoa butter for confectionery applications. Generally, the continuous fat phase is required to maintain the other nonfat ingredients (primarily sugar and cocoa solids) in an acceptable form prior to consumption. Upon consumption, the fat phase must melt quickly and completely so that the other

Table 9.7 Typical chemical and physical properties of cocoa butter (30)

	Fatty Acid Composition (%)		Triglyceride Profile (%)	
C16:0 (P)	25		POSt	36.3–41.2
C18:0 (St)	36		StOSt	23.7–28.8
C18:1 (O)	34		POP	13.8–18.4
C18:2 (L)	3		StOO	2.7–6.0
C20:0 (Ar)	1		StLP	2.4–6.0
Specific gravity at 15°C	0.970–0.998		PLSt	2.4–4.3
Refractive index at 40°C	1.4565–1.4570		POO	1.9–5.5
Melting point	30–35°C		StOAr	1.6–2.9
Saponification value	188–195		PLP	1.5–2.5
Iodine value	35–40		StLSt	1.2–2.1
Unsaponifiable residue	0.3–0.8		OOAr	0.8–1.8
Titer	48–50°C		PStSt	0.2–1.5
			POL	0.2–1.1
			OOO	0.2–0.9

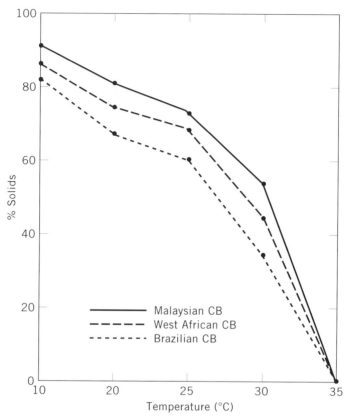

Figure 9.4 Typical melting profile of origin cocoa butters: Malaysian, West African, and Brazilian.

ingredients contained therein become available to contribute to sweetness, chocolaty flavor and aroma, and the other sensory experiences associated with chocolate.

Any solid fat remaining above 35°C is undesirable as it will be recognized as waxy and chewy and will not give up much of the aromas and flavors entrapped within. Generally, more flavor has to be added to fat-rich products in order to deliver the desired impact. Consider, for example, (*1*) chocolate, (*2*) chocolate-flavored ice cream, and (*3*) chocolate-flavored skimmed milk. The fat content for each is approximately 35, 10, and 0.2%, respectively, while cocoa solids represent 16, 2.5, and 1.3%. Fat masks the bitter element (theobromine) of cocoa as well as the sour note, and thus alkalized powders are not often used in higher-fat-content products. On the other hand, aqueous systems tend to highlight bitter notes, and the typically lower fat contents of such systems lack the masking effect. Consequently, alkalized powders are usually selected for chocolate-flavored milk, for example (31).

Triglycerides are believed to conform to a chair or tuning fork configuration, regardless of whether in the liquid or solid state. Furthermore, in the crystalline state, triglycerides likely exist as dimers. The longitudinal packing is determined by the degree of unsaturation and chain lengths of the constituent fatty acids. Most saturated triglycerides containing a single acid, for example tristearin (StStSt), pack into a double chain length structure. However, those triglycerides containing fatty acid moieties of varying chain length (more than four carbon atoms difference between longest and shortest fatty acid) tend to exhibit triple chain length packing (Figure 9.5). This is also the case for triglycerides that contain an unsaturated fatty acid at the β position. Hence, the packing arrangement for crystallized cocoa butter closely approximates the triple chain length configuration (32).

Fundamentally, three crystal forms are typically discussed, each exhibiting different stability. The α form is normally the first to develop upon cooling from the melt and is often only transitory because as its melting point is approached it transforms rapidly into the more stable β' form. This transition may require a matter of seconds to several hours. Under the appropriate conditions, the β' form transforms to the most stable β crystal. This change

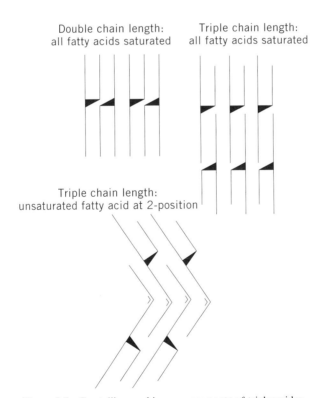

Figure 9.5 Crystalline packing arrangements of triglycerides.

may require a matter of hours to a matter of months. Cocoa butter exhibits complex polymorphism because of the small differences among the constituent symmetrical triglycerides. Because cocoa butter is a product of nature, its origin and climatic conditions within the growing region will alter its triglyceride profile to a degree. An additional variable is introduced by the process selected for extraction of butter from the bean. Consequently, a range of melting points have been reported over the years as each polymorph was identified. Today, six forms are generally recognized (Table 9.8) (32–34).

The crystal forms become progressively more stable moving from form I through form VI. The γ form (form I) crystallizes below 17°C and rapidly transforms spontaneously into α (form II). This crystal remains viable about 1 h and melts around 23°C. All transitions between forms I and IV occur in the liquid state while the conversion from form V to VI is a solid-state transition.

Classification schemes are further complicated by the fact that forms III and IV have been proposed to be β' polymorphs and that forms V and VI are thought to be similar to one another and both β. Dr. S. V. Vaeck studied cocoa butter extensively and much of his work was summarized in a Swiss publication, the *International Chocolate Review*, between 1951 and 1955 (35). The melting points he reported (Table 9.9) for the α, β' and β polymorphs agree closely with those published by Johnston and Willie and Lutton (Table 9.8).

1.7 Tempering

Given the foregoing, it is clear that cocoa butter must be "encouraged" into the appropriate crystal form so that products formulated with this fat exhibit the physical properties desired. Attributes affected in chocolate include appearance (gloss and gloss retention), snap, contraction, texture, and overall product acceptability. The technique of tempering is applied to promote the

Table 9.8 *Cocoa butter polymorphs and their melting ranges reported by various researchers (32–34)*

Form	Systemic Nomenclature	Melting Point (°C)	
		Johnston	Wille/Lutton
I	γ'_3 (sub α) (gamma)	16–18	17.3
II	α-2	21–24	23.3
III	β'_2-2	25.5–27.1	25.5
IV	β'_1-2	27–29	27.5
V	β_2-3	30–33.8	33.8
VI	β_1-3	34–36.3	36.2
VII(?)	—	38–41	—

Table 9.9 Properties of cocoa butter polymorphs reported by Vaeck (35)

Polymorph	Melting Point (°C)	Latent Heat of Fusion	Approximate Life	Contraction from Liquid
α	21–24	19 cal/g	1 hour	0.060 mL/g
β'	27–29	28 cal/g	1 month	0.080 mL/g
β	34–35	36 cal/g	Stable	0.097 mL/g

development of stable crystals. Tempering involves the basic steps of

1. Complete melting
2. Cooling to initiate crystallization
3. Warming to melt out unstable crystals
4. Crystal development

Molten chocolate is typically stored at 45–50°C, a temperature range that is sufficiently high to preclude the development of any fat crystals. If previously crystallized chocolate is remelted, once melted it must be maintained within this temperature range for at least 30 min to 1 h to ensure all traces of cocoa butter crystals have been melted out prior to any further processing. The mass is then cooled, with agitation, to 28–29°C over a period of about 20 min. During this time, sensible heat is removed from the cocoa butter, the continuous phase, which in turn cools the cocoa and sugar and any other solids; thermal movement of the molecules slows. Eventually, the molecular motion slows enough to permit triglycerides to interact as they approach one another. Ultimately, they begin to pack together in a triple chain length chair configuration. As ordering progresses, the viscosity begins to build, thus impeding mobility of molecules, consequently prolonging the time required for a triglyceride to orient and mesh into a packing site.

Ideally, a proper balance must be imposed such that molecular motion is slowed enough to permit individual triglycerides to orient properly to establish lattices and yet not impede movement to the extent that reorientation into a more favorable conformation is precluded (36). Cooling can be accommodated with either about 18°C circulating air or a 25°C water jacket. Lower temperatures are to be avoided as they may result in localized shock cooling with the subsequent formation of unstable crystals. The chocolate is maintained at 28–29°C under agitation for several minutes to permit development of seed crystals. Only a very small proportion of the cocoa butter, 1–4%, needs to be crystallized at this stage if very small crystals are formed. As cocoa butter crystallizes, the total amount of solids present in the mass increases and consequently the viscosity increases.

The objective of tempering or precrystallization then is to introduce the proper quantity, size, and polymorph of seed crystals such that substantial and unmanageable increases in viscosity are avoided. Small seed crystals are also advantageous from a crystallization kinetics perspective. Attraction of a triglyceride to a growing crystal matrix is favored over the association and interaction of free triglycerides to form new nuclei. Therefore, a large number of smaller crystals is more conducive to progression of the crystallization process than is a much smaller quantity of substantially larger aggregates. Unstable β' (form IV) as well as the preferred β crystals will have formed over this crystallization range. Hence, the temperature is increased 2 or 3°C to promote transition to the stable β (V) form and melt out unstable β' crystals. It is critical that hereafter the temperature of the chocolate mass is never permitted to exceed 33°C, even that portion in direct contact with tank walls, because the stable crystals formed as a result of tempering will be melted out. The process required for milk chocolate is exactly the same except that all temperatures are reduced about 1°C (or more depending on the milk fat content) to accommodate the diluent effect of milk fat. Other approaches to introduce temper include (*1*) partial melting, (*2*) seeding, and (*3*) the mush method. The first two techniques require properly tempered solid chocolate.

The "partial melt" technique involves melting the solid chocolate, preferably broken into chunks first, and heating to 37–38°C. Thereafter, a solid chunk of chocolate representing about one fourth the weight of the molten chocolate is added and melted into the main mass with stirring. As the chunk melts, the temperature of the melted chocolate falls, and when it reaches 31–32°C, any remaining unmelted chocolate is removed. The molten chocolate is now in temper.

The "seeding" method, on the other hand, requires only a small portion of tempered solid chocolate shavings. The chocolate to be tempered is melted completely to 45–50°C and then cooled with stirring to 32–34°C. The tempered chocolate shavings, representing about 5% of the quantity of molten chocolate to be tempered, are stirred into the cooled main mass. As the shavings melt, the temperature of the chocolate falls to the desired range of 31–32°C and is tempered and ready to use.

The time-honored "mush" method was traditionally used by hand dippers. Chocolate is melted completely to 45–50°C and then cooled with stirring to 35°C. About one fourth of the chocolate to be tempered is poured onto a marble slab at room temperature. The puddle of melted chocolate is worked back and forth with a spatula on the slab as it cools. As crystallization progresses, the puddle thickens and becomes dull and mushy. This portion is then added back into the molten mass with stirring to produce tempered chocolate (37–39).

As was mentioned earlier, proper tempering will develop approximately 1–4% of the available cocoa butter into the β polymorph. Criteria such as appearance, viscosity, and temperature are reasonable indicators of how well tempered a chocolate mass is. However, considerable opportunity for error persists, and consequently a number of instruments have been developed over the years in an effort to quantify the degree of temper. Generally, these instruments evaluate the cooling curve (Figure 9.6) generated when a tem-

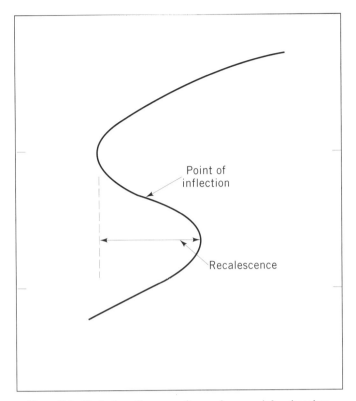

Figure 9.6 Typical cooling curve (tempering curve) for chocolate.

pered chocolate sample is cooled under prescribed conditions (a relationship exists between degree of temper and the cooling curve). Initially, the chocolate cools from the melt as sensible heat is removed and the temperature of the sample falls. Eventually, nucleation and crystallization occur and the resultant exotherm is evidenced by an increase in temperature. Ultimately, the capacity of the cooling medium to extract heat from the sample overcomes the exotherm and the temperature again falls.

Natural variability in the composition of cocoa butter can have a marked effect upon its temperability. Large amounts of StOSt, for example, enhance tempering because this triglyceride has a high crystallization rate in the β form. On the other hand, minor components (especially free fatty acids and monodiglycerides) tend to retard crystallization. Wennermark (40) evaluated the influence of digylcerides and tristearin upon the crystallization rate of cocoa butter having the following composition:

Free fatty acids, 1.0%
1,3-Diglycerides, 1.1%
1,2-Diglycerides, 0.7%

Triglyceride composition (as % of total triglycerides):
StOSt, 25.5%
POSt, 40.5%
POP, 17.6%
Trisaturates, 1.6%
Others, 14.8%

To the reference cocoa butter, 5% of (*1*) 1,2-StSt, (*2*) 1,3-StSt, or (*3*) StStSt was added and the rate of β development evaluated by x-ray diffraction. The crystallization rate measured at 26°C was markedly retarded by 1,2-StSt and by StStSt but not affected at all by 1,3-StSt. Another crystallization experiment wherein only 1% of 1,2-StSt was added also significantly retarded the rate of β crystal development. Wennermark ascribes the differences between the impact of 1,2- and 1,3-diglycerides in this model system to their crystallization behaviors. Pure 1,2-diglycerides crystallize into an α form and may transform into β' prime; 1,3-diglycerides immediately crystallize into the β polymorph, and consequently do not influence β development in cocoa butter. Given the foregoing, it is clear that even minor differences in the content of trisaturates and 1,2-diglycerides might be expected to impose a significant influence upon the temperability and crystallization properties of cocoa butter.

1.8 Cooling

Molten, tempered chocolate, whether applied to a center, molded into a solid form or hollow shell, deposited to form a chocolate chip, or applied as a stripe or swirl to a confection or baked good, must be further cooled to continue the crystallization process initiated during the tempering step. Properly tempered, properly cooled chocolate will transform into a brittle, glossy, stable, homogenous form as the cocoa butter crystallizes. Ideally, chocolate is cooled and crystallized within a sanitary, dry environment that provides well circulated 18°C air. Very cold temperatures are to be avoided because they can (*1*) crystallize unstable, lower melting polymorphs, and (*2*) cause a skin of crystallized fat to form on the surface of pieces being cooled; the resultant insulation effect will greatly hamper heat removal from the interior and thus delay crystallization of deeper lying liquid cocoa butter. In either case, the end result will be unsatisfactory product and the eventual formation of fat bloom, a white haze of recrystallized surface fat that severely detracts from eye appeal and represents a quality defect. Bloomed chocolate products are certainly safe to consume; they just are not very attractive and to many consumers recrystallized fat has the appearance of mold. On the other hand, cooling chocolate at higher temperatures also compromises product quality. In this case, the crystallization process proceeds very slowly and consequently very large, grainy crystals develop. The finished chocolate will have a crumbly texture and mouthfeel, dull surface, and will eventually bloom.

Typically, insulated tunnels are employed for cooling, this final and very important stage of chocolate production. Air temperature, air velocity, and air direction relative to product travel and tunnel transit time are the parameters typically controlled in commercial cooling operations. The laws of nature apply to the heat exchange process that occurs and include conduction, convection, and radiation.

Two types of cooling tunnels prevail in the industry—convection and radiant. The concept of convection tunnels involves primarily guiding cooling air down the tunnel. Product is placed onto a belt that traverses the length of the tunnel at a constant rate. Cool air is often introduced into the middle of the tunnel and flows toward either end. Thus, air moves countercurrent to product in the first half of the tunnel and cocurrent to product in the back half. This scheme has the advantage of gradually ramping temperatures down and then up again as product traverses the tunnel length. The air in the first stage or section of the cooler should range between 16 and 20°C to ensure the formation of considerable quantities of stable crystals and preclude the formation of unstable crystals. As product nears the middle of the tunnel, the temperature gradually falls to 10–16°C. This is typically the point in the crystallization process where the most heat is released. In the initial cooling period, primarily sensible heat is removed; but as product continues further into the tunnel latent heat of crystallization must be removed, and this represents a substantially greater content of energy than does sensible heat. Air velocity is important and higher velocities are capable of moving away greater quantities of heat at a given air temperature. The last third of the tunnel transit offers a gradual increase in temperature to typically 15–18°C. This warming is necessary to keep surface temperatures above the dew point of the air in the packing room, thereby preventing condensation from forming upon the surface of cooled product when exiting the tunnel. Surface moisture dissolves sugar and forms a syrup. Later, as the water evaporates, sugar recrystallizes to form a gray, rough surface similar in appearance to fat bloom. Radiant cooling relies primarily upon black, water-cooled radiation absorbers installed above the product as it travels the entire length of the tunnel. In addition, 6–12°C cooling water is generally circulated through tubes located beneath the belt to provide bottom cooling. The air within the tunnel remains relatively still. Although properly tempered and properly cooled chocolate products will exit from the cooling tunnel with a dry, glossy appearance and be firm and dry to the touch, the crystallization process is not complete. Some 15–20% liquid cocoa butter remains and will crystallize over the next 2 or 3 days. Product is therefore stored in a cool (16–21°C) dry environment to complete crystallization (41–43).

1.9 Standards of Identity

Any product labeled chocolate in the United States must comply with the criteria described in Food and Drug Administration (FDA) Title 21 CFR Part

163. The FDA adopted the original standards of identity for chocolate in 1944, and they have changed only relatively little over the years. On December 2, 1985, FDA solicited comments concerning the desirability of and need for amending the U.S. standards of identity for chocolate products to achieve consistency with the Codex Standard for Chocolate (Codex Standard 87-1981) developed by the Codex Alimentarius Commission. The commission, sponsored jointly by the Food and Agriculture Organization (FAO) and the World Health Organization (WHO), conducts a program to develop worldwide food standards. Because the United States is a member of the Codex Alimentarius Commission, it is obliged to consider all Codex standards for acceptance. The Codex Standards for Chocolate (Codex Stan 87-1981) define 14 products (*Worldwide Standard*):

2.1.1 Chocolate with the addition of sugars (3.1.1)

2.1.2 Unsweetened Chocolate without the addition of sugars (3.1.20)

2.1.3 Couverture Chocolate with the addition of sugars (3.1.3), which is suitable for covering purposes

2.1.4 Sweet (Plain) Chocolate with the addition of sugars (3.1.4)

2.1.5 Milk Chocolate with the addition of sugars and milk solids (3.1.5)

2.1.6 Milk Couverture Chocolate with the addition of sugars and milk solids (3.1.6), which is suitable for covering purposes

2.1.7 Milk Chocolate with High Milk Content with the addition of sugars and milk solids (3.1.7)

2.1.8 Skimmed Milk Chocolate with the addition of sugars and skimmed milk solids (3.1.8)

2.1.9 Skimmed Milk Couverture Chocolate with the addition of sugars and skimmed milk solids (3.1.9), which is suitable for covering purposes

2.1.10 Cream Chocolate with the addition of sugars and cream and milk solids (3.1.10)

2.1.11 Chocolate Vermicelli with the addition of sugars (3.1.11), which is in the form of grains

2.1.12 Chocolate Flakes with the addition of sugars (3.1.12), which is in the form of flakes

2.1.13 Milk Chocolate Vermicelli with the addition of sugars and milk solids (3.1.13), which is in the form of grains

2.1.14 Milk Chocolate Flakes with the addition of sugars and milk solids (3.1.14), which is in the form of flakes

Table 9.10 (excerpted from the Codex standards) lists the ingredients required for each product.

U.S. standards in 21 CFR Part 163 also define 14 cacoa products, six of which have similar, but not identical, counterparts in the Codex standard. The six U.S. standards are:

Table 9.10 Composition (% calculated on the dry matter in the product) of the chocolate products defined in Codex Standard 87-1981

	Constituents Product	Cocoa Butter	Fat-Free Cocoa Solids	Total Cocoa Solids	Milk Fat	Fat-Free Milk Solids[a]	Total Fat	Sugars
3.1.1	Chocolate	>18	>14	>35	—	—	—	—
3.1.2	Unsweetened chocolate	>50–<58	—	—	—	—	—	—
3.1.3	Couverture chocolate	>31	>2.5	>35	—	—	—	—
3.1.4	Sweet (plain) chocolate	>18	>12	>30	—	—	—	—
3.1.5	Milk chocolate	—	>2.5	>25	>3.5	>10.5	>25	<55
3.1.6	Milk couverture chocolate	—	>2.5	>25	>3.5	>10.5	>31	<55
3.1.7	Milk chocolate with high milk content	—	>2.5	>20	>5	>15	>25	<55
3.1.8	Skimmed milk chocolate	—	>2.5	>25	<0.5	>14	>25	<55
3.1.9	Skimmed milk couverture chocolate	—	>2.5	>25	<0.5	>14	>31	<55
3.1.10	Cream chocolate	—	>2.5	>25	>7	>3–<14	>25	<55
3.1.11	Chocolate vermicelli	>12	>14	>32	—	—	—	—
3.1.12	Chocolate flakes							
3.1.13	Milk chocolate vermicelli	—	>2.5	>20	>3.5	>10.5	>12	<66
3.1.14	Milk chocolate flakes							

Source: from Codex Standards for Chocolate (World Standard) formerly CAC/RS 87-1976.
[a] In their natural proportions.

§163.111 Chocolate liquor (also chocolate, baking chocolate, bitter chocolate, cooking chocolate, chocolate coating, bitter chocolate coating) is the solid or semiplastic food prepared by finely grinding cacao nibs. To such ground cacao nibs, cacao fat or a cocoa or both may be added in quantities needed to adjust the cacoa fat content of the finished chocolate liquor.

§163.123 Sweet chocolate (also sweet chocolate coating) is the solid or semiplastic food the ingredients of which are intimately mixed and ground, prepared from chocolate liquor (with or without the addition of cacao fat) sweetened with one of the optional saccharide ingredients. . . . One of the optional emulsifying ingredients or combinations of ingredients. . . . One or more of the optional dairy ingredients . . . may be used in such quantity that the finished sweet chocolate contains less than 12 percent by weight of milk constituent solids. . . . The finished sweet chocolate contains not less than 15 percent by weight of chocolate liquor. . . . Bittersweet chocolate (also semisweet chocolate, semisweet chocolate coating, and bittersweet chocolate coating) is sweet chocolate which contains not less than 35 percent by weight of chocolate liquor.

§163.130 Milk chocolate (also sweet milk chocolate, milk chocolate coating, sweet milk chocolate coating) is the solid or semiplastic food the ingredients of which are intimately mixed and ground, prepared from chocolate liquor (with or without the addition of cacao fat) and one or more of the optional dairy ingredients . . . , sweetened with one of the optional saccharide ingredients. . . . One of the optional emulsifying ingredients or combinations of ingredients. . . . The finished milk chocolate contains not less than 3.66 percent by weight milk fat, not less than 12 percent by weight of milk solids, and not less than 10 percent by weight of chocolate liquor.

§163.135 Buttermilk chocolate (also buttermilk chocolate coating) conforms to the standards of identity . . . prescribed for milk chocolate by §163.130, except that:

(a) The dairy ingredients used are limited to sweet cream buttermilk, dried sweet cream buttermilk, or any combination of two or all of these.

(b) The finished buttermilk chocolate contains less than 3.66 percent by weight of milk fat and, instead of milk solids, it contains not less than 12 percent by weight of sweet cream buttermilk solids.

§163.140 Skim milk chocolate (also sweet, skim milk chocolate, skim milk chocolate coating) conforms to the definition and standard of identity . . . prescribed for milk chocolate by §163.130, except that:

(a) The dairy ingredients used are limited to skim milk, concentrated skim milk, evaporated skim milk, sweetened condensed skim milk, nonfat dry milk, and any combination of two or more of these.

(b) The finished skim milk chocolate contains less than 3.66 percent by weight of milk fat and, instead of milk solids, it contains not less than 12 percent by weight of skim milk solids.

§163.145 Mixed dairy product chocolate conforms to the definition and standard of identity . . . prescribed for milk chocolate by §163.130, except that:

(1) The dairy ingredient used in each such article is a mixture of two or more of the following four components:
(i) Any dairy ingredient or combination of such ingredients specified in §163.130 . . .
(ii) One or more of the five skim milk ingredients specified in §163.140.
(iii) One or more of the three sweet cream buttermilk ingredients specified in §163.135.
(iv) Malted milk.

(2) Each of the finished articles may contain less than 3.66 percent by weight of milk fat and, instead of milk solids, it contains not less than 12 percent by weight of milk constituent solids of the components used.

COCOA BUTTER ALTERNATIVES

The fat phase of chocolate, cocoa butter, is largely responsible for its desirable properties, which include a brittle nongreasy texture at and below room temperature, excellent storage qualities, and characteristically rapid melting near body temperature. However, chocolate exhibits other properties that challenge certain applications:

1. Chocolate is sensitive to temperature fluctuations; therefore the environment of distribution systems must be carefully controlled to maintain finished product appearance and integrity.
2. Chocolate is characterized by a strong contraction and very brittle texture upon setting. Such properties are less than optimal if a soft, spongy matrix is to be coated; cracking and flaking off are a certainty.
3. The presence of other fats in a coated center elicits shelf life concerns. Compatibility of center fats with cocoa butter in the chocolate coating must not be overlooked.
4. Chocolate requires careful tempering to ensure good set, gloss, and gloss retention.
5. Cocoa butter and chocolate liquor command premium prices; cocoa butter alternatives typically provide the opportunity to reduce total formulation cost not only as a function of their discount vis-à-vis cocoa butter but also basis production streamlining (e.g., the conche step can often be eliminated or significantly shortened and many cocoa butter alternatives do not require tempering—such options represent the opportunity to save production time and energy).

Chocolate analogues, formulations within which either some or all of the cocoa butter is replaced with an alternative fat, have been developed to address these issues and hence, in certain applications, offer technical advantages over chocolate. Analogues produced today are far superior to the cheap replacements for chocolate, known in the trade as grease coatings, that were common in the 1930s. This improvement can be attributed primarily to technological advances in the field of fats and oils processing. The concept and utility of "tailor-made" fats gained acceptance in the 1950s as evidenced by the number of patents describing the manufacture of cocoa butter alternative fats with specific physical and chemical properties. The dramatic increase in cocoa butter prices in 1953–1954 likely fueled this enthusiasm as demand for confectionery coatings reached an all-time high for that period. The confectionery industry became acquainted with and took advantage of the uniformity and versatility represented by cocoa butter alternative fats. These factors, coupled with economic advantages, firmly established a market for cocoa butter alternatives (44). Cocoa butter alternatives, often referred to generically as hard butters, exhibit varying degrees of similarity to cocoa butter in terms of percent solid fat at room temperature, rapid complete melting at or near body temperature, and high stability. Major market segments for such fat systems include confectionery, biscuit and cracker, and industrial chocolate companies (convertors) that manufacture coatings and drops for user industries.

Three chocolate coating analogues, known generically in the industry as compound coatings, are described in 21 CFR Part 163:

§163.150 Sweet cocoa and vegetable fat (other than cacao fat) coating is subject to the requirements ... prescribed for sweet chocolate by §163.123, except that:

(a) In its preparation cocoa is used, instead of chocolate liquor, in such quantity that the finished food contains not less than 6.8 percent by weight of the nonfat cacao portion of such cocoa ...

(b) In its preparation is added one or any combination of two or more vegetable food oils, vegetable food fats, or vegetable food stearins, other than cacao fat, which oil, fat, stearin, or combination has a melting point higher than that of cacao fat. Any such oil or fat may be hydrogenated.

(c) The requirement of §163.123(a) that the milk constituent solids be less than 12 percent by weight does not apply.

§163.153 Sweet chocolate and vegetable fat (other than cacao fat) coating (a) conforms to the definition and standard of identity ... prescribed for sweet chocolate by §163.123, except that:

(1) In its preparation there is added one or any combination of two or more vegetable food oils or vegetable food fats, other than cacao fat, which oil, fat, or combination may be hydrogenated and which has a melting point lower than that of cacao fat.

§163.155 Milk chocolate and vegetable fat (other than cacao fat) coating (also sweet milk chocolate and vegetable fat [other than cacao fat] coating) conforms to the definition and standard of identity ... prescribed for milk chocolate by §163.130, except that:

(1) In its preparation there is added one or any combination of two or more vegetable food oils or vegetable food fats other than cacao fat, which oil, fat, or combination may be hydrogenated and which has melting point lower than that of cacao fat.

Cocoa butter alternatives are used to either replace or extend the cocoa butter component present in traditional chocolate. These fats find utility far beyond the coatings described in 21CFR. They are also widely used in molded products (solid as well as shell), centers, and for deposits like drops and inclusions. Hard butters are also used in the formulation of nonchocolate coatings, more commonly referred to as pastels, which are available in various colors and flavors. The very desirable properties of cocoa butter provide the quality standards against which alternative fats are evaluated. The fat phase, just as is the case for cocoa butter in chocolates, provides a continuous matrix that holds the other ingredients contributing to flavor, aroma, and color in an acceptable form prior to consumption. Relatively high levels of solid fat are required at room temperature, generally at or near 90%, to preclude a greasy or tacky feel on handling. Upon consumption, the fat must melt away rapidly and completely to maximize flavor release. The cocoa butter alternative selected will largely determine the flavor release and flavor stability, initial gloss and gloss retention, hardness and snap, compatibility with other fats present in the formulation, and rate of crystallization and contraction. The nonfat solids (primarily cocoa solids, sugar, and milk solids) will contribute significantly to the flavor, color, and palate perception of fineness. Together, these phases will establish the overall quality, rheological properties, price, and consequently acceptability of the finished goods.

One useful approach to the classification of cocoa butter alternatives is to consider the dominant properties of the source oils present. Three families emerge: cocoa butter equivalents and extenders (CBEs), nonlauric cocoa butter replacers (CBRs), and lauric cocoa butter substitutes (CBSs). Each family can be further resolved into a variety of subcategories of specialty fat classes. Cocoa butter equivalents and extenders are composed of the same types of triglycerides as cocoa butter and consequently must be tempered; CBRs and CBSs, on the other hand, are formulated with triglycerides quite different from those in cocoa butter, and they crystallize spontaneously (without tempering) into their stable β' polymorph upon cooling. Table 9.11 summarizes the distinguishing characteristics for each category.

With deference to the regulatory constraints concerning the use of the term "chocolate," the following terminology will be employed to discuss formulations/applications that contain cocoa butter alternatives:

Chocolate—contains only cocoa butter and dairy fat.
CBE chocolate—some or all of the cocoa butter is replaced with a CBE.
CBR chocolate—most or all of the cocoa butter is replaced with a CBR.
CBS chocolate—all of the cocoa butter is replaced with a CBS.

2.1 Cocoa Butter Equivalents

The terms cocoa butter equivalent and cocoa butter extender are often used interchangeably; however, technically they describe separate product catego-

Table 9.11 Distinguishing characteristics of cocoa butter and cocoa butter alternatives

Percent	Cocoa Butter	CBE	CBR	CBS
C8	—	—	—	3
C10	—	—	—	3
C12	—	—	—	54
C14	—	—	—	20
C16	25	30	12	9
C18	36	30	14	10
C18:1	34	35	67	—
C18:2	3	3	6	—
Tempering required	Yes	Yes	No	No
Stable crystal	β	β	β'	β'

ries. Equivalents are fats that behave like and are compatible with cocoa butter in any proportion. They do not alter the melting, processing, and rheological properties of cocoa butter, and they have physicochemical characteristics similar to cocoa butter. Extenders, on the other hand, can be mixed with cocoa butter to a limited extent without significantly altering its melting, processing, and rheological properties. They do not necessarily have physicochemical characteristics similar to cocoa butter. The degree of compatibility, a function of triglyceride profile (Table 9.12), determines the quality of the extender. Equivalents may also be used as extenders. Source oils for the production of CBEs are all of tropical origin and include:

Palm (*Elaeis guinneensis*) fat from Malaysia, Indonesia, West Africa, Papua New Guinea, and South America
Illipe (*Shorea stenoptera*) fat, sometimes referred to commercially as green butter, from the island of Borneo.
Shea (*Butyrospermum parkii*) fat from the savanna regions of West and Central Africa
Sal (*Shorea robusta gaertn* f.) fat from India
Kokum (*Garcinia indica choisy*) fat from India

The triglyceride composition and melting properties of illipe fat are similar enough to cocoa butter to permit its inclusion, without any special processing techniques, at significant levels. Kokum fat is rich in StOSt triglycerides, and therefore is also a valuable raw material, without fractionation, for the production of CBEs. Cocoa butter typically contains some 80% symmetrical 2-oleo disaturated triglycerides. The positioning of the fatty acids with respect to one another is paramount to the unique melting properties of cocoa butter. However, palm and shea fats, for example, although good sources of the

Table 9.12 Typical fatty acid compositions and triglyceride profiles for selected CBE feedstocks (30)

	Cocoa Butter	Palm	Illipe	Shea	Sal	Kokum
Palmitic (P)	25	45	16	4	5	2
Stearic (St)	36	5	46	43	44	57
Oleic (O)	34	38	35	45	40	40
Linoleic (L)	2	10	—	7	2	1
Arachidic (Ar)	1	—	2	—	7	—
PPP	—	5	—	—	—	—
POSt	39	3	35	5	11	5
StOSt	26	—	45	40	42	72
POP	16	26	7	—	1	—
StOAr	2	—	4	2	13	—
StLP	4	2	—	—	—	—
PLP	2	7	—	—	—	—
StLSt	1	2	—	—	—	—
PPO	—	5	—	—	—	—
StOO	4	3	3	27	16	15
POO	4	19	—	2	3	—
StOL	—	—	—	6	1	—
OOO	—	3	—	5	3	2

desirable SUS triglycerides, are significantly softer than cocoa butter. This is due to the high levels of di- and triunsaturated glycerides also present. In addition, palm fat contains significant levels of PPO, a triglyceride much less compatible in crystalline behavior with cocoa butter. Such undesirable triglycerides can be separated via fractionation processes including dry pressing, solvent, and detergent. Selective blending of whole fats and/or fractions yields hard fats suitable for CBEs.

Biotechnology provides another avenue for the production of CBEs. Genetic manipulation of oil-producing plants, harvesting lipids from single-cell organisms, and enhancing traditional process capabilities with the specificity offered by enzymes represent the biologically based technologies currently under development. High-oleic varieties of sunflower and safflower oils, for example, represent excellent sources of oleic feedstocks for the production of SUS triglycerides obtained by introduction of saturated fatty acids at the 1,3-position (45). Such transesterification reactions are facilitated by lipases. This group of enzymes catalyzes the hydrolysis of triglycerides; the reaction is reversible and thus lipases will also encourage the formation of acylglycerols from fatty acids and glycerol. Enzymic modification permits the production of pure and specific products under mild conditions thereby severely limiting the extent of side reactions. A regiospecific lipase, for example, 3A from fermentation of a selected strain of *Mucor miehei* fungus, will confine the exchange of fatty acid groups to the α positions on the triglyceride. The process can be made continuous by immobilization of the enzyme; this also provides the opportunity to reuse the lipase (46).

Macrae has described one such example that involves the directed interesterification of palm midfraction (rich in POP) with either stearic acid or tristearin to produce the following new triglycerides: POSt, StOSt, PStP, and PStSt. Fractionation typically follows to further tailor the triglyceride families to mirror the distribution characteristic of cocoa butter (47). Single-cell oils represent another potential source of oils and fats that might be suitable elements for CBE formulations. Yeasts and molds (both eukaryotic organisms) can accumulate substantial quantities of triglycerides, up to 70% of their dry cell weight; these lipids are generally similar in both fatty acid distribution and triglyceride profile to vegetable oils and fats (48). Bacteria, on the other hand, tend not to accumulate substantial quantities of lipids; or if they do, those products generally are other than triglycerides. Consequently, bacteria are of less commercial interest as a potential source of specialty lipids than are yeasts and fungi (49). The cost of the carbon source to sustain single-cell oil producers is a major impediment to widespread commercialization of the process, and therefore their niche is in value-added areas like CBEs (50). It is also possible to alter, *in vivo,* the composition of the lipids accumulated by single-cell organisms. Cells can be grown on a medium rich in stearic acid, for example, to encourage its uptake and incorporation directly into triglycerides as they are synthesized. This approach tends to circumvent the natural tendency to preferentially desaturate stearic acid over elongating palmitic acid (51).

The additional cocoa butter required for the formulation of a dark sweet chocolate could be replaced with a completely compatible CBE. No changes are necessary in the manufacturing process since the cocoa butter–CBE blend contains the same symmetrical triglycerides as cocoa butter and hence requires tempering. The term "supercoating" has often been used to describe a CBE chocolate so formulated. It is in fact even possible to completely replace the chocolate liquor with the appropriate proportions of CBE and cocoa solids (Table 9.13). From a formulation perspective, it is important not to overlook the fact that the final processing step applied during the production of CBEs (and any other cocoa butter alternative) is deodorization, and consequently a bland and odor-free CBE replaces aromatic cocoa butter. On the other hand, for recipes that contain deodorized cocoa butter, replacement of that butter with a CBE will not, of course, alter the flavor impact or intensity.

The symmetrical triglycerides POP, POSt, and StOSt are key to the performance of CBEs. Their β polymorphs melt at 38, 37, and 43°C, respectively; consequently the hardness of a given CBE is determined to a great extent by the quantity of each of these triglycerides present. The possibility to alter the relative proportions of POP, POSt, and StOSt in a CBE offers the opportunity to incorporate a harder cocoa butter alternative into a CBE chocolate that (*1*) contains a softer cocoa butter, (*2*) is severely softened by high amounts

Table 9.13 Typical recipes for chocolate, CBE chocolate, and supercoating[a]

	Chocolate (%)	CBE Chocolate (%)	Supercoatings (%)	
			Cocoa Butter Replaced	Cocoa Butter and Liquor Replaced
Chocolate liquor	42.0	42.0	42.0	—
Cocoa butter	8.4	2.0	—	—
Cocoa powder (10/12%)	—	—	—	20.8
CBE	—	6.4	8.4	29.6
Sugar	49.6	49.6	49.6	49.6
Fat content	31.9	31.9	31.9	31.9

[a] Lecithin added to each formula at 0.2–0.5% (salt and other flavors as required).

of milk fat (a function of the diluent effect of milk fat upon cocoa butter), or (3) that requires some tolerance to warmer temperatures. In general, SUS triglycerides are highest in Malaysian cocoa butters, somewhat less in African butters and lowest in South American cocoa butters. These natural variations can be largely dampened with the incorporation of an appropriate CBE into the recipe. Mixing, tempering, and crystallization of CBE chocolate is fundamentally as described for chocolate. Certainly, it is possible to include milk products in formulations if milk CBE chocolate is desired. Or, cocoa solids can be completely eliminated to produce a "white" CBE chocolate.

2.2 Cocoa Butter Replacers

Nonlauric cocoa butter replacers (CBRs) are derived from partially hydrogenated or partially hydrogenated and fractionated blends of primarily soybean, cottonseed, canola, and palm oils. These source oils are essentially composed of triglycerides containing 16 and 18 carbon atom fatty acids. In the United States CBRs are also often referred to as "domestic hard butters" because many of the currently available products are formulated with oils produced domestically. The triglycerides of the source oils typically utilized for the production of CBRs have relatively high levels of unsaturated fatty acids and thus are far too soft and unstable in their native forms. Soybean and cottonseed oils, for example, are liquid at room temperature and melt at around -17 and $-2°C$, respectively. Selective partial hydrogenation techniques including elevated temperatures, reduced hydrogen gas pressure, and partially inactivated nickel catalyst such as previously used catalyst or commercially available sulfur-promoted catalysts are required to manufacture these products.

The selective hydrogenation process favors the production of trans oleic acid and minimizes the formation of stearic acid. Isomerization of naturally occurring cis oleic acid to the trans form dramatically improves solid fat content at room temperature while minimizing the level of solid fat remaining at body temperature. The reaction can be terminated at melting ranges slightly higher than body temperature with a resultant acceptable level of solid fat at room temperature. Oxidative stability is improved as a result of reduction in the number of carbon-to-carbon double bonds and the superior resistance of trans isomers to oxidation. Partially hydrogenated CBRs fall short of the melting properties of cocoa butter because of the triglyceride families available for tailoring via isomerization and hydrogenation of their constituent fatty acids.

Compared to chocolate, CBR chocolates formulated with these alternative fats generally are characterized by poor eating quality, poor snap, and a low coefficient of contraction resulting in poor mold release. The limited compatibility of CBRs with cocoa butter generally precludes the inclusion of chocolate liquor in most formulations. Hence, low-fat cocoa powder is required and is the source of chocolaty color, aroma, and flavor. The inherently poor flavor release characteristics of these fats obviously works against the percep-

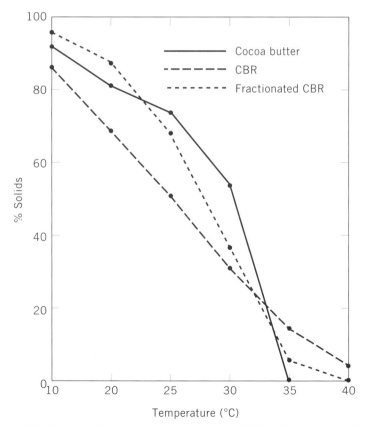

Figure 9.7 Typical melting profiles for cocoa butter, CBR, and fractionated CBR.

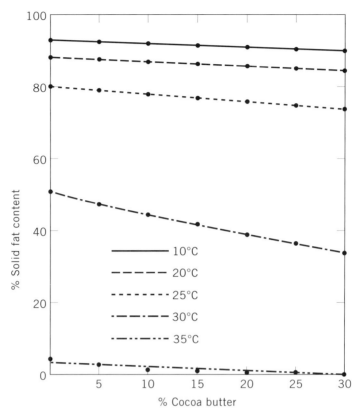

Figure 9.8 Influence upon solid fat content of incremental additions of cocoa butter into CBR-isosolids diagram.

tion of flavor notes in CBR chocolates. However, these hard butters are associated with fair to good gloss and good shelf life. Cocoa butter replacer chocolate handling is relatively simple in that it is merely brought to a fully molten state and crystallized without tempering. This is possible due to minimal tendencies toward polymorphism in crystal structure. These fats crystallize in the β' form, which practically speaking, is the desired stable polymorph (ultimately, CBR fats will recrystallize to β). Thus, CBR chocolate products tend to maintain acceptable gloss following heat stressing and subsequent recovery to ambient temperature. Market applications include biscuit and cracker coatings, economy baking chips, and high-bulk/low-cost candy bar coatings.

The physical and functional properties of CBRs can be improved by controlled crystallization and separation techniques. Some fractionation processes involve the use of food-grade solvents while others accomplish separation simply by cold, dry pressing. Figure 9.7 illustrates that the resultant selective concentration of desirable triglycerides significantly improves solid fat content at room temperature and narrows the melting range while effecting only a marginal change in the melting point.

Some of these products reportedly tolerate up to 25% cocoa butter in the total fat phase of a CBR chocolate. Admixture with cocoa butter up to this limit slightly softens CBRs and thus has the important effect of improving eating quality by reducing the amount of solid fat present at mouth temperature (Figure 9.8). Maximum improvement of mouth melt occurs at about a 10% addition of cocoa butter (fat basis).

The limited compatibility with cocoa butter is a function of the similarity in chain length shared between fatty acids associated with cocoa butter and CBR triglycerides (Table 9.14).

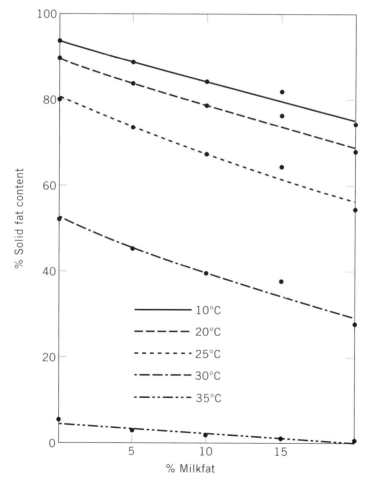

Figure 9.9 Influence upon solid fat content of incremental additions of milkfat—isosolids diagram.

Table 9.14 *Typical composition of soy-based CBR*[a]

Fatty Acid Composition			Triglyceride Profile			
C16 (P)	12%	(25%)	OElEl	10%	PElEl	20%
C18 (St)	13%	(36%)	POEl	6%	PPEl	2%
C18:1 (O)	69%	(34%)	ElElEl	25%	StOEl	10%
C18:2 (L)	5%	(2%)	PPO	2%	StElEl	10%

[a] Cocoa butter fatty acid composition is given in parenthesis for reference.

Cocoa butter admixture prolongs the setting time (see Table 9.16) and reduces the gloss and gloss shelf life of CBR chocolate. Milk fat is often a component in CBR chocolate and it interacts with the CBR fat as well. Like cocoa butter, it softens the CBR, but in a predictable linear manner and thus also can be used to advantage for improvement of mouthmelt. Again, like cocoa butter, milk fat delays setting time but it conveys the benefit of improved gloss stability. The severe softening resulting from the addition of milk fat limits its inclusion to between 10 and 15% of the fat phase (Figure 9.9).

Cocoa butter replacer chocolates (Table 9.15) generally are described as having good mouthfeel and flavor release properties, good gloss and shelf stability, and are nontempering. Applications include enrobing better quality bakery products and high bulk candy bars as well as in the formulation of cookie drops. Cocoa butter replacers are especially attractive center fats in complex products that may contain moisture in the center (to preclude the risk of hydrolytic rancidity) or are coated with chocolate or CBE or CBR chocolate. Because they are less brittle and do not contract as severely as chocolate, CBRs are especially well suited to coatings for covering jellies, marshmallows, cakes, and other pliable or spongy substrates.

Cocoa butter replacer chocolate represents several advantages over chocolate and CBE chocolate:

1. Conching is generally eliminated or at least the time substantially minimized to a few hours.

Table 9.15 *Typical CBR chocolate formulas*[a]

Ingredient	Milk (%)		White (%)		Dark (%)	
	9.0	—			8.0	—
Chocolate liquor	—		5.5		—	
Cocoa butter	—		—		16.0	20.0
Cocoa powder (10–12%)	12.0		—		—	
Whole milk powder (24% fat)	13.0		30.5		—	
Skim milk powder (<1% fat)	26.0		28.0		30.0	34.0
Sugar	40.0		36.0		46.0	46.0

[a] Lecithin added to each formula at 0.2–0.5% (salt and other flavors as required).

2. Tempering is generally not required; the stable β' polymorph forms spontaneously.
3. Provides a flexible, more elastic coating for soft or spongy substrates and hence precludes cracking and flaking off.
4. Tolerance to higher temperatures.
5. Lower cost.

2.3 Cocoa Butter Substitutes

Lauric cocoa butter substitutes (CBSs) represent a wide range of alternative fat systems. These products are predominantly lauric fats obtained from the oil palm and coconut palm. World supply is mainly of palm kernel and coconut origin. These differ from the nonlauric fats and oils in that their fatty acid compositions are 40–50% lauric acid. Hydrogenation to near saturation results in fats that melt relatively rapidly and cleanly upon heating as opposed to the gradual softening behavior of partially hydrogenated nonlauric fats. Furthermore, the inherently low levels of unsaturates, such as oleic and linoleic fatty acids, imparts a high degree of oxidative stability to these fats. Such characteristics make the lauric fats particularly attractive to the confectionery industry. Processing techniques such as hydrogenation, interesterification, and fractionation are applied to provide a wide range of melting points and melting curves suitable for confectioners' applications. Interesterification is often combined with hydrogenation to manufacture "rearranged" or "modified" lauric hard butters. The interesterification process permits the random rearrangement of the fatty acids on the triglyceride molecules with the aid of a catalyst such as sodium methylate. The interesterification process does not change the degree of unsaturation or isomerization. This technique permits an increase in solid fat content at room temperature to required levels at melting points consistent with good melting properties (Figure 9.10). By employing a variety of oil blends, a wide range of solid fat contents and melting ranges are possible. Generally, these products are made from palm kernel and/or coconut fats in combination with lesser quantities of nonlauric fats such as palm, cottonseed, or soybean.

Formulation flexibility provides the ability to produce compound coatings suitable for almost any climate or season. It is not uncommon for a bakery item to be enrobed with a compound coating containing an interesterified CBS with a melting point of 37 or 38°C during winter months and 45°C or higher during summer months. Cocoa butter substitute chocolate formulated with an interesterified CBS typically exhibits good mouthfeel and flavor release, good gloss, good mold and belt release, and is hard and dry to the touch at ambient temperature. It also displays excellent oxidative stability. Coatings may be run either tempered or nontempered; however, tempering (or some form of conditioning) is generally advisable to maximize gloss and gloss retention. They crystallize quickly and therefore often represent a manufacturing advantage over CBR-based products (Table 9.16).

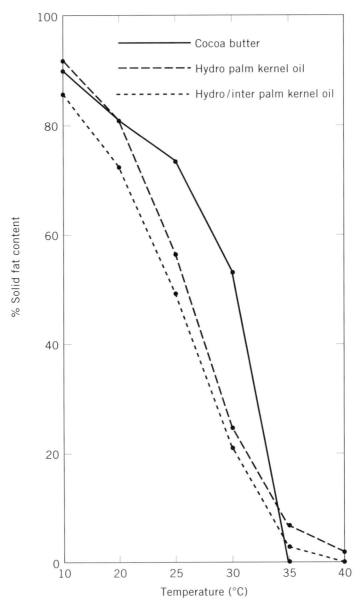

Figure 9.10 NMR curves for cocoa butter, hydro palm kernel oil, and hydro/interesterified palm kernel oil.

Table 9.16 *Relative crystallization times for CBR and CBS chocolates as influenced by crystallization temperature and by cocoa butter content (of CBR)*[a]

	Crystallization Temperature		
	5°C	10°C	12.5°C
CBS	1.0	1.2	1.5
CBR (6% cocoa butter in recipe)	1.2	1.3	1.7
CBR (17% cocoa butter in recipe)	1.5	1.6	2.2
CBR (25% cocoa butter in recipe)	2.7	2.9	3.9

[a] CBS crystallized at 5°C is assigned a value of unity.

Applications include coatings for bakery products and confectionery products in which the coating is important but is not the dominant factor in the eating quality of the product. The predominantly short-chain fatty acid composition of lauric fat triglycerides is largely responsible for their very limited tolerance to cocoa butter. Admixture of these fats with cocoa butter in excess of only a few percent results in a severe eutectic (Figure 9.11) with subsequent softening and dulling of a coating so formulated. Therefore, chocolate liquor is replaced with low-fat cocoa powder in most lauric based coatings (Table 9.17).

The highest quality CBS alternatives are those derived from fractionated palm kernel oil. Palm kernel stearins closely parallel the melting profile of cocoa butter. They offer the firmness, snap, and steep melting profile typical of cocoa butter but at a lower price. The manufacturing process involves crystallizing whole palm kernel fat under very specific conditions. The softer

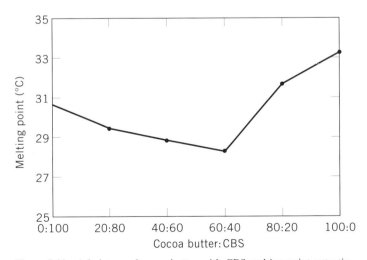

Figure 9.11 Admixture of cocoa butter with CBS-melting point eutectic.

Table 9.17 Typical CBS chocolate formulas[a]

	Dark CBS (%)	Milk CBS (or Pastel) (%)
Cocoa powder (10–12%)	16.0	7.0
CBS	35.0	34.0
Sugar	49.0	40.0
Skim milk powder	—	19.0

[a] Lecithin added to each formula at 0.2–0.5% (salt and other flavors as required).

fractions, referred to as olein, are removed via centrifugation, solvent extraction, or cold dry pressing. Some hard fractions, or stearins, nearly duplicate or even exceed the steep melting profile typical of cocoa butter (Figure 9.12).

Inspection of fatty acid compositions (Table 9.18) shows that the stearin is primarily lauric, myristic, and palmitic-based triglycerides while the short-chain and unsaturated fatty acid triglycerides are partitioned into the olein.

Subsequent partial hydrogenation of stearin fractions can be employed to further improve the melting properties. These fats are selected when quality is a dominant factor in determining overall product performance and appeal. Fractionated CBSs are especially well suited to applications involving extreme levels of diluent fats and/or oils, e.g. peanut oil. Products so formulated eat extremely well and tend to remain bloom free for long periods of time, perhaps as a function of entrapment of nut oil within the crystal matrix. Cocoa butter substitute chocolates formulated with these fats exhibit excellent eating properties, texture, mold and belt release, oxidative stability, and gloss. Fractionated CBSs are typically the hard butter of choice in the formulation of pastel coatings (Table 9.17) for molding and enrobing as well as in the production of butterscotch, peanut butter, and chocolate-flavored baking chips.

In general, many of the same advantages described for CBR chocolate also apply to CBS chocolate. These fats, however, do yield products that have more snap and brittleness than CBR chocolates, and they crystallize more rapidly. Because of their high content of shorter chain length fatty acids, the triglycerides of CBS fats represent the potential for the development of severe off-flavors. Oils and fats, by nature, are designed to liberate energy upon decomposition (cleaving fatty acids). However, it is very undesirable for this reaction to proceed in CBS chocolate. The organoleptic effect is often called hydrolytic rancidity, lipolytic rancidity, or simply soapy rancidity. In the case of shorter chain length fatty acids, many are associated with unpalatable flavors and odors at extremely low levels when freed. One of the predominant notes produced is that of soap, obviously a flavor that does not belong in any food product. This reaction is promoted by lipase, alkali, and acids, and water must be available. Milk powder, cocoa powder, egg albumen, and some spices

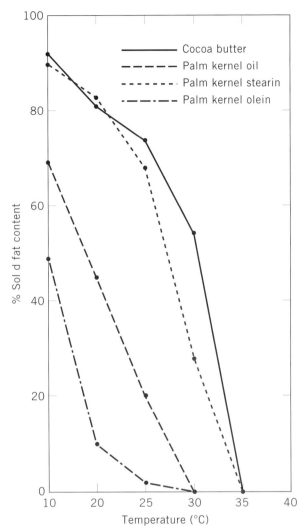

Figure 9.12 NMR curves for cocoa butter, palm kernel oil, palm kernel stearin, and palm kernel olein.

represent potential sources of lipase activity. Alkali can possibly be introduced if dutched cocoa powder is used in the recipe. It is not practicable to consider lack of available moisture as an option for complete control over the hydrolytic potential. As little as 0.1% water is all that is required to convert lauric triglycerides to diglycerides and 1.1% lauric acid. To put the impact of the free lauric acid formed in perspective, flavor thresholds of 0.02 and 0.07% have been reported for capric and lauric acids, respectively (52). The best prevention includes hygienic processing conditions, lipase-free raw materials and restricting as much as possible free moisture.

Table 9.18 Typical fatty acid distribution for palm kernel (PK) and PK fractions

Distribution	Whole PK	PK Stearin	PK Olein
C8	3.5	2.5	4.5
C10	3.5	3.0	4.0
C12	48.0	55.0	42.0
C14	16.0	20.0	12.5
C16	8.0	8.0	8.5
C18	2.5	3.5	2.5
C18:1	15.5	7.0	22.0
C18:2	2.5	0.8	3.5
Iodine value	13–18	6–9	25–31

EMULSIFIERS

An emulsifier is almost always a key component in chocolate and CBE, CBR, and CBS chocolate recipes. Surface-active agents provide the following benefits (53):

1. Reduce viscosity thereby sparing fat and thus reduce formulation costs.
2. Reduce thickening due to moisture and thus improve stability of viscosity.
3. Reduce thickening due to high temperatures (caused by formation of sugar syrup) thus conveying greater process temperature latitude.
4. Modification of crystallization behavior thereby providing greater latitude in tempering, improved gloss, and snap and prolonged gloss.

Emulsifiers can be broadly divided into two categories—those that affect rheological properties and those affecting crystallization. Lecithin, synthetic phospholipids (YN, e.g.), and polyglycerol polyricinoleate primarily influence rheological properties while sorbitan esters, monodiglycerides, and ethoxylated sorbitan esters are incorporated to steer crystallization. Heemskerk (54) summarized (Table 9.19) the impact on viscosity of emulsifiers that might be added to influence crystallization.

There is disagreement regarding the rheological impact of some of these emulsifiers. DuRoss (55) reported that polysorbate 60 at 0.1, 0.2, and 0.3% increased the yield value of a 30% fat chocolate and that 0.8% sorbitan monostearate decreased both the plastic viscosity and yield value of a 30.8% fat dark sweet chocolate. Even though a particular emulsifier might be selected solely based on its ability to maneuver crystallization tendencies in the desired direction, because it is a surface-active agent it can also be expected to alter rheological properties. Depending on the formula and process, this may not be a problem at all; or, under other circumstances, it might become a major

Table 9.19 Rheological impact of emulsifiers in chocolate and coatings (54)

Emulsifier	Plastic Viscosity	Yield Value
Monodiglycerides	Increase	Increase
Acetylated glycerides	Increase	Decrease
Lactic acid esters	Increase	Increase
Citric acid esters	Increase	Increase
DATEMs	Slight increase	Slight increase
Sucrose esters	Increase	Increase
Propylene glycol esters	Increase	Decrease
Sorbitan esters	Increase	Increase
Na stearylactylate	No change	Increase
Ca stearylactylate	Increase	No change
Polysorbates	Increase	No change

issue. DuRoss (55) demonstrated that not only the emulsifier(s) selected but also the order of addition of emulsifiers can influence rheology. A light chocolate paste to which 6.0% cocoa butter (CB) and 0.4% lecithin was added served as the control. The effect of order of addition of lecithin, sorbitan monostearate (SMS), and polysorbate 60 (POES) upon rheology was evaluated. All samples, including the control, were mixed a total of 10 h at 60°C and thereafter viscosity (PV = plastic viscosity [dynes/cm^2]; YV = yield value [Poise]) determinations made. The results are summarized in Table 9.20.

Sorbitan esters, particularly sorbitan tristearate, are especially effective additives to improve gloss and gloss retention in CBR and CBS chocolate. Sorbitan tristearate also has the benefit of slightly elevating the solid fat content up to about 35°C and decreasing it at higher temperatures thus giving the opportunity to improve mouthmelt and also to provide some tropicalization effect (Figure 9.13).

Garti and co-workers (56) showed that small incremental additions over a range of a few tenths to 5% of sorbitan tristearate (1) lowered the melting point of cocoa butter in form V and (2) increased the liquid fraction at 29°C from 27% to about 34%. Because transformation from the less stable form IV to the preferred stable form V is liquid mediated, sorbitan tristearate facilitates that transition. Garti's studies also showed that the solid-state form V to form VI transition is delayed by sorbitan tristearate, thus preserving gloss.

Pastel coatings based on CBS fats occasionally exhibit a defect Timme (57) calls mottling—a disorganized crystallization visible as tiny, starlike holes in the surface. Sometimes this defect appears as a dull haze, other times surface irregularities are seen. Timme claims that the addition of 1–3% of a distilled glycerol monostearate will prevent this phenomenon.

Musser (58) has previously found that emulsifiers function effectively in certain types of fats in confectionery coatings:

Table 9.20 Influence of order of addition of emulsifiers upon chocolate rheology (55)

Order	PV	YV
1. Control	5049	514
2. CB + 0.5% SMS 　0.3% POES 　0.4% Lecithin 　Mix 1 h between additions	1930	168
3. CB + 0.4% lecithin 　0.5% SMS 　0.3% POES 　Mix 1 h between additions	2006	290
4. CB + 0.4% lecithin + 0.5% SMS + 0.3% POES	3275	1234

1. As an aid in improving the impact of chocolate flavor.
2. As an aid to the dispersion and maintaining the shelf life of color in pastel coatings.
3. To modify the rate of crystallization during tempering, improving crystal size and gloss.
4. To improve the maintenance of gloss.

He also evaluated the impact of various emulsifiers upon the molding temperature, crystallization rate, and viscosity of both CBR and CBS chocolate. The following emulsifiers were investigated at 0.5, 1.0, and 1.5% (formula basis) levels of addition:

Distilled (fully hydrogenated) soybean monoglyceride
Distilled (fully hydrogenated) cottonseed monoglyceride
Distilled palm monoglyceride
Distilled diacetyl tartaric acid esters of monoglycerides (DATEM)

Compared to the control (0.5% addition of a 60/40 blend of sorbitan monostearate/polyoxyethylene sorbitan monostearate):

1. Both CBR and CBS chocolates generally required progressively higher molding temperatures at increasing levels of emulsifier; the higher temperatures reportedly produced superior gloss and shrinkage on all samples.
2. All of the emulsifiers modified the rate of crystallization significantly. Musser says the fine-grained crystalline structure required for maximum gloss and hardness is promoted by (*1*) an increase in the rate of crystallization at nucleation and (*2*) a nucleation temperature consistent with the normal (without added crystal modifiers) nucleation temperature of the fat. DA-

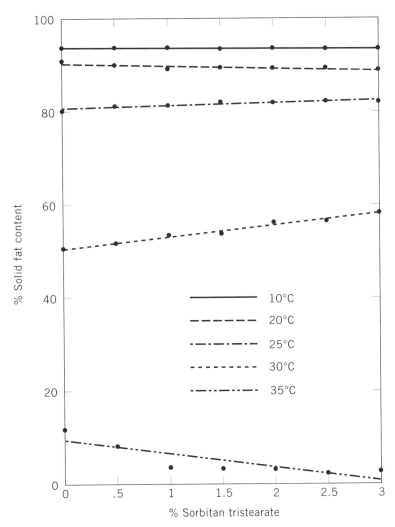

Figure 9.13 Influence upon solid fat content of incremental additions of sorbitan tristearate (STS)-isosolids diagram.

TEM worked especially well as a seeding agent at all levels evaluated in both recipes. The CBS chocolate was also favorably affected by the other emulsifiers with the exception of the soybean at 1.5% and the cottonseed at both 1.0 and 1.5%; the CBR chocolate, on the other hand, benefitted only from cottonseed at 0.5% and the palm at all levels.

3. DATEM had an especially interesting influence upon viscosity. This work showed that DATEM has the ability to reduce viscosity beyond that achievable with lecithin, giving the possibilities to (*1*) correct for lecithin overdoses and (*2*) reduce viscosity in high-moisture situations. When added in the

presence of lecithin, DATEM primarily affects yield value; a blend of two parts lecithin to one part DATEM is recommended to achieve the maximum viscosity reduction.

Clearly, it is possible to influence the physical properties of confectionery coatings via the addition of emulsifiers, but it is difficult to predict in general terms what the outcome might be. Thus, it is important to evaluate each new situation on its own merit. The sorbitan monostearate–span 60 blend mentioned in the Musser work has been promoted as a bloom inhibitor for many years (typical level of use is 1% on a formula basis). Cross and co-workers (59) published work in 1952 demonstrating the efficacy of this blend for bloom prevention in chocolate, and in 1965 DuRoss and Knightly (60) presented their work showing that the inclusion of sorbitan monostearate and polysorbate 60, in the proper ratio and at the proper level, improves the resistance of chocolate to bloom when chocolate is properly tempered. They cycled product [12 h at 15.6°C (60°F), 12 h at 29.4°C (85°F)] until samples bloomed; control chocolate required 8 cycles while chocolate containing the emulsifier package did not bloom until 26–28 cycles. However, results have proven to be inconsistent in commercial situations and there seems to be much disagreement regarding the efficacy of this combination for preservation of gloss in chocolate as well as compound coatings.

Additional functions of emulsifiers in confectionery products include the following: polysorbate 60, polyglycerol esters, and sucrose esters can improve the palatability of coatings formulated with higher melting point fats. It is likely that the reduction in perceived waxiness occurs as a result of emulsification of the fat with saliva during mastication. Mono- and diglycerides are added to caramels, nougats, and fudges at 0.4–1.0% to improve eating quality, reduce stickiness, improve shelf life, control spread of caramels, and add firmness; they also improve the chewing quality and shelf life of chewing gums. Mono- and diglycerides are also effective stabilizers for peanut butter and are beneficial in the production of starch-molded jellies (they keep the starch soft and compressible). Polyglycerol esters are good aerators for nougats, fudges, and whipped centers.

TROPICALIZATION OF CHOCOLATE

Moderately warm temperatures can compromise the integrity of many chocolate products. Partial melting of the crystal matrix quickly results in loss of structure, form, and texture. Food technologists have worked for many years to improve the heat stability of chocolate; and the inherent paradox is obvious. The objective is to physically alter the system such that it maintains its form, structure, and snap at elevated temperatures, i.e., it does not melt, while simultaneously maintaining chocolate's characteristic rapid and pleasant mouthmelt.

One approach to this problem is to add small portions of harder fat to the recipe. In the case of chocolate, perhaps harder cocoa butter; for CBE chocolate, higher melting SOS fractions. Sato and co-workers (62) demonstrated that the heat resistance of cocoa butter systems can be somewhat improved by the addition of a few percent seed crystals. This material, BOB powder, is composed largely of symmetrical triglycerides (especially 1,3-dibenoyl-2-oleoylglycerol) plated onto sucrose crystals; the particles are mostly smaller than 20 μm (61,62). The heat stability of CBR and CBS chocolates can be enhanced via the addition of small amounts of fully hydrogenated vegetable fat. Unfortunately, this approach provides, practically speaking, opportunity for only marginal improvement because mouthmelt is quickly compromised by relatively small additions of higher melting components.

An alternate approach essentially ignores the fat crystal matrix but rather strives to establish another network capable of providing structure. This secondary network is established between and among the sugar and polysaccharides in cocoa particles. Both are hydrophilic and tend toward strong interaction with one another. This interaction is largely eliminated during the chocolate manufacturing process wherein the objective is to coat each particle with a continuous layer of fat. Thus, reintroduction of water, at quite low levels, into the system again establishes strong interaction between the nonfat solids and represents the possibility to develop a new network.

The challenge with this approach is to introduce the water at a point beyond which it no longer interferes with process requirements (because of the thickening) or in such a fashion that it is not immediately captured by the hydrophilic solids. A patent issued to General Foods in 1956 describes spraying an emulsion of lecithin and water onto the surface of chocolate mass (to provide 2–3% water) followed by mixing in under prescribed conditions (63). Cadbury has described the preparation of an emulsion of chocolate ingredients in water wherein the water is evaporated under reduced pressure to below 5% (64) to give a foamed chocolate that maintains its form up to 104°C. A 1984 Battelle Memorial Institute patent teaches incorporation of water in the form of a solidified milled emulsion of cocoa butter, water, and lecithin (65). In 1991 this assignee disclosed an alternate approach that introduced water via an aqueous foam of, for example, egg white and water (66). Mars, Inc. proposed adding moisture to chocolate via a stable water-in-oil emulsion, e.g., hydrated lecithin (67). Another structuring approach to improve the thermal stability of chocolate as well as imitation chocolate, the addition of a polyol such as glycerine or sorbitol at about a 1% level, is described by Finkel (68).

OTHER APPLICATIONS

Oils and fats find utility in the confectionery industry beyond the important role they play in chocolate (including CBE, CBR, and CBS chocolate) and pastel (both white and colored) bars, coatings, baking chips, and seasonal novelties. In general, the oils and/or fats are expected to contribute nothing to flavor and aroma but rather provide attributes such as structure, adhesion, suspension, release properties, lubricity, richness, and texture to convey the desired sensory experience when the confection is consumed. The percentage of oils and fats introduced, on a finished formulation basis, span a very wide range—less than 0.1% to beyond 60%. The most critical criteria to be considered are stability and melting properties of the oil/fat candidate.

The surfaces of belts, forming equipment, cutting devices, cooling tables, and wrapping materials that come into direct contact with confections often require a food-grade lubricant coating to ensure clean release. Such situations typically present large surface areas and at times elevated temperatures, consequently oxidative stability of the lubricant is a concern. Mineral oils have traditionally satisfied this role but have come under scrutiny in recent years because of some toxicity concerns. Spain, South Africa, Israel, and the United Kingdom are among countries that have either banned or announced plans to ban the use of mineral oils in foods and food processing. In the United Kingdom, the Ministry of Agriculture has stipulated the use of vegetable oils instead (69). Suitable vegetable-oil-based alternatives must be stable against oxidation and possess low melting points and good flow properties. Possibilities include coconut fat, medium-chain triglycerides, selectively hydrogenated or hydrogenated/fractionated nonlauric oils, and high-oleic/low-polyunsaturate genetically modified oils. These oils are also suitable vehicles for polishing agents, glazes, pigments, and flavorants. In addition to the organoleptic issue, excellent oxidative stability is required to preclude the potential for degradation products to interact with flavors, fade colors, or inactivate vitamins in fortified products.

Another interesting processing-aid-oriented application has been described by Minifie. Ingredients such as milk powder, cocoa powder, and spices represent potential sources of lipase. These can be suspended in a high stability oil and the slurry heated to 110°C to destroy lipase. This process also thoroughly disperses cocoa powder and provides a lump-free product for ease of incorporation into formulations (70). High stability oils are also required in the production of gums and pastilles. These products are prepared from a low boiled mixture of sugar, glucose syrup, and water. Their high moisture content necessitates the incorporation of a setting agent, a gum or gelatine, so that the product can be formed into individual pieces. The cooked mixture is deposited into starch molds and allowed to set. Thereafter, product is demolded, brushed to remove excess starch, and then steamed lightly to remove any remaining traces of starch and to give a bright sparkling appearance. Gums that are made with gelatine are then oiled to prevent the pieces from sticking together as a result of the gelatine being partially melted during the steam bath (71). A blend of medium-chain triglycerides and beeswax represents one option for a very stable oiling agent. Tabletted candies represent another application requiring a lubricant. These confections are produced by either (*1*) direct compression or (2) granulation followed by compression of the formula ingredients. In either case, a lubricant is required to prevent product from adhering to the punch face or to the die wall. Suitable lubricants, added up to about 2% of the formula, include magnesium stearate, stearic acid, powdered emulsifier, or powdered fully hydrogenated vegetable fat (typically soybean or cottonseed).

The properties of some cooked confections rely heavily upon their fat content. Cutting, forming, and machining during processing are facilitated by the lubricating properties of fats and oils. Lubrication is often also important to the chewing properties of confections. In addition, fats might provide body, retard crystallization, and promote moisture retention and a smooth texture. For the lubrication effect, a stable oil is often acceptable. On the other hand, for formulations where structure and body are largely dictated by the lipid component, hard fats exhibiting appropriate melting characteristics are required. Furthermore, confections that are cut or subjected to some form of shear stress require fats with sufficient body and melting point to preclude oiling off during machining. And because the ingredients of cooked confections are largely hydrophilic in nature, an emulsifier such as lecithin or mono- and diglycerides plus vigorous agitation are necessary elements to ensure thorough dispersion of the oil or fat added to the recipe. Generally, lauric fats (including 76° coconut, hydrogenated or hydrogenated/interesterified blends, and fractionated palm-kernel-based systems) are preferred because of their inherent stability, steep melting curves, and clean flavor release. Dairy butter is unquestionably an absolutely necessary ingredient of quality caramels, toffees, and butterscotch. These products' appeal is largely a function of the unique flavor notes developed as a result of the Maillard reaction. The partial decomposition and caramelization that occurs during boiling are a result of interaction among milk solids, water, and sugars. Furthermore, the native lecithin in butter promotes emulsification of the fat into the syrup. No confectionery fat or oil by itself is capable of delivering the unique flavor profile of butter.

Traditionally, butter and cocoa butter were the fat sources for most confections. The development and evolution of cocoa butter alternative fats has provided confectioners the opportunity to take advantage of the very desirable sharp melting properties inherent to cocoa butter, but at a much lower cost. In addition, the flexibility represented by cocoa butter alternatives provides the option to deliver an even greater variety of textural experiences as well as the ability to adapt recipes to climates and equipment. Interestingly, confectionery boilings have traditionally been conducted in copper kettles. It is, of course, a fundamental tenet of oil and fat chemistry that contact with copper must be avoided to preserve oxidative integrity. Yet, the confections are quite acceptable; it is likely that the sugar provides some protection for the lipids. However, it has been demonstrated that products cooked in stainless steel kettles have superior shelf lives compared to those from copper pans (72).

Caramel, toffee, and butterscotch are similar products. Their distinguishing characteristics primarily result from (1) milk and fat content, (2) moisture content, and (3) fat type. These chewy confections all generally contain between a few percent to 15 or 20% fat. Generally, higher boiled sweets (lower final moisture content) require lower levels of fat because they are inherently firmer and have better stand up. Furthermore, lower melting fats often are suitable since they are not required to contribute to structural integrity but rather primarily provide lubricity. Thus toffees, cooked to much lower moisture contents than caramels, typically contain a few percent fat. A harder toffee, butterscotch, for example, might contain 3 or 4% butter. Caramels, on the other hand, generally contain more than 10% of a hardened fat (typically lauric) that melts at 32–35°C. In addition, glycerol monostearate might also be added to improve the chewing quality (72). Fudge is a related product and has been described as a cross between fondant and caramel (73). Fondant is prepared by boiling a mixture of sugar, glucose syrup, and water to about 88% solids. The resultant supersaturated solution is cooled under agitation to yield a suspension of very fine sugar crystals in a saturated sugar solution (74). Fat, accounting for about 10% of the recipe, is important to provide the proper standup in fudge and generally partially hydrogenated laurics are selected. The melting points of these hard fats typically range between 35 and 45°C. Nougats, essentially a combination of marshmallow and caramel, usually contain 1 or 2% of a hard fat (35–45°C melting point) to improve texture and chewing quality (75).

Chocolate is not the only high-fat confection. Ice cream coatings, sometimes called frozen novelty coatings or pail coatings, typically contain some 60% fat. A very basic recipe is 60% coconut oil, 33.5% sugar, 6% cocoa powder, and 0.5% lecithin. The keys to performance of these products include their ability to quickly crystallize during the production process and to melt rapidly and completely upon consumption. Mouthmelt is especially challenging given the low palate temperature that results from consumption of a frozen product. It is common to blend 10–20% of a stable nonlauric oil with coconut fat to create a eutectic that depresses the melting point and improves flavor release. This also has the benefit of dampening some of the brittleness inherent in such a coating formulation, consequently reducing the tendency to fracture and break off during consumption. The challenge is to balance these organoleptic improvements against the delayed coating crystallization and drying times that result. Recently, nonlauric fats (often based on partially hydrogenated and fractionated components) have been successfully incorporated into such formulations. The meltaway, or American truffle, is another example of a high-fat confection. The severe eutectic that occurs when incompatible fats are blended together is the basis for this delightful confection. A blend of 75% chocolate with 25% lauric fat (coconut, hydrogenated coconut, or fractionated palm kernel) is the traditional approach. The melting point of the resultant blend is substantially lower than either of the components, and the mouthmelt is extremely rapid and cooling (see Figure 9.11). Many examples of the overlapping of bakery and confectionery technologies are present on the market today. Perhaps the most common marriage is represented by the tremendous number of enrobed baked products. Filled confections (nut pastes, creams, jellies) and filled baked goods (cream horns, snack cakes) have been around for years. This concept has provided another interesting hybrid—a cookie that surrounds a pocket of soft, creamy confectionery filling. Such bakeable fillings contain about 40% fat; the balance of a typical recipe is sugar, milk solids, lecithin, flavoring, and possibly some cocoa powder. To deliver the desired textural properties, the fat system must remain fluid over a broad temperature range and also be stable against oxidation. Selectively hydrogenated nonlauric oils are especially well suited to the required melting profile.

The reader interested in an in-depth discussion of confectionery technology and especially recipes is referred to References 1, 71, 76, and 77.

OIL MIGRATION

One of the greatest challenges to maintenance of product quality and appeal is oil migration. The classic bane of chocolate is fat bloom, the result of recrystallization. The potential for this quality defect is inherent within most fat-containing matrices although it has traditionally been discussed in relation to chocolate. Oils and fats contained within confections, bakery products, or any other foods for that matter, will migrate throughout the matrices within which they are entrapped to achieve an equilibrium between solid and liquid fractions (78). Redistribution of triglycerides is often most severe within compound products; in other words those that are composed of several elements each of which may contain a fat or oil as an ingredient. The physical appearance of a fat can be deceiving if one thinks of oil available for migration as only any salad oil in the recipe. A typical household all-purpose shortening, for example, is largely a liquid phase (perhaps 80%) at room temperature entrapped within a hard crystalline matrix. Similar statements can be made for harder fats, like cocoa butter, that although substantially firmer still contain considerable amounts of entrapped liquid triglycerides.

One of the most challenging combinations is a peanut-butter-based center overlayed with chocolate or compound coating—substantial oil migration is inevitable. Significant amounts of oil are released when peanuts are ground to make peanut butter, and gravitational separation of peanut solids from the freed oil will occur shortly thereafter (case in point, "natural" peanut butters are always biphasic—peanut solids at the bottom of the jar overlayed with an oil layer). Commercial peanut butters are usually stabilized with about 1–3% hard fat or emulsifier whose role it is to establish a crystalline matrix capable of entrapping and suspending the peanut solids. Peanut oil will rapidly move from an unstabilized ground peanut-based center toward the coating chocolate or shell in an effort to establish a solid–liquid equilibrium throughout the entire confection. Although the coating also contains liquid triglycerides, the movement will largely be from the center outward because the center is so rich in liquid fractions. As the center loses oil, it becomes drier and its textural properties and mouthfeel are altered. Unfortunately, the coating is not impervious to the oil and therefore takes it in. The consequences for the coating are typically softening, dulling, and perhaps bloom.

Oil migration is a fact of life for the confectioner, and over time, equilibration between phases will occur. This phenomenon is very much temperature dependent because at increasingly warmer temperatures, more solid triglycerides are melted and are added to the pool of oil seeking equilibrium. A higher fat content represents a higher migration potential simply because there is more fat to melt and mobilize. The source oils and fats are an important element as well because (1) lauric fats tend to migrate more rapidly than nonlaurics of similar hardness and (2) dissimilar fats, for example, cocoa butter and laurics, interact more severely (stronger eutectics) than do more compatible fats. The rate of migration is rapid initially and then slows as equilibrium is reached. Work conducted within the author's laboratory showed this to be an exponential progression that slows significantly after about 3 weeks storage.

Several possibilities exist to delay and/or reduce oil migration and thus dampen its impact. As was previously mentioned in the peanut butter example, a crystal network can be established in an effort to entrap oil. In addition to the compatibility (e.g., center fat versus coating fat) consideration, migration tendencies are significantly reduced when the solid fat content exceeds about 35% (79). Nonfat ingredients can also be manipulated to deal with free oil. For example, some of the sugar can be replaced with cocoa solids or milk solids to present a larger surface area to the oil and thus delay migration. Similarly, the particle size of all nonfat solids can be reduced (longer milling, finer grinding, e.g.) to present more surface area to absorb fat. Furthermore, smaller particles can pack more tightly and therefore present a less porous matrix as well as interact with one another to form networks that might trap or impede oil mobility.

FAT REDUCTION

Because the fat is so important to eating quality, rheology, and structure for many confections, it is extremely challenging to remove significant quantities of it in the quest for fat-reduced products. A chocolaty chip containing less than 3% cocoa butter has been produced by blending a microparticularized cocoa dispersion with approximately an equivalent portion of confectionery sugar (10X) and defatted cocoa powder to form an extrudable damp mix that can be formed into chips. The microparticularized cocoa dispersion is made by blending 13% cocoa powder (10–12% fat), 60% sugar, and 27% skim milk. This dispersion is ground in a wet mill to reduce at least 75% of the particles to between 2 and 10 μm (80).

Fat continuous emulsions have been proposed to reduce the fat content of confectionery centers. One invention describes emulsion-based fillings and centers wherein the fat content has been reduced to 23%. The water phase contains acidity regulator, thickener, bulking agent, emulsifier, sweetener, flavor, colorant, humectant, and preservative (81). The patent literature abounds with chemical structures claiming to deliver the physical properties of fats and oils, but at a reduced caloric density. One of the few to actually see commercialization thus far is Caprenin. It was offered as a functional alternative to cocoa butter for milk chocolate and other confectionery fats used in soft candies, like caramels and nougats, and confectionery coatings (82,83). Caprenin is claimed to be a reduced-calorie fat whose triglycerides contain predominantly behenic, capric, and caprylic fatty acids. Poor absorption of behenic acid is largely responsible for the reduced caloric density while the heat loss due to metabolism of the medium-chain fatty acids contributes to a lesser degree; Peters and co-workers determined a value of 5 kcal/g (84,85).

REFERENCES

1. B.W. Minifie, *Chocolate, Cocoa, and Confectionery*, 3rd ed., Van Nostrand Reinhold, New York, 1989, pp. 1–33.
2. S. Crespo, *Cacao Beans Today*, Silvio Crespon, Lititz, Penn., 1986.
3. W.J. Shaughnessy, *Manuf. Confect.* **72**(11), 51 (1992).
4. A.S. Lopez, Chemical Changes Occurring During the Processing of Cacao, *Proceedings of the Cacao Biotechnology Symposium,* Pennsylvania State University, University Park, Penn., 1986.
5. D.L. Zak, *Manuf. Confect.* **68**(11), 69 (1988).
6. E.A. Niediek in S.T. Beckett, ed., *Industrial Chocolate Manufacture and Use*, Van Nostrand Reinhold, New York, 1988, pp. 90–93.
7. E.H. Meursing, *Cocoa Powders for Industrial Processing*, 3rd ed., Knijnenberg B.V., Krommenie, Holland, 1983, pp. 66–73.
8. pp. 76–78 of Ref. 1.
9. p. 33 of Ref. 7.
10. p. 32 of Ref. 7.
11. pp. 101–105 of Ref. 6.
12. E.A. Niediek, *Manuf. Confect.* **71**(6), 91 (1991).
13. pp. 98–191 of Ref. 6.
14. p. 161 of Ref. 1.
15. K. Müntener, *Confect. Prod.* **56**(7), 414 (July 1990).
16. W.H. Pratt-Johnson, *Manuf. Confect.* **67**(5), 52 (1987).
17. Anon., *Confect. Prod.* **51**(11), 626 (Nov. 1985).
18. J.M. Hoskin and P.S. Dimick in Proceedings of the 33rd Annual Production Conference of PMCA, 1979, pp. 23–31.
19. F.Y. Maniere and P.S. Dimick, *Lebensm.-Wiss. u.-Tech.* **12,** 102 (1979).
20. T.A. Rohan and T. Stewart, *J. Food Sci.* **31,** 202 (1966).
21. D. Ley in Ref. 6, pp. 122–141.
22. E.H. Reimerdes and H.-A. Mehrens in Ref. 6, pp. 53–55.
23. pp. 144–148 of Ref 1.

24. pp. 116–122 of Ref. 1.
25. Brit. Pat. 1,032,465 (to Cadbury Brothers Ltd.) (1966).
26. J. Chevalley in Ref. 6, pp. 151–152.
27. E.S. Seguine, *Manufact. Confect.* **73**(11), 37 (1993).
28. T.L. Harris, *SCI Monograph* **32,** 108 (1968).
29. p. 153 of Ref. 6.
30. *Cocoa Butter Alternatives,* Karlhsmans Oils & Fats Academy, 1991, pp. 8–9.
31. pp. 28–30 of Ref. 7.
32. P.S. Dimick, *Manufact. Confect.* **72**(5), 109 (1991).
33. G.M. Johnston, *J. Am. Oil Chemi. Soc.* **49,** 462 (1972).
34. R.L. Wille and E.S. Lutton, *J. Am. Oil. Chem. Soc.* **43,** 491 (1966).
35. Anon., *Technology of Chocolate,* p. 78.
36. E.S. Seguine, *Manuf. Confect.* **71**(5), 117 (1991).
37. pp. 183–198 of Ref. 1.
38. T. Papoutsis and E.S. Seguine in J. Stitley, ed., *AIB Technical Bulletin—Selection, Specification, and Use of Chocolate in Confectionery and Baking,* Vol. 19, AIB Research Dept., Manhattan, Kans., 1987.
39. Anon., "Making Chocolate—Tempering (Part II)," from the RCI Short Course in Candy Making, Erie, Penn.
40. B. Wennermark, Tempering of Cocoa Butter and CBE Fats, Paper No. 15, presented at the Karlhamns 1993 International Specialty Fats Seminar, Scheveningen, Holland, May 12–15, 1993.
41. K. Muntener, *Manuf. Confect.* **68**(6), 91 (1988).
42. R.B. Nelson in Ref. 6, pp. 172–226.
43. pp. 212–221 of Ref. 1.
44. V.K. Babayan, *J. Am. Oil Chem. Soc.* **55,** 845 (1977).
45. Jpn. Pat. 62-155048 (to Fuji Oil-Refining Inc.) (1987).
46. T.T. Hansen and P. Eigtved in A.R. Baldwin, ed., *Proceedings—World Conference on Emerging Technologies in the Fats and Oils Industry,* American Oil Chemists' Society, Champaign, Ill., 1986, pp. 365–369.
47. A.R. Macrae in Ref. 46, pp. 7–13.
48. C. Ratledge in C. Ratledge, P. Dawson, and J. Rattray, eds., *Biotechnology for the Oil and Fats Industry,* American Oil Chemists' Society, Champaign, Ill., 1984, pp. 119–127.
49. C. Ratledge in Ref. 46, p. 318
50. U.S. Pat. 4,032,405 (to Tatsumi, Hashimoto, Terashima & Matsuo) (1977).
51. pp. 324–325 of Ref. 49.
52. J.B. Rossell in J.C. Allen and R.J. Hamilton, ed., *Rancidity in Foods,* 2nd ed., Elsevier Science Publishers, Essex, England, 1989, pp. 25–27.
53. T.L. Harris, Surface Active Lipids in Chocolate, SCI Monograph No. 32, April 22, 1968, pp. 108–121.
54. R. Heemskerk, How to Adapt the Interrelationship Between Viscosity and Yield Value in Compounds and Coatings, Paper No. 2, presented at the 1987 Friwessa International Specialty Fats Seminar, Zaandijk, Holland.
55. J.W. DuRoss, *Manuf. Confect.* **67**(6), 105 (1987).
56. N. Garti, J. Schlichter, and S. Sarig, *J. Am. Oil Chem. Soc.* **63,** 230 (1986).
57. E. Timme, *Manuf. Confect.* **65**(5), 65 (1985).

58. J.C. Musser in *Proceedings of the 34th Annual Production Conference of PMCA,* Pennsylvania Manufacturing Confectioner's Association, Drexel Hill, Penn., 1980, pp. 51–59.
59. S. Cross and co-workers, *Food Tech.* **6**(11), 1952.
60. J.W. DuRoss and W.H. Knightly, Relationship of Sorbitan Monostearate and Polysorbate 60 to Bloom Resistance in Properly Tempered Chocolate, presented at PMCA Conference, Lancaster, Penn., April 30, 1965.
61. H. Mori, *Manuf. Confect.* **69**(11), 63 (1989).
62. K. Sato, and co-workers, *J. Am. Oil Chem. Soc.* **66,** 664 (1989).
63. U.S. Pat. 2,760,867 (to General Foods) (1956).
64. U.S. Pat. 4,081,559 (to Cadbury Ltd) (1978).
65. U.S. Pat. 4,446,166 (to Battelle Memorial Institute) (1984).
66. U.S. Pat. 5,004,623 (to Battelle Memorial Institute) (1991).
67. U.S. Pat. 5,149,560 (to Mars, Inc.) (1992).
68. G. Finkel U.S. Pat. 4,664,927 (1987).
69. Anon., *Oils & Fats International*, Issue Two, 1989.
70. p. 577 of Ref. 1.
71. R. Lees, *A Course in Confectionery,* 2nd. ed., Specialised Publications Ltd., Surrey, United Kingdom, 1980, pp. 98–106.
72. pp. 530–538 of Ref. 1.
73. Anon., *Confect. Prod.,* November, 1985, pp. 633–635.
74. pp. 506–507 of Ref. 1.
75. pp. 134–138 pg Ref. 71.
76. W. Richmond, *Choice Confections,* 3rd ed., Manufacturing Confectioner Publishing Company, Glen Rock, N.J., 1977.
77. N. Harris, M. Peterson, and S. Crespo, *A Formularly of Candy Products,* Chemical Publishing Company, Inc., New York, 1991.
78. J. Wacquez and R. Dallemagne, Fat Migration Into Enrobing Chocolate, presented at the 1975 European Conference on Chocolate and Cocoa Products in Chateau Thierry.
79. p. 53 of Ref. 30.
80. U.S. Pat. 5,190,786 (to Kraft General Foods) (1993).
81. EP App. 92203725.4 (to Unilever) (1992)
82. U.S. Pat. 5,066,510 (to The Procter & Gamble Company) (1991).
83. EP App. 88202854.1 (to The Procter & Gamble Company) (1988).
84. D.R. Webb and R.A. Sanders, *J. Am. Coll. Toxicol.* **10**(3), 325 (1991).
85. J.C. Peters and co-workers *J. Am. Coll. Toxicol.* **10**(3), 357 (1991).

10

Oils and Fats in Snack Foods

The primary function of fats and oils in fried snack foods is to serve as a heat transfer mechanism while adding flavor and desirable eating qualities to the finished product. Fats and oils used in slurry or spray oil applications are not subjected to the high temperatures of frying; however, they are still required to possess high stability characteristics. An oil's resistance to oxidation and hydrolysis during the frying operation are crucial to the oil's performance and contribute to the production of high quality snack foods. Such issues as oil turnover rate and fryer design must be considered to obtain the best possible performance from the frying fat or oil. Proper care and maintenance of the oil handling/storage equipment and frying systems are essential in producing consistant quality snack foods.

Oil selection depends on the type of snack food being produced, with special attention being given to the specific eating characteristics desired. A liquid oil will produce a potato chip with an oilier surface, whereas a solid shortening will produce a potato chip that has a dryer appearance and texture. Often the type of potato chip produced by smaller manufacturers is a function of regional preference. Fats and oils are used in snack foods to provide a wide variety of eating characteristics.

MAINTAINING OIL QUALITY PRIOR TO USAGE

To ensure the highest quality oil is received at the snack manufacturing facility, each oil seed processing step through growing, crushing, and refining must be properly carried out. If the oil has not been properly processed by the crusher or refiner, it will be difficult for the snack manufacturer to obtain the optimum performance from that oil. Quality indicators such as color, free fatty acid, peroxide value, and AOM are common on oil specifications included in oil formulas. The snack food manufacturer should hold the oil supplier accountable for supplying the highest quality oil possible on a consistent basis.

Oil can be packaged for delivery to the snack manufacturer in a variety of ways. Depending on the usage level and storage capacity at the production facility, the amount and frequency of oil purchased must be determined to ensure timely usage of the oil. Because oil begins its degradation process immediately after processing at the refinery, use of the oil as quickly as possible is in the best interest of the snack manufacturer. Oils can be packaged in a variety of sizes, ranging from 15.8-kg (35-lb) containers to 68,000-kg (150,000-lb) railcars. Again, the oil use rate will determine the size of the shipping container. Most larger snack food manufacturers will order 20,400-kg (45,000-lb) tank trucks or 68,000-kg (150,000-lb) railcars.

To help protect oil from the oxidation process in shipment, nitrogen protection is required (1). Nitrogen sparging of the railcars or tank trucks is often done at the refinery. To maintain this protection at the snack food manufacturing facility a nitrogen sparging system should be installed at the oil unloading site. Without nitrogen protection, the oil in a storage tank should be used within 3–4 weeks. Under ideal conditions and with nitrogen protection the oil quality can be maintained 6–8 weeks. Sealed containers such as 181-kg (100-lb) drums and 907-kg (2000-lb) totes will maintain oil quality for up to 6 months in properly controlled temperature conditions.

On receipt of a bulk oil shipment by a snack food manufacturer, he or she must consider several details to maintain oil quality. The oil must be properly sampled before unloading into the storage tank. Several methods are used to sample a bulk oil shipment, of which a sampling bomb is most common and provides the most representative sample. Proper handling and analysis of the sample is the joint responsibility of the receiving and quality assurance departments. A decision to accept or reject the oil should be based on a combination of analytical and organoleptic results as well as visual inspection.

Once the decision has been made to accept the oil into the manufacturing facility, transporting the oil from the truck or railcar into the plant requires special attention to prevent unnecessary incorporation of air into the oil. Pumps should be checked to make sure no air is sucked into the oil as it passes through the pump. Storage tanks must be bottom filled to eliminate the free falling of oil through the air and reduce the potential for incorporation of oxygen into the oil. Liquid oils stored outside during winter months in colder climates require heat to preclude crystallization, especially if nonwinterized oils are used (cottonseed oil has a melting point of about 0°C). Partial hydrogenation does not necessarily dictate external heat, because this reaction can be selective in that the melting point is only minimally affected in an effort to improve oxidative stability; in addition, selectively hydrogenated oils might be winterized or fractionated to remove higher melting components. However, any external heating requirement depends on the environment:

outside versus inside storage in cold climates. The heat source should be hot water rather than steam, due to the excessive surface temperatures associated with steam piping. Because heat serves as a catalyst in degradation reactions of oils, it is recommended fats be stored at no more than -12.4--$9.4°C$ (10–15°F) above the respective melting point of the fat. Slight, periodic agitation or circulation is recommended for storage of fats and oils, which prevents the settling out of higher melting fractions in partially hydrogenated oils and reduces the potential for oil separation. The separating of nonagitated partially hydrogenated fats results in solid, more stable fat settling to the bottom of the tank and the liquid, less stable oil portion remaining on top. This phenomenon leads to inconsistent oil stability as the oil level in the tank decreases as it is used in the plant. In addition to affecting oil stability, separation alters the oil's physical properties; thus performance will change and finished product quality will be inconsistent.

1.1 Oil Breakdown during Frying

The most obvious reactions occurring to the oil during the frying process are the result of three sources: heat, air, and moisture. The degradation of the oil from these sources leads to polymerization, oxidation, and hydrolysis. Heat not only serves as a catalyst accelerating the formation of free fatty acids during the frying process but also is responsible for polymerization of the oil. Overheating the oil, usually due to hot spots in the fryer, often leads to thermal polymerization. This is evidenced by a gummy varnishlike buildup on the frying equipment. To maintain the proper frying temperature range during the frying process, a constant source of heat is applied to the oil.

Putting room temperature, or cooler, raw product such as potato slices into the hot oil reduces the oil temperature significantly at the inlet end of the fryer and requires constant reheating of the oil. The outlet end of the fryer will typically maintain a higher oil temperature, usually -9--$1°C$ (15–30°F) hotter than the inlet end. This difference in oil temperature, often referred to as delta T (ΔT) is not as great in continuous fryers as it is in the batch type fryers commonly used in the production of kettle chips. A drop in oil temperature of as much as 10°C (50°F) over a 4–6-min frying cycle is common in kettle chip production. This drop in temperature and subsequent reheating are obviously extremely stressful on the frying oil and contribute to the thermal breakdown of the oil. The method used to reheat this oil is critical and directly related to the stress applied to the oil during the reheating process. Heat exchangers typically subject the oil to less heat stress then direct fired fryers. This is due to the reduced surface temperature contact of the oil in heat exchangers.

Other common causes of thermal breakdown in oils are improper melting procedures and thermostat malfunction. Partially hydrogenated or solid frying fats should be heated slowly in an effort not to burn the fat. Thermostats should be checked periodically to ensure the readout temperature on the panel is within -17--$-18°C$ (1--2°F) of the actual oil temperature in the fryer.

Heating oil to frying temperatures without actually frying in the oil subjects the oil to unnecessary abuse. The frying process actually aids in the distilling off of free fatty acids through the escape of steam up the exhaust stack. When the oil is heated without frying, this steam distilling process does not occur and free fatty acids tend to increase in the frying oil. Excessive heating of the oil not only results in polymerization but may also cause foaming of the oil. Typically the oil and product fried in the oil will exhibit a bitter, acrid, burned flavor. Proper oil handling by protecting the oil from excessive heat will help prevent the deterioration of the oil through polymerization.

Unless protected from exposure to air, oils will react with oxygen molecules at the points of unsaturation. Oxidation can occur during storage, frying, or in the package of finished product. The more unsaturated the oil the more reactive it is with oxygen. Polyunsaturated fatty acids are more reactive with oxygen than monounsaturated fatty acids (Table 10.1) (2). In general, the less saturated an oil, the more reactive the oil is with oxygen; therefore, the more reactive the oil, the less stable the oil. It is not, however, always true that the less saturated an oil, the more reactive it is with oxygen. Consider soybean oil (about 15% saturates) versus canola, brush hydrogenated canola, high oleic canola, and high oleic sunflower: these all are less reactive than soybean and are also less saturated than soybean. The oxidation reaction can be increased by metal ions, especially those found in copper (3). Copper serves as an oxidation catalyst and should never be allowed to come into contact with the oil. Special care must be taken to eliminate copper fittings and brazings in the fryer equipment and oil handling lines. Periodic checks of the equipment must be conducted, and all maintenance personnel must be properly educated with respect to oil's sensitivity to copper and copper containing materials.

In addition to high temperatures, oxygen, and metals, the exposure of oils to uv light should be avoided. Uv light exposure will also tend to accelerate the oxidation process, exhibited by a rise in peroxide value. There are precautions that can be taken to inhibit the oxidation reaction of oil. Such measures as using nitrogen protection in the storage of unused oils and nitrogen flushing packaged product will aid in protection against the oxidation process. Antioxi-

Table 10.1 Reactivity of fatty acids

Fatty Acid	Relative Oxidation Rate
Stearic (C18:0)	1
Oleic (C18:1)	10
Linoleic (C18:2)	100
Linolenic (C18:3)	150

dants such as TBHQ, BHT, BHA, and propyl gallate can be added to the oil immediately after processing to help protect the oil from premature oxidation.

Antioxidants will aid in the protection of oils against oxidation in storage. Metal scavengers, of which citric acid is most common, assist in removing trace metals, thus inhibiting oil deterioration (4). Ascorbyl palmitate has been shown to protect the quality of frying fats during the frying process (5). Perhaps a more important point is the ability of antioxidants like ascorbyl palmitate and tocopherols to survive the frying process and consequently provide good carrythrough protection to fried foods.

Moisture contributes to the degradation of oils during the frying process. This process of deterioration is referred to as hydrolysis. During the hydrolysis reaction, water breaks the chemical bonds holding the triglycerides together. The result is the creation of free fatty acids, monoglycerides, diglycerides, and in some extreme cases, glycerine (Figure 10.1) (6).

A common belief is that the greater the moisture contained in the raw product, the greater the hydrolysis reaction during frying. This appears to be not entirely true. Jacobson (7) reports that free fatty acid development is not appreciably influenced by the moisture content of the food fried. Dornseifer and co-workers (8) reported that the total volatile carbonyl (TVC) development was not proportional to moisture content in cottonseed oil. Over the moisture range studied, TVC reached a maximum at 2.5% water and fell off at higher levels (highest moisture level evaluated was 10%). Frying operations that do not turn the oil over quickly, such as a kettle chip operation, tend to have hydrolysis reactions that, if not properly managed, affect the taste and quality of the finished product. Continuous frying operations tend to maintain oil breakdown in an optimum stage, giving the finished product a desirable fried flavor. In some kettle chip frying operations, the oil breaks down into some of the secondary products of degradation (6). This is primarily due to the lower oil turnover rate typical in kettle chip frying operations. *Turnover rate* refers to the length of time required to replace the volume of oil in a fryer (oil removed through oil uptake in the fried product) with fresh or makeup oil. The oil turnover rate can be calculated by

$$\text{pounds of finished product (per hour)} \times \% \text{ oil content of finished product (average)} \times 8 \text{ h} = \text{total oil use over 8 h}$$

Compare this number to the actual oil volume in the frying system. Ideally, the entire volume of oil in the frying system should be replaced every 8 h. Frying systems with higher oil turnover rates tend to take advantage of the steam-releasing phenomenon and use the process to strip free fatty acids from the oil. Proper handling of fryer dripback generated from steam going up the fryer stack is essential to maintaining lower free fatty acids and contributes to good flavored finished product.

Additional factors contributing to oil breakdown are foreign material and

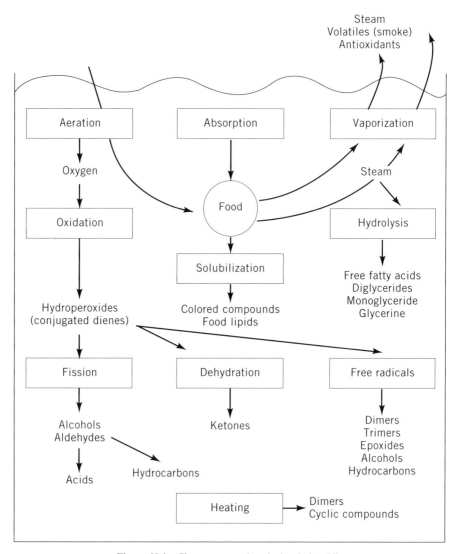

Figure 10.1 Changes occurring during frying (6).

extraneous matter. Maintaining the purity of frying oils is important in achieving the optimum performance of the oil. Especially in potato chip frying, such materials as starch, dirt, sand, and other foreign materials find their way into the oil. Crumbs and burned food particles tend to accumulate in the fryer. This buildup of particles increases the rate of development of free fatty acids. Frequent or continuous filtration during frying is essential in maintaining consistency in the quality of the frying oil. The method of filtration used is important when trying to eliminate particular types of foreign material. Care must be taken not to aerate the oil during filtration (9). If continuous filtration is not possible, daily filtration is the next viable option. In this procedure, after the oil has cooled to approximately 93°C (200°F), the oil is pumped from the fryer through a filtration system and back into the fryer after crumbs and residue have been removed from the kettle. Crumbs and debris left in frying oil for extended periods of time at elevated temperatures will cause the oil to darken and exhibit off-flavors. Free fatty acids will increase causing a lowering of the smoke point (10). Contamination of frying oils usually results from soap residue left from improper cleaning procedures. Alkaline materials are also produced from the reaction of free fatty acids with metals either leached from the food being fried or from the processing equipment as well as from traces carried over from the oil refining process. If an alkaline solution is used in the cleanout procedure it is critical to rinse the fryer with an acidified solution until the pH of the rinse solution is between 6.0 and 7.5. This will neutralize any remaining alkaline residuals (11).

THE FRYING OPERATION

2.1 Fryer Design

There are two types of fryers used in snack food production: continuous and batch. These fryers differ in the type of product produced and their affect on the oil. Batch-type fryers are much more detrimental to oil quality. Such factors as ΔT, oil turnover rate, and aeration of the oil make batch-type frying more of a challenge to maintaining the integrity of the frying oil. Fryers can be heated by gas-fired immersion tubes as well as external heat exchanges (12). Heat exchanger-type fryers range in size from 340 kg (750 lb)/h to 2268 kg (5000 lb)/h.

When designing a frying system, it is wise to involve the fryer manufacturer first to ensure the fryer will produce the type of product desired. All too often companies buy a fryer based on price or availability without addressing the capabilities of the particular fryer. Several issues must be considered when choosing a fryer to fit your specific needs: maximum and minimum production rate, oil heating source, heat exchanger or direct fire, gas fired or electric, oil filtering capabilities, cleaning accessibility, turnover rate (based on your desired output) and oil circulation. Be sure also to consult the oil supplier to ensure the fryer purchased is capable of producing the product desired.

It is also important at this stage to select a frying system that will adequately remove condensation and prevent the likelihood of dripback into the fryer. Condensation buildup on the underside of the fryer hood is high in free fatty acids and should be properly caught by catchpans and channeled away from the frying oil.

Other factors influencing oil absorption can be directly related to fryer design. It is true that oil absorption is influenced by the surface area of the product being fried (13). If the fryer is too large for the desired output, the product will tend to remain in the oil too long and result in the higher than desired oil content of the finished product. Frying conditions have the most impact on the amount of oil absorbed in the finished product. The frying conditions in potato chip production that affect oil absorption are listed below.

2.2 High Oil Absorption

1. *Oil temperature too low.* Low oil temperature frying will produce a potato chip that is oil soaked and raw. The potato slice absorbs oil from a soaking effect into the potato slice and not frying.
2. *Fryer overload.* Fryer overloading results in an extreme drop in the oil temperature. In an overloaded fryer, the heat recovery potential of the fryer cannot reheat the oil to the desired frying temperature. When the fryer is overloaded, the potato slices are subjected to low temperature frying, resulting in raw, oil-soaked potato chips.
3. *Broken-down or abused oil.* Broken-down or abused oil tends to result in an increase in the heat recovery time during frying. As a result, the potato slices are exposed to lower temperature frying and will produce a product with a higher oil content and a raw texture.
4. *Extended dwell time.* Each type of snack food has a particular frying time that produces the desired flavor, texture, and mouthfeel. Exceeding this fry time will result in excessive oil content of the finished product.
5. *Slicer blades rough or dull.* Slicer blades must be properly maintained and not allowed to wear down or become damaged. A dull blade will produce a rough and uneven potato slice. This rough slice will tend to be more porus and allow more oil absorption during frying.
6. *Low specific gravity, high moisture potato.* Potatoes with a lower potato solid (less than 25%) and higher moisture will produce a finished potato chip with a lower yield and higher oil content. The higher the specific gravity, the higher yield of potato chips (14).

2.3 Low Oil Absorption

1. *Oil temperature too high.* Frying potato chips in excessively high oil temperatures will tend to sear the exterior of the potato slice and prevent adequate cooking of the inner portion of the potato slice. The potato chip will look fully cooked on the outside, but have a raw texture in the center.
2. *Short dwell time.* If the potato slice is rushed through the fryer and not given adequate time to cook, the result is a raw, undercooked potato chip with high moisture content.

3. *High specific gravity, low moisture content.* Potatoes with high solids (more than 25%) will result in a higher chip yield and lower oil content. This is the ideal condition because chip manufacturers are paying for more potato solids and less moisture. As a result, less oil is absorbed in higher specific gravity potatoes.

2.4 Type of Oil Used

Choosing the type of oil to be used for snack food frying can be influenced by many factors. It is important to consider the following finished product attributes when choosing a frying oil: flavor, mouthfeel, texture, and product appearance. If the frying oil is liquid at room temperature, the finished chip will have an oilier appearance than if the oil is partially hydrogenated. Liquid oils tend to give a quicker flavor release of the oil in the mouth than do partially hydrogenated oils. Heavy-duty frying shortenings give doughnuts a pleasant creamy mouthfeel; however, using the same shortening for potato chips will produce a bland flavor with a waxy mouthfeel. Recently, consumers have been more concerned with the nutritional aspects of oils. The public has shown concern in such issues as saturated fat, cholesterol, and trans fatty acids and their effects on cholesterol in the blood. Some studies show trans fatty acids act similarly to saturated fatty acids by raising levels of low density lipoproteins ("bad" cholesterol), and decreasing levels of high density lipoproteins ("good" cholesterol) in the blood (15).

SNACK FOOD PRODUCTS AND PROCESSING SYSTEMS

3.1 Potato Chips

There are two types of potato chips produced by snack food manufacturers: chips made from fresh sliced potatoes and chips processed from potato dough and formed into potato chip shapes. There are two basic methods used to process fresh potato slices: a continuous frying system and a batch-type system, or kettle chip frying. These fried potato products are distinctly different, with each having their own market segment.

Because potatoes are at the mercy of weather variations during their maturation in the ground, care must be taken by the potato chip manufacturer to ensure the highest quality potatoes are purchased. Consideration must be given to tuber size, shape, defects, peel, and eye depth. Processing potatoes with minimum variation allows the chip manufacturer to produce potato chips that are consistant from bag to bag. Any variations in the chips may be seen by the consumer as inconsistant and not typical of a quality product. In addition, the chip manufacturer must pay special attention to specific gravity.

The specific gravity of potatoes can be measured by several methods. The most widely used method in snack food production is the SFA hydrometer method, which measures the relationship between the weight of potatoes in air and the weight of potatoes in water. Total solids in potatoes can be predicted by measuring the specific gravity content of the potato (Table 10.2) (16). It is important for the chip manufacturer to select potatoes with high specific gravity, as they will produce a higher yield and lower oil absorption.

Another attribute affecting the finished product quality of the potato chip is the amount of reducing sugars (glucose and fructose) in the potato. If the amount of reducing sugars in a tuber is higher than 0.1%, dark brown or black chips could result from the frying process (17). Reducing sugars are formed naturally during photosynthesis and vary, depending how the tuber is handled before frying.

Common oils used in the production of potato chips are liquid cottonseed, partially hydrogenated soybean/cottonseed blends, peanut, and corn. The typical processing line for potato chips will use the following pieces of equipment: destoner, peeler–washer, inspection table, slicer, washer–tumbler, fryer, and seasoner–salter.

Table 10.2 Relationship of specific gravity, water content, dry matter, and starch in potatoes

Specific Gravity	Percent Water	Percent Dry Matter	Percent Starch
1.040	86.4	13.6	7.80
1.045	85.4	14.6	8.75
1.050	84.5	15.5	9.60
1.055	83.6	16.4	10.46
1.060	82.6	17.4	11.41
1.065	81.7	18.3	12.26
1.070	80.8	19.2	13.11
1.075	79.8	20.2	14.06
1.080	78.8	21.2	15.00
1.085	78.0	22.0	15.76
1.090	77.0	23.0	16.71
1.095	76.1	23.9	17.56
1.100	75.1	24.9	18.51
1.105	74.2	25.8	19.37
1.110	73.3	26.7	20.22
1.120	71.4	28.6	22.01
1.125	70.5	29.5	22.87
1.130	69.6	30.4	23.72
1.135	68.6	31.4	24.67
1.140	67.8	32.2	25.43

Destoner, Peeler–Washer, and Inspection Table. The purpose of the destoner is exactly what the name implies. Coming directly from the fields, the potatoes are often covered with dirt and/or sand. The shipment may also contain stones or other debris that must be removed before processing. The destoner removes any stones or debris from the potato shipment. After the destoning step, the potatoes are washed to remove excess dirt or sand just before they go into the peeler. The peeler consists of several rotating course cylinders covered with grit that grind away the skin of the potato. Sometimes, potatoes are received that have tender or less durable skin. These are usually new crop potatoes harvested early in the spring. When such potatoes are received, coarse peeler rollers may be replaced with brush-type rollers to reduce potato loss during the peeling operation. If the coarse rollers were not replaced, excessive losses in yields would occur. From the peeler–washer station, the potatoes pass over an inspection table where visible defects such as rotting, bruising, and blackspots are removed with a knife or, in severe cases, the entire potato is separated and discarded.

Slicer and Washer–Tumbler. Depending on the type of potato chip desired (flat, wavy, thick, thin) the slicer can be adjusted to achieve a wide variety of potato chips. The peeled potatoes are sliced by a centrifugal force pressing the potatoes against a stationary circular ring that holds several gauged blades. As the potatoes come into contact with the knives, the slices separate and fall into the fryer. Thickness and shape can be varied to meet marketing needs. Flat slices are generally 0.13 cm (0.050 in. thick), or 18 slices/in. Wavy slices can be as much as 0.4 cm (0.070 in. thick), or 13 slices/in. Slices produced from rough or dull slicer blades tend to absorb more oil during frying. The style or shape of the potato slice determines the oil content of the chip as the amount of oil is affected by the ratio of the surface area of the potato slice to the total volume of the slice (14). Because the wavy potato chip has a larger surface area, it will have more surface oil than a flat potato chip. Thicker potato slices will tend to absorb less oil than thinner slices. The relationship between fryer temperature and slice thickness and their effect on oil content is illustrated in Figure 10.2 (16).

Figure 10.2 shows that the lowest oil content was achieved by thicker slicing and higher temperature frying. To remove excess starch, the potato slices are washed in a perforated tumbler with spray nozzles. After passing through the bath, the slices are dried on a flat wire conveyor with air pressure to remove as much surface starch and water as possible. Some chip processors, especially kettle chip manufacturers, do not wash the slices, which gives added flavor to the finished potato chip. Since the starch is not washed off, care must be taken to prevent clustering of the potato slices going into the fryer.

Fryer. Potato chip frying can be broken down into two methods currently used in the industry: continuous frying and batch frying. As previously men-

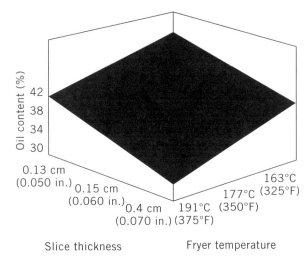

Figure 10.2 Effects of slice thickness and fryer temperature on the oil content of potato chips.

tioned, batch frying is generally more stressful on the oil than continuous frying. Most manufacturers of potato chips use continuous fryers. In the frying process, moisture in the potato slices is fried out and partially replaced by oil (Figure 10.3).

As the dried potato slices enter the fryer, a rotating paddle wheel pulls the slices into the fryer while pushing the floating slices under the oil. The slices are then periodically pushed under the oil surface by a series of paddle wheels. After the paddle wheels, the slices are pulled under the surface of the oil by means of the submerger. The slices may remain under the oil surface for as long as 1–2 min. This time can be adjusted to achieve the desired browning as the color of the potato chip is typically set during this stage of the frying process. This final stage of moisture removal is critical and goes hand in hand with the browning effect. Immediately after the submersion, the slices are removed on a takeout conveyor. There is an opportunity during this step to remove as much surface oil as possible by slowing the conveyor and allowing the excess oil to drain off.

Seasoning–Salter. Potato chips are usually seasoned in a rotating tumbler where flavored topical seasonings adhere to the fried potato slice. The chips

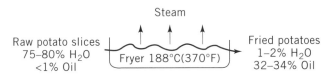

Figure 10.3 Water and oil content of raw and fried potatoes.

are cooled on a conveyor and sorted by size before being seasoned and salted and packaged in bags ranging from 1 to 20 oz. The oil temperature for potato chip frying will range from 177 to 191°C (350 to 375°F).

The finished potato chip should meet the following quality control points:

Moisture, %	<1
Oil, %	30–40 range, 32–34 average
Salt, %	1 average
Seasoning, %	6–8
Wet centers, %	0
Broken chips, %	<10
Defects, %	major (1.3 cm/0.5 in. diameter) < 3
	minor (0.65 cm/0.25 in. diameter) < 5

3.2 Tortilla Chips

Tortilla chips and corn chips are made from alkali treated cooked corn. Most producers use a soybean-based oil in the frying process. Tortilla chips are made from cooked yellow, white, or blue corn. Oils commonly used for frying are partially hydrogenated soybean and partially hydrogenated and winterized soybean oils.

Although more expensive, use of a partially hydrogenated and winterized soybean oil typically will provide a more stable oil than the originally partially hydrogenated soybean oil. Research has suggested that impurities contributing to poor flavor stability tend to stay with the "stearine" portion filtered out of hydrogenated and winterized oil. The remaining clear liquid fraction of the hydrogenated and winterized oil tends to be more stable than the partially hydrogenated soybean oil before the winterization step (18).

Tortilla chip production uses the following steps. For chips made from cooked corn: cook–steep, wash–drain, grind–mill, sheeting, oven baking, frying, and season–salter. For chips made from a masa mix: formulation, mixing, floor time, sheeting, oven baking, frying, and season–salter.

Cook–Steep. Raw corn is cooked in a steam-jacketed kettle in a water and lime solution. Cook time will vary between 4 and 10 min. at 93–99°C (200–210°F). Special care must be taken to stop the cooking cycle at the precise time by the addition of cool water to the kettle. At this point, the moisture of the corn has increased from 8–12% on receipt into the plant to approximately 30%. The lime imparts a desirable flavor into the corn kernel. The steeping step serves to equalize the moisture within the corn kernel while raising the kernel moisture to approximately 45%. The steeping process allows the cooked corn to soak in the cooking water for a specified number of hours, usually between 8 and 24 h. This soaking also allows the husk to loosen from the kernel, making its removal much easier during the washing step. The temperature of the cooked corn in the soaking vats will range from 46 to 60°C (115 to 140°F).

Wash–Drain. The corn is now transferred to the processing area for rinsing and draining, which take about 4–7 min. The temperature of the corn kernels after washing is between 10 and 21°C (50 and 70°F).

Grind–Mill. Recent advances in technology have virtually eliminated the use of lava stones for corn grinding. Aluminum oxide stones give a much more even grind of the corn (or masa) and do not wear as quickly. Proper care and alignment of the stones is essential in producing a consistant quality tortilla chip. As the cooked corn is auger fed into the center of the stones (one stationary and one rotating), the corn is forced out between the stones through decreasing swirled grooves. Water addition may be necessary for lubrication. This masa should be slightly grainy, below 38°C (100°F) and have a moisture content between 52 and 54%. Overcooked corn will tend to produce a hot masa that sticks to the sheeter rollers and die cutters.

Sheeting. From the hopper below the mill, the masa is pumped to a sheeter, consisting of two rollers and a die cutter. The die cutter can be of various shapes, such as round or triangular to produce the desired shape tortilla chip. Special care must be taken to ensure proper adjustment of the sheeter to eliminate such defects as foldovers and clusters.

Oven Baking. The sheeted tortilla chip then enters a triple-pass oven to toast the exterior of the tortilla chip for appearance and provide strength during frying. The oven is maintained between 399 and 482°C (750 and 900°F). The tortilla chip is exposed to this heat for 12–18 s bringing the moisture down to 40–45%. Some tortilla chip producers condition the baked chip in a humidity chamber for as long as 30 min to allow the moisture to equalize throughout the chip.

Frying. Tortilla chips are fried in a continuous fryer between 171 and 185°C (340 and 365°F): for approximately 2 min, bringing the finished moisture down to less than 1.5%.

Seasoner–Salter. Before being seasoned, the fried tortilla chips are sometimes sprayed with a fine oil mist to improve seasoning adherence. For convenience sake, the oil used is usually from the storage tanks feeding the fryer.

3.3 Corn Chips

Corn chip processing is similar to tortilla chip production with some changes in corn cooking and grinding. Corn chip masa is extruded and not sheeted.

Corn chips go through the following production steps: cook–steep, wash–drain, grind–mill, masa extruding, frying, and season–salting.

Cook–Steep. Corn chip corn is cooked between 15 and 45 min in a water and lime solution at between 93 and 99°C (200 and 210°F). The moisture after the cook cycle is 42–44%. The corn is steeped 8–24 h. After steeping, the moisture is between 50 and 52%.

Wash–Drain. The corn undergoes the same wash–drain procedure as described for tortilla chips before entering the grind–mill step.

Grind–Mill. With a kernel moisture of 50–52% after cooking, water is usually not added as the corn enters the stones. The masa for corn chips is typically finer than that for tortilla chips. This finer grind tends to produce a chip with greater oil uptake during frying.

Masa Extruding. Masa from the grind–mill step is pumped or manually carried to an extruder equipped with a die head for shaping strips of corn chips. The desired length of corn chip is achieved by adjusting the speed of the rotating knife that cuts the strips of corn chips. The cut strips are dropped directly into the fryer.

Frying. Corn chips are fried for approximately 90 s at 199–204°C (390–400°F). Yellow corn, typically a softer corn, is fried at a lower temperature than white corn. A finished moisture of less than 1.5% must be achieved.

Season–Salting. Corn chips are typically seasoned or salted directly out of the fryer before going to a cooling conveyor and on to packaging.

3.4 Extruded Snacks

Extruders are used in the snack industry to produce what is commonly referred to as puffed snacks. Most extruded products are produced from a dry milled cornmeal with the hull and germ removed. The two most common extruded snacks are baked and fried collets. During the process of extrusion, cornmeal with a desired moisture content is fed into an extruder. The pressure and heat generated in the extruder by the auger causes the cornmeal moisture to vaporize. The super heated cornmeal is then forced through a die (baked collets) or through a stationary and rotating head (fried collets). The exposure of the cornmeal mixture to the atmosphere allows the seam vapor to release, resulting in the expansion of the collet. The resulting collet is then processed further, depending on which type of extruded finished product is desired, baked or fried.

Fried Extrusion. Oils commonly used for fried extrusion are partially hydrogenated soybean oil and partially hydrogenated soybean and cottonseed oils with a melting point between 32 and 37°C (90 and 98°F). Extruded snacks are usually packaged in packaging with a clear window. If a liquid oil is used for frying, or in the slurry application, a greasy film will form on the inside of the packaging material. Oils that are basically solid at room temperature are required to prevent this from occurring.

The oil content of the cornmeal when received is normally less than 1%. Particle size or cornmeal granulation is important. Many extruded snack manufacturers have granulation standards for incoming cornmeal (Table 10.3) (19).

If the cornmeal is too coarse or is out of specification on the larger particle sizes, the producer will not be able to get the desired moisture into the cornmeal pieces. In addition, processing problems such as poor expansion and rough texture will result. If the cornmeal is ground too fine, the water added in the blending step will tend to be absorbed by (flourlike) cornmeal and result in poor extrusion.

Moisture. The moisture content of the cornmeal is important. A moisture content of 11–13% as received is recommended for good expansion during extrusion. Although the manufacturer does not want to pay cornmeal prices for water, the moisture in the cornmeal at the time of delivery from the mill should be as close to 11–13% as possible. This is because it is difficult to get moisture into the cornmeal pieces during blending. Water can be added during blending, but the water primarily serves as lubrication during extrusion. The moisture content should be raised to 14–15% in the blender. Proper mixing is critical during this stage, as moisture variation within the batch will result in collet size and shape variations. Cornmeal with too much moisture will inhibit expansion during extrusion and produce a thin, smooth collet. These collets tend to absorb less oil during frying. Too little moisture will result in collets that are tough and rough in texture and tend to absorb more oil during

Table 10.3 **Typical cornmeal specification**

Granulation	Percent Retained on Screen
16 mesh	0
20 mesh	0–2
25 mesh	0–10
30 mesh	25–50
40 mesh	45–65
50 mesh	0–8
60 mesh	0–2
Moisture	11–13%
Fat	1%

frying. Proper moisture content of the cornmeal is also related to the bulk density of the raw unfried collets. Ideal bulk density is 1.70 kg (3.75 lb)/ft^3 with a range between 1.6 and 2.0 (3.5 and 4.5) (16). After the cornmeal and water have been mixing in the blender for 8–10 min, the blend is ready to be transferred to the extruder. The extruder consists of a rotating auger enclosed in a sleeve designed to increase pressure and thus raise the temperature of the cornmeal. During this stage, the cornmeal undergoes a transformation called gelatinization, whereby the cornmeal changes into a gelatin-like paste. This paste is forced through the gap created between the faces of the stationary and rotary plates. Swirls and groves in the plates create the collet. The collet is then severed by rotating blades. Alignment and gap distance of the plates are crucial in producing consistent collets.

In summary, friction and compression provide heat on pressure to superheat the moisture, cooking the cornmeal. The pressure forces the dough through the opening of the heads after preventing the water from vaporizing inside the extrude sleeve. As the collets are forced out from between the heads and collected on a conveyor the moisture is between 6 and 8%.

After the extrusion step, the collets pass over a perforated vibrating conveyor to remove small bits and pieces. These pieces are removed before frying as these pieces tend to absorb excessive oil and usually end up in catch pans under conveyors. The perforations in the vibrating conveyor are usually 0.625 cm (0.25 in.) in diameter. During the frying process moisture in the collets is removed going from 13–15% to 1%, while oil is introduced, increasing from less than 1% to approximately 22–24%. Frying time is between 17 and 20 s. Oil content is a function of collet moisture, bulk density, dwell time, and oil temperature. After frying, the collets are seasoned with a cheese slurry made of cheese powder, oil, and salt. The slurry is typically made up of 45 kg (100 lb) cheese powder, 79–91 kg (175–200 lb) oil, and 0.4536–4.536 kg (1–10 lb) of salt. The ingredients are blended in a stream-jacketed kettle and held at a temperature between 38 and 49°C (100 and 120°F). When the collets are coated in a rotating rumbler for 70–90 s, the oil content is increased to 32–38%. The seasoned collets are cooled and sent to packaging.

Baked Extrusion. Baked extruded collets are produced using the same extrusion principle as fried extrusion. The difference between the two is the gelatinized cornmeal is forced through a fixed plate die. The baked extrusion process produces a much more symmetrical collet with little or no size variation. The baked collets tend to be lighter than fried collets and baked in an oven to lower the moisture content. The cheese slurry is applied topically to the baked collets to bring the finished oil content up to approximately 38%.

3.5 Popcorn

There are two types of commercially produced popcorn: hot air popped and hot oil popped. When the popcorn kernel is superheated, the moisture inside the kernel changes to steam, causing an explosion of the kernel or hull. Shapes of popped kernels are affected by seed variety and moisture content of the kernel. The ideal shape of the popped kernel is similar to the shape of a mushroom. These popped kernels tend to be more durable than elongated shaped kernels. Dry popped popcorn typically incorporates a spray oil for flavor addition and seasoning application. Common oils are partially hydrogenated soybean oil, cottonseed oil, and corn oil. Wet poppers, until recently, nearly always used coconut oil for popping. This preference was due to stability and flavor characteristics. Partially hydrogenated soybean and cottonseed oils are commonly used today. Proper care and adjustment must be taken in operating the hot air popper to reduce the amount of scorched and unpopped kernels. High temperatures [greater than 227°C (440°F)], slow drum speed, and slow feed rates are all causes of scorched kernels. Low temperatures [less than 213°C (415°F)], excessive drum speed, and fast or heavy feed rates will result in unpopped kernels. After popping, the kernels pass through a rotating wire tumbler to remove unpopped kernels or scrap.

3.6 Nut Roasting

Other than peanut oil, a majority of nut producers roast in cottonseed oil or partially hydrogenated soybean or cottonseed oils. After they are sorted for defects, the peanuts are blanched to remove the skin. Once the skins are removed, the peanuts are separated by size and conveyed to the roaster.

Nut roasting oil temperatures range from 154 to 182°C (310 to 360°F). The oil content of the peanut will increase nearly 2% during frying. Oil absorption is affected by oil temperature, frying time, and fryer or roaster load.

3.7 Pork Rinds

Pork rinds or pork skins are produced from green hog belly skins. The raw skins, after removal from the hog are submerged into a sugar and salt brine solution at 100°C (212°F). The skin is thin sliced into small pellet pieces approximately 0.2688 kg (0.5 in.) square.

The pork skin pellets are then rendered in prime steam lard between 110 and 116°C (230 and 240°F) (20). After rendering, the pellets are packed for frying. The pellets are fried in oil at temperatures of 193–204°C (380–400°F). Oil content will range from 34 to 38%. Oil temperature, frying time, and fryer load all affect the oil absorption in the pork rinds.

REFERENCES

1. G.R. List and D.R. Erickson, *Handbook of Soy Oil Processing and Utilization,* 4th reprint, AOCS Publication, Champaign, Ill., 1987.

2. H.J. Beckman, *J. Am. Oil Chem. Soc.* **60**(2), 282 (1983).
3. N.O.V. Sonntag in D. Swern ed., *Bailey's Industrial Oil and Fat Products,* Vol. 1, 4th ed., John Wiley & Sons, Inc., New York, 1979, 152.
4. T.J. Weiss, *Food Oils and Their Uses,* 2nd ed., AVI Publishing Co., West Point, Conn., 1983, pp. 109–113, 157–180.
5. Y.Y. Gwo, G.J. Flick, Jr., H.P. Dupuy, R.L. Ory, and W.L. Baran, *J. Am. Oil Chem. Soc.* **62**(12), 1666–1671 (Dec. 1985).
6. C.W. Fritsch, *J. Am. Oil Chem. Soc.* **58**, 272 (1981).
7. G.A. Jacobson, *Food Technol.* **21**(2), 45 (Feb. 1967).
8. T.P. Dornseifer, S.C. Kim, E.S. Keith, and J.J. Powers, *J. Am. Oil Chem. Soc.* **42**(12), 1073–1075 (Dec. 1965).
9. H.W. Lawson, *Standards for Fats and Oils,* AVI Publishing Co., Inc., Westport, Conn., 1985, p. 81.
10. M.W. Formo in Rd. 3, p. 212.
11. L. Boucher, *Snack Food Association Edible Oils Manual,* Snack Food Association, 1990, p. 37.
12. S.A. Matz, *Snack Food Technology,* AVI Publishing Co., Inc., 1984, p. 269.
13. R. Guillaumin in G. Valera, A.E. Bender, and I.D. Morton, eds., in *Frying of Food: Changes, New Approaches,* VCH Publishing, New York, 1988, pp. 82–89.
14. O. Smith, *Am. Potato J.* **38**, 265–272 (1961).
15. R.P. Mensink and M.B. Katan, *N. Engl. J. Med.* **323,** 439 (1990).
16. W.A. Gould, *Chipping Potato Handbook,* Snack Food Association, Alexandria, Va., 1989, pp. 18–29.
17. E.C. Lulai and P.H. Orr, *Am. Potato J.* **56**(7), 379–390 (1979).
18. H.A. Moser and H.J. Dutton, *Food Technol.* **4**(3), 105 (1950).
19. R. Vandaveer, *Corn Quality Assurance Manual,* Snack Food Association, Alexandria, Va., 1987, pp. 49–50.
20. p. 131 in Ref. 12.

11
Frying Technology

INTRODUCTION

Commercial deep-fat frying is estimated to be a $75 billion industry in the United States and at least double that for the rest of the world. These figures include the oils, fats, and shortenings used and the commercial and retail value of foods fried at the industrial and food service or restaurant levels. Supporting figures must be assembled from various sources. The Institute of Shortening and Edible Oils (20 member companies that together produce 90% of the edible oils and fats marketed in the United States) reported member companies shipped 12.245 billion pounds of product in 1985 (1). The categories were 4.684 billion pounds of shortening (solid and semisolid cooking and baking fats), 5.114 billion pounds of salad or cooking oil, 1.998 billion pounds of margarine oil, and 448 million pounds of oil used in other products such as confectionery, toppings, and candies. That major segment of the snack food industry that uses frying oils in continuous or batch fryers produced 2,896,600,000 pounds of products with a value of $7,128,300,000 in 1987 (2). These snack foods were mostly potato chips, corn/tortilla chips, nut meats, and pork rinds. These figures have grown at an annual rate of 4–5% according to yearly U.S. Department of Agriculture (USDA) statisics. Similar figures are available from meat and poultry groups, potato groups, and other trade associations, usually on an annual basis. Assembly of various data led to the $75 billion estimate.

Such a large industry has to consider regulation, quality control, and significant process and product development. Scientific principles are studied to both improve the process and the foods produced. The unit operation known as deep-fat frying is more a technology than a science. The focus of frying is to produce uniquely cooked foods. Deep-fat frying operations are the concern of food scientists, engineers, and technicians. Continuous and batch fryers are modified to obtain more efficiency and higher throughputs for the local process. Safety and ease-of-use issues are of concern to operators. Frying oil usage rates, energy costs, and product salability are in the domain of cost accountants. In practice, fried-food quality and process parameters are generally set in ranges of acceptability rather than absolute units.

The deep-fat frying cooking process uses fats and oils as a heat transfer medium to convey energy from heated fryer surfaces to the heating fluid (hot oil) and from the hot oil to the cooler surfaces of immersed foods. During frying, water is evaporated, and the exterior (the crust) becomes differentiated from the interior of the food. The heat transfer medium is also incorporated into the frying food or its coating by adsorption, absorption, and chemical interaction to form a crust that has its own unique properties. The amount of oil taken up by the food, whether it is simply absorbed/adsorbed onto the food or incorporated into the crust or a coating, is dependent on a number of factors including the food being fried, the time and temperature of frying, and the chemistry of the oil itself.

Some of the heat transferred to the exterior of the immersed food is in turn transferred by conduction to the interior of the food. The oil used to transfer heat energy is affected by various degradation processes during use that yield both the delicious odors, flavors, and textures associated with fried foods, plus a myriad of less desirable products that affect heat transfer efficiency and also the nutritional value of the incorporated oil.

The frying process exists solely to produce high-quality, uniform fried food for eventual or immediate food service. A process engineer regards the equipment and its components, the oil, and the food involved as a single system contributing to the acceptability of finished product. The system defines the deep-fat frying process. Chemical degradation of frying oils causes the frying process to become complex with respect to production of uniform food quality, regardless of set points. Degradation is brought about by a combination of heat, interaction with oxygen, interactions with the food, interactions with the fryers themselves, and reactions between the degradation chemicals. The degraded frying oil has an altered nutritional and toxicological profile when compared to fresh oil. The changed frying oil composition is reflected in the oil incorporated into/onto fried food, which is ingested by consumers. The ingested oil is studied to assess its potential for creating health hazards.

To understand deep-fat frying, the whole system must be methodically examined. One objective of study is to provide fryer operators, oil chemists, researchers, and others with an understanding of how and why food fries. The ultimate goal of current frying research is to shift from simply looking at the chemistry of heated oils to looking at how the quality of the fried food being produced is affected by the overall system, including the hot oil.

1.1 Benefits of Deep-Fat Frying

In direct contradiction to widely perceived needs by the public to limit dietary intake of fatty products, processing companies spend thousands or hundreds of thousands of dollars on new fryers from which they expect increased sales. Why make the investment? Thompson has provided a list of benefits to food processors achievable by frying (3):

1. *Taste and texture.* Fried foods taste good. Properly prepared foods have a good flavor, an excellent mouthfeel, and an agreeable texture. Three-dimensional items are crunchier on the outside than on the inside, and there are not many ways to obtain that combination.
2. *Set a coating.* Battered and breaded products are frequently only partially cooked at the plant. Frying, or flash (par or pre-) frying will set the crisp coating for future reheating.
3. *Impart color.* Frying will impart a pleasant golden brown color to coated products and can be used to brown all sorts of items from potatoes to turkeys.
4. *Oil addition.* Immersion in the frying oil results in the pickup of oil by the foods being fried. Oil pickup ranges from 1 to 2% with nuts to up to 40% with potato chips. In most products the added oil acts as a lubricant imparting a desirable mouthfeel.
5. *Convenience.* Fried products can be easily reconstituted in a fryer, a conventional oven, or a microwave oven.
6. *Blanch.* The temperatures of frying [(usually in excess of 177°C (350°F))] will blanch products. Traditionally, blanchers are used to inactivate enzymes, reduce intercellular air, reduce volume, and destroy some microorganisms. French fry producers consider parfrying to be a blanching operation.
7. *Inactivate microorganisms/destroy pathogens.* Frying temperatures will destroy some microorganisms, and some processes are designed to destroy pathogens. The USDA has required that hamburger patties be heated to an internal temperature of 71°C (160°F) to assure the destruction of enteropathogenic *E. coli.*
8. *Heat transfer.* The hot oil is an excellent heat transfer medium.

Consumers seem almost addicted to the flavor of fried foods and buy them in ever-increasing quantities. Logical arguments aside, the consuming public wants fried foods to be easily available, which clouds the issue of perceived benefits. The production, distribution, shelf life, variety, and ease of service of fried foods undoubtedly make them beneficial for a given economy.

The question of benefits to the consumer in eating a particular type of oil on/in fried food revolves around somewhat subjective assessments. Whether saturated oils, or hardened oils containing substantial amounts of trans acids, or oils containing high levels of monounsaturates or tocopherols are better or worse for an individual consumer are hard to pin down. However, the effects on consumer health, aside from obesity, due to consumption of frying oil degradation products is seen as a quantifiable issue of consequence to regulatory authorities in many countries. This is because individual isolated compounds may be shown to be toxic in animal models at specific, measurable levels.

The potential conflict between producer and consumer benefits of deep-fat frying has been brewing for nearly 100 years and currently remains unresolved.

1.2 How Frying Differs from Other Cooking Processes

From an engineer's viewpoint, frying differs from other cooking processes in a number of ways. These differences help to make frying an efficient and practical means to cook food (3).

1. Frying is accomplished in a very short time. Processes generally range from a few seconds in flash-fried products to a few minutes.
2. There is a large temperature difference between the oil and the product being fried. This makes for efficient cooking once surfactant levels have increased sufficiently to assure optimal contact between food and oil.
3. The size of fried products is usually quite small. Consider typical fried products such as snacks, nuts, nuggets, fish pieces, etc. The largest sizes are usually only 4 ounces or so. Large items may be fried, but these are generally cooked to impart color (browning).
4. The cooking oil becomes a significant part of the end product. Water and oil or oil and fat will be exchanged.
5. Greater degrees of crispness may be achieved through frying than with other cooking processes. Crispy texture or bite is very desirable.

1.3 What Happens During Frying

What happens to the food during deep-fat frying plays a major role in how fryer systems are designed and built (3–5).

1. Moisture is evaporated from the food. The product surface temperature is raised. Frying is in essence a dehydration process carried out in hot oil instead of hot air. The moisture must be removed from the fryer system and environment.
2. The product is heated to a desired temperature to achieve desired characteristics. This information must be established for each product.
3. The surface temperature of the product is raised to achieve browning and crispness. Degree of browning is affected by surfactant levels in the oil.
4. The products change dimensions. They can shrink, expand, or remain the same size.

5. Fat is removed from the fryer as product is processed and picks up oil, or in a few cases, fat renders out of the product (such as chicken) into the frying oil.
6. The system must be designed to replace the oil removed by the product, or drain excess oil rendered into the fryer by the product.
7. The products not only change in size but also in density, causing some products to float.
8. The oil chemistry and heat transfer capabilities change, resulting in changes in product quality (oil pickup, degree of browning, flavor, etc.). Oil chemistry also affects equipment by causing difficult-to-remove polymers to build up on equipment surfaces.

FRYING PROCESS AND SYSTEMS

There are several basic components of the frying system. These are the oil, the food, and the equipment, the latter of which includes the fryers, filters, ventilators, fryer feeds, and others. However, the most important issue that needs to be addressed from a commercial viewpoint (before marketing a new fried food) is to define the food product that is to be produced. That is, the processor or the foodservice operator must establish descriptive and quality parameters for the products he or she wishes to sell (6). For products such as frozen, prefried french fries or fish sticks bought by the military, there are well-established government purchasing specifications; but for the retail or institutional market, quality standards are usually set internally and perhaps with buyer input. Products are conceived by marketing, developed by research and development, and manufactured in the production facilities. It is the responsibility of research and development in concert with production to define those parameters that allow the product to be produced to those specifications established by marketing.

Without first describing the food quality arising from the process, the study of deep-fat frying becomes merely the study of the qualitative nature of the degradation products of heated and oxidized oils. The fascination with deep-fat frying is that it is a process that can be understood quantitatively and be made into an engineering application. Process engineering can optimize the processing equipment, the food, the oil, and the process parameters into an efficient system. Even the natural variations from food to food can be quantified.

If there is a first law of quality control for deep-fat frying it is "focus on the product being manufactured." This is ultimately accomplished by continuously monitoring the frying process and the products being produced and relating observed changes in fried-food quality to specific changes in the frying oil as it degrades within the process with continuing use. This empirical database will then describe, and place in a hierarchical order, the relevant chemistry and physics of the frying oil and of the food, so that meaningful quality control measurements can be made on both. Stauffer (7) lists some of these parameters as crispness, fat pickup, and color.

2.1 The Fryers

A fryer is an integral part of a whole system. The fryer manufacturer must understand the amount of food that is to be fried, how it is conveyed to the fryer, the type and condition of the food(s) being fried, and how it will be handled after frying. Designers need to understand the system. Reputable manufacturers will work closely with the client to assure that they get the right fryer for the work they want to do. An oversized fryer can be a disaster. Too little volume through a system results in poor efficiency, will severely damage oil, create cleanup problems, compromise product quality, and create operational problems. It is better to use two small systems than one large one that is not used to capacity.

Using industrial fryers as an example, consider the several basic components of frying equipment. These components are similar for both continuous and batch systems (3).

Oil holding kettle. Each system must have a source of oil. Most operations add fresh oil to the fryer during operation to assure that oil levels are proper.

Conveying system. There needs to be a means to bring food to and from the fryer. In continuous operations, the conveyor system moves product through the fryer. The type of conveyor needed is product dependent.

Heating and control system. Each fryer has a means to heat and control oil temperature. These may be direct or indirect heated systems or ones that utilize an external heat exchanger to heat the oil directly.

Hood or canopy. All continuous fryers are fully covered to conserve energy. Within the hood are ducts and blowers to remove steam and volatiles.

Hoist system. There is a means to lift the hood and conveyor from the fryer. These can be manual or mechanical.

Filter/treatment system. A system to remove solids from the oil is a standard feature on all fryers. Solids may be sinkers, suspended, or floaters.

Frying vessel. The hot oil is contained in a vessel of some sort in which food is fried. Fryers are either continuous or batch systems.

One point that must be emphasized involves multipurpose systems. Many processors try to get the most for their dollar, and want a system that can fry a wide variety of products. Whenever a fryer system is used for multiple products, that system will be compromised. A manufacturer can build a very efficient system. Once a system is asked to produce multiple products or product types, efficiency drops. It is nearly impossible to optimize a system for multiple products, but it very possible to do it for one product or product type.

2.2 Conveyance Systems

Foods to be fried may be classified into four groupings:

1. Surface fried—doughnuts.
2. Submerged fried—pellets
3. Partially floating—chips or tempura
4. Nonfloating—meat products or some french fries.

Each type of food requires a different type of conveyor system.

Conveyance systems must be designed to fit the product (8). With batch systems, a measured amount of product is metered into or dumped into the fryer or placed in a basket and submerged in the fryer. The product is then cooked for a set period of time and pushed from the fryer or the basket is dumped. Kettle-style potato chips and nuts are cooked in batch systems. Batch systems are also very common in small operations and in most foodservice operations, with the exception of fresh tortilla chip production in some Mexican restaurant chains.

In continuous operations, the conveyance system through the fryer must fit the product. For a product like doughnuts, which float, the product moves on a chain or in pockets and is flipped part way through the process. Potato chips are a similar case. Their buoyancy with increasing moisture loss causes them to rise, so turning is a necessity.

There are products, such as battered and breaded shrimp, that need to be fully cooked and must, therefore, remain in the oil. The problem is that these products may tend to float. Because of this feature the products are held under the oil as they pass through the fryer.

There are also products such as tempura or extruded potato coproduct that both sink and bob to the surface. These products are conveyed in a system that allows the products to rise and forces them under the surface part way through the process.

There are also multizone fryer systems that provide a different set of cooking parameters at points along the fryer. Multizone cooking is only possible with externally heated systems.

The products are fed into fryers as viscous fluids or as solids (or semisolids). Examples of fluid feed products are doughnuts, corn chips, or formulated potato products. These products are extruded and drop directly into the oil. The most common type of feed is when solid or semisolid products are conveyed into the fryer. These kind of products include french fries, fish sticks, meat patties, potatoes for chips, and most coated products. The solid products may be conveyed into the fryer or dropped directly into the system, which is what happens with kettle-style chips. (This latter product is sliced directly into the oil, and the presence of starches and sugars on the surface, plus the design of the fryer give these products their unique flavor and bite).

2.3 Heating Systems

There are three major design systems used in achieving heating and temperature control of the cooking oil. There are many determining factors that come into play when deciding what type of fryer may be selected. Among these factors are the type of product being produced, the equipment cost, space availability, manufacturer's experience, and environmental issues. There can be many more (3).

The three basic frying oil heating designs are direct heating, indirect heating, and external heating systems.

In *direct heating systems,* energy is provided by combustion of gas (natural or propane) or, to a lesser extent, electricity is used to heat the heat transfer surfaces. On rare occasions fuel oil may be burned for energy. The heating elements are immersed in the cooking oil directly below the product zone. Heating is achieved by firing these heating tubes (flues) or resistance elements. The heat is transferred directly into the oil by conduction from the heating elements. Temperature control is achieved by modulating the fuel input to the burners or switching the electrical elements on and off. A thermocouple or temperature probe in the oil is used to sense temperature. Immersion elements or flues can be run across the fryers or lengthwise (multiple burners) or be single-element S-shaped tubes. Foods processed in directly heated systems include kettle-stye chips, battered and breaded products (fish, meat, poultry, and vegetables), oil (dry) roasted nuts, doughnuts, and other snack foods.

Advantages and disadvantages of directly heated systems may include the following (3,5):

Advantages	Disadvantages
Compact package	No control of overall temperatures
Generally lower initial cost	Relatively high oil volume
Some temperature zoning	Limited heat capacity
Flexibility (multipurpose)	Lower thermal efficiency
	Fines settle on fire tubes
	Difficult removal of fines
	Lack of temperature uniformity
	Limited product packing depth

Indirect heating systems utilize an external heat source. A heater or boiler is fired with gas, oil, or electricity to heat a thermal fluid. Thermal fluids are usually chlorinated hydrocarbons. This thermal fluid is controlled at a set temperature above the target temperature of the frying oil. The oil is heated by circulating the thermal fluid using the same immersed heating tubes concept as in direct heating systems. Temperature control is achieved by controlling the heated thermal fluid entering the fryer's circulating fluid system. A three-way valve is used to modulate the fluid temperature in the heaters. Feedback from an oil temperature probe tells the valve to release thermal fluid into the fryer tubes or divert back to the heat exchanger. This allows a constant flow of fluid through the circulating pump.

Products processed in this indirect kind of system include battered and breaded products and oil roasted nuts. These systems are popular in Europe but are not used extensively in the United States. It is believed that there are concerns in the United States that oil may accidently become contaminated with the thermal fluid (3–5).

Advantages and disadvantages of indirectly heated systems may include the following:

Advantages	Disadvantages
More uniform temperature control	No control of overall temperature
Good temperature control	Limited heat capacity
No gas equipment in the cooking area	Fines settle on heater tubes
Very quiet	Difficult fines removal
High thermal efficiency	Limited product packing depth
Some temperature zoning	May be difficult to clean
	Chlorinated hydrocarbon thermal fluid

External heat exchangers use the frying oil itself to maintain fryer temperature. The frying oil is passed through a heat exchanger and returned directly to the fryer. These systems are variable. They range from those using gas, fuel oil, or electric heating to utilizing steam or thermal fluid to heat the oil in the heat exchanger, with steam to thermal fluid shell-in-tube or tube heat exchangers. Temperature control is achieved by inserting a thermocouple on the inlet side of the cooker thus controlling the fuel, steam, or thermal fluid to the heat exchanger.

Advantages and disadvantages of external heat exchangers may include the following:

Advantages	Disadvantages
Uniformity of temperature control	High initial cost
Extremely high heat capacity	Require more floor space
No gas equipment in process area	Pump/motor noise
Generally low oil volumes	
Highest thermal efficiencies	
Temperature zoning possible	
Auto C.I.P. systems usually included	
Able to match oil–product velocity	

The following two methods are in general use for controlling the temperature of cooking oil by the various heating systems:

1. *On/Off (or Hi/Lo/Off).* This is a simple system adopted where accurate temperature control is not required. Temperature control in the range of ±5.5°C (±10°F) is possible. The heat source is generally switched on or off by means of a thermostat switch connected to a thermocouple located in the oil to achieve temperature control.
2. *Fully modulating controllers.* These are much more common in high-production operations. A sensing thermocouple is installed in the fryer at a carefully selected location. The control source modulates the heat source to match the system's heat demand. These control devices are usually electronic and are capable of controlling to within a few degrees Celsius or Fahrenheit. These systems are more expensive, but as with many higher technology systems, the costs are dropping with increased utilization.

There are two methods for determining the *net heat load* of a product. The first is by calculation. To do this parameters such as cook time, oil temperature, oil absorbance, moisture loss, product input and output temperatures, plus other variables must be known. Information on similar products held in manufacturer databases can be used to make these calculations. Knowing these parameters, a calculation can be made to establish heat exchanger sizing (3,4).

The second method is to conduct a frying test. The goal is to simulate actual frying conditions in a controlled environment. Manufacturers use an already used oil for frying in these studies, which is closer to the real world. A fryer with a "window" is filled with the oil and the weight of oil determined. For greater control, the fryer may be mounted on a scale. The unit is then brought up to frying temperature and held at that temperature to bring the system into equilibrium. A test food of known weight is then placed into the fryer. Before placing the food into the fryer, the heater is switched off. The food is usually fixed with a thermocouple to measure center temperature. The food is held in the oil for the specified cook time, and ΔT is measured. The amount of water lost can be determined by monitoring weight loss of the system with the food in the oil. The food is then removed and weighed. The weight of the fryer is again checked to determine oil pickup. This simple study is repeated several times. This technique provides the manufacturer with the follow information: approximate heat load requirement, oil absorption, product yield, whether the product floats or sinks, the action of boil, and texture of the finished product, which can affect the design of the conveyance system.

Once the net heat load is established, the fryer size needs to be determined. The best pan loading (weight/ft^2) and cook time are then selected. Using the fryer dimensions, a gross heat load can be established. The style of fryer used allows determination of the heat transfer efficiency from which the firing rate of the system is determined. Manufacturers try to determine these values early on, so prospective customers can establish appropriate utilities in their plants.

This is a thumbnail sketch of a process that is integral to custom fryer design and manufacture. Computer programs allow rapid determination of these values. All manufacturers design some reserve into their systems for product variability and startup needs.

2.4 Styles of Frying

There are different styles of frying. Flash frying is a quick fry used to parfry or prefry materials. Frying times are as short as 20–30 s on products such as pellets or small breaded products. Cook times may range up to 50–60 s to flash fry a piece of "bone-in" product. Final frying may be carried out in a finishing plant or in a restaurant or institutional operation. Battered and breaded products prepared from frozen blocks (fish) are often coated and flash fried. Rapid handling allows the coating to be set without thawing the fish, allowing more efficient use of energy downstream in the freezer.

Fully fried products are completely cooked. These include snack items and many meat products, such as hamburger patties. The size and weight of the products and maintenance of these parameters are essential to assure proper cooking, not only to assure food quality but also to assure microbiological food safety (9). Many of these products are prepared for reconstitution in the microwave oven, in the convection oven, or in a fryer.

2.5 Frying System Care and Maintenance

Maintaining the quality of the oil, and therefore the product, is a function of several factors. Robertson (10,11) proposed the following principles for frying oil maintenance. These principles, when consistently applied, will assure proper frying system care.

Proper Design, Construction, and Maintenance of Equipment. As with any process, frying is most efficient when using the proper system. Most industrial fryers are built to specific design specifications. Foodservice fryers may also be designed to produce specific products for retail sale. To ensure the best use of a system, it should be configured to the product to be manufactured and maintained according to the manufacturer's instructions. Purchasing equipment that can be used for multiple products or styles of products may appear to be less expensive initially, but it is extremely difficult to operate such systems at maximum efficiencies.

Proper Operation of Equipment. All fryers should be operated according to the manufacturer's instruction. They should also be operated according to established specifications. If the process development programs have determined that optimum product is produced by operating the system within set parameters, those parameters must be followed. A common error by fryer operators is to increase the oil temperature as a means to counteract problems. In most cases, all this does is further damage the oil and subsequent product quality. One of the major problems that occurs in industrial frying is caused by operating fryers at less than their designed capacity. In cases such as this, too much heat is pumped into the oil resulting in the formation of large quantities of thermal polymer. The resulting polymer formation compromises oil life, increases the time and energy required to clean a system, and can jam the system.

Proper Cleaning of Equipment. Failure to clean fryers properly can damage oil and thereby affect product quality. It can also reduce the oil's usable life and create worker safety problems. Cleaning protocols are usually recommended by the fryer manufacturer, but they can also come from suppliers of the cleaning materials. The recommended cleaning protocol for a deep-fat fryer entails the following steps (12):

1. Drain the fryer.
2. Rinse the fryer to remove particulates.
3. Boil out the unit with a caustic cleaner.
4. Drain and neutralize the caustic cleaner.
5. Rinse the fryer with an acidic solution to remove residual caustic.
6. Rinse with water to remove residual salts and soaps.
7. Be sure the fryer has been drained after cleaning.

If necessary, workers may have to scrub residual polymer from the side walls, heater tubes, and other areas taking care not to scratch the surfaces. If the fryer design is such that there are parts of the system that are not cleanable (unreachable), oil quality could be compromised.

A serious problem with fryer cleaning is improper removal or neutralization of residual caustic remaining on or in the fryer. In the presence of water, the metals (cations) in the caustic react with free fatty acids in the oil forming alkaline soaps. The soaps dissolve most readily in oils, which are already slightly oxidized and polar. Soap left in the oil causes water evolving from the food to become emulsified in the frying oil. The emulsions change the heat transfer properties of the oils, and affect the sensory properties of the fried food with respect to taste and color spotting.

Soaps are strong surfactants and act to accelerate incorporation of oxygen into the frying oil. Their presence increases the rate of oil degradation, and, in high enough concentrations, adversely affects the quality of the food being produced. These materials bind water in a reverse micelle (polar cation heads inward, and nonpolar lipid tails outward around a spherical cavity holding the water) as described by Vold and Vold (13), which serves to enhance the rate of chemical reactions.

Minimum Exposure to Ultraviolet (UV) Light. Ultraviolet light will catalyze the degradation of triglycerides. Double bonds of the unsaturated fatty acids are attacked to produce by-products that act as prooxidants and can adversely affect product quality. The resulting free radical reactions produce undesirable off-flavors and can compromise shelf life (14). Frankel (15) shows some of the oxidation reactions that occur as a result of exposure to UV light.

Elimination of Salt and Other Metal Sources from the Oil. Like UV light, metals are strong catalysts and prooxidants. They also accelerate oxidation of the double bonds, producing the same and often more undesirable compounds. Sedlacek (16) reported that frying in sunflower seed oil in the presence of heavy metals showed the following reactivities:

$$Zn < Ni < Pb < Co < Cu < Mn < Fe$$

Thomson (3) indicated that the primary metals of concern with fryers are copper and brass, particularly if workers accidentally replace steel fittings with those made of brass. Looking at the practical effects of both heavy and transition metals on oil degradation (6), the progression of most to least reactive determined by in-field consulting is

$$Cu > Brass > Fe > Zn > stainless\ steel > Mg > Ca > Na$$

Regular Filtering. Jacobsen (17) describes the importance of removing charred pieces and breading materials, because these materials darken oil, contribute bitter flavors to food, impede heat transfer, and ruin the appearance of fried foods. Filtration can also help to extend oil life, as these particles may also act as prooxidants, speeding the rate of oil degradation.

Use of Chemical Markers to Measure Oil Degradation. Each of the six previous principles has been established to maintain oil quality. As has been stated, the key to success for a food processor, restaurant operator, or foodservice operation is the food quality. These processors need to monitor certain oil parameters as a means of assuring the quality of the food they are producing, or in other words, producing food to predetermined quality specifications. Quality control in deep-fat frying is a series of activities aimed at monitoring and controlling a dynamic organic system (the frying oil) to produce a given end product. To assure finished product quality, there are many different chemical markers and physical markers that have been and may be used. These vary for the product and frying system and will be discussed at length later.

2.6 Frying Oil

Almost all varieties of fats and oils have been or are used for frying. These include vegetable oils, hydrogenated vegetable oils, animal fats (tallows and lards), animal–vegetable blends, and margarines or shortenings. The type of frying fats used once depended almost exclusively on where people lived. For example, residents of the Mediterranean basin used olive oil or other liquid vegetable oils, whereas the residents of Northern Europe utilized fats of animal origin. Typical characteristics of two often-used cooking oils are summarized by Krishnamurthy (18).

Analytical Test	Cottonseed Oil	Peanut Oil
Iodine number	108	95
Refractive index at 60°C	1.4572	1.4550
Free fatty acids expressed as percent oleic acid	0.03%	0.03%
Smoke point °C (°F)	226.6 (440)	226.6 (440)
AOM hours (to PV = 125)	10	12
Lovibond color (Y/R)	20/2.5	25/2.0
Cold test [hours to cloud at 0°C (32°F)]	<1	<1
Cloud point (ASTM) °C (°F)	3.3 (38)	4.4 (40)
Solid point (ASTM) °C (°F)	−2.7 (27)	−1.1 (34)
Titer °C (°F)	36.0 (96.8)	31.3 (88.3)

Lawson (19) emphasized the importance of selecting the proper shortening for deep-fat frying. He noted that a good frying fat should have the following features:

1. No contribution of off-flavors to the food.
2. Long frying life, resulting in economical operation.
3. Ability to produce an appetizing, golden brown, nongreasy surface on the foods during its frylife.

4. Resistance to excess smoking after continued use.
5. Ability to produce foods with excellent taste and texture.
6. Resistance to gumming (polymer formation), which helps keeps equipment clean.
7. Resistance to rancidity.
8. Uniformity in quality.
9. Ease of use, which includes both form (fluid, solid, etc.) and packaging.

Not all frying fats can be used for all applications. Product development must focus on finished product quality when selecting frying fats. Schaich (20) addressed this issue in a discussion of heat capacity, crystallinity, and relative unsaturation of the frying media. Methods for determining the quality of fresh (unheated) frying oils can be found in analytical compendia such as produced by the Association of Official Analytical Chemists (AOAC) International (21).

2.7 Passive and Active Filters/Treatments

One of Robertson's quality principles was to filter oil regularly (10,11). Jacobsen (17) described how it is essential to remove charred batter and breading materials, commonly referred to as cracklings, because these materials can darken oil, contribute bitter flavors to food, impede heat transfer, and ruin the appearance of fried foods. These particles also continuously leach their components into the oil, which chemically degrade the frying medium. Finally, the particulates serve as reactive sites for degradation reactions.

Jacobsen (17) also described the role of passive filter aids, which are used for "polishing" a frying oil. These materials or systems, which perform a polishing function, remove particles with still reactive surfaces but do not remove specific oil-soluble chemical entities, which promote oil degradation. The frying oil quality curve (Figure 11.1) shows the five stages of oil degradation and relates them to food quality (22,23).

For the moment, assume that this curve represents changes in oil food quality over time for a pierogi fried in a continuous fryer with no filters. Simple filtration will remove particulates, such as flour and charred materials, and the optimum frying region width will be increased.

By using an active filter aid or system, the optimum frying area may be even further extended (Figure 11.2), pushing the zones further out at the degrading end of the curve, thus giving the producer a wider window to produce high-quality products (24).

2.8 Filtration

Most filter systems that are in use or have been used in the industry over past years are systems that are now considered passive filters or systems. These

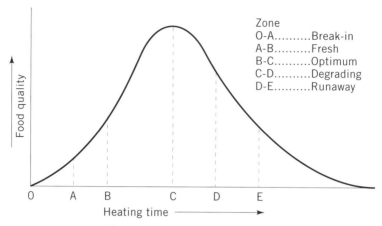

Figure 11.1 Frying oil quality curve.

systems include the metal screens, rolling (indexing) paper filters, paper cones, plastic cloths, canvas on plate and frame systems, systems using diatomaceous earth, and leaf filters (25). All of these systems are routinely used in frying operations. They remove only the insoluble particulates of certain sizes and have little, if any, effect on the actual chemistry of the frying oil.

Kubose (25) describes three of the basic processes by which frying oils are filtered as absolute filtration, depth filtration, and absorption.

Absolute filtration (sieving) works by sieving or screening. Particles of a certain size are trapped on the sieve and retained. They will not pass through the filter, unless the filter ruptures or is damaged. Filter papers work by sieving and are considered passive filters capable of removing particles down to about 5 μm.

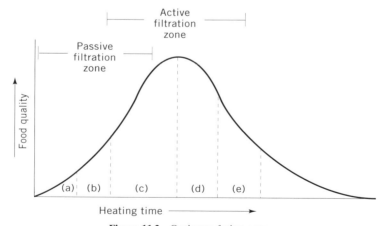

Figure 11.2 Optimum frying area.

Depth filters consist of a relatively thick layer of porous materials containing irregular and torturous channels. Particles are entrapped within these matrices. Depth filters have greater porosity than absolute filters and are not "blinded" (reducing or shutting off pores), thereby shutting off flow through the filter (26) as easily as absolute filters. Systems using diatomaceous earth may be considered depth filters. These media can remove particles down to about 1 μm.

There is a "new" breed of systems that have been developed recently. They remove specific oil-soluble chemical compounds from heated oils. These are called *active* filters or systems, and they remove or trap not only the particulates but will alter the chemistry of the frying oil by removing soluble components in the oil, thereby decreasing the overall rate of degradation.

Absorption is the final type of filtering for frying oils described by Kubose (25). With these systems, an interaction occurs between the filter material or component of the filter material and a particulate or compound within the frying fat. The materials are then bound to the filter. This allows the removal of minute particulates and even specific molecules. An example of this type of system is a molecular sieve-like clay or zeolite, which can entrap molecules. Another type of system is a filter pad designed to remove soaps. These media can selectively remove specific molecules from frying oil.

There are also two other kinds of reactions that occur in filtering oil; adsorption and phase separation/emulsion breaking. Adsorption may be described as a "microadhesion" system. In these systems, the materials being removed are trapped on fine particles, such as loose silicas or aluminas. Phase separation/emulsion breaking may be described as acidifying an oil so a microemulsion of water becomes destabilized and separates out as water at the bottom of a vat of oil.

2.9 Benefits of Filtration

Jacobsen's work (17) highlighted some of the potential benefits of filtration and Figure 11.2 shows how the installation of active filters may improve the performance of a frying oil by extending the optimum frying region, and the food produced in that oil will be uniformly good for an extended period. These are not the only benefits processors can obtain when they install systems with active filter capabilities. Blumenthal (23) has cited other areas where active filter materials may potentially enhance operations.

Benefits of Active Filtration or Treatment.

1. Reduced energy usage.
2. Improved food quality.
3. Oil life extension.

446 Frying Technology

4. Reduced oil usage.
5. Enhanced shelf life.
6. Reduced downtime.
7. Reduced cleanup time and costs.

Jacobsen (17) mentioned that one of the roles of filtration was removal of particulates, which could impede heat transfer. Active treatment can remove surfactant materials, which attach to heater surfaces and are carbonized (become coke) on these surfaces. Coke on heater surfaces reduces heat transfer by acting as an insulating blanket.

Improved food quality may also be achieved. Figure 11.2 demonstrates how this occurs. By simply maintaining oil quality at a high level for a longer time, the window for producing top-quality food is enlarged. Potential quality improvements for the industrial fryer are enhanced shelf life, reduced oil pickup, improved adhesion of coatings, and reduced stickiness of formulated products. For restaurant fryers, beneficial quality factors are reduced oil pickup, better adhesion of coatings, and longer life under heat lamps

Maintaining oil at a higher quality level not only will improve food quality, it can reduce oil usage. Chain restaurants, in-store bakeries, and delicatessen operators have installed active filter systems in their fryers and have reported to our laboratory that they are both using and purchasing less oil, with no adverse effects on food quality. These savings come from a combination of longer oil life and reduced addback.

Industrial fryer operators, particularly those producing battered and breaded products, are often plagued by product sticking to the conveyor belts, which results in product being dragged back through the continuous fryer. Once sticking occurs, the system must be shut down and the belts cleaned. Active filters have the potential to remove the materials (surfactants) that contribute to this expensive problem. For an industrial producer, the loss of even a day or two of production in a year makes this problem terribly expensive.

The last potential benefit is reduced time and expense for cleanup. With tightening environmental regulations, which limit what and how much can be used (much less can be put down the sewer), cleanup procedures are becoming more and more expensive. To this, add the usual push from management to produce more product in a shorter time, and there is a situation that appears to be an unresolvable conflict. Use of active filter materials cannot only reduce the frequency of cleanups, but they can reduce the amounts of caustic fryer cleaning compounds needed. By eliminating surfactants and other precursors of polymers (the sticky and rubbery materials that plate out onto fryer surfaces from the oils), active systems reduce the amount of polymer formed from the frying oils. It is the polymer that is so hard to clean, requiring caustics for removal.

2.10 Disadvantages of Active Treatment Systems

The active filtration systems or treatment products have great potential for enhancing operations, but there are a number of problems associated with them. These concerns include (27):

Disadvantages of Active Treatments.

1. Leaching of, or presence of, free-form foreign powders into oil.
2. Leaching of metals into oils.
3. Lack of good filtration equipment.
4. Legal issues.
5. Capital costs.

Many of the active treatments are free-form powders. These materials include diatomaceous earth, magnesium and calcium silicates, rhyolitic and pumice materials, zeolites, and aluminas (22). These materials may be added directly to the frying oil, after which they must be filtered out; or they are used in a filter system, whereby used oil passes through the material and is "polished." These materials may not be totally removed from the system or they may leach into the fryer during filtration. The presence of these free-form materials in oil can compromise food texture and could be construed as an undeclared additive on served portions of food.

Several of these first-generation active filter materials described by Blumenthal (22) were designed to lower the concentration of free fatty acids in oils. This was accomplished by the use of filter materials containing metal ions, such as calcium or magnesium. In the presence of the small quantities of water found in frying fats, these metals reacted with free fatty acids to form alkaline soaps (8,22,28), which are more detrimental to the oil than the free fatty acids themselves. The metals also catalyze other reactions, which can further degrade oils. The residual silicic acid is believed by our laboratory to support undesired foaming of oil.

Reaction of Free Fatty Acids with Metals to Form Alkaline Soaps (28).

$$CH_3(CH_2)_7CH=CH(CH_2)_7\underset{\underset{O}{\|}}{C}-OH + Na^+ \text{ silicate}$$

(oleic acid)   (sodium silicate)

(in the presence of water)

$$\Downarrow$$

$$CH_3(CH_2)_7CH=CH(CH_2)_7\underset{\underset{O}{\|}}{C}-ONa + \text{free silicate}$$

(sodium oleate)   (silicic acid)

448 Frying Technology

One of the major roadblocks to efficient filtration of oil, independent of the type of filter medium, has been the available equipment. Most of the filter equipment in use today for oil filtration (e.g., plate-and-frame or pressure leaf systems) were originally designed for aqueous systems. They were not designed to filter oil at 190.5°C (375°F). When one sees these systems in industrial operations, they are all too often leaking. There is a great need in the industry for equipment that has been designed to filter nonaqueous materials at high temperatures. Filter systems, old or new, are also rather expensive. Many companies are unwilling either to install or to spend any money for filtration equipment. Others, who have systems in place, do not want to spend the money to retrofit or modify a system to handle the active filter materials or systems (27).

The final disadvantage of active filter systems, especially those employing free-form powders is the legal one. If the materials or their breakdown products get into the oil, they can be considered an additive or an adulterant. These materials may also be considered as not being food grade in the context of their use with hot oils. Finally, because many of the materials are silicates, there is a fine line between these materials and what may be construed as being glasslike. For example, regulatory agencies in both Japan and France have taken the position that these materials are, indeed, glass and will not allow them to be used in fryer systems. In fact, all countries are unhappy with chemical leachates, such as phenolic binders from paper and nonwoven materials, and plasticizers from cloth.

2.11 Selection of an Active Treatment System

Whether a company decides to simply passively filter its oil or use an active filter material or system is an individual decision. The benefits of filtration have been well established, but the type of approach selected must be weighed against the costs of the system and the benefits it will bring to the user (26). If a company decides that active filtration or treatment would be useful, it should not simply jump into the first system that comes along. Active systems can be very beneficial, but selection of the wrong system or materials can create even greater problems.

Before buying and installing any filter system, and in particular an active system, potential users should follow the path described now. The first step is to gain some understanding of the existing oil system and operations. This will help the user determine potential benefits, both operationally and financially. Potential users should understand their oil usage, their fryers and controls, the economies of operation, and how their oil degrades. The last point, oil degradation, is the key to the program. Potential users should fully characterize their fresh oils, then embark on a program to determine how the frying oil degrades over time during normal operations. The following analyses should be conducted on the fresh oils (27):

Percent polar materials
Percent free fatty acids
Alkaline [water emulsion titratables (WET)] materials (ppm)
Fatty acid profile [gas chromatography (GC)]
Metals (ppb)
UV/visible spectrum
Polymers [high-pressure liquid chromatography (HPLC)]
Lovibond color
Peroxide value
Anisidine value
Iodine value
Infrared spectrum [Fourier transform infrared (FTIR)]

Once the fresh oil has been characterized, and frying is initiated, samples of the heated oil should be collected at specific, meaningful intervals that are related to transitional events in the heated frying oil.

Having this information on how an oil degrades chemically and merging it with data on amount of food fried, oil added to the system, and operating parameters will provide the fryer with a baseline for oil degradation. Once this information is in hand, the effects of oil treatment can be properly determined. The food producer should also take note of food quality, because one of the potential benefits of treatment is just that, enhanced food quality.

Once baseline data have been developed, pilot studies or controlled plant trials to evaluate the product or systems should be initiated. These studies should be conducted in a similar fashion as described earlier. Using the same format, the cumulative effect of the system can be determined and compared directly to the baseline data for oil and food quality. These studies will also help potential users determine if and how the product may benefit their operation.

One thing that potential users of active treatment materials or systems must realize is that these will not rejuvenate severely degraded oil. They can clean it up somewhat, but that is not why they were developed, or how they perform.

NEW PARADIGM

Many authors have written about the effects of frying processes on foods and oils. There is a great body of work in which researchers have isolated and identified countless compounds in oil, yet very few of these hundreds of compounds have been shown to play any role in the actual frying process. Major reviews by Artman (29) and Belitz and Grosch (30) describe the organic chemistry and reaction mechanisms of lipids, fats, and oils and their degradation products. Contributors to the symposium collection edited by Varela and co-workers (31) documented many details about frying and fried foods.

Individual descriptions and analyses of these effects of frying processes can be found in articles on potato chips/crisps (32), flavor volatiles (33), polymers (34,35), oil quality control techniques (36,37), and oil oxidation products (15,38).

A few authors (39,40) have discussed causes of changes in foods due to changes in the oils. However, there has been no working model that explains why food fries, which is really the key to understanding frying. Frying is considered by many to be more of an art than a science or technology (41). Addressing this issue using research programs employing the essentially mechanical, though mathematically rigorous, disciplines of physical chemistry and engineering approaches can provide dramatic insights as to how and why foods cook, rather than char or burn, in hot oil; how heated oils and their degradation products interact with fried foods; and how to optimize the process of frying. Once the actual mechanics of the frying system have been established, it becomes possible to more easily control that system, yielding higher-quality products and improved operational efficiencies. This is the new paradigm described by Blumenthal (22).

Model systems, conceptual thinking, and simplistic drawings are helpful in understanding the important features of laboratory or commercial frying. A water-laden sponge may be used as an analogous system for food undergoing frying. The sponge has an interior volume, edges, walls, and a uniform distribution of water in an evenly divided matrix of cells. The simplest real-life analogy to this idealized model system is a french fried (chip) potato, which will be used in the following discussion as a model food undergoing frying.

3.1 How and Why Foods Cook in Frying Oils

A discussion of how and why foods fry in hot oil involves both mass transfer and heat transfer

Mass Transfer. Water in a frying (dehydrating) french fry migrates from the central portion radially outward to the walls and edges to replace that which is lost by dehydration of the exterior surfaces. The term "pumping" is used to describe this migration; i.e., water is pumped by an imaginary machine from the interior to the exterior. This pumping phenomenon is described by Lydersen (42) and Keller and Escher (43) and is easily observable in the laboratory by examining cross sections of food pieces as they fry over a period of time.

Therefore, it is fair to use mass transfer calculations to derive an initial simulation for the frying of food, based on the overall rate of loss of water, as well as the relative ease of migration of water through a dehydrating sponge matrix and from walls and edges. These conditions define the kinetics and dynamics of the mass transfer of water in and from a frying food, an exercise beyond the scope of this Chapter.

Three quantitatively smaller mass transfer phenomena will be addressed later: the uptake of frying oil into the food "sponge"; the small (but crucial to food product quality) movement of water-soluble food materials to the exterior as water is pumped out during frying; and the leaching/rendering of liquefied food components from the food.

Heat Transfer. Water plays a number of roles in the transfer of heat into the sponge. First, it carries off thermal energy from the hot frying oil surrounding the frying food. This removal of energy from the food's surface prevents charring or burning caused by excessive dehydration. The conversion from liquid water to steam as the water leaves the food carries off the bulk of the contacting oil's energy. As long as the water is leaving, the food will not char or burn. Therefore, although the temperature of the oil may be 180°C (356°F), the temperature of the food frying is only about 100°C (212°F), representing the temperature of the change in phase from water to steam. This, too, can be easily demonstrated by locating microthermocouples in the food, the oil, and at the oil–food interface. Subsurface water also conducts heat energy from the surfaces contacted by hot frying oil to the interior. (Water is a better conductor than the fat, protein, and carbohydrate portions of food.)

The final function is to cook the interior of frying food. In the french fry model system, this means gelatinizing the starchy interior. Sufficient heat (thermal energy) must be transferred to bound water to begin swelling the starch, but not enough to strip the water from the starch gel, which leads to collapse of the interior structure, a phenomenon that may be seen in a food that has been overcooked or that has been fried in a badly abused oil. Hallstrom and co-workers (44) describe models for understanding heat transfer in food systems.

Cooking foods that are more complex than the french fry model often requires delivery of a large amount of energy to a core area by conduction through water and food solids. The energy flux is provided by long immersion times in hot frying oil, though efficiency in heat transfer from the hot oil to the frying food is not simply related to time and temperature.

3.2 How Heated Oils Interact with Foods

The factors affecting how heated oils and their degradation products interact with food include the food, the oil, water and oil interactions, surfactants, and oxygen.

Food. As an example, all french-fried potatoes should be cut to the same dimensions, i.e., with a fairly square cross section and uniform length. Each unit will have the same fixed, average surface-to-volume ratio, with the same chemical composition. In the real world, there are differences between lots and batches and so forth (e.g., storage and nonstorage products), but for the simulation of the frying process, we can assume that the chemical composition and the cross section are the same for each piece. Then each of these volume, mass, and compositional characteristics can be considered constant for a given batch of potatoes.

It follows that the specific heat (otherwise known, for water systems, as heat capacity) and the thermal conductivity (the rate at which energy will move through the section) are constants. Therefore, for a large lot of french-fried potatoes, the variations in finished food quality, for statistically reliable samples, will be due only to changes in the frying oil.

If there are no changes in the processing equipment, and no changes in the type or volume of oil in the fryer, all the changes that are observed in cooked (finished) food quality are really due to the changes (degradation) in the frying oil. In actual frying operations, this assumption is very close to real-world operating parameters.

This concept explains why raising the temperature in an oil does not cook the food to set specifications more quickly. Higher temperatures and longer immersion times simply distort the desired heat and mass transfer effects necessary for the production of food of a specified quality. Lowering the frying temperature to balance the rate of surface water loss with the inherent thermodynamics of food and oil systems is most efficient. In fact, excess energy in the oil is converted into crosslinks, leading to case hardening of food and polymer formation in the oil.

Oil. Frying oils should be studied to learn to control the process and the resulting quality of the fried food simultaneously. This, unfortunately, has not been the case in the past. Oil chemists and other researchers must remain aware of the fact that it is the food that people are buying, and consuming, not the oil itself. Frying oils have bulk properties that change with use (45,46). Initially, the oil has a high heat capacity that is diminished with use. This means that there is a "thermal flywheel," or energy sink, at a high level in a fresh (unused) oil. As the oil degrades, the heat capacity decreases and the flywheel "slows down" to a low value.

Other properties start low and end up with high values; an example is viscosity (47), which increases as a result of dimer and polymer formation. The polar material content of an oil, the sum total of those materials that are not triglycerides, also goes from low to high (37). In addition, the initially low color or ultraviolet absorption spectrum dramatically increases, although the quantitative amount of chemicals causing the increase in color is very small. And there is an initially low thermal conductivity that rises drastically with use.

The quality of the oil as a frying medium and the quality of the food produced in it are intimately bound. Blumenthal (8,23,24,48) described five phases that an oil passes through during the degradation process, which was illustrated in Figure 11.1:

1. *Break-in oil* White product; raw, ungelatinizated starch at center of the fry; no cooked odors; no crisping of the surface; little oil pickup by the food.
2. *Fresh oil* Slight browning at the edges of the fry; partially cooked (gelatinized) centers; crisping of the surface; slightly more oil absorption.

3. *Optimum oil* Golden-brown color; crisp, rigid surfaces; delicious potato and oil odors; fully cooked centers (rigid, ringing gel); optimal oil absorption.
4. *Degrading oil* Darkened and/or spotty surfaces; excess oil pickup; product moving toward limpness; case-hardened surfaces.
5. *Runaway oil* Dark, case-hardened surfaces; excessively oily product; surfaces collapsing inward; centers not fully cooked; off-odor and off-flavors (burned notes).

These results have been confirmed by other researchers (49).

Water and Oil. The interplays of water and oil near a food can be explained, as in Miller and Neogi (50). Water converted to steam carries off the latent heat of vaporization, an effect moderated by the rate of water migration from the interior to the exterior. The faster the water leaves the surface, the faster the water will move through the matrix, limited by the mechanics and capillarity of the matrix itself.

Oil transfers sensible heat by surface contact and then by capillarity into the open pores of the surface from which steam has rushed out. This effect is moderated by the viscosity and surface tension of the oil. The contact between the food and the oil progresses through the different stages already described. A new oil has to be "broken-in." This means that there is very little in the oil that will allow the oil to cling to the exterior of the food against the gradient of steam escaping. The actual hot oil's contact time at the surface is only about 10% of the food immersion time (unpublished data), meaning that only a limited contact time is available for the energy in the oil to transfer into the heat sink of the colder food. As the oil continues to be used, the contact time increases to about 20%, then optimally to about 50%, then degrading 70–80%, and then in runaway oil, where all the systems that are normally in place for producing a food are out of control, nearly continuous. This, too, can be demonstrated using microthermocouples.

Surfactants. The causes of increasing oil contact with the watery food are twofold: water-activated surfactants and lipid-activated surfactants (45). The derivation of this "surfactant theory of frying" may be more easily seen by applying the following assumptions, some of which have already been discussed.

1. Frying is basically a dehydration process. When a food is fried, water and materials within the water are heated and pumped from the food into the surrounding oil (31).
2. The heat transfer medium, i.e., the frying oil, is a nonaqueous material, and food, for all intents and purposes, is almost all water (42,43). Oil and water are immiscible.
3. For frying or cooking to occur, heat must be transferred from the nonaqueous medium, the oil, into the mostly aqueous medium, the food.

4. Any changes in heat transfer or cooking ability of the oil, including at the oil–food interface, must result from degradation products formed as a result of breakdown or interaction of the oil.
5. The food materials leaching into the oil, breakdown of the oil itself, and oxygen absorption at the oil–air interface all contribute to change the oil from a medium that is almost pure triglyceride to a mixture of hundreds of compounds (29,36).
6. Those materials that affect the heat transfer at the oil–food interface must act to reduce the surface tension between the two immiscible materials. These materials act as wetting agents and are regarded as surfactants (50,51).
7. As the oil degrades, more surfactants are formed, causing increased contact between food and oil. This causes excessive oil pickup by the food and an increased rate of heat transfer to the surface of the food. Eventually, excessive darkening and drying of the surface occur, while conduction of heat to the interior is a constant and cannot be sped up by changes in the oil.

3.3 The Theory of Frying

This theory of why food fries may be stated as: "Surfactants are responsible for the surface and interior differences in fried foods as induced by aging oils" (22).

Stern and Roth (52) surmised that breakdown products of oils could be forming and acting as surfactants, but they did not have the analytical capabilities to follow up on this observation. Blumenthal and Stockler (53) isolated and identified a water-activated surfactant, sodium oleate (soap), in a very polar fraction of frying fat. Water-activated surfactants are soaps, phospholipids, and inorganic salts. The lipid-activated group includes low-polar thermal polymers, which are carbon-to-carbon linked, and high-polar oxidative polymers.

Soaps (metal salts of fatty acids) form inverted micelles (13), i.e., a water-in-oil emulsion, exactly the opposite of mayonnaise, which is an oil-in-water emulsion. Because of the formation of these micelles, used frying oil may contain 0.5–1.5% entrained water at frying temperatures (24).

A french fry, frying in an oil bath, releases water that steams and forms bubbles at the air–oil interface. In a break-in oil, there is a very high interfacial tension, and small bubbles form and break rapidly on the oil's surface. Polymers formed in the oil are at very low concentration and cannot help strengthen the oil film of the steam dome. In fresh oil, there are also small steam bubbles forming when frying any food, but they persist longer than in oils with only a trace of polymer. In optimum oil, with a moderate interfacial tension, medium-sized bubbles result. In degrading oil, with low interfacial tension, large persistent bubbles are observed. In runaway oil, with very low interfacial tension, very persistent, stacked bubbles, both large and small, are formed; i.e., foaming occurs. The presence of large amounts of polymer in well-used oils provides the elastic, sheet-forming, strong component of the steam domes or bubbles. It is conjectured that oxygenated polymers behave more like soaps in that they are also wetting agents.

At low concentrations of surfactants in frying oil, little oxygen is introduced into the oil. At moderate surfactant concentrations, oxygenation produces chemicals, such as oxidized fatty acids, which produce good heat transfer properties in the oil and desirable volatiles. At high surfactant concentrations, incorporation of oxygen proceeds at relatively high rates, so oil degradation dynamics and kinetics are forced to production of short-chain fatty acids. At this stage, flammable ketones and ethers are formed, and polymer deposition onto the fryer vat walls and conveyor or basket surfaces increases noticeably.

Low surfactant concentrations produce low absorption of oil into the food, and little cooking of the exterior or interior of food. Moderate concentrations produce normal absorption of oil into food and proper cooking of the exterior and the interior. High concentrations produce oil-soaked food with over-cooked exterior and undercooked interior.

Oxygenation. A gas such as oxygen does not stay dissolved in a fat. The higher the temperature, the less soluble the gas. Oxygen apparently reacts with oil components in the thinly stretched (activated) films of the bubbles and at the oil meniscus with air. A measurement of the absolute volume or the absolute surface area of the bubbles should suffice to calculate the actual kinetics of oxygen uptake in frying oil. When the steam domes break and collapse, they drag oxygen into the bulk of the oil, where the oxygen is free to react with various chemical species.

3.4 Relationship Between Food Quality and Oil Quality

There is a direct relationship between food and oil quality, and understanding this relationship is one of the keys to process optimization. As an oil degrades, the quality of the food being produced in that oil changes. It improves initially and then begins to degrade. This may be seen in Figure 11.1, the frying oil quality curve (22,23). As an oil degrades from "break-in" to "runaway," its ability to produce high-quality food is altered. These changes in food quality are reflected in the changing chemistry of the frying fat. With the goal being to produce high-quality food, the processor or foodservice operator should strive to keep oil at the top or optimum part of the quality curve for as long as possible. This curve may be applied to all kinds of frying systems as may be seen in Figure 11.3 (23). The differences between the curves is primarily due to the rate of addition and total percentage of makeup oil added to the fryer.

456 Frying Technology

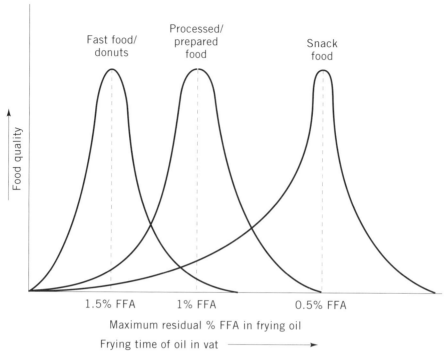

Figure 11.3 Family of frying quality curves.

3.5 How to Optimize Frying

Optimization of the frying process allows simplification of control and documentation, which can yield benefits such as increased frying oil life and reduced oil pickup into food, reduced product reject rates through tightened process control specifications, greater efficiencies in cleanup (reduced use of caustics and shortened times), energy conservation, and ultimately reduced operating costs.

Simulation of the process to be optimized requires that the most important elements in the process be modeled:

Model Foods. There are basically three surface-to-volume ratios for foods being fried. The first model looks like cotton candy or a ball of cotton, which is all interior volume, with a wispy external surface and no possibility of crust differentiation; an example is the food beneath batter and breading coatings, such as chicken or fish (muscle tissue). The second model has a significant interior volume, with a significant external surface and good crust differentiation; an example is the french fry. The third model has no significant interior volume but a very high exterior surface area, approximating an all-crust and no-center product; an example is a potato chip (crisp). The equations for calculating frying simulations for these different kinds of products are quite disparate.

Model Oils. All triglyceride-based frying oils seem to behave alike with respect to how much abuse they can take before becoming unfit for making good-quality food. Even with reasonable turnover of oil in a vat and reasonable care in handling and use, oil fails as a quality component of fried food and as a heat transfer medium. Oils that appear to overcome this hurdle often become laden with surfactants; this causes extra oil to be dragged out on the food, or large amounts of oil to be absorbed by the food, therefore requiring more fresh oil to be added to the vat. Thus, the oil does not last longer, but is just diluted with fresh oil more frequently. This is what occurs in the production of snack foods, in which the oil "turns over" or is fully replaced at intervals of 8–12 h (3).

Oils that fail early are often subjected to excessive heating and low turnover of the oil in the vat; oils that are heated continuously without fluctuations in temperature will last longer (54). However, all frying oils can be treated alike when developing a frying simulation. The simulation does not take into account the formation of flavors or volatiles (55) but only considers the bulk properties of the oil itself.

Heat and Mass Transfer. As described earlier, the theory of frying suggests that as the surfactant increases with increasing use and/or breakdown, or as surfactants are incidentally or purposely introduced to the oil, all of the effects in the food can be described in terms of the surfactant concentration (22).

The residence time of food in oil can moderate "soakage" (absorption) for a single surfactant level, but the frying time must not be arbitrarily changed to control the amount of oil soakage because the surfactant level is always changing. Real-world operations, and the equipment used, are not designed to deal with changing oil quality.

More surfactant means more polymer deposition. As the interfacial tension drops at the air interface, and as more oxygen is dragged in, more polymer forms in an oil until the oil is saturated. More polymer allows more stable steam domes, and the rate of oxygenation goes up; this can be observed and measured. In degrading oils, foaming may be directly observed or videotaped. These observed physical characteristics roughly correlate with increases in polymer and other surfactant materials.

Increased surfactant levels cause longer contact times between the hot oil and aqueous food surfaces. The increased contact times transfer more heat from the oil to the food in a fixed period of time. The enhanced heat transfer causes heightened dehydration at the surface, which usually translates to increased water migration from the core to the exterior of the frying food. These phenomena also can be observed and quantified.

The rate of breakdown of an oil is increased by higher surfactant concentrations, as increased surfactancy causes the oil to wet heated metal surfaces.

The oil "cokes" (carbonizes) at the high temperature of the heater's surfaces. This coking produces red products that account for much of the color in the used oil, as well as most of the cyclic acids and other products felt to be undesirable.

Wetting of the heater surfaces ultimately leads to complete carbonization of an oil layer and the formation of an insulating blanket around the heater elements. Insulation leads to higher temperatures on the heater surfaces as the controllers call for more sensible heat from the sources, even though the oil temperature is still quite high.

Oil Degradation Model. In conducting 30,000 analyses on 6000 samples of frying oils at Libra Technologies, we have found that on the average the amount of triglycerides in bulk fresh oil from the factory is somewhere around 96%. As this oil is used, the amount of triglycerides surviving in the oil drops in direct proportion to the total polar materials (the nontriglyceride material) that have formed or accumulated in the oil.

Before, or at, the point at which fast-food chains ordinarily discard their oil, about 25% polar materials have formed in the oil. That means that the oil being consumed as a part of fried food is only about 75% triglycerides. Several European nations have established that point as regulatory limits for polar materials (56–58). If the oil is only 75% pure and 10% oil has been picked up on the food, the consumer eats 2.5% total polar materials by weight of fried food. The polar fraction contains all the degradation chemicals in an oil that may be toxic and that account for the cooking and eating qualities of an oil.

The effect on the thermodynamic properties of the oil and on food quality of such a large amount of conversion products easily accounts for changes in the bulk properties and cooking efficiencies of frying oils. This area requires more work, as the changing heat and mass transfer capabilities of a degrading oil have yet to be established.

It has long been known that filtration to remove particulates may be used to enhance oil life and improve food quality (31). Most industrial and institutional fryers are designed today with built-in filters (3–5,41). The soluble chemicals, such as surfactants, that play a major role in the rate of oil degradation are not filterable, however. New technologies have provided the means of removing these materials by selective ad/absorption, thereby extending the amount of time that the oil stays in its optimum condition, producing optimum foods. Shaw (59) describes the movement and properties of particles to be removed.

3.6 Surfactant Control Can Lead to Better Fried Foods

The surfactant theory of frying described provides industrial and university researchers with a model to better understand the chemistry and physics of how and why food fries. This model can ultimately lead to enhanced production efficiencies and higher-quality, more healthful foods. Active filter aids and treatments offer a route to controlling the level of surfactants left in a frying oil. Foods produced in the proper range of surfactant concentration are optimal.

CHEMICAL DEGRADATION OF FRYING FATS AND OILS

Refineries prepare oils to be light colored, of low odor, and free of oxidation and other products that would spoil their neutral, bland flavor. Some cultures have come to expect that there will be some "bite" to their oils, reflecting a modest concentration of free fatty acids. The excess acidity is usually due to incomplete deodorization of the oil. Deodorization (vacuum steam stripping) does cause a small amount of polymer to form in the oil, which accounts for about one-half of the polar materials found in fresh frying oils.

Fritsch provided a diagram (36) that summarizes the routes through which triglycerides degrade and interact during the frying process. To that scheme must be added the formation of acrolein by the dehydration of the glycerol remaining after three fatty acid residues have hydrolyzed from the triglyceride molecule.

Funding for intellectual studies at institutions to support the deep-fat frying industry has resulted in the publishing of long lists of identified, volatile degradation compounds. Much shorter lists of nonvolatile degradation products have been more recently developed. The proportion among the lists reflects the earlier advances in gas chromatography compared to the later advances in liquid chromatography and ancillary detectors and spectroscopy equipment.

Recent frying research has focused on determining the primary relationships between volatile and/or nonvolatile degradation products, the efficiency of frying, health and human nutrition, worker safety, and heat transfer. As the body of knowledge from this applied work increases, the ultimate result will be greater understanding of the process of frying. With understanding will come greater chances for optimization of the process, which will lead to more efficient, economical, and safer frying.

4.1 Determination of Degradation Products

The "index" used to determine the degree of degradation of an oil is percent total polar materials. The polar material fraction in an oil contains all the compounds that influence the cooked quality of fried food and that linearly relate to oil usage over time with respect to the amount of food fried. All other chemical and physical measurements determine "markers," which note the presence and concentration of individual component compounds or groups of related materials. Some suppliers of testing equipment supply mathematical correlations of specific marker components to percent total polar materials.

The amount or rate of formation of specific marker compounds may, in some instances or ranges, coincide with the formation of total polar materials, but no single marker compound or method is a satisfactory index for all oils under all use conditions in all ranges of use or abuse. Chemical methods commonly used to evaluate degradation products in frying oils are discussed in following sections.

4.2 The Index of Degradation

Polar Materials. By definition, total polar materials (TPM) are those materials that remain on the column after the first elution when a heated oil is tested for polar materials using AOAC method 28.074 (21). They include all partially oxidized triglycerides, nontriglycerides, lipids, and other materials soluble in, emulsified in, or suspended particulates in the frying oil. A refinery strives to produce frying oils that contain 99+% triglycerides. Our laboratory typically finds virgin frying oils (for other than the most delicate snack food production) contain 2–4% polar materials. The impurities are largely small amounts of oxidized and dimerized triglycerides, water, free fatty acids, mono- and diglycerides, sterols, carotenoids, antioxidants, antifoamers, crystal inhibitors, bleaching earth, filteraid, and hydrogenation catalyst residues, soaps, and residues of chlorophyll and phospholipids.

Once an oil is exposed to frying temperatures and food product, a portion of the triglycerides is converted to a myriad of degradation products. Total polar material is considered to be a chemical *index* of oil degradation. Since it also includes conversion products, %TPM measures cumulative degradation of the frying oil. There are many who consider the polar materials measurement to be the single most important test for assessing degrading oil. Regulatory agencies in Spain, Portugal, France, Germany, Belgium, Switzerland, Italy, and the Netherlands have established regulatory limits for polar materials in frying oils (56–58).

The main concern with the use of polar materials in regulatory and quality operations is that the official method is time and solvent intensive. The analysis may take 2.5–3.5 h to fully complete. There are several rapid methods on the market claiming to correlate markers somehow to polar materials, and one officially accepted for %TPM.

Polar materials are an excellent predictor of food quality for many operations. Pokorny (55) demonstrated that increases in the polar fraction resulted in an eventual decline in food quality. His data yielded a food quality curve that resembles that shown in Figure 11.1. Referring to similar data has been shown to be useful in the production of fried snacks such as nuts, doughnuts, and other products. Because of the amount of oil removed by the food in the production of chips, other quality indices, such as free fatty acids and WET materials, may be more sensitive tests for indicating oil quality while the values are quite low. Oil whose quality is degraded by about 5% or more, and any abused oil, is usually better measured and controlled by polar materials.

4.3 The Markers of Degradation

Free Fatty Acids (FFA). "Everyone" tests for %FFA. Free acidity is determined using an American Oil Chemist's Society (AOCS) or AOAC method titration procedure (21) with a difficult end point when testing highly colored or emulsified oils, and there are also quick tests for this parameter. It is frequently the only frying oil marker used. Free fatty acids (also expressed as acid value outside North America) is a measure of the amount of fatty acid chains hydrolyzed off the triglyceride backbone that have not yet further degraded to nontitratable species, or may be formed through oxidative processes. The %FFA is the titration of acidic material in an oil, usually expressed as percent oleic, but sometimes as percent palmitic for palm oil. It can be a useful chemical marker for the degraded oil on the surface of a fried food, but it is not a good frying oil quality indicator for several reasons.

1. FFA are transient. They both volatilize and are changed to oxidized fatty acids and other decomposition products even as they form. After some use, frying oils show an accelerating increase in total acidity and then after considerable accumulation, a drop.
2. FFA does not affect frying. Good-quality foods can be fried in pure FFA under laboratory conditions that largely exclude oxygen.
3. FFA can be diluted out using fresh or partially degraded "run-around" oils, but foods also adsorb FFA differentially depending on their crusts, coatings, surfaces, or available surfactants and the overall state of degradation of the frying oil (6,24).

A number of researchers (36,60,61) have stated that there is no direct relationship between FFA and the quality of a used frying oil. Therefore, basing quality decisions on this one parameter may be a mistake. This statement usually rubs experienced food producers the wrong way until a "mystery" process upset occurs on the line. Our experience at Libra Technologies suggests that %FFA is found to linearly increase until approximately 0.5%, but then upon dilution or other treatment the oil becomes nonideal with respect to this chemical marker. The %FFA in the oil on the surface of a food offers a "sink" for oxygen in a package and can be used to predict (with experience) the shelf life of the food with respect to the development of rancidity. Thus, by knowing %FFA and percent oil pickup, and package permeability, it is possible to design a shelf stability model. In this way high levels of FFA have been related to reduced shelf life and the development of rancidity in fried snacks.

Water Emulsion Titratables (WET). Alkaline soaps, other surfactants, and water are other chemical markers of oil degradation. They are determined using AOCS method Cc 17-79 (62) or by using a commercially available quick test. Like FFA, this parameter alone is not adequate to predict food quality.

Alkaline soaps (sodium oleate, etc.) are formed by the reaction of metals and free fatty acids in the presence of water. They are fairly stable once formed. Precursors are derived, as discussed earlier, from residual caustic cleaners, from coatings or breading containing metals left over from leavening agents (in cake doughnuts, e.g.), or from rupturing plant and animal cells, and from animal blood and bone cells. As the concentrations of soaps and other entirely organic surfactants such as phospholipids increase in an oil, they can affect heat penetration, oil incorporation into the crust of a food, and the makeup oil required to replace oil ad/absorbed by a food. Many commercially available active filtration systems have been developed to reduce or eliminate surfactants and water. Snack foods produced in oils with high levels of WET may have their shelf life compromised, as the presence of these prooxidants on such products accelerates the development of rancidity.

Peroxide Value (PV). Peroxides are unstable organic compounds formed from triglycerides. They are determined using an AOCS or AOAC method (21) or a variant using isooctane instead of chloroform as solvent. This is a standard test for fresh oils but has limited value for frying oils. One reason is that the test itself is very sensitive to oil temperature. Users must take great care to standardize their sampling and testing procedures. Peroxides are destroyed at frying temperatures but begin to reform during cooling, and "staircase" upward with each additional temperature cycle. Fritsch (36) has stated that this is not a good test for measuring the degree of abuse of a frying oil. One area where it has been found to be useful is snack foods. Processors extract oil from the fried snack and measure the PV. A PV over 2 is an indicator that product has a high rancidity potential and could fail on the shelf. Although the snack food industry utilizes this test regularly, it is not popular because it requires chloroform, which is an extremely hazardous solvent and is expensive to dispose of. A rapid, solventless test for low peroxide values is now available.

Anisidine Value. Aldehydes are products of the decomposition of peroxidized fatty acids. The anisidine value quantitates important aldehydes (AOCS method) (62). The aldehydes can be used as a marker to determine how much peroxidized material has already broken down. In conjunction with current peroxide levels, the past and future degradation profile of an oil can be mapped—especially for reprocessed "fresh" oils and reheated frying oils. By using totox values, where totox = [anisidine value + 2(peroxide value)] a useful single value can be calculated to put a scalar value to oxygen-directed oil degradation.

Polymers. Polymers are usually the largest single class of degradation products in a frying oil. They include dimers, trimers, tetramers, etc., and may be formed through oxidative and thermal reactions. The dark "shellacs" that form on fryer walls, heater tubes, and belts are polymeric, sometimes sticky materials. They are an excellent chemical marker of oil degradation but not applicable in food quality monitoring. The official method (21), originally an International Union of Pure and Applied Chemists (IUPAC) method, utilizes high-pressure liquid chromatography (HPLC). Although pure frying oil polymers are not something that can be readily measured as part of the quality control program, understanding the rate of polymer formation is extremely important toward understanding how an oil degrades and optimizing the system.

Physical test methods were once widely used to assess frying oil degradation when fewer instruments and rapid tests better related to the operational chemistry were available. They are not, however, precise enough to be considered reliable for research, nor do they seem to be repeatable between operators or locations when the somewhat subjective readings are analyzed with rigorous statistical methods. Recently, some automation has been applied to these tests, but they still suffer from sampling and handling differences between samples.

Oil Color. Oil color has been used by many as a quality index over the years. The belief was "dark oil yielded dark food." This is a complete myth. High-quality food may be produced in dark oil, and poor-quality food in apparently "clean" oil. It is valuable, however, to measure Lovibond color (62) when evaluating a frying system and developing quality standards. Researchers should measure Lovibond Red, Yellow, and Blue. Oil color should not be used as an index for state of degradation, oil quality, or for discard. It is too subjective and really bears no relation to the thermodynamic capacity of that oil to produce quality food.

The red color in a finely filtered (<5 μm) oil loosely correlates to combined oxidized fatty acids and pyrolytic condensation products. The yellow color usually relates to the combined peroxides and aldehydes in an oil that may be weakly correlated to the oil's totox value (see anisidine value), and is also related to carotenoid and other compounds. The blue color is related to the haze created by water and fine particulates suspended/emulsified/dispersed in the oil and is removed by fine filtration. This complex color test can sometimes be useful as a marker when screening samples for research. Color often varies with the temperature for the same sample. Color also depends on tocopherol and carotene content of the original oil, which varies due to processing.

Oil color measured at a single wavelength in a spectrophotometer or against color standards in a fast-food shop is often claimed by frying oil producers/marketers to relate to used frying oil quality. Our experience is that this measurement is highly dependent on the batch of oil, the amount and type of food fried, the completeness of filtration of the oil, and the specificity of active treatment, the temperature and type of fryer, and in the case of visual estimation the extent of colorblindness and type of lighting used. (Also, the layer of oil on the surface of fried food is so thin that it does not contribute to the color of the food.) Since the darkness (redness) of an oil relates very strongly to finely divided suspended matter in the oil, the use of a visual color reference can cause the oil to be misevaluated and discarded before the polar compounds cause the oil to fail as a cooking medium. It is also a truism in analytical chemistry that a colorimetric measurement should be made at a wavelength far removed from the perceived color of the analyte to assure that the determination is not confused with the background sample. Measuring a reddish sample in the red to yellow portion of the spectrum is to be avoided.

Suspended Solids. The silt left suspended in an oil varies with the amount of emulsified water and surfactants, the temperature of the oil, charge transfer properties, size, shape, and source. Filtration technology has reduced the need for this observation.

Viscosity. Viscosity varies with temperature. Active filter aids and treatments produce oils with dissimilar cooking properties and polar contents even though the viscosity remains the same as before the treatment.

Foam/Foam Height. When an oil foams, it is already abused or contaminated with a large amount of surfactant. Good oil bubbles, it does not foam.

Smoke/Smoke Point. White smoke over frying oils is mostly steam. Blue to gray smoke contains organics codistilling with steam. This latter condition can indicate the oil is unfit for use but cannot be used for quantitation.

Capacitance. The electrical properties of a frying oil depend on water and electrolytes suspended/emulsified in the oil. The rise in conductivity or drop in capacitance parallel the formation of polar materials in the oil until the oil is contaminated with an electrolyte such as salt or citric acid. Removal of small amounts of water (FTIR determination) by filtration can alter capacitance readings. Particles caught in an electrode can cause low and zero readings. The technique is temperature sensitive, and hot oil from a fryer has to reach equilibrium with the measurement cell before a reading is taken.

4.4 Summary Discussion

The degradation of frying oils due to cooking stresses forms a variety of chemical substances through different and sometimes competing reaction routes, and at varying rates sometimes influenced strongly by trace materials. The degradation products formed from triglycerides (triacylglycerides) are both volatile and nonvolatile. These various degradation chemicals may be found free in the oil, may combine or interact among themselves, may combine or interact with chemicals leached from the food being cooked, may distill from the oil with or without codistillation with steam from the cooking food, or may precipitate from oil solution.

Though the potential number of chemical entities formed in such a complex matrix is huge, specific lines of inquiry may raise the significance of just a few star performers from among the degradation products. These important products can be placed in a simplified, comprehensible framework for study. When the framework itself is arranged to be used to teach others about what happens when food is cooked in an oil, a hierarchy of topical assignment and relative importance develops among the remaining identified degradation products. The whole of the degradation products can then be categorized within one or another topic being taught in the logical framework of study.

REGULATION

Research into whether heated and/or abused fats and oils are safe has been underway for over 50 years. The majority of the work in this area has focused on the effects of abused fats on laboratory animals. The two general areas of concern have been (63):

1. Health and longevity—overall health of the animals and their organs, and survival.
2. Growth and food utilization—absorption of nutrients and acceptance of food.

It is crucial that readers examine and critique the test conditions employed in the laboratory studies. Many of the test parameters are ones that would be considered extremely abusive and would never be encountered when consuming heated fats and oils in the real world. For instance, this author has not seen food being fried in oil that was ever heated above 215°C (419°F), even for a short time. However, 205°C (401°F) for frying has been encountered in over 100 food processing locations.

This laboratory research on heated oils has been, and will continue to be, extremely important. It is the data gathered from toxicological and/or nutritional studies, such as the work conducted with heated fats and oils, that provide regulatory agencies with the background for enacting laws and establishing guidelines to protect the health and welfare of their constituency.

5.1 Heated Oils Research—The First Generation

Among the first workers in the area were Morris and his colleagues (64). They reported that they were able to induce vitamin E deficiencies in rats fed heated lard. Their work also showed that rats fed diets of 50% lard, which had been heated at 300°C (572°F) for 120 min suffered growth failure and loss of weight.

Roffo (65) indicated that rats fed diets in which sunflower or olive oil, lard, tallow, or cholesterol that had been heated to 350°C (662°F) for 30 minutes developed tumors. Later workers refuted these findings on the basis that the tumors failed to meet the established criteria for cancer.

In 1946, Crampton and Millar (66) observed that rats fed a diet consisting in part of linseed oil polymerized at high temperatures [275°C (527°F) using CO_2 to exclude air for 6–15 h] suffered high mortality. Harris (67) reported that laboratory rats that consumed fish oils heated at 280°C for 8–12 h suffered retarded growth and ill health.

Experiments by Lassen and co-workers (68) showed that sardine oil that had been heated to high temperatures was less digestible, and Lane, Blickenstall, and Ivy (69) observed no tumor formation in rats fed oil that had been browned for 30 min at 350°C (662°F). These animals reportedly had higher incidence of papillon and ulcers, however.

The work of Crampton and his colleagues (70,71) has been frequently cited. As reported in a series of papers in 1951, rats received a diet consisting of 10–12% oil (linseed, soy, cottonseed, rapeseed, corn, peanut, or herring) that had been heated at 275°C (527°F) for up to 30 h in some cases. These oils were then incorporated into the meal by baking for 20 min at 190.5°C (375°F). All of the abused oils at 20% of the diet produced weight loss in the test animals, with linseed being the worst. They observed that as heating time increased, percent weight gain decreased. In a second study, esters from heated linseed oil were separated by distillation into distillable and nondistillable fractions. Heavy mortality and overall poor health were observed in the group receiving the nondistillable fraction. The distillable fraction was noted to be more deleterious than the unheated oil.

Deuel and co-workers (72) conducted a study in which rats were fed a diet of oils containing citric esters heated for 8 h at 205°C, (401°F), and used to fry chips. Neither incorporation of the oils or the chips into the diets had any adverse effects on the test animals. Frahm, Lembke, and Von Rappard (73) observed deleterious effects in mice fed heat-polymerized whale oil.

Crampton and co-workers (74) continued along the same lines established 2 years earlier. This study again showed reduced weight gain in animals whose diets consisted of greater percentages of polymerized oils. They also fractionated these oils into six fractions. At 20% in the diet, the fraction that contained high levels of cyclized monomers (non-urea-adducting fraction) resulted in high mortality, whereas those fractions containing straight chains did not.

Raju and Rajagopalan (75) conducted work in which sesame, peanut, and coconut oils were heated in open pans at 270°C (518°F). The thermal polymers that were formed resulted in depressed growth rates. Individuals familiar with commercial or restaurant frying operations will quickly observe that the conditions that the oils were exposed to, particularly temperature, far exceed those reached by fryers used in industry or food service.

Kaunitz and co-workers (76) published a study in which oils were oxidized by aeration at 95°C (203°F) for 200 h. The resulting oils contained between 17 (lard) and 40% (cottonseed) polymers. These highly oxidized oils were incorporated into diets at levels up to 20%. Rats receiving the diet with 20%

oxidized fat suffered high mortality. Adverse effects were also noted in animals fed diets with 10% oxidized fat. The adverse trends were observed to be reversible if the diets included fresh oils. Crampton and co-workers (77) published yet another paper in 1956 reporting on feeding studies with rats receiving nonurea adduct-forming components from linseed, soy, and sunflower oils. The fractions from linseed were found to be toxic, with those from soy less so, and the sunflower fractions only slightly injurious to the test animals.

Johnson, Sakuragi, and Kummerow (78) demonstrated that thermally oxidized corn oil depressed growth in rats, a trend that could be reversed by a return to a normal diet. The oils used in these studies were oxidized by blowing air through them.

5.2 Heated Oils Research—The Second Generation

The experimental conditions used in much of the early (first-generation) work were not those to which frying fats would normally be exposed. This point was made very clear by Melnick (79) in his 1957 paper in which he also presented information gathered from oils used in commercial frying operations. He called much of the early work "impractical." Oils obtained from 89 potato chippers indicated that there was little change in iodine value and that polymers were not a problem. Rice, Mone, and Poling (80) continued work along this practical line. Their studies examined fats from restaurants, bakeries, and chip producers. They observed that feeding rats even the most abused oils generated during normal operations resulted in only slight decreases in caloric value and slight increases in liver size. Alfin-Slater (81) noted that growth, reproduction, lactation, and longevity were not impaired when cottonseed oil is heat polymerized to effect a drop of 5% iodine value.

Witting, Nishida, Johnson, and Kummerow (82) suggested that the incorporation of a single peroxide group into fatty acid molecules was enough to cause toxicity in rats and mice.

In 1958, Nishida, Takenaka, and Kummerow (83) reported that cholesterol levels in test animals fed heated oils dropped. Perkins and Kummerow (84) collaborated in 1959 on a study in which corn oil was continuously heated for 48 h with agitation. The abused oil was then fractionated. Weanling rats fed the nondistillable residue from the non-urea-adducting fatty acids died within 7 days.

Alfin-Slater and co-workers (85) reported on a study in which oils were heated under vacuum for 60–100 min at 321°C (610°F). In their feeding studies, no evidence of nutritional impairment was noted, with the exception of a highly polymerized soy oil. With this one oil, there was a slight decline in digestibility. This same year, Custot (86) warned that improper frying produced severe damage in oils. Keane, Jacobsen, and Krieger (87) reported on a study conducted using heated oils from commercial operations. No toxicity was observed in the test animals, and in fact, increased weight gain was observed in those fed the heated fats. The same study reported that oils heated and oxidized under laboratory conditions produced lower growth rats in the test animals but not toxicity.

Two excellent review articles were published in 1960. Rice and co-workers (88) reported that there was "no reason to believe that fats were nutritionally damaged by normally accepted good practices in present-day food preparation." They also observed that undesirable changes in viscosity, browning, and flavor of the oil usually preceded any biological effects. They described a situation in which an oil used to fry potato chips became unusable due to foaming at a point in which biological activity would be noted. Perkins (89), presented his review in *Food Technology*. He reviewed past work in the area and posed three questions:

1. Are polymers formed in unsaturated oils during deodorization, processing, and use?
2. Are polymeric materials absorbed on food products and if so, to what extent?
3. What are the physiological and nutritional effects of these materials?

In 1960 and 1961, workers from the U.S. Food and Drug Administration (FDA) (90,91) reported on studies in which cottonseed oil was heated at 225°C (437°F) for 190 hr. The work indicated that heated oils had greater viscosities and more higher molecular weight compounds. There was also a drop in iodine values. They observed that rats fed abused oils suffered a loss of ability to absorb those oils and hypothesized that such oils could interfere with absorption of other nutrients. The force feeding of urea filtrate monomers and dimers to weanling mice resulted in death. The adverse changes appeared to be caused by the non-urea-adducting fatty acids. What is interesting and important about this work was that these researchers were the first to propose using this class of compounds as an indicator of the quality of heated oils.

In 1962, Poling and co-workers (92) studied the influence of temperature, heating time, and aeration on the nutritive value of heated cottonseed oil. They found that the changes in the oil were proportional to severity of the conditions and that conditions more severe than those encountered in normal processing operations or cooking are necessary to yield detectable damage. This study confirmed the fact that abused oils are less utilized and, therefore, provide less energy. Livers tended to be larger in animals fed the damaged oils, but these workers questioned whether increased organ weights were really harmful.

Fleischman and co-workers (93) recommended against both the reuse of oils, as a second heating would cause additional hydrolysis and oxidation, and overheating of oils for the same reason. They also commented that frying appears to decrease the hypocholesteremic effect of the high concentration of polyenoic acids in raw oils.

Raju, Rao, and Rajagopalan (94) conducted another study in which they heated peanut, sesame, and coconut oils in an open iron pan for 8 h at 270°C (518°F). Rats receiving a diet in which these fats were included at a 15% level had reduced growth rates, reduced food intake, increased liver weight and fat, reduced B vitamin storage, reduced carbohydrate absorption, and higher blood glucose and cholesterol levels. This same year, Perkins and Van Akkeren (95) demonstrated that intermittent heating and cooling of cottonseed oil caused the oil to break down more rapidly. The oil that was heated intermittently had polar material levels at 62 h, which were equivalent to an oil that had been continuously heated for 166 h. They concluded that oils in small operations, where the turnover is low, may be damaged more rapidly than in large operations, where turnover is more rapid. This is, of course, the case in actual practice.

Kaunitz, Johnson, and Pegus (96) conducted long-term feeding studies in which rats were given a diet that included up to 30% fat, fat that had been oxidized for 40 h at 60°C (140°F) with an airflow of 1–2 L/min. Cottonseed, chicken fat, beef fat, and olive oils were used. They found that death rates and lesions observed after death were greater in animals that had received the oxidized oils, with the exception of the oxidized olive oil. A year later, Kaunitz (97) participated in an IFT-sponsored symposium on the chemistry and technology of deep-fat frying. In his summary with regards to effects of cooking and storage he noted that "intelligent cooking or even storage procedures may even improve the quality of some dietary fats." This was based on their work in which the rats receiving oxidized olive oil had longer survival rates.

One of the most exemplary pieces of work on the nutrition of frying fats was conducted by Nolen, Alexander, and Artman (98). Included were hydrogenated soybean oils, cottonseed oil, and lard, which were used under restaurant conditions until they were judged to be unfit for use in this feeding study. These fats made up 15% of the diet. The used fats were found to be slightly less absorbable and gave correspondingly slower growth rates. There were no clinical, pathological, or metabolic criteria that suggested that the used fats adversely affected the population consuming them. Distillable non-urea-adducting fractions from used fats proved somewhat toxic when high concentrations were administered to weanling rats. Their conclusion was: "Although heating of fats under actual frying conditions does cause the formation of substances which can be shown to be toxic, the level of such substances and the degree of their toxicity are so low as to have no practical dietary significance."

Nolen published additional studies (99,100) that supported this work. His target this time was the effects of hydrogenation, and his conclusion was that hydrogenated fats, either fresh or used, do not affect reproduction in rats.

Poling and co-workers (92) also conducted long-term feeding studies on laboratory animals fed heated [182°C (359.6°F) for 120 h] cottonseed salad oils, corn oils, lards, and hydrogenated vegetable shortenings. They found no significant differences between the heated and unheated oils. In their abstract they state that "the absence of adverse effects attributable to the heated fats during the life span of the rats is further evidence of the safety of these fats of the quality customarily consumed by the human population."

In 1972 and 1973, Ohfuji and his colleagues (101–103) in Japan conducted a series of studies on the nutritive value of heated oils. They found a toxic dimer in abused oils and concluded that this was found in the most polar fraction of the oils. This dimer was found to be more digestible by itself than the thermally oxidized oil. The oils were heated for 90 h at 275°C (527°F) in the presence of air or nitrogen. They did report that they were able to recover small amounts of the compound in some commercial cooking oils.

In 1977 at the AOCS meeting in New York and at the DGF Symposium in 1979, Billek, Guhr, and Sterner (104,105) presented work that they had conducted on sunflower oil obtained from the commercial production of fish fingers. They isolated different fractions from the oil, before and after it had been used, and used these for feeding studies. They found that rats consuming the polar (P) fraction had significantly lower weight gains than the other test animals. As a result of this study, they proposed that the polar fraction could be used as an index of oil quality. They felt that 30% polar material was a good number, a value that was reduced to 27% for regulatory purposes. This fraction is the non-urea-adducting fraction, so 18 years later, the proposals of Firestone and co-workers (90,91) were resurrected. Polar materials are, in fact, now used in several European nations as a quality index. At the same DGF meeting, Wolfram (106) noted the difficulties in setting up experimental models that allow exact quantitative statements on the possible harmful effects of fat. At the AOCS meeting in 1978, Clark (107) presented a review on the state of heated oils, reporting that they are not a health hazard, if they are not abused. He reiterated and updated these comments in 1991 (108).

Fong and co-workers (109) used the Ames test (microsomes) as a means of evaluating Chinese peanut oils for aflatoxins. They detected mutagenic activity in fresh oils, an activity that decreased with repeated cooking. This activity was reduced by cooking. This study demonstrated that heating can remove certain undesirable components from fresh oils, that is, the aflatoxins. The presence of aflatoxins in oils produced in Western countries would not be a problem, where they are more highly refined and would be removed. On the subject of mutagens, Taylor and his colleagues (110,111) conducted two studies to determine if mutagens were present in fried foods and the effect of frying conditions on mutagen formation. Their work agreed with that of many others, that is, mutagens are not a problem, if oils are not abused.

Goethart and co-workers (112) conducted a study on frying oils and foods from canteens in the Netherlands and conducted their own frying studies. Their work indicated that there was a slight decline in weight gain and some differences in liver and kidney size in animals fed more abused oils (frying up to 14 days). They concluded, however, that mutagens are not present in fried foods or heated oils.

Hageman and co-workers (113–115) also conducted work using the Ames test on heated oils and foods. These workers observed increased mutagenicity in the polar fraction of the oil with increasing abuse. They were unable to isolate the mutagenic components within the polar fraction.

Addis (116) has postulated that cholesterol oxides and other oxidative materials formed as a result of heating oil may act as initiators in arterial damage. These materials allegedly damage the arterial walls, setting the stage for plaque deposition. This work is, however, still in its infancy, but has prompted a great deal of popular interest in incorporating antioxidants in the diet.

Perhaps the greatest concern of the food industry is irresponsible journalism, such as with questions regarding the toxicity of heated oils. An article in *Science Digest* (117), entitled "Here's the Beef: Fast Foods Are Hazardous to Your Health," is one such example. The author has basically accused the fast-food restaurants of misleading the public and knowingly feeding them foods fried in tallow.

5.3 Industry and Food Service Practices

Those individuals who have conducted work on heated oils generally agree that only abused oils are a concern. The next question is whether oil abuse is a problem in the food processing and food service industries. Several of the studies mentioned earlier utilized oils collected from commercial operations (63,73,79,80,88,107,108). These studies concluded that the oils that were utilized were not a health concern.

In actual practice, the degree of oil abuse varies with the type of industry and the products they produce. As oils are abused, they break down into literally hundreds of compounds, many of which are detrimental to shelf life and sensory qualities of the food and shorten usable life of the oil (22,23,33,55,118).

Blumenthal (22,23) has described industrial and food service frying practices. The oils used in those industries involved in the production of chips and other snack items abuse their oils the least, due to high takeout of oil by the food and constant replenishment with fresh oil. Industrial fryers producing materials such as french fries and battered and breaded products are the next most abusive. The most abusive frying operations are those used in fast-food and institutional frying and in the production of coffee-shop donuts. This agrees with the observations of Perkins and Van Akkeren (95) who observed that repeated heating and cooling were the most damaging, precisely the situation seen in these kind of operations.

As described earlier, Firestone and co-workers (90) and Billek and co-workers (104,105) have recommended that the polar fraction be used as an index of oil quality. This belief is supported by a large number of researchers throughout the world (22,23,36,40,60,61,119,120). Using the 27% polar level as being an abused oil as recommended, Blumenthal compares average values for these industries (23).

It may be seen that the industry of concern is the fast-food and restaurant business. In actual practice, very few restaurant operations allow their oils to reach such high polar concentrations. Foods produced in such oils are of poor quality and could well be rejected by the consumer (22,23) and cause an operator to suffer economic loss.

5.4 Heated Oils and the Regulatory Environment

Food laws and regulations are enacted to protect the consumer and ensure that foods and the food preparation environments are sanitary and free of filth, chemicals, and other contaminants, including those from processing and handling operations (56–58). In the United States, laws or acts enacted by Congress and signed by the President become the law of the land.

Laws, such as the Food, Drug, and Cosmetic Act, mandate what is or is not allowed and frequently include penalties for violation, but the laws do not describe how they will be enforced. Enforcement falls to the appropriate regulatory agency. These agencies pass regulations (which are administrative actions or interpretations of the act or law) defining how the act or law should be enforced. Regulations may be substantive, which means they hold the weight of a law and are neither subject to interpretation nor interpretive.

To enact regulations to control an operation or process, an agency must perceive a concern and either gather data or review current literature to support its position, set limits to control and regulate the issue, and establish an inspection or monitoring program to ensure compliance with the established regulation. Recent events in Europe have indicated that there are many nations that perceive a concern with frying fats.

5.5 Establishing the Issue

As has just been reviewed, there has been a great deal of laboratory and statistical work conducted over the years evaluating the safety of frying oils. Studies have focused on health and longevity, that is, overall health of test animals, their organs, and growth and food utilization. The bottom line appears to be that frying oils do not present a health concern unless they have been abused. Even that finding has been questioned.

Nolen and co-workers (98) observed that "although heating of fats under actual frying conditions does cause the formation of substances which can be shown to be toxic, the level of such compounds are so low as to have no practical dietary significance."

Some of the compounds that form are extremely toxic, however, hence the concern. Among these are the cyclic monomers (74) and fractions containing the non-urea-adducting fatty acids (90).

Although researchers regularly acknowledged that abused fats contained potentially dangerous compounds, there was no cause to suspect that they might cause human health problems. This changed when German health authorities received a number of reports from individuals suffering gastrointestinal distress (121) after consuming foods fried in restaurants. Epidemiological studies failed to provide a link between abused oils and these complaints, but the studies did show that many, perhaps one-half, restaurant operators were abusing their oil, especially those frying meat items. Despite not being able to find an epidemiological link, the German scientists felt that the magnitude of oil abuse in restaurants was significant and a public health concern. After analyzing hundreds of samples from different restaurant operations, they proposed the first standards for frying oils, specifically for restaurant frying oils (122,123). Scientists in other countries also found that a significant percentage of restaurants and food service locations served food that had been cooked in abused oils.

5.6 Establishing Regulatory Guidelines or Limits

There are a wide range of tests used to monitor or measure oil quality. These include physical parameters such as foam height, color, capacitance, or viscosity; and chemical markers, like free fatty acids, oxidized fatty acids, polymer content, and iodine value; and the chemical index total polar materials. Which test is the best indicator of oil quality has been a subject for debate among oil chemists and quality control people for many years. The food industry, especially the snack food industry, tends to use free fatty acid as a chemical marker for predicting oil stability on the product as it passes through distribution to the consumer, but more and more individuals are becoming convinced that the total polar materials may be the best oil quality or end point index, as originally proposed by the Germans (123).

The pioneering work by German scientists led to the establishment of the first recommended guidelines for restaurant frying oils. The first set of recommendations (56,122) stated that frying fats were considered unacceptable if either "the flavor or taste was unacceptable; if the smoke point was less than 170°C, and the oxidized fatty acid content was 0.75% or higher." Six years later, these recommendations were revised following studies that indicated that a value of 27% polar materials correlated with the 0.7% oxidized fatty acid value proposed earlier (123). One of the reasons for the switch was the development of a column chromatographic procedure for polar materials, a method that was simpler than that for oxidized fatty acids. In 1990, the upper limit for polar content was reduced to 24–25% for restaurant oils (124). This change was based on work conducted by Gertz (125).

Since the publishing of the first recommendations, a number of nations have followed suit setting their own standards for restaurant frying oils. Firestone and others (56–58) have summarized the regulatory situation. The criteria for determining what constitutes an abused oil varies for the different nations that have set guidelines or established regulations.

The regulatory criteria established include chemical markers of oil quality, such as free fatty acids (or acid value), peroxide value, anisidine value, and polymer concentration; physical characteristics such as viscosity, color, and foam height; and sensory parameters such as organoleptic qualities. Notice that these markers of oil quality include both objective or measurable parameters, and subjective qualities.

For regulatory purposes, and in the scientific community, objective values are preferred. The most common oil degradation index, however, is the polar material fraction. Those countries using total polar materials as an end-point index for frying oil are Portugal, Germany, Belgium, Austria, the Netherlands, Italy, France, Spain, and Switzerland.

What is interesting is that the Germans, who did the pioneering work in this area, do not have a national regulation. The national guidelines are, however, enforced with fines by the individual states within the country. Switzerland's cantons provide enforcement in a similar fashion.

5.7 Inspection and Monitoring to Ensure Compliance

All food regulations establish protocols to allow the agency responsible for enforcement to ensure compliance. Enforcement protocols include inspection of facilities in question, sampling or examination of operations or product to ensure compliance, and record review. For example, to enforce the regulations or guidelines for restaurant frying oils in Europe, inspectors will enter the facilities and examine the frying operations. As these inspections are conducted by public health authorities, they will also check the restaurant's sanitary operations. If the inspector believes the oil is being or has been abused, he or she will collect a sample, which is then shipped to a laboratory that has been sanctioned by the government for analysis. In most cases, the inspector must rely on his senses, looking at physical indicators of oil degradation. These indicators include oil color and odor, smoking, and foaming. There are several countries whose programs for controlling operations at the restaurant level allow inspectors to use rapid tests for preliminary evaluation on-site.

Regulatory agencies are extremely interested in rapid tests. They allow suspect samples to be screened in the field or in the laboratory by inspectors and technicians, respectively. By screening samples and resampling only those that fail the quick test screen, the number of samples submitted to laboratories can be greatly reduced. The U.S. FDA interest in quick tests for frying oils has been publicly stated by Firestone (126).

Having such a quick test would allow restaurants, commissaries, or industrial producers to do their own "at fryer" oil testing, thereby helping to ensure compliance and provide for an upgrade to existing operations. Older quick tests have not gained greater acceptance because they employed toxic or flammable solvents or glass that should not be used in a food processing environment; did not provide equivalent results in multiple locations; did not have the backing of a technical support group to help start or support the testing program; provided only fugitive results that were not available for rechecking without retesting; failed in the hands of colorblind or handicapped people; or were not objectively readable in, for example, a digital meter to suggest the action to be taken at a particular location for continued use or discard of an oil.

If an inspector feels that a restaurant operator has been abusing oils, a sample will be collected for further testing. In countries such as Germany, Belgium, and the Netherlands, these samples are sent to private laboratories designated by the government. These laboratories test the suspect oils using recognized IUPAC procedures. North American practice generally follows the approved methods of AOAC International (formerly the Association of Official Analytical Chemists). Results are reported to the appropriate enforcement agency, which will take any necessary action.

For those nations having regulations for restaurant frying fats, frying in oils that have been abused can be expensive. Once an oil has reached the end point, be it 24–25% polar matcrials, or whatever, it cannot be used to produce food for human use. In Germany, operators caught using abused oils may be warned, fined, and closed down, depending upon whether the operator is a repeat offender. In 1990, Germany issued 7000 fines to operators, for the equivalent of over $250 each, for frying in abused oils. They also issued 50,000 warnings (124). In a worst-case scenario, a repeat offender may be closed down for a set period and not allowed to reopen at the same location. For a vendor who has a prime location, such as one near a train station, such an event could be catastrophic.

5.8 Regulating Frying Oils and Fats in the United States

The U.S. Department of Agriculture (USDA/FSIS) has established guidelines for frying oils used to produce meat and poultry products in factories but not restaurants, which are outside its jurisdiction. To cite the guidelines: "Large amounts of sediment and free fatty acid content in excess of 2% are usual indications that frying fats are unwholesome and require conditioning or replacement" (127).

What is interesting is that the USDA guideline for the upper limit of free fatty acid content (2%) is higher than what is commonly used in industrial practice. Also, the guidelines totally ignore any other problem moieties in the oil, which are crucial for food quality, and perhaps food safety as well. The FDA is interested in the issue (56–58,127) but to this point has established no regulations. Both the USDA and FDA are resource limited and are unlikely to take drastic action unless an emergency occurs or consumer activists become vocal in the media. Without serious guidelines and enforcement the issue could be quickly forced.

Local health inspectors may look at frying operations during their inspections, but there are not yet specific limits on oils, such as in Europe. It now seems likely that if U.S. regulatory attention is turned to frying oils, guidelines will be set by federal agencies, but enforcement will be through state and local agencies.

5.9 Future of Frying Oils Regulation

If the Common Market nations do establish such universal regulations, the effects will be felt in the United States and Canada, and very likely around the world. The fast-food and restaurant operations that have worldwide operations will be forced into compliance in the regulated part of the world. They may deem it prudent to establish similar standards throughout their whole operations before consumer activist groups bring a dichotomous situation to the press. For these operations, a program to control oil quality could also be used as a marketing tool.

The U.S. FDA has been looking at frying oils for a long time, so it may eventually move to establish regulatory limits based on food safety. Because of the inherent complexity of frying oil chemistry, and the unknown properties of potential frying fat replacers, the FDA will probably look to quick tests as a means for on-site monitoring. The burden on the FDA labs is great enough now though, without inviting more complex analytical work to ensure compliance. Thus, state laboratories may pick up the task.

There may well be some regulatory activity in the industrial arena also. The USDA/FSIS could well change its guidelines to requirements, particularly in keeping with its belief in the HACCP system as a means for assuring the safety of meats and meat products, including meat pies and pastries.

The issue of heated fats and public health has not been addressed here in the United States, but, as noted earlier, is an issue in Europe. Firestone and co-workers (56) presented a review of the regulatory situation pertaining to heated fats and oils at the 1990 meeting of the Institute of Food Technologists. In this work, Firestone and co-workers noted that Belgium, France, Germany, Spain, and Switzerland have established regulatory limits for polar materials in restaurant frying oils, the segment most likely to abuse an oil. It was, in fact, abuse of frying fats at the restaurant level that led to the promulgation of these regulations in Germany. The other nations mentioned then followed the lead of Germany. The belief is that similar regulations will probably be established for all European Common Market nations.

5.10 Discussion

The basic belief among the scientific community is that heated oils, particularly those that have not been abused, are not a health hazard. There are compounds found in abused oils that are potential mutagens, but the levels at which they are found are so low that they are not considered significant. Test animals fed large quantities of abused oil or fractions do tend to gain weight at a slower rate than those given fresh or less abused oils, apparently due to the formation of indigestable compounds (polymers). When conducting animal feeding studies using heated oils, it has been shown that it is essential to ensure that diets are balanced so the actual effects of the oils may be observed. Some of the early work has been questioned on this point. The most deleterious compounds have been found in the non-urea-adducting fraction of abused oils, which contains highly polar materials. It was proposed as early as 1961 (90) that these polar compounds be used to index, and set a 30% limit to, the cooking use of abused oil, an event that has already come to pass in several European nations. This polar material limit discard point generally coincides with the best practices of cooks in the most abusive of frying operations (the restaurant operation) in the United States.

Claims have been made that animal- and vegetable-based frying oils should be replaced with synthetic materials. The claims are that synthetics can be more durable than natural, triglyceride-based oils, and that the synthetics, being nondigestible, would provide less calories into the diet while still preparing delicious fried foods. Controversies have arisen about these claims and about the nutritional and toxicological properties of frying oils in general.

It is clear that frying oils used to fry different foods form polar materials at different rates in relation to the food, and not the frying oil. A common mistake is to believe that all frying oils should be discarded at some arbitrary value such as 25%, which has been suitable for simple, home-style french fry oil in Europe. However, sensory perceptions and analyses show that different foods are spoiled for food service or later consumption at much lower polar material contents. For instance, industrially prepared potato chips must be produced in oils below 13% polar content, and battered and breaded products must be produced in oils below 18% polar content. Therefore, even for the same oil, the food controls the type(s) of polar materials formed in the oil, either in their relative ratios of subcomponents or more specifically in the types of subcomponents. Without attention to this reality, regulatory agency attempts to limit polar material contents may actually encourage overuse of some frying oils by setting a single high limit for the content of polar materials in all frying oils. As mentioned, it is better to assess polar material content(s) by means of tests that are sensitive to the leading chemical indicators among the polars than to depend on tests correlated to, or coincident with, the simple gravimetric presence of nontriglycerides in frying oils.

Experiments with single fryers used to cook a mixture of foods as is common in laboratories, test kitchens, family-style restaurants, and counter-service shops produce similar results. With similarly unsaturated frying oils, and similar throughputs of food, the polar material contents found in the oils are about the same through time and average out to be about the same at the discard point, which is roughly 20%. This is related to the fact that about 75% of the food fried is french fries or other potato products, and the rest is battered and breaded meat, fish, poultry, or vegetables, with a smattering of seafood, filled roll-ups, or pastries. Fryers devoted to donut production generally have oils that are out of specification at about 18% polars, but often the oils are "never" discarded and contain in excess of 25% polars.

The polar contents below 25% at discard due to food quality failure in the above cases are due to the averaging out among the polar subfractions produced by the individual foods fried. Therefore, the less a food contributes to an oil qualitatively and quantitatively by leaching or other process, the longer an oil can be used at equivalent throughput of food, before the polar material content reaches about 25%. At about 25% polar content, any given frying oil apparently produces gastric distress and initiates bowel activity to eliminate even small amounts of ingested abused frying oil.

REFERENCES

1. R. Reeves, Institute of Shortening and Edible Oils, personal communication (1987).
2. Anon., *Snack World Magazine,* (June 1989).
3. J. Thomson, Chemistry and Technology of Deep-Fat Frying, Short Course at University of California at Davis, May 16–18, 1990.
4. R. Swackhamer, Chemistry and Technology of Deep-Fat Frying, Short Course at University of California at Davis, Apr. 3–5, 1989.
5. R.F. Stier, Deep Frying of Foods: Theory & Practice, Short Course at Rutgers University, New Brunswick, N.J., Mar. 25–27, 1992.
6. R.F. Stier and M.M. Blumenthal, *Baking Snack,* **15**(2), 67 (1993).
7. C.E. Stauffer, *Baking Snack,* **13**(9), 18 (1991).
8. M.M. Blumenthal, Short Course, The Chemistry & Technology of Deep-Fat Frying, University of California at Davis, 1989
9. D.A. Corlett and R.F. Stier, *A Practical Application of HACCP,* ESCAgenetics Corp., San Carlos, Calif., 1990.
10. C.J. Robertson, *Food Technol.* **21**(1), 34 (1967).
11. C.J. Robertson, *Can. Inst. Food Technol.* **1**(3), A66 (1968).
12. D. Bogart, Short Course, Deep Frying of Foods: Theory and Practice, Rutgers University, New Brunswick, N.J., Feb. 25–27, 1992.
13. R.D. Vold and M.I. Vold, *Colloid and Interface Chemistry,* Addison-Wesley Publishing, Reading, Mass., 1983, p. 633.
14. R.F. Stier and M.M. Blumenthal, *Baking Snack,* **13**(3), 29 (1991).
15. E.N. Frankel, L.M. Smith, C.L. Hamblin, R.K. Creveling, and A.J. Clifford, *J. Am. Oil Chem. Soc.* **61,** 87 (1984).
16. B.A.J. Sedlacek, *Nahrung* **15,** 413 (1971).
17. G.A. Jacobsen, *Food Technol.* **45**(2), 72 (1991).
18. R.G. Krishnamurthy in D. Swern, ed., *Bailey's Industrial Oil and Fat Products,* Vol. 2, 4th ed., John Wiley & Sons, New York, 1982, pp. 315–342.
19. H.W. Lawson, *Standards for Fats and Oils,* Avi Publishing Company, Inc., Westport, Conn., 1985.
20. K. Schaich, Short Course, Deep Frying of Foods: Theory and Practice, Rutgers University, New Brunswick, N.J., Feb. 25–27, 1992.

21. AOAC, *Official Methods of Analysis,* 15th ed., Association of Official Analytical Chemists, Arlington, Va., 1990.
22. M.M. Blumenthal, *Optimum Frying: Theory and Practice,* Libra Laboratories Monograph Series, Piscataway, N.J., 1987.
23. M.M. Blumenthal, *Food Technol.* **45**(2), 68, 94 (1991).
24. M.M. Blumenthal and R.F. Stier, *Trends Food Sci. Technol.* **2**(6), 144 (1991).
25. D. Kubose, Short Course, Deep Frying of Foods, University of California at Davis, May 16–18, 1990.
26. H.W. Ballew, *Basics of Filtration and Separation,* Nucleopore Corp. Pleasanton, Calif., 1978.
27. R.F. Stier and M.M. Blumenthal, *Baking Snack,* **13**(5), 15 (1991).
28. R.F. Stier and M.M. Blumenthal, *Baking Snack Sys.* **12**(9), 15 (1990).
29. N.R. Artman in D. Kritchevsky and A. Paoletti, eds., *Advances in Lipid Research,* Vol. 7, Academic Press, New York, 1969, p. 245.
30. H.-D. Belitz and W. Grosch, in *Food Chemistry,* English translation by D. Hadziyev, Springer-Verlag, New York, 1986, Chapters 3 and 14.
31. G. Varela, A.E. Bender, and I.D. Morton, *Frying of Food,* Ellis Horwood Ltd., Chichester, U.K., 1988.
32. G.P. Mottur, *Cereal Foods World* **34,** 620 (1989).
33. S.S. Chang, R.J. Peterson, and C.-T. Ho, *J. Am. Oil Chem. Soc.* **55,** 718 (1978).
34. A.E. Waltking, W.E. Seery, and G.W. Bleffert, *J. Am. Oil Chem. Soc.* **52,** 96 (1975).
35. C.N. Christopoulou and E.G. Perkins, *J. Am. Oil Chem. Soc.* **66,** 1338 (1989).
36. C.W. Fritsch, *J. Am. Oil Chem. Soc.* **58,** 272 (1981).
37. A.J. Paradis and W.W. Nawar, *J. Food Sci.* **46,** 449 (1981).
38. W.W. Nawar in D.B. Min and T.H. Smouse, eds., *Flavor Chemistry of Lipid Foods,* American Oil Chemists Society, Champaign, Ill., 1985, p. 39.
39. L.M. Smith, A.J. Clifford, C.L. Hamblin, and R.K. Creveling, *J. Am. Oil Chem. Soc.* **62,** 996 (1985).
40. S.G. Stevenson, M. Vaisey-Genser, and N.A.M. Eskin, *J. Am. Oil Chem. Soc.* **61,** 1102 (1984).
41. J. Grob, presented at Extension Short Course, Short Course on Deep Fat Frying of Foods, University of California at Davis, May 16–18, 1990.
42. A.L. Lydersen, *Mass Transfer in Engineering Practice,* John Wiley & Sons, New York.
43. C. Keller and F. Escher, presented at International Congress of Engineering and Food V, Cologne, Germany, 1989.
44. B. Hallstrom, C. Skjoldebrand, and C. Tragardh, *Heat Transfer and Food Products,* Elsevier Applied Science, New York, 1988.
45. R. Ohlson in E.G. Perkins and W.J. Visek, eds., *Dietary Fats and Health,* Am. Oil Chem. Soc., Champaign, Ill., 1983, p. 44.
46. M.W. Formo, in D. Swern, ed. *Bailey's Industrial Oil and Fat Products,* Vol. 1, 4th ed. John Wiley & Sons, New York, 1979, p. 177.
47. S.P. Rock and H. Roth, *J. Am. Oil Chem. Soc.* **43,** 116 (1966).
48. M.M. Blumenthal, presented at Stephen S. Chang Symposium at Annual Meeting of the Am. Oil. Chem. Soc., Phoenix, AZ, 1988.
49. M. Love, Personal communication, Iowa State University, Ames, 1989.
50. C.A. Miller and P. Neogi, *Interfacial Phenomena: Equilibrium and Dynamic Effects,* Surfactant Science Series, Vol. 17, Marcel Dekker, Inc., New York, 1985, p. 184.
51. L.R. Fisher, E.E. Mitchell, and N.S. Parker, *J. Food Sci.* **50,** 1201 (1985).
52. S. Stern and H. Roth, *Cereal Sci. Today* **4,** 176 (1959).

53. M.M. Blumenthal and J.R. Stockler, *J. Am. Oil Chem. Soc.* **63,** 687 (1986).
54. E.G. Perkins and L.A. Van Akkeren, *J. Am. Oil Chem. Soc.* **42,** 782 (1965).
55. J. Pokorny in D.B. Min and T.H. Smouse, eds., *Flavor Chemistry of Lipid Foods,* Am. Oil Chem. Soc., Champaign, Ill., 1989, p. 39.
56. D. Firestone, R.F. Stier, and M.M. Blumenthal, *Food Technol.* **45**(2), 90 (1991).
57. R.F. Stier and M.M. Blumenthal, *Baking Snack* **13,** 11 (1991).
58. D. Firestone, *Inform* **4,** 12 (1993).
59. D.J. Shaw, *Introduction to Colloid and Surface Chemistry,* 3rd ed., Butterworths, Boston, 1980, p. 19.
60. J. Castang, *Ann. Fals. Exp. Chim.* **74,** 701 (1981).
61. A. Mankel, *Fette Seifen Anstrichm.* **72,** 677 (1970).
62. AOCS, *Official Methods of the American Oil Chemists' Society,* 16th ed., American Oil Chemists Society, Champaign, Ill., 1979.
63. R.F. Stier and M.M. Blumenthal, *Baking Snack* **13**(9), 27 (1991).
64. H.P. Morris, C.D. Lassen, and J.W. Lippincott, *J. Natl. Cancer Inst.* **4,** 285 (1943).
65. A.H. Roffo, *Biol. Inst. Med. Exp.* **21**(64), 1 (194-).
66. E.W. Crampton and J. Millar, unpublished data, cited in Ref. 71.
67. P. Harris, unpublished data, cited in Ref. 71.
68. S. Lassen, E.K. Bacon, and H.J. Dunn, *Arch. Biochem.* **23,** 1 (1949).
69. A. Lane, D. Blickenstall, and A.C. Ivy, *Cancer* **3,** 1044 (1950).
70. E.W. Crampton, F.A. Farmer, and F.M. Berryhill, *J. Nutrition* **43**(3), 431 (1951).
71. E.W. Crampton, R.H. Common, F.A. Farmer, F.M. Berryhill, and L. Wiseblatt, *J. Nutrition* **44,** 177 (1951).
72. H.J. Deuel, S.M. Greenberg, C.E. Calbert, R. Baker, and H.R. Fisher, *Food Res.* **16**(3), 258 (1951).
73. H. Frahm, A. Lembke, and G. Von Rappard, *Michwirtsch. Forschungsber,* **4,** 443 (1953), cited in Ref. 77.
74. E.W. Crampton, R.H. Common, F.A. Farmer, A.F. Wells, and D. Crawford, *J. Nutrition* **49**(2), 333 (1953).
75. N.V. Raju and R. Ragopalan, *Nature* **176,** 513 (1955).
76. H. Kaunitz, C.A. Slanetz, R.E. Johnson, H.B. Knight, D.H. Saunders, and D. Swern, *JAOCS* **33**(12), 630 (1956).
77. E.W. Crampton, R.H. Common, E.T. Pritchard, and F.A. Farmer, *J. Nutrition* **60**(1), 13 (1956).
78. O.C. Johnson, T. Sakaragi, and F.A. Kummerow, *JAOCS* **33**(10), 433 (1956).
79. D. Melnick, *JAOCS* **34**(11), 578 (1957).
80. E.E. Rice, P.E. Mone, and C.E. Poling, *Fed. Proceed.* **16,** 398 (1957).
81. R.B. Alfin-Slater, A.F. Wells, L. Aftergood, and H.J. Deuel, *J. Nutrition* October (1957).
82. L.A. Witting, T. Nishida, O.C. Johnson, and F.A. Kummerow, *JAOCS* **34**(9), 421 (1957).
83. T. Nishida, F. Takenaka, and F.A. Kummerow, *Cir. Res.* **6,** 194 (1958).
84. E.G. Perkins and F.A. Kummerow, *J. Nutrition* **68**(1), 101 (1959).
85. R.B. Alfin-Slater, S. Auerbach, and L. Aftergood, *JAOCS* **36**(12), 638 (1959).
86. F. Custot, *Ann. Nutri. Aliment.* **13,** 417 (1959).
87. K.W. Keane, G.A. Jacobsen, and C.H. Krieger, *J. Nutrition* **68**(1), 57 (1959).
88. E.E. Rice, C.E. Poling, P.E. Mone, and W.D. Warner, *JAOCS* **37**(11), 607 (1960).
89. E.G. Perkins, *Food Technol.* **14**(10), 508 (1960).

90. D. Firestone, W. Horwitz, L. Friedman, and G.M. Shue, *JAOCS* **38**(5), 253 (1961).
91. L. Friedman, W. Horwitz, G.M. Shue, and D. Firestone, *J. Nutrition* **73**(1), 85 (1961).
92. C.E. Poling, E. Eagle, E.E. Rice, A.M.A. Durand, and M. Fisher, *Lipids* **5**(1), 128 (1970).
93. A.I. Fleischman, A. Florin, J. Fitzgerald, A.B. Caldwell, and G. Eastwood, *J. Am. Diet. Assoc.* **42**(5), 394 (1963).
94. N.V. Raju, M.N. Rao, and R. Rajagopalan, *JAOCS* **42**(9), 774 (1965).
95. E.G. Perkins and L.A. Van Akkeren, *JAOCS* **42**(9), 782 (1965).
96. H. Kaunitz, R.E. Johnson, and L. Pegus, *JAOCS* **42**(9), 770 (1965).
97. H. Kaunitz, 1967, *Food Technol.* **21**(3), 278 (1967).
98. G.A. Nolen, J.C. Alexander, and N.R. Artman, *J. Nutrition* **93**(3), 337 (1967).
99. G.A. Nolen, *JAOCS* **49**(12), 688 (1972).
100. G.A. Nolen, *J. Nutrition* **103**, 1248 (1973).
101. T. Ohfuji, K. Sakurai, and T. Kaneda, *Yakagaku* **21**(2), 68 (1972).
102. T. Ohfuji, H. Igarashi, and T. Kaneda, *Yakagaku* **21**(2), 21 (1972).
103. T. Ohfuji and T. Kaneda, *Lipids* **8**(6), 353 (1973).
104. G. Billek, G. Guhr, and W. Sterner, paper presented at 69th Annual Meeting of the AOCS, New York, 1977.
105. G. Billek, G. Guhr, and W. Sterner, *Fette Seifen Anstrichm.* **81**, 562 (1979).
106. G. Wolfram, *Fette Seifen Anstrichm.* **81**, 559 (1979).
107. W. Clark, Paper presented at the 70th Annual Meeting of the AOCS, Chicago, Ill., 1978.
108. W.L. Clark and G.W. Serbia, *Food Technol.* **45**(2), 84, 94 (1991).
109. L.Y.Y. Fong, C.C.T. Ton, P. Koonanuwatchaidet, and D.P. Huang, *Fd. Cosmet. Toxicol.* **18**, 467 (1980).
110. S.M. Taylor, C.M. Berg, N.H. Shoptaugh, and V.N. Scott, *Food Chem. Toxicol.* **20**, 209 (1982).
111. S.M. Taylor, C.M. Berg, N.H. Shoptaugh, and E. Traisman, *JAOCS* **60**(3), 576 (1983).
112. R.L.D. Goethart, H. Hoekman, E.J. Sinkeldam, L.J. Van Gamert, and R.J.J. Hermus, *Voeding* **46**(9), 300 (1985).
113. G. Hageman, R. Kikken, F. ten Hoor, and J. Kleinjans, *Mutation Res.* **204**, 593 (1988).
114. G. Hageman, R. Kikken, F. ten Hoor, and J. Kleinjans, *Lipids* **24**(10), 899 (1989).
115. G. Hageman, R. Hermans, F. ten Hoor, and J. Kleinjans, *Food Chem. Toxicol.* **28**, 83–92 (1990).
116. P.B. Addis, *Nutrition News* **62**(2), 7 (1990).
117. M.H. Brown, *Sci. Digest* **31**, 76 (Apr. 1986).
118. J.A. Thompson, M.M. Paulose, B.R. Reddy, R.G. Krishnamurthy, and S.S. Chang, *Food Technol.* **21**(3), 405 (1967).
119. U.P. Buxtorf, W. Manz, and M. Schupbach, *Gebiete Lebensm. Hyg.* **67**, 429 (1976).
120. G. Guhr and J. Waibel, *Fette Seifen Anstrichm.* **80**(3), 106 (1978).
121. A. Seher, personal communication, Federal Institute for Fat Research, F.R.G. Munster, 1980.
122. DGF, Meeting Summary, DGF (German Society for Fat Research), Symposium of Frying and Cooking Fats, Fette Seifen Anstrichm., Vol. 76, 49, p. 1973.
123. DGF, DGF Symposium of Frying and Cooking, *Fette Seifen Anstrichm.*, Special Issue, **81**, 493 (1979).
124. C. Gertz, Chemisches Untersuchungsant, Hagen, Germany, personal communication, 1991.
125. C. Gertz, *Fette Seifen Anstrichm.* **88**(12), 475 (1986).
126. Anon., *Food Chem. News,* July 9, 1990, pp. 19–21.
127. USDA/FSIS, Part 18.40 *Meat and Poultry Inspection Manual,* September, Food Safety and Inspection Services, USDA, Washington, D.C., 1985.

12
Emulsifiers for the Food Industry

EMULSIFIERS AS AMPHIPHILES

Surfactant is a coined word (from *sur*face *ac*tive ag*ent*) that is applied to molecules that migrate to interfaces between two physical phases and thus are more concentrated in the interfacial region than in the bulk solution phase. The key molecular characteristic of a surfactant is that it is amphiphilic in nature, with the lipophilic (or hydrophobic) part of the molecule preferring to be in a lipid (nonpolar) environment and the hydrophilic part preferring to be in an aqueous (polar) environment (Figure 12.1). By the word *preferring* it is actually meant that the thermodynamic free energy of the system is at a minimum when the lipophilic part is in an oil (or air) phase and the hydrophilic part is in water (1). If a surfactant is dissolved in one phase of an ordinary mixture of oil and water, some portion of the surfactant will concentrate at the oil–water interface, and at equilibrium the free energy of the interface (called interfacial or surface tension, γ) will be lower than in the absence of the surfactant. Putting mechanical energy into the system (e.g., by mixing) in a way that subdivides one phase will increase the total amount of interfacial area and energy; the lower the amount of interfacial free energy per unit area, the larger the amount of new interfacial area that can be created for a given amount of energy input. The subdivided phase is called the discontinuous phase, and the other phase is the continuous phase.

As shown in Figure 12.1 surfactants have a lipophilic (fat-loving) and a hydrophilic (water-loving) part; for this reason they are sometimes called amphiphilic (both-loving) compounds. The lipophilic part of food surfactants is usually a long-chain fatty acid obtained from a food-grade fat or oil. The hydrophilic portion is either nonionic (e.g. glycerol); anionic (negatively charged, e.g., lactate); or amphoteric, carrying both positive and negative charges (e.g., the amino acid serine). Cationic (positively charged) surfactants

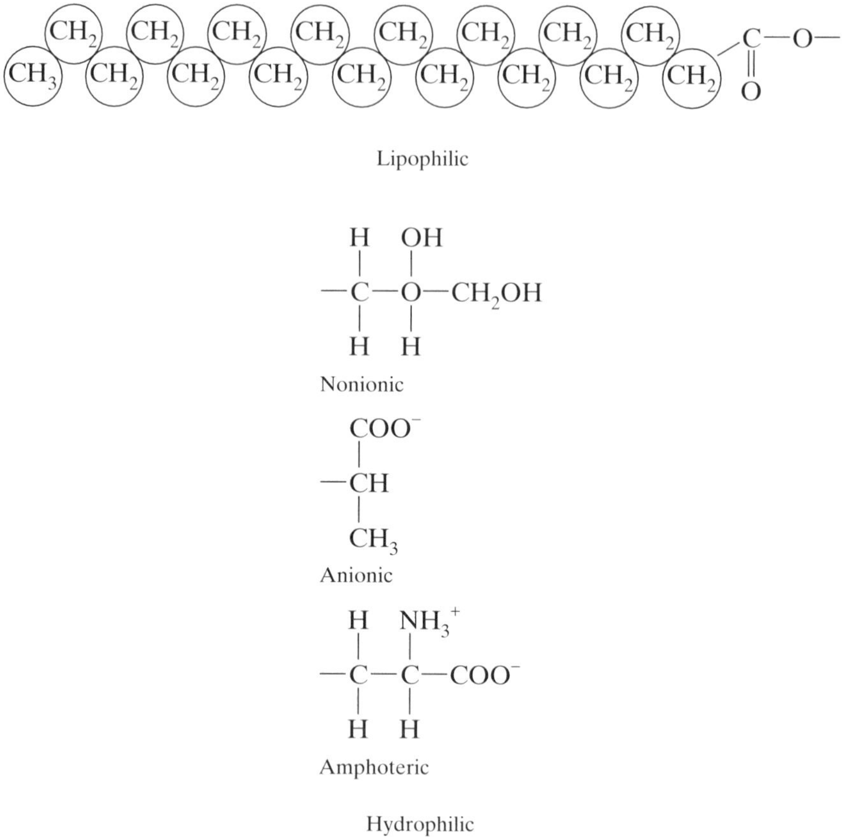

484 Emulsifiers for the Food Industry

$$CH_3-(CH_2)_{16}-C(=O)-O-$$

Lipophilic

$$-\underset{H}{\overset{H}{C}}-\underset{H}{\overset{OH}{O}}-CH_2OH$$

Nonionic

$$-\underset{CH_3}{CH}-COO^-$$

Anionic

$$-\underset{H}{\overset{H}{C}}-\underset{H}{\overset{NH_3^+}{C}}-COO^-$$

Amphoteric

Hydrophilic

Figure 12.1 Generalized amphiphilic molecular structure. The lipophilic portion is the long-chain fatty acid stearic acid.

are usually bactericidal and somewhat toxic, and they are not used as food additives. Examples of surfactants are monoglyceride (nonionic), stearoyl lactylate (anionic), and lecithin (amphoteric). The nonionic surfactants are relatively insensitive to pH and salt concentration in the aqueous phase, while the functionality of the ionic types may be markedly influenced by pH and ionic strength.

Numerous books have been written on the subject of emulsions, but three good sources are Adamson (1), Larsson and Friberg (2), and Becher (3). They summarize and discuss all phases of emulsifier technology for the practical working scientist.

SURFACES AND INTERFACES IN FOODS

Table 12.1 lists five systems whose properties are influenced by interfacial tension. Usually emulsions are of the oil-in-water (o/w) type, where the continuous phase is polar (an aqueous solution of salts, sugars, proteins, etc.) and

Table 12.1 Food systems involving interfaces (4)

System	Continuous Phase	Divided Phase
Emulsion	Liquid	Liquid
Foam	Liquid	Gas
Sol (dispersion)	Liquid	Solid
Fog	Gas	Liquid
Aerosol	Gas	Solid

the discontinuous or divided phase is nonpolar (lipid, i.e., fat or oil). The suspension of whey in melted fat, used in margarine manufacturing, is a water-in-oil (w/o) emulsion. In foams, the discontinuous phase is usually air; the continuous phase may be either aqueous or lipid. The interface between the continuous aqueous phase of a dough and insoluble components (starch granules, gluten proteins) typifies a sol or dispersion. Fogs are usually found in processing steps (e.g., a spray application of an ingredient such as oil); aerosols are formed during spray drying of food ingredients. In these two instances, particle sizes are usually larger than sizes generally associated with the terms fog and aerosol.

SURFACE ACTIVITY

3.1 **Energetics**

Surface Tension, Surface Free Energy. Consider a soap bubble film contained in a rectangular wire frame (Figure 12.2). One side of this frame, of length l, is moveable. The work necessary to move this side to the right by a distance dx is given by:

$$\text{work} = \gamma l \, dx = \gamma \, dA \quad (1)$$

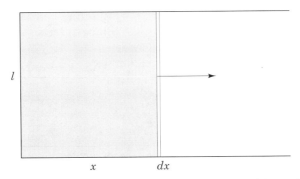

Figure 12.2 Sketch of a wire frame with one movable side, containing an interior film.

where γ is the free energy of the water–air interface. The total surface free energy equals $\gamma l x$. The usual units for γ is milli-Newtons per meter (mN/m), which is numerically equal to the older (non-SI) units of dynes per centimeter or ergs per centimeter squared. The concept is general; the surface could be the interface between two condensed phases, e.g., water and oil, in which case the term is interfacial tension (free energy).

If the surface is curved, the radius of curvature plays a role. Given an air bubble of radius r the total surface energy is $4\pi r^2 \gamma$. Decreasing the radius by the amount dr, decreases total surface energy by $8\pi r \gamma dr = 4\pi r^2 \gamma dr$. This change must be balanced by a pressure increase, ΔP, otherwise the bubble would be compressed to nothingness. This pressure difference times the change in surface area equals the change in total surface energy:

$$\Delta P 4\pi r^2 dr = 8\pi r \gamma dr \qquad (2)$$

and

$$\Delta P = 2\gamma/r \qquad (3)$$

Equation 3 indicates that the internal pressure of a small bubble is greater than that of a large bubble. This has practical consequences in aerated food systems. In a cake batter, for example, time-lapse photography shows that small bubbles (containing carbon dioxide) disappear and large bubbles increase in size (5). The carbon dioxide in the small bubbles dissolves into the aqueous phase because of a higher pressure, then enters the larger bubbles, which represent a region of lower internal pressure. Similar effects are expected in other foods where the continuous phase may act as a conduit for dissolved gases.

The surface tension of a solution of a surfactant is lower than that of the pure solvent. Surface tension is roughly a linear function of ln(surfactant concentration) up to the critical micelle concentration (CMC) (Figure 12.3). Above the CMC the thermodynamic activity of the surfactant does not increase with the addition of more surfactant, and the surface tension remains constant. Interfacial tension also decreases with the concentration of an emulsifier dissolved in one of the phases. In Figure 12.4 the decrease in γ does not level off, because the emulsifier (PGMS) does not form micelles in the organic solvent phase (heptane). The changes in the slope of the plot are attributed to changes in orientation of emulsifier molecules at the interface (7).

3.2 Concentration at the Interface

Surface Excess of Emulsifier. Surfactant molecules concentrate at the interface, with the lipophilic portion being in the nonpolar phase (air, organic

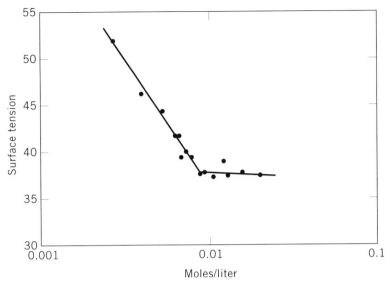

Figure 12.3 Surface tension of laurylsulfonic acid solutions (6).

solvent) and the hydrophilic portion being in the polar (water) phase. This migration of some surfactant lowers the free energy of the total system. The result is a higher concentration of surfactant in the region that includes the interface (Figure 12.5). The difference between this concentration and the bulk concentration is called the surface excess, Γ.

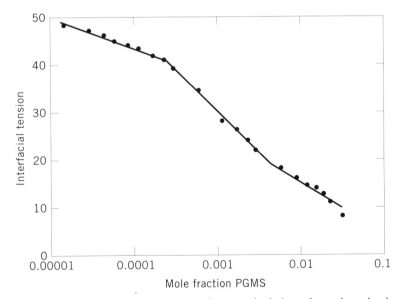

Figure 12.4 Interfacial tension at the interface of water and solutions of propylene glycol monostearate in heptane (7).

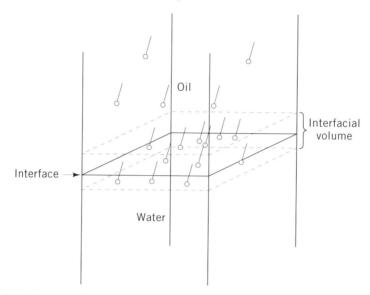

Figure 12.5 Demonstration of the excess concentration of a surfactant at a water–oil interface.

The surface excess is calculated from a plot of γ versus the thermodynamic activity, a, of the surfactant:

$$\Gamma = -(a/RT)(d\gamma/da) \tag{4}$$

For dilute solutions, surfactant activity equals concentration, and Γ is calculated from the plot of γ versus molar concentration.

When surfactant molecules concentrate at the interface, some solvent molecules are displaced, so the surface solvent concentration is lower than the bulk solvent concentration. The Gibbs convention defines the dividing line between the two phases so that the (negative) surface excess of solvent equals zero. Then equation 4 gives the surface excess of (say) laurylsulfonic acid at the air–water interface. When the actual interfacial concentration of surfactant is needed, the situation is more complicated. Methods for handling these complications have been discussed (1,7).

3.3 Interactions between Surfaces

When two surfaces approach each other, two forces exist: one repulsive and one attractive. Whether or not the surfaces touch and coalesce depends on the relative sizes of the two forces. This is equally true for liquids (e.g., oil droplets in an emulsion), solids (e.g., finely divided $CaCO_3$), and films (air bubbles in a foam). The description of these interactions is stated in the DLVO theory (8).

Electrical Double Layer. Electrical repulsion exists when the surfaces carry net charges of the same sign and the continuous phase is water. For example, if an o/w emulsion is stabilized by an anionic surfactant, the oil droplets have a negative charge on their surface. Electrical repulsion then tends to keep the droplets from making contact. At the oil surface, the electrical potential is ψ_o, and the potential decreases as the square of the distance from the surface, because cations are attracted into the region, partially neutralizing the surface negative charge. This change is shown in Figure 12.6.

The rate of decrease of ψ is directly related to the ionic strength of aqueous phase. Ionic strength, μ, is related to the concentration of individual salt ions and the square of the ionic charge (z) of each ion:

$$\mu = \tfrac{1}{2}\Sigma c_i z_i^2 \tag{5}$$

Divalent ions are four times as effective as monovalent ions in decreasing ψ. Thus 0.25 M zinc sulfate, for example, is as effective as 1 M sodium chloride in promoting emulsion flocculation or coalescence. If gravity (creaming) is the only force bringing the droplets together, they will approach to a distance

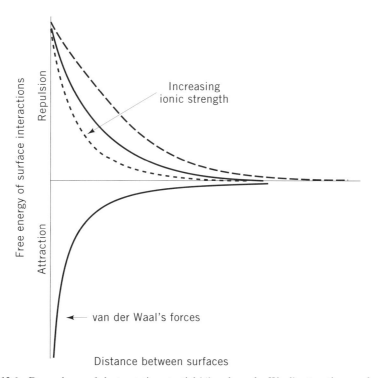

Figure 12.6 Dependence of electrostatic potential (ψ) and van der Waal's attraction as a function of distance between two surfaces.

where repulsion due to ψ is just balanced by gravitational effects, and the emulsion will then be stable.

Attractive Forces. Attractive forces, collectively called van der Waal's forces, exist between two oil droplets. Simplistically, these forces may be thought of as the attraction between oil molecules at the o/w interfaces that have lower energy when in contact with each other than when in contact with water. Several phenomena are involved: hydrophobic interactions and London dispersion forces are most commonly considered. These are effective as (roughly) the fourth power of the distance between the surfaces and are unaffected by ionic strength. The attraction due to van der Waal's forces is shown in Figure 12.6. Suspensions of solids (cellulose fibers, finely divided $CaCO_3$, etc.) are stabilized in the same way. Ionic surfactants are used that selectively adsorb to the solid surface, generating a ψ potential and making possible a stable suspension.

Drainage of foams is governed by similar principles. In a soap foam, for example, carboxylate molecules are concentrated at the air–water interface, and the surface has a net negative charge. Water drains from between two air bubbles until the ionic repulsion between the two surfaces equals the gravitational force on the water; at that thickness, the film becomes stable. The presence of salt in the water phase decreases the thickness of the final drained film, until the ionic strength is high enough to allow the surfaces to touch, when the foam collapses. Stabilization of protein foams (e.g., meringues) is due to a somewhat different mechanism (discussed below).

EMULSIONS

4.1 Formation

Division of Internal Phase. Simply adding oil to water does not result in emulsion formation. Input of mechanical energy subdivides the droplets of internal phase until they reach the final average droplet diameter, in the range 1–100 μm. A cylinder of liquid whose length is more than 1.5 times its circumference is unstable and tends to break up into droplets. Mechanical stirring of an oil–water mixture forms drops of liquid that are then distorted into cylinders (along the lines of flow) and that break up into smaller droplets (Figure 12.7). The process is repeated until the droplets are so small they cannot be further distorted, and further subdivision ceases.

A suspended liquid drop forms a sphere, because this shape has minimum surface area (hence minimum interfacial free energy) for a given volume; area is related to the cube of droplet radius. Distortion is a flow shear effect, depending on droplet cross-section, related to the square of the radius. At large diameters, shear forces are greater than interfacial tension forces, droplets are distorted into cylinders, and subdivision occurs. Droplet radius decreases, until

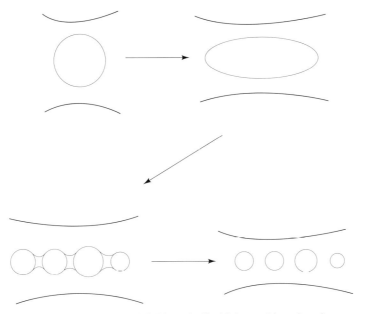

Figure 12.7 Stages in subdivision of a liquid drop subjected to shear.

the interfacial tension forces balance (or exceed) shear forces, and further division stops. In emulsification experiments in which the amount of mixing energy is constant and γ is changed by adding emulsifier, it is found that the average oil droplet diameter parallels γ, i.e., as more emulsifier is added, γ decreases and so does average droplet size. If γ is unchanged but mixing energy is increased, droplet size is also decreased. This is due to the change in the balance of shear and interfacial forces, allowing cylindrical distortion of smaller droplets. Equipment design that enhances shear is more effective at dividing droplets.

Oil–Water Versus Water–Oil Emulsions. If oil and water are vigorously shaken, they form a dispersion of water droplets in oil and oil droplets in water. When shaking is stopped the phases start to separate; small water drops fall toward the interface, and oil drops rise. The "emulsion" quickly breaks. Adding an emulsifier to the system changes the outcome; after standing, one phase becomes continuous, while the other remains dispersed. The nature of the emulsion is determined by the emulsifier. As a general rule, the continuous phase is the one in which the emulsifier is soluble. Thus sodium stearate promotes an oil-in-water (o/w) emulsion, while zinc distearate promotes a water-in-oil (w/o) emulsion. Several qualitative theories have been advanced to explain this empirical rule.

The oriented wedge theory states that the emulsifier at the interface is wedge shaped. The ionized end of a sodium soap has a wider (effective) radius than the hydrocarbon chain, hence the oil–water interface should be curved with the convex side oriented toward the water phase. This favors oil droplet formation, hence gives an o/w emulsion. The polar end of zinc distearate, on the other hand, is smaller than the two hydrocarbon chains, the interface is convex toward the water phase, and a w/o emulsion is formed.

A second theory considers the relative ease with which the two types of droplets can coalesce. Upon shaking, drops of both phases are formed. Sodium stearate ionizes, and the electrical potential hinders approach and coalescence of oil droplets; water droplets, on the other hand, experience no such hindrance and readily touch and coalesce. Zinc distearate, being un-ionized, does not interfere with the mutual approach of oil droplets, whereas van der Waal's forces favor subsequent coalescence. Thus the type of emulsion formed depends on the relative kinetics of oil–oil and water–water coalescence.

HLB System. The concept of hydrophilic–lipophilic balance (HLB) as a number for characterizing surfactants is an extension of the general rule of thumb stated above (4,9). The HLB scale as originally proposed ranged from 0 to 20, with the low end signifying an emulsifier that is much more soluble in oil than in water, and the high end meaning the reverse. A listing of HLB values for some food surfactants is given in Table 12.2. HLB is recommended as a guide for selecting emulsifier systems to give high emulsion stability to manufactured foods such as salad dressings, whipped toppings, and similar oil–water mixtures that must retain their emulsified character during a shelf life of up to 1 year.

The basic idea is that the HLB of a blend of two emulsifiers is equal to their algebraic sum, i.e., the weight fraction of *A* times its HLB plus the weight fraction of *B* times its HLB value. Boyd and co-workers (10) used a test system comprising equal volumes of water and Nujol (food grade mineral oil) and made blends of various Span (sorbitan fatty acid esters) and Tween (polyoxyethylene$_{20}$ sorbitan fatty acid esters) surfactants to cover the HLB ranges 8.5–16.5 (for o/w emulsions) and 2.0–6.5 (for w/o emulsions). After

Table 12.2 Hydrophilic–lipophilic balance

Surfactant	HLB
Sodium stearoyl lactylate	21.0
Polysorbate 80 PE(20) sorbitan monooleate	15.4
Polysorbate 60 PE(20) sorbitan monostearate	14.4
Sucrose monostearate	12.0
Polysorbate 65 PE(20) sorbitan tristearate	10.5
Diacetyl tartaric ester of monoglyceride	9.2
Sucrose distearate	8.9
Triglycerol monostearate	7.2
Sorbitan monostearate	5.9
Succinylated monoglyceride	5.3
Glycerol monostearate	3.7
Propylene glycol monostearate	1.8

the emulsions were made under controlled conditions, the rate of oil globule coalescence was used to measure emulsion stability. It was found that o/w emulsions were most stable at an HLB of about 12 and w/o emulsions were most stable when HLB was about 3.5. In a practical system that includes other ingredients such as sugar, salt, protein, and other typical food components, the value for this optimum might shift, and a series of tests are necessary to determine the exact blend of surfactants that gives the best results. As a rule of thumb, w/o emulsions are favored by HLBs in the 3–6 range, o/w emulsions are stabilized with HLBs in the 11–15 range, and intermediate HLBs give good wetting properties but not good emulsion stability.

Microemulsions. It is possible to make emulsions in which the diameter of the oil droplets is in the range 1.5–150 nm. The droplet size is less than the wavelength of visible light, and the emulsion appears transparent because no light scattering occurs. Several different strategies for making microemulsions have been tried (11), but a simple example using mineral oil and water demonstrates the principles involved. With pure liquids, γ is 41 mN/m, but the inclusion of 0.001 M oleic acid in the water phase reduces γ to 31 mN/m; a reasonably stable emulsion may be formed. Neutralization of the acid with 0.001 M NaOH lowers γ to 7.2 mN/m and gives a stable emulsion. Making the water phase 0.001 M in NaCl further lowers γ to less than 0.01 mN/m. This system will spontaneously form an emulsion; Brownian movement provides sufficient shear forces to elongate droplets into cylinders and cause further subdivision.

Spontaneous emulsions such as this one are frequently opalescent, because some particles have a diameter approaching the wavelength of light: 400 nm. Transparent microemulsions generally require a surfactant plus a cosurfactant, for example, acetyl monoglyceride plus hexanol. For use in food, various polyglycerol esters have shown some promise for making w/o microemulsions (12). The technology is promising but needs further refinement before it is readily applied to food systems.

4.2 Flocculation

Creaming and Adhesion. Oil droplets in an o/w emulsion float to the top, because the density of vegetable oil is about 0.91 g/mL, 0.08 g/mL less than that of water. The rate at which they rise depends on particle diameter. A drop having a 1 μm diameter rises at a rate of 4 cm/day, while one with a 10 μm diameter rises 4 m/day. Obviously, reducing the average droplet size reduces the rate of creaming. Fat globules in raw milk have an average diameter of 3 μm; after homogenization the average diameter is 0.5 μm. In raw milk, the average flotation rate is 36 cm/day, while in homogenized milk it is 1 cm/day. Creaming brings the oil droplets closer together, and if contact is not prevented (say, by ionic repulsion) coalescence occurs. A simple creamed layer of oil drops is readily redispersed by inverting the container a few times.

In some circumstances the oil drops actually adhere to each other and are not readily redispersed. This occurs when the emulsifier is polymeric (e.g., protein, gum, or polyoxyethylene derivative). By one mechanism, different segments of a polymer molecule adsorb to the surfaces of two drops, thus forming a bridge that holds them together. Another mechanism obtains when polar parts of two polymer molecules (adsorbed to separate drops) approach each other and intertwine. This "tangle," say of long polyoxyethylene chains, then holds the drops in proximity, although they are not in actual contact. The creamed layer is said to be flocculated, although it is also stable against coalescence.

4.3 Stabilization and Coalescence

Ionic Repulsion. As discussed above, two surfaces carrying an identical net charge repel each other. The thickness of the electrical double layer (the region where $\psi > 0$) is affected by ionic strength. As long as ionic strength is low, electrical repulsion is greater than van der Waal's attraction, and the droplets remain suspended. If, by addition of salt (particularly divalent or trivalent salts) the ψ potential is markedly suppressed, the surfaces can approach so closely that van der Waal's forces override repulsion, and the droplets can touch and coalesce. At some intermediate ionic strength, the two forces are approximately equal. There is actually a small free energy minimum, and the droplets will remain separated by about one droplet diameter (Figure 12.8). The practical conclusion to be drawn from this is as follows: if the emulsifier used is ionic in nature, the salt concentration of the aqueous phase markedly affects emulsion stability. Low salt concentration enhances stability, while high salt concentration increases flocculation and/or coalescence.

Steric Hindrance. Another form of stabilization is relatively independent of ionic strength: the oil droplets are prevented from making contact by simple steric hindrance. This may take two forms, either an immobilized water layer at the interface or a solid interfacial film. Emulsion stabilization by proteins, gums, and polyoxyethylene derivatives occurs by the first mechanism. Hydrophobic parts of the stabilizers adsorb at the oil surface, but adjacent large hydrophilic segments are hydrated and form an immobilized layer on the order of 10–100 nm thick (Figure 12.9). As mentioned, these hydrated segments frequently interact to cause flocculation, while coalescence of the oil drops themselves is prevented. Such emulsions are frequently used as carriers for oil-soluble flavors, essences, and colorants.

The α-tending emulsifiers such as propylene glycol monostearate are oil soluble. The emulsifier adsorbs at the oil–water interface, but under certain conditions (low temperature, presence of a free fatty acid) the emulsifier forms a solid interfacial film (Figure 12.10). While the oil droplets may make contact, the film prevents coalescence. The interfacial layer actually appears to be crystalline, with a well-defined melting point (14).

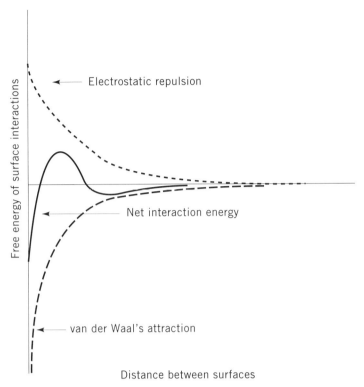

Figure 12.8 Net free energy of interaction between two droplets.

FOAMS

5.1 Formation, Film Drainage

For the purposes of this Chapter, air is a nonpolar medium. Surfactants concentrate at the air–water interface, with the hydrophobic portion extending into the gas phase. When the gas phase is finely divided, a foam is formed. In regard to the energetics involved, a foam is nearly identical to an o/w emulsion.

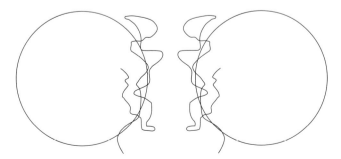

Figure 12.9 Hydrophilic polymeric surfactant at the interface in an oil-in-water emulsion.

496 Emulsifiers for the Food Industry

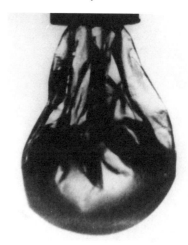

Figure 12.10 The interfacial film formed by an α-tending emulsifier. A water drop was suspended in vegetable oil containing 10% propylene glycol monoester; after a few minutes some of the water was withdrawn. Reprinted by permission from Ref. 13.

The terminology is somewhat different, but the results are the same: a foam is stable or else the gas bubbles coalesce and the foam breaks. Rather than referring to "creaming," a foam is said to "drain"; the effect is the same, with the water phase concentrating at the bottom and the dispersed phase concentrating at the top of the container. The volume fraction (ϕ) of gas in a foam is usually much higher than the ϕ of oil in an emulsion. Whipping egg white, for example, may easily give a 10- to 15-fold expansion ($\phi = 0.9$–0.93), while mayonnaise ($\phi = 0.7$–0.8) is the food emulsion with the highest oil content.

The mechanism of air incorporation and subdivision in a foam is the same as for an emulsion; large bubbles are elongated, and the unstable cylinders spontaneously divide. In a wet foam ($\phi < 0.7$), the initial drainage of liquid from the regions between the bubbles is due to gravity and is governed primarily by viscosity. Figure 12.11 shows how the film between three bubbles meets at a 120° angle, after drainage. In three dimensions, four touching

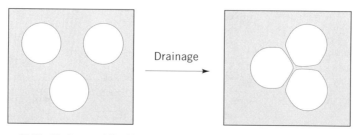

Figure 12.11 Drainage of liquid from between bubbles, forming the plateau border.

bubbles meet at the tetrahedral angle of 109°28′. In the real situation, the bubbles assume the shape of regular polyhedra, but the contact angles remain fairly close to the ideal value. The thicker liquid in the corners, known as the plateau border, has a lower pressure relative to the straight contact films, and liquid moves into these regions. Drainage of a dry foam ($\phi > 0.7$) is probably via these borders, connected throughout the foam.

5.2 Stabilization

Foam stability is governed by similar factors as emulsion stability. Thus in a soap foam the negative charges located at the air–water interface lead to repulsion as the two surfaces of the film approach each other, and drainage stops when an equilibrium film thickness is reached. This thickness is influenced by the ionic strength of the aqueous phase; increasing the ionic strength gives foams of lower stability. Protein stabilizes foams by a combination of steric hindrance and surface viscosity. When egg whites are whipped, the protein molecules unfold, with the hydrophobic side chains entering the air phase and the hydrophilic chains remaining in the water phase. In Figure 12.12 the heavy lines in the magnified section represent unfolded albumin proteins adsorbed at the air–water interface. The portion of the proteins located in the aqueous phase hold water, preventing it from draining away from this region and hence preventing the air bubbles from coalescing and destabilizing the foam.

Film breakage is thought to be due to random fluctuations (e.g., Brownian movement) that momentarily bring the two surfaces into contact, allowing the air bubbles to merge. These fluctuations are minimized when the surface viscosity is increased. The addition of an alcohol to a soap solution (e.g., dodecanol added to sodium laurate) increases surface viscosity and also increases foam stability. Surface viscosity of some (but not all) protein solutions is quite high, and there is some correlation between this property and the

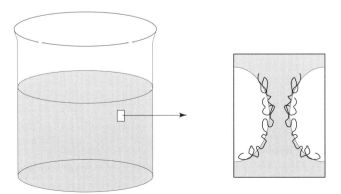

Figure 12.12 A protein foam, showing unfolded protein molecules at the air–water interface.

498 Emulsifiers for the Food Industry

ability of the protein to give a stable foam. Bulk viscosity does not correlate with stability of drained films, but if a wet foam is desired, increasing bulk viscosity (say, by adding a high viscosity gum) extends the usable life of the foam.

WETTING

6.1 Liquid Spreading on Solid Surface

As mentioned above, surfactants with HLBs in the 8–10 range are not good emulsion stabilizers but are good wetting agents. This property is useful in many circumstances: enhancing dispersion of dry mixes in liquid, improving spreadability of chocolate and cocoa-based coatings, and incorporation of dietary fiber materials in dressings. Qualitatively, a drop of water is placed on the solid surface. If the contact angle $\theta > 90°$ (Figure 12.13), the water does not spread; it is said that the solid is not wetted. If $\theta < 90°$, the water spreads, and the solid is wetted.

The angle θ is determined by the surface tension at the three interfaces involved:

$$\cos \theta = (\gamma_{SV} - \gamma_{SL})/\gamma_{LV} \tag{6}$$

The spreading coefficient is defined as:
$$S_{L/S} = \gamma_{SV} - \gamma_{LV} - \gamma_{SL} \tag{7}$$

When $S_{L/S} > 0$, wetting occurs, and the liquid spreads. An efficient wetting agent is one that minimizes the surface tension of the air–water and solid–water interfaces, while leaving the air–solid surface tension unchanged. This is the situation, for example, when dry beverage powder is added to water. Sodium lauryl sulfate lowers the air–water and solid–water interfacial tensions and enhances dispersibility. In the absence of the surfactant, the contact angle at many of the (irregular) solid surfaces is $> 90°$, and the powder with its entrapped air floats on the top of the water.

In chocolate coating, the liquid phase is an oil (cocoa butter). The addition of lecithin aids the wetting of solid cocoa particles by this oil, most probably

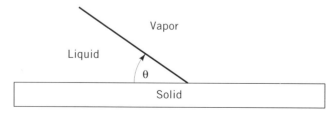

Figure 12.13 The contact angle between liquid, solid, and air phases.

by lowering γ_{SL}. This lowers the viscosity of the heterogenous mass as well as giving a smoother mouthfeel to the final product.

PHYSICAL STATE OF EMULSIFIER PLUS WATER

7.1 Phase Diagrams

Mixtures of surfactant and water form a number of different physical structures, depending on the surfactant: water ratio and the temperature. These mixtures are frequently opalescent dispersions, called liquid crystals, but are more properly termed mesophases. In publications on the subject these are usually shown as phase diagrams with temperature as the y-axis and percent water on the x-axis; the interior of the graph is divided into regions that represent the various mesophases (Figure 12.14) (15). Phase diagrams provide guidance to researchers who are trying to produce surfactant systems that are functional under use conditions. They are used to characterize detergent systems and industrial applications of surfactants as well as for food emulsifiers, which are the present concern.

7.2 Mesophase Structures

The main mesophase structures of interest in the baking industry are depicted in Figure 12.15 (15,16). A monoglyceride such as GMS crystallizes in bilayers, with the thickness of each bilayer being defined by the length of two monoglyceride molecules end to end. When heated in water, the crystals melt (the fatty acid chains gain thermal mobility and lose their ordered structure) and water begins to intrude between the bilayers along the plane defined by the glycerol head groups. Under the proper conditions of temperature and water content, this intrusion results in the formation of the *lamellar* mesophase. The thickness of the water layer may be from 60 nm to 1.75 μm, depending on various conditions (16), while the distance between the water layers due to the lipid bilayer remains relatively constant at around 400 nm. This material is rather fluid, but when the mixture is cooled, the lipid layers solidify in the α-crystalline state, and the material becomes a gel with a lipid bilayer about 550 nm thick; the water layer also may increase in thickness, and some of the water may become free. This phase is of particular interest to bakers, because there is evidence that the lamellar mesophase is the most efficient in promoting the interaction between monoglyceride and starch, producing the antistaling effects (17). If this gel is cooled even further and allowed to come to equilibrium, all the water is expelled from the gel and the α-crystalline layers of monoglyceride transform into the more stable β-crystalline form, yielding a suspension of β crystals in water. At higher water levels and within a limited temperature range, the lamellar mesophase is transformed into spherical multilamellar vesicles (liposomes), which is sometimes called a lamellar dispersion.

500 Emulsifiers for the Food Industry

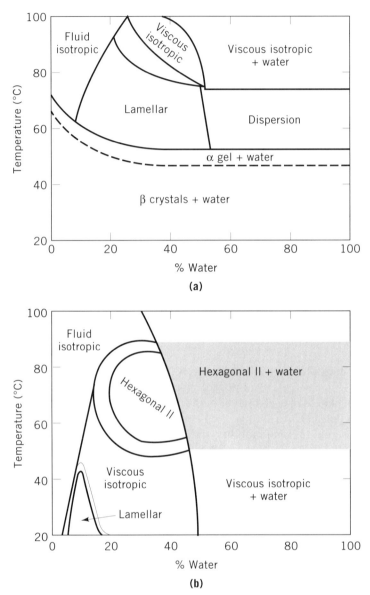

Figure 12.14 Phase diagrams of monoglycerides in water. (**a**) Refers to a distilled monoglyceride made with hydrogenated lard; (**b**) shows distilled monoglyceride made from sunflower oil (15).

At higher temperatures and water concentrations, the system may shift into the cubic mesophase structure (see Figure 12.15). The water is present as spheres totally surrounded by monoglyceride. This phase has a high viscosity and is sometimes called viscous isotropic in the literature; the two terms refer to the same structure. In the presence of more water than can be accommodated in the internal spherical phase, one obtains a mixture of lumps of this

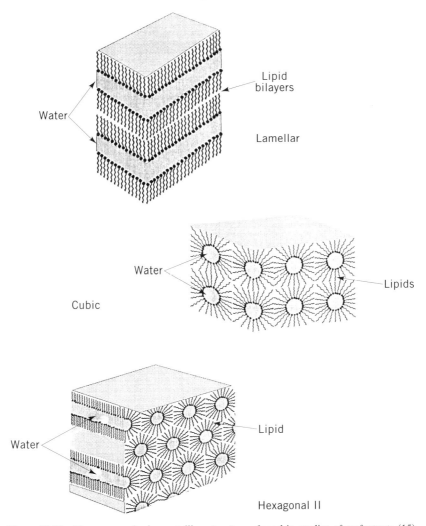

Figure 12.15 Three mesophasic crystalline structures found in studies of surfactants (15).

cubic structure dispersed in excess water. With a saturated monoglyceride such as GMS, the lamellar structure is the main mesophase found under practical conditions, while with unsaturated monoglycerides this cubic phase is the predominant one at lower temperatures. At lower water concentrations, the spherical water "micelles" are farther apart, so the viscosity of the mixture becomes lower, approaching that of melted pure surfactant. This is the fluid isotropic mesophase, sometimes referred to as the L2 phase.

The third important mesophase structure is formed by unsaturated monoglycerides and also by most of the other surfactants of interest to bakers. The hexagonal structure (see Figure 12.15) consists of rods of the internal phase arranged hexagonally within a matrix of the external phase. In the case of monoglycerides and SSL, the internal phase is water, forming the hexagonal II structure as shown in Figure 12.15. With highly water-soluble surfactants of the polyoxyethylene type (polysorbates, ethoxylated monoglyceride), the internal phase is the lipophilic tail of the material and the external phase is water, giving the hexagonal I mesophase (15).

The lamellar mesophase of monoglyceride is stabilized if an anionic surfactant, such as SSL or a neutralized free fatty acid (soap), is part of the lipid fraction. The lipid–water interface takes on a negative charge, and electrostatic repulsion prevents the collapse and expulsion of water at the lower temperatures described above. As expected, this electrostatic stabilization can be counteracted by adding salts, and low concentrations (0.3%) in the aqueous phase will counteract the favorable effects of the anionic surfactant. Anionic monoglyceride derivatives, the succinate and diacetyltartrate esters, form lamellar mesophases under most conditions. This penchant is enhanced if the carboxyl group is partially neutralized so that the pH of a water dispersion of the surfactant is in the typical pH range for dough of 4–6.

Finally, the inclusion of a nonpolar triglyceride (oil) in the system tremendously complicates the phase diagrams. As an example, GMS forms a lamellar mesophase with water, as was noted above, but when soybean oil is added, the system converts into the hexagonal II structure (18). The ternary phase diagrams from combinations of flour lipids, water, and salt have been studied (19,20). The structures listed above were all found in some region of the phase diagram. The studies are interesting, but as of this writing the direct relevance to the practical application of surfactants in dough systems is not clear.

EMULSIFIERS FOR FOOD APPLICATIONS

The actual food emulsifier of commerce is seldom exactly like the organic chemical structures that are discussed in this section but is a mixture of similar compounds derived from natural raw materials. As a simple comparison, sodium chloride is one single, relatively pure chemical entity, accurately described by the formula NaCl. Glycerol monostearate, on the other hand, is made from a hydrogenated natural fat or oil, and the saturated fatty acid composition may well be something like 1% C_{12}, 2% C_{14}, 30% C_{16}, 65% C_{18}, and 2% C_{20}, reflecting the chain length distribution in the source fat. In addition, the monoglyceride will be approximately 92% 1-monoglyceride and 8% 2-monoglyceride, which is in chemical equilibrium. Keep in mind, then, that the chemical structures shown here represent the major components in the commercial material and that related molecular species will also be present.

8.1 Monoglycerides and Derivatives

The manufacture and application of monoglycerides and derivatives have been reviewed by several authors (21–23). Roughly 18 million kg (40 million lb) of monoglyceride are used annually in the United States in yeast-raised goods for retarding staling (23). At least an equal amount finds its way into cakes, icings, and other applications. The third major use of monoglycerides is in the manufacture of margarine. Overall, this group of surfactants is the single most important one for food uses, representing about 75% of total emulsifier production.

The use of monoglycerides in baking first began in the 1930s, when "superglycerinated shortening" became commercially available. Glycerin was added to ordinary shortening along with a small amount of alkaline catalyst, the mixture was heated, causing some interesterification of triglyceride with the glycerin, and the catalyst was removed by neutralization and washing with water. The resulting emulsified shortening contained about 3% monoglyceride and was widely used for making cakes, particularly ones containing sugar at high levels. The effectiveness of monoglyceride in retarding staling (crumb firming) in bread became known at about the same time, and bread bakers sought a more concentrated source of monoglyceride. This need was met by suppliers of plastic monoglyceride, which is made by altering the ratio of glycerin to fat to achieve a final concentration of 50–60% monoglyceride, with most of the remainder being diglyceride. When industrial-scale molecular distillation processes became available it was logical to subject the plastic monoglyceride to this step, producing distilled monoglyceride containing a minimum of 90% monoglyceride, the rest of the mixture being diglyceride and small amounts of fatty acids, glycerol, and triglyceride. The next stage was to make a lamellar mesophase product from this distilled monoglyceride, adding some anionic surfactant (usually SSL) to stabilize the hydrated monoglyceride, which contains roughly 25% monoglyceride, 3% SSL, and 72% water. More recently, manufacturers have developed a powdered distilled monoglyceride in which the composition of the original feedstock fat is balanced between saturated and unsaturated fatty acids, so that the resulting powder is hydrated fairly rapidly during the process of dough mixing and is functional in complexing with gelatinized starch. Today, bakers use all three monoglyceride types (plastic, hydrated, and powdered distilled) with about equally good results.

The monoglyceride structure shown in Figure 12.16 is for 1-monostearin, also called α-monostearin. If the fatty acid is esterified at the middle hydroxyl the compound is 2-monostearin, or β-monostearin. Both isomers are equally effective at retarding bread staling (23). In technical specifications, manufacturers usually give the monoglyceride content of their product as percent of α-monoglyceride. The routine analytical method for monoglyceride (AACC method 58-45) detects only the 1-isomer; quantitation of the 2-isomer is much more tedious. The total monoglyceride content of a product is about 10% higher than the reported α-monoglyceride content. In a practical sense, however, when various products are being compared for functionality and cost effectiveness, the α-monoglyceride content is a useful number, since for all products this equals about 92% of the total monoglyceride present.

The fatty acid composition of monoglyceride reflects the makeup of the triglyceride fat from which it is made. Commercial GMS may contain as little as 65% stearate if it is made from fully hydrogenated lard, or as much as 87%

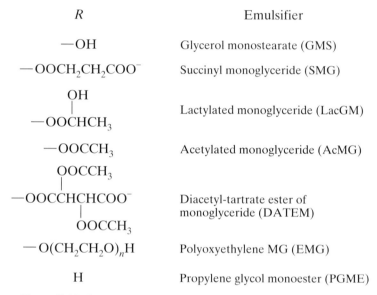

Figure 12.16 Structure of monoglyceride and several related compounds.

stearate if it is made from fully hydrogenated soybean oil. The other major saturated fatty acid will be palmitic, and because complete hydrogenation (to iodine value of 0) is not practical, a few percent of unsaturated (oleic and/or elaidic) acid are also usually present. A typical commercial GMS will have an iodine value of about 5. Iodine values for powdered distilled monoglycerides are in the range of 19–36, and for plastic monoglycerides, a typical range is 65–75. The unsaturated fatty acids are a mixture of oleic, linoleic, and the *trans* isomers of these acids. The phase diagram of a highly unsaturated monoglyceride is quite different from that of a saturated one (see Figure 12.14), but the higher melting point of the hydrogenated monoglyceride makes its phase behavior much more like that of GMS than that of the sunflower monoglyceride. To the extent that mesophase behavior governs monoglyceride functionality, the plastic monoglycerides are quite adequate.

The derivatives of monoglyceride shown in Figure 12.16 are of two types: (*1*) dough strengtheners (SMG, EMG, and DATEM) and (*2*) α-tending emulsifiers (LacGM, AcMG, and PGME). Details of the functionality of these two groups in baking are discussed below.

The succinate and diacetyltartrate half esters are fairly discrete compounds, produced by reacting the respective acid anhydride with monoglyceride in the presence of alkaline catalyst. The product of the manufacture of ethoxylated monoglyceride is somewhat more random in structure. Monoglyceride is treated with ethylene oxide gas under pressure in the presence of alkaline catalyst and at elevated temperatures. Ethylene oxide is polymerized via a series of ether linkages and also forms ether bonds with the free hydroxyl groups on monoglyceride. The average chain length comprises about 20 units ($n = 20$ in Figure 12.16). Chains may be attached to hydroxyls at both the number 2 (β) and number 3 (α) positions of the monoglyceride, although many more chains will be located at the α position than at the β spot because of the difference in their chemical reactivities. The exact distribution of polymer chain lengths and distribution between α and β positions are functions of reaction conditions, e.g., catalyst type and concentration, gas pressure, temperature, agitation, and length of reaction time.

The second group of monoglyceride derivatives, the α-tending emulsifiers, find their main use in cake production. These emulsifiers are dissolved in the shortening phase of the cake formulation, and they contribute to the emulsification of the shortening in the water phase as well as promoting incorporation of air into the fat phase. The particular property of these emulsifiers that makes them valuable in liquid shortening cakes is that they form a solid film at the oil–water interface, not only stabilizing the emulsion but also keeping the lipid phase from preventing air incorporation (protein-stabilized foam formation) during cake batter mixing.

8.2 Sorbitan Emulsifiers

When the sugar alcohol sorbitol is heated with stearic acid in the presence of a catalyst, two reactions occur: sorbitol cyclizes to form the five-membered sorbitan ring and the remaining primary hydroxyl group is esterified by the acid. The resulting sorbitan monostearate (Figure 12.17) is oil soluble, with a rather low HLB value, and it is the only one of the many sorbitan esters presently approved for food use in the United States. Other sorbitan esters of importance are the monooleate and the tristearate. Any of the three esters may be reacted with ethylene oxide to give polyoxyethylene derivatives (see Figure 12.17). The monostearate derivative is known as polysorbate 60, the tristearate is polysorbate 65, and the monooleate is polysorbate 80. The remarks made in connection with EMG regarding the length and location of the polyoxyethylene chains apply to these compounds. The average number of oxyethylene monomers is 20 ($n = 20$), and in the case of the monoesters, chains may be located on more than one hydroxyl group of the sorbitan

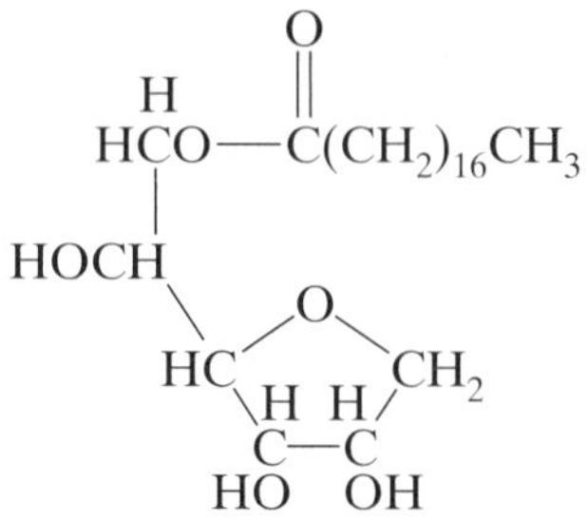

Sorbitan monostearate

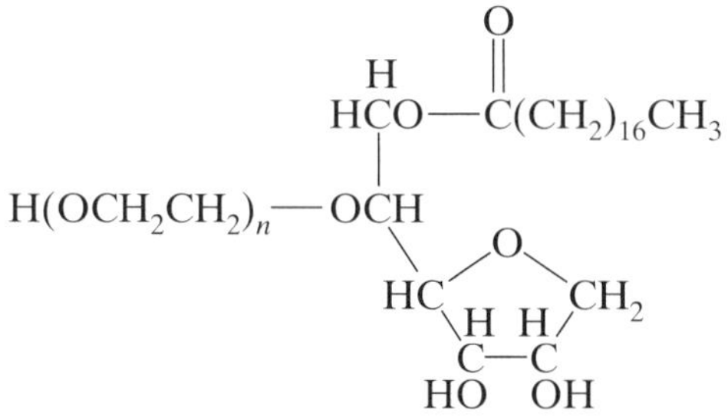

Polyoxyethylene (20) sorbitan monostearate (Polysorbate 60)

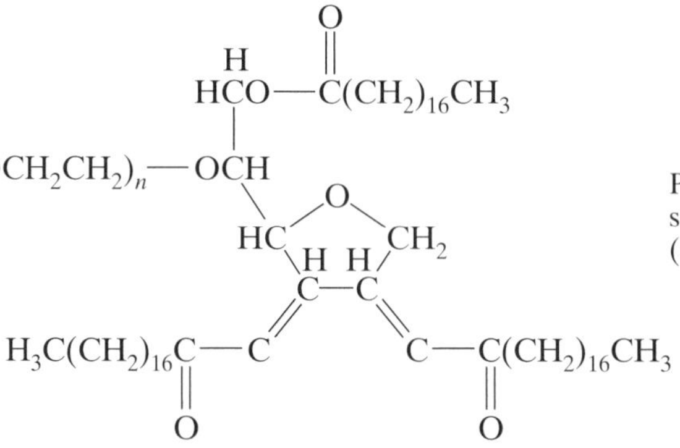

Polyoxyethylene (20) sorbitan tristearate (Polysorbate 65)

Figure 12.17 Structures of food-grade sorbitan derivatives.

ring (with the triester, of course, only one hydroxyl group is available for derivatization).

Sorbitan monostearate finds use as a good emulsifier for making icings that have superior aeration, gloss, and stability characteristics. It is also used as part of the emulsifier system in whipped toppings and in coffee whiteners. The polyoxyethylene derivatives have found more acceptance, with the monostearate polysorbate 60 being the most widely used of the group. At a level of 0.25% (flour basis) the ability of polysorbate 60 to strengthen dough against mechanical shock is greater than that of SMG and about equal to those of EMG and SSL. Polysorbate 60 has also been used in fluid oil cake shortening systems, generally in combination with GMS and PGMS.

8.3 Anionic Emulsifiers

In addition to SMG and DATEM, some other anionic surfactants that have been tried as dough strengtheners are shown in Figure 12.18. Presently SSL is the one most widely used in the United States; sodium stearyl fumarate did not find acceptance, and sodium lauryl sulfate is used mainly as a whipping agent with egg whites.

Lactic acid, having both a carboxylic acid and a hydroxyl function on the same molecule, readily forms an ester with itself. In commercial concentrated solutions almost all the acid is present in this polylactylic form, and to get free lactic acid it must be diluted with water and refluxed for a period of time. When stearic acid is heated with polylactic acid under the proper reaction conditions and then neutralized with sodium hydroxide, a product having the structure shown in Figure 12.18 is obtained. The monomer lactylic acid shown represents the predominant product; the dilactylic dimer is also present as well as lactylic trimers and tetramers (24). As with all compounds based on commercial stearic acid derived from hydrogenated fats, some percentage of the fatty acid is palmitic, with small portions of myristic and arachidic acids also being present. Sodium stearoyl lactylate is readily water soluble, but the calcium salt is practically insoluble. In this respect, it mimics a soap e.g., sodium stearate is water soluble but calcium stearate is oil soluble. Either form may be used, depending on the details of the intended application, but as a dough strengthener, the sodium salt is more commonly used. As a stabilizer for hydrated monoglyceride the sodium form is used, because ionization in the water layer is necessary.

Stearyl fumarate is a half ester of fumaric acid with stearyl alcohol (octadecanol). Although stearyl fumarate might be expected to have dough-strength-

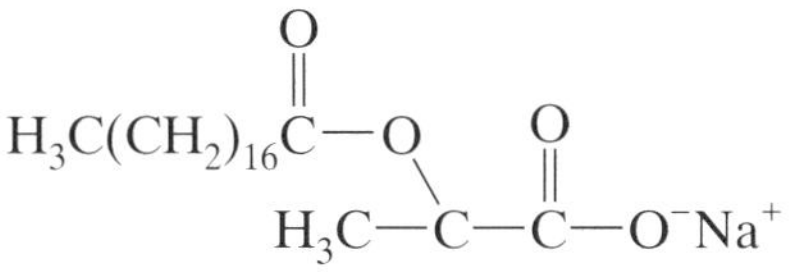

$$H_3C(CH_2)_{16}\overset{O}{\underset{\|}{C}}-O-\underset{H_3C}{\underset{|}{C}}H-\overset{O}{\underset{\|}{C}}-O^-Na^+$$

Sodium stearoyl lactylate

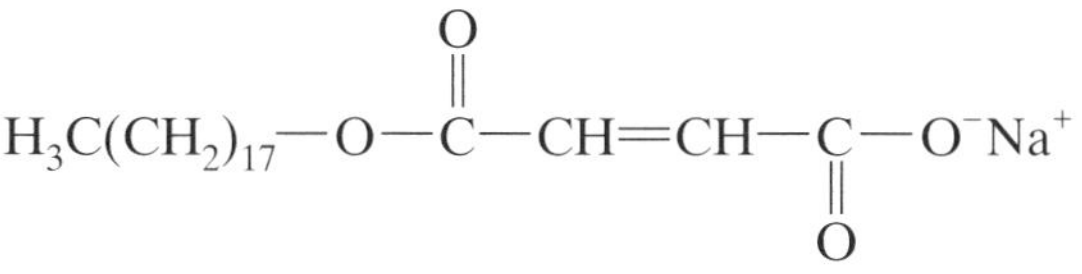

$$H_3C(CH_2)_{17}-O-\overset{O}{\underset{\|}{C}}-CH=CH-\overset{O}{\underset{\|}{C}}-O^-Na^+$$

Sodium stearyl fumarate

$$H_3C(CH_2)_{11}-O-SO_3^-Na^+$$

Sodium lauryl sulfate

Figure 12.18 Three food-grade anionic surfactants.

ening properties similar to those of SSL, this was not found to be so in practice, and the product was not a commercial success. Stearyl fumarate is still approved by the FDA for use in bread.

The third structure shown in Figure 12.18 is sodium dodecyl sulfate (SDS), often used by research workers for solubilizing proteins. It is a sulfate ester of the C_{12} alcohol dodecanol. Commercially, this alcohol is produced by reduction of coconut oil, and the resultant mixture is called lauryl alcohol (from lauric acid, the predominant fatty acid in coconut oil). The alcohol portion of sodium lauryl sulfate is a mixture of chain lengths, the approximate composition being 8% C_8, 7% C_{10}, 48% C_{12}, 20% C_{14}, 10% C_{16}, and small amounts of longer chains. In bakeries the most common use of sodium lauryl sulfate is as a whipping aid. The compound is added to liquid egg whites at a maximum concentration of 0.0125%, or to egg white solids at a level of 0.1%. It promotes the unfolding of egg albumin at the air–water interface and the stabilization of the foam.

8.4 Polyhydric Emulsifiers

Polyglycerol Esters. Polyglycerol esters (Figure 12.19) have a variety of applications as emulsifiers in the food industry. The polyglycerol portion is synthesized by heating glycerol in the presence of an alkaline catalyst; ether linkages are formed between the primary hydroxyls of glycerol. In Figure 12.19 n may take any value, but for food emulsifiers the most common ones are n = 1 (triglycerol), n = 4 (hexaglycerol), n = 6 (octaglycerol), and n =

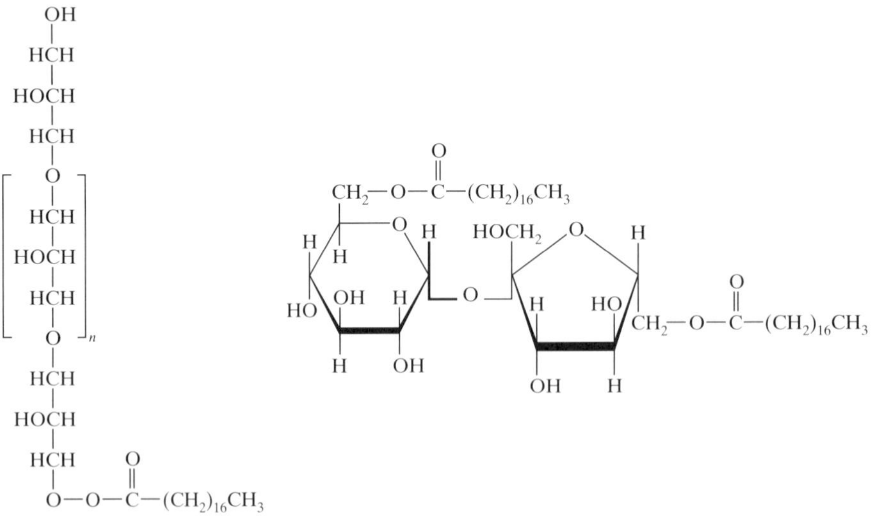

Polyglycerol monostearate

Sucrose diester

Figure 12.19 Surfactants based on polyglycerol and sucrose.

8 (decaglycerol), remembering that in all cases n is an average value for the molecules present in the commercial preparation. The polyglycerol backbone is then esterified to varying extents, either by direct reaction with a fatty acid or by interesterification with a triglyceride fat. Again, the number of acid groups esterified to a polyglycerol molecule varies around some central value, so an octaglycerol octaoleate really should be understood as an (approximately octa)-glycerol (approximately octa)-oleate ester. By good control of feedstocks and reaction conditions, manufacturers manage to keep the properties of their various products relatively constant from batch to batch.

The HLB balance of these esters depends on the length of the polyglycerol chain (the number of hydrophilic hydroxyl groups present) and the degree of esterification. For example, decaglycerol monostearate has an HLB of 14.5, while triglycerol tristearate has an HLB of 3.6. Intermediate species have intermediate HLB values, and any desired value may be obtained by appropriate blending, as described earlier. The wide range of possible compositions and HLBs makes these materials versatile emulsifiers for food applications.

Sucrose Esters. Sucrose has eight free hydroxyl groups, which are potential sites for esterification to fatty acids. Compounds containing six or more fatty acids per sucrose molecule have been proposed for use as noncaloric fat substitutes under the name Olestra; this material acts like a triglyceride fat and has no surfactant properties. Compounds containing one to three fatty acid esters, on the other hand, do act as emulsifiers and are approved for food use in that capacity. They are manufactured by the following steps:

1. An emulsion is made of fatty acid methyl ester in a concentrated aqueous sucrose solution.
2. The water is removed under vacuum at elevated temperature.
3. Alkaline catalyst is added, and the temperature of the dispersion is raised slowly to 150°C under vacuum, distilling off methanol formed on transesterification.
4. The reaction mixture is cooled and purified.

The degree of esterification is controlled by the reaction conditions, especially the sucrose: methyl ester ratio, and the final product is a mixture of esters (Table 12.3). The HLB value of a particular product is smaller (more lipophilic) as the degree of esterification increases, as would be expected.

8.3 Lecithin

Lecithin is a by-product of the processing of crude soybean oil; it is the "gum" that is removed during the degumming step of oil refining. The crude gum is treated and purified to give the various commercial lecithin products that are available to the food processor today (26,27). Oil-free lecithin is a sticky plastic

Table 12.3 Sucrose ester surfactants (25)

Manufacturer's Designation	Percent Monoester	Percent Diester	Percent Triester	Percent Tetraester	HLB
F-160	71	24	5	0	15
F-140	61	30	8	1	13
F-110	50	36	12	2	11
F-90	46	39	13	2	9.5
F-70	42	42	14	2	8
F-50	33	49	16	2	6

material, but blending with half its weight of soy oil sharply reduces the viscosity to give the product known as fluid lecithin. The crude material is dark, almost black (mainly because of the high temperatures during processing), so it is bleached to give a more acceptable light brown color. Treatment with up to 1.5% hydrogen peroxide gives the product known as single-bleached lecithin, and further addition of up to 0.5% benzoyl peroxide yields double-bleached lecithin (26). Reaction with hydrogen peroxide at even higher levels, plus lactic acid, hydroxylates unsaturated fatty acid side chains at the double bond (yielding, e.g., dihydroxystearic acid from oleic acid). Hydroxylated lecithin is formed, which is more dispersible in cold water than the other types and is more effective as an emulsifier for o/w emulsions.

Figure 12.20 shows the structure of the main surface-active components of lecithin. The phosphatidyl group is a phosphate ester of diglyceride. The fatty

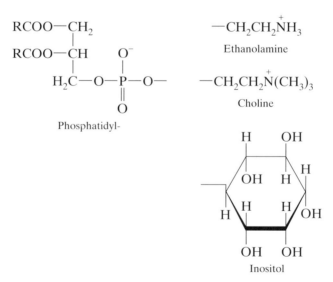

Figure 12.20 Lecithin structure and components.

acid composition of the diglyceride is similar to that of the basic oil so a number of different fatty acids are found, not just the stearic and oleic acids depicted. Little phosphatidylserine is present in soybean lecithin, and the other three derivatives are found in approximately equal amounts. Phosphatidylethanolamine (PE) and phosphatidylcholine (PC) are amphoteric surfactants, while phosphatidylinositol (PI) is anionic. The HLB values of the three species are varied, with PC having a high HLB, PE an intermediate HLB, and PI having a low value. The HLB of the natural blend is in the range of 9–10, and emulsifier mixtures having values in this range will tend to form either o/w or w/o emulsions, although neither type is highly stable. On the other, hand intermediate HLB emulsifiers are excellent wetting agents, and this is a major application for lecithin.

The emulsifying properties of lecithin can be improved by ethanol fractionation (28). Phosphatidylcholine is soluble in ethanol, PI is rather insoluble, and PE is partially soluble. Adding lecithin to ethanol gives a soluble and an insoluble fraction. The phosphatide compositions of the two are (*1*) ethanol soluble, 60% PC, 30% PE, and 2% PI; and (*2*) ethanol insoluble, 4% PC, 29% PE, and 55% PI (29). The soluble fraction is effective in promoting and stabilizing o/w emulsions, while the insoluble portion promotes and stabilizes w/o emulsions. At the present time several European companies are using this process to produce industrial food-grade emulsifiers.

INTERACTIONS WITH OTHER FOOD COMPONENTS

9.1 Starch

Starch molecules are polymers composed of α-D-glucopyranosidyl residues joined primarily by 1,4 acetal linkages (1,6 linkages occur at the branch points in amylopectin). Glucopyranoside is a six-membered ring, not flat, but puckered in what is called the "chair" configuration (Figure 12.21). The bond angles from each carbon are such that the hydrophilic hydroxyl groups project outward to the side of the plane of the ring, while the hydrogen atoms project either above or below this plane; the perimeter of the ring is hydrophilic, while the two faces are hydrophobic. The bond angle of the α-1,4 acetal linkage is such that the starch chain coils to form a helix, with about six residues per turn (see Figure 12.21). It is difficult to draw the details of this helix, but from molecular models it is apparent that the plane of the residue ring lies parallel to the wall of this helix and the hydrogen atoms on carbons 3 and 5 project into the interior of the helix. The result is a hollow cylinder that has a hydrophilic outer surface and a hydrophobic inner surface. This inner space is about 45 nm in diameter, and straight-chain alkyl molecules such as stearic acid will fit into it (see Figure 12.21), as will other molecules,

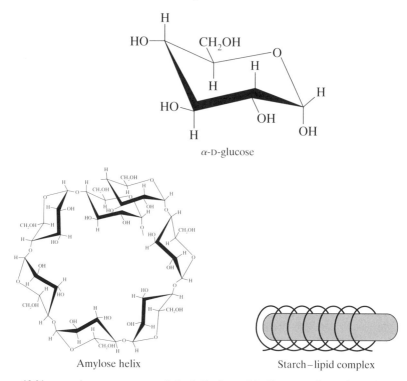

Figure 12.21 α-D-glucose, one turn of the helix formed in linear portions of starch, and the complex between a linear alkane and a starch helix.

e.g., iodine. The blue color of an iodine–starch complex is caused by the presence of iodine in this nonpolar environment; iodine dissolved in chloroform is blue, but in water, it is brown. The complex of amylose with n-butanol crystallizes much more readily than does amylose alone, and it has been used to separate amylose from amylopectin.

The n-alkyl portion of emulsifiers such as glycerol monostearate form a complex with helical regions of starch, a phenomenon that is thought to be responsible for the ability of GMS to retard starch crystallization in bread crumb, slowing the process of staling. Several workers have measured the stoichiometry of complex formation by using various methods. Some of these studies are mentioned later in connection with the discussion on bread staling; here three typical reports are summarized.

Lagendijk and Pennings (30) heated amylose or amylopectin with excess monoglyceride, cooled the heated material, and measured the amount of uncomplexed monoglyceride. With amylose, the maximum complexation occurred with monopalmitin, with both longer and shorter saturated fatty acid monoglycerides reacting to a lesser extent. If the average molecular weight of amylose is taken as 150 kDa, then 1 mole of amylose bound 10.6 moles of monomyristin (23.4 mg/g starch), 16 moles of monopalmitin (37.4 mg/g), 14 moles of monostearin (33.9 mg/g), 10.4 moles of monoolein (24.6 mg/g), and 5.1 moles of monolinolein (12.2 mg/g). These results show the effect of unsaturation in the alkyl chain; the cis double bond makes the chain bend so that it is more difficult to form a complex with the supposedly straight cavity of the amylose helix. Amylopectin bound increasing amounts of the saturated monoglycerides, the weight ratios being 5 mg monopalmitin, 8.3 mg monostearin, and 11 mg monoarachidin per gram of amylopectin.

Batres and White (31) reacted monoglyceride with amylopectin at 60°C, isolated the precipitate that separated on cooling, and analyzed this complex. They found significantly more interaction than Lagendijk and Pennings, obtaining values of 370 mg of monomyristin, 580 mg of monopalmitin, and 250 mg of monostearin per gram of amylopectin. While there were procedural differences between the two studies, these differences would not seem to be enough to account for such large discrepancies. Nevertheless, it is clear that monoglyceride complexes with amylopectin, probably by forming clathrates with the helices in the linear portions of the molecule.

Eliasson and Ljunger (32) reported that the cationic surfactant cetyl trimethylammonium bromide (CTAB) slowed down the rate of formation of amylopectin crystallites in gelatinized waxy maize starch, as measured by differential scanning calorimetry (DSC).

Krog (33) reacted an excess of amylose with various surfactants at 60°C, removed the precipitated complex after cooling, and measured the amount of uncomplexed amylose remaining in the supernatant. Taking as a standard ratio 5 mg of emulsifier per 100 mg amylose he calculated the amylose complexing index (ACI), which is a measure of the amount of amylose precipitated by 5 mg of the surfactant. Monomyristin had an ACI of 100, with monopalmitin and monostearin giving values of 92 and 87, respectively. Unsaturated fatty acid monoglycerides gave results of about 30, several other stearate esters (of propylene glycol, sorbitan, or sucrose) yielded ACIs in the range of 10–25, but sodium stearoyl lactylate had an ACI of about 75. The low results with the stearate esters were thought to be due to the inability of these materials to form the particular lamellar mesophases that promote the interaction of the alkyl chain with the amylose helix. Several papers from his research group have appeared since 1971, relating the complex formation between emulsifiers and starch to bread making and antistaling properties. The overall picture is that under the right conditions, the straight-chain hydrophobic portion of emulsifiers will complex with helical sections of amylose and amylopectin as shown in Figure 12.21, and this complexation has advantageous consequences for bakery foods.

Surfactants modify the gelatinization behavior of starch. Figure 12.22 shows the changes in amylograph gelatinization curves for wheat starch caused by the inclusion of 0.5% of various emulsifiers (34). Of the three emulsifiers shown, DATEM is the least interactive, raising the swelling temperature by about 5°C but not changing the viscosity of the gelatinized starch. Glycerol monostearate is the most effective, raising swelling temperature by about 18°C

514 Emulsifiers for the Food Industry

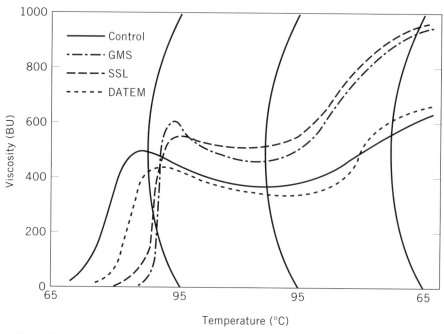

Figure 12.22 Amylograph curves obtained with wheat starch and various emulsifiers (34).

and also increasing the paste viscosity. Sodium stearoyl lactylate is less effective in regard to inhibiting swelling, but it increases paste viscosity to about the same extent as does GMS. Monolaurin effectively prevents gelatinization of potato starch when it is heated for one hour at 64°C (35). Sodium stearoyl lactylate and GMS reduce the extent of solubilization of starch molecules on heating in excess water (36). At 85°C, 10% of the starch from the control sample was soluble, but in the presence of 2% SSL or GMS, only 1.7% of the starch dissolved. Measures of swelling of the samples followed the same pattern. The interaction between emulsifier and starch takes place at the surface of the granules, and the starch–surfactant complex apparently serves to stabilize the granule, retarding water penetration and swelling as the temperature is raised.

9.2 Protein

Some of the amino acid side chains in proteins are hydrophobic, generally buried in the interior of the folded protein molecule but exposed if the protein is unfolded. Sometimes these hydrophobic regions are partially exposed even in the native folded protein, and they are often referred to as hydrophobic patches on the protein surface. The lipophilic parts of surfactants interact with these hydrophobic regions, sometimes contributing to unfolding (denaturation) of the protein and further binding of surfactant. Much work has been done on the binding of surfactants (sodium dodecyl sulfate in particular) to soluble proteins such as bovine serum albumin (37) and also on interactions of lipids with food proteins (38,39). The adsorption of protein at the oil–water interface is frequently altered by the presence of surfactant and vice versa. As a general rule, surfactant contributes to protein unfolding, enhancing interfacial absorption and emulsion stabilization. Sodium lauryl sulfate promotes unfolding of egg white albumin, thus increasing its ability to form a stable foam. Given the disparate nature of proteins found in various foods, generalizations about the effects to be expected by adding a surfactant to the food system are tenuous at best.

Wheat gluten protein contains about 40% hydrophobic amino acids and interacts with lipid-type materials such as the surfactant SSL. Figure 12.23 depicts schematically the effect of this interaction on gluten characteristics. The addition of acid to a flour–water dough solubilizes some of the protein. Dough pH (about 6) is roughly the isoelectric point of the gluten proteins. As the pH is lowered many of the anionic carboxylates are protonated, i.e., become unionized, and the protein molecule takes on an overall positive net charge. At pH 6 the hydrophobic patches can interact, and the protein aggregates via hydrophobic interaction (37). At pH 3, the net positive charges cause the molecules to repel each other, and solubilization occurs. This situa-

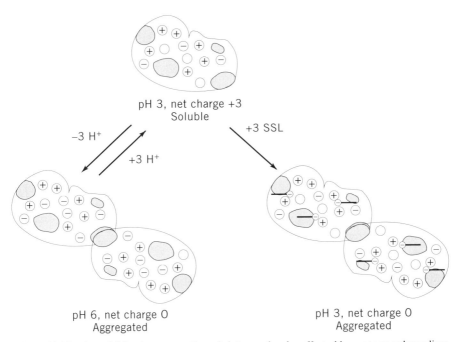

Figure 12.23 A model for the aggregation of gluten molecules affected by protons or by sodium stearoyl lactylate.

tion has many similarities to that of emulsified oil droplets stabilized by an ionic surfactant, where the surface charge prevents droplet contact and coalescence. As in that case, salt represses the electrostatic repulsion, and protein aggregation is favored. Most dough strengtheners are anionic surfactants, and when the lipophilic tail of the surfactant binds to the protein hydrophobic patches, it incorporates this negative charge into the complex, moving the overall charge closer to zero and promoting aggregation in the dough (see Figure 12.23). Salt and SSL have similar effects on dough mixing properties. Salt suppresses the electrostatic repulsion while SSL neutralizes it; the final effect in both cases is the same, namely hydrophobic aggregation of the gluten protein and an increase in dough strength.

An excess amount of surfactant can solubilize proteins, presumably by additional adsorption to the surface and generation of a large net charge at the surface of the protein molecule. This is the basis for estimation of protein molecular weight by gel electrophoresis in the presence of sodium dodecyl sulphate (SDS-PAGE). Such effects are usually found only at surfactant concentrations well in excess of those normally found in foods, and have more use in laboratory investigations than in ordinary applications.

SOME FOOD APPLICATIONS

The use of emulsifiers in foods is regulated in nearly all countries. Table 12.4 lists emulsifiers approved in the United States (by the Food and Drug Administration) and in Europe (by the European Economic Community). Several emulsifiers that have FDA sanction are not legally usable in Europe and vice versa. In some cases, specific limits are placed on the amount of emulsifier that a manufacturer may include in a food; in other instances the limitation is "that amount sufficient to produce the intended functional effect." If there is any question the appropriate regulation (in the United States, chapter 21 of the *Code of Federal Regulations, 21 CFR*) should be consulted.

One emulsifier, triethyl citrate, is approved only for use as a whipping agent for egg whites, and this is also the main use for sodium lauryl sulfate. Dioctyl sodium sulfosuccinate is used only as a wetting agent for various dry desert and beverage bases. Some emulsifiers may be classified as good for "general use," whereas others are emulsion promoters or stabilizers in certain limited applications or have effects such as the "dough conditioners" discussed above.

10.1 Baked Goods

Bread Antistaling. Monoglycerides added to the dough delay bread staling of baked bread during storage. During baking, starch gelatinizes when the internal temperature reaches a temperature of 60–65°C. The amylose (a linear polymer having a molecular weight of 100–200 kDa) is partially solubilized

Table 12.4 Regulatory status of emulsifiers

Emulsifier	U.S. FDA (21 *CFR*)	EEC Number
Monoglycerides and diglycerides (GRAS)	182.4505	E 471
Succinyl monoglyceride	172.830	—
Lactylated monoglyceride	172.852	E 472
Acetylated monoglyceride	172.828	E 472
Monoglyceride citrate	172.832	E 472
Monoglyceride phosphate (GRAS)	182.4521	—
Stearyl monoglyceride citrate	172.755	E 472
Diacetyl-tartrate ester of monoglyceride (GRAS)	182.4101	E 472
Polyoxyethylene monoglyceride	172.834	—
Propylene glycol monoester	172.854	E 477
Lactylated propylene glycol monoester	172.850	—
Sorbitan monostearate	172.842	E 491
Polysorbate 60	172.836	E 435
Polysorbate 65	172.838	E 436
Polysorbate 80	172.840	E 433
Calcium stearoyl lactylate	172.844	E 482
Sodium stearoyl lactylate	172.846	E 481
Stearoyl lactylic acid	172.848	—
Stearyl tartrate	—	E 483
Sodium stearyl fumarate	172.826	—
Sodium lauryl sulfate	172.822	—
Dioctyl sodium sulfosuccinate	172.810	—
Polyglycerol esters	172.854	E 475
Sucrose esters	172.859	E 473
Sucrose glycerides	—	E 474
Lecithin (GRAS)	184.1400	E 322
Hydroxylated lecithin	172.814	E 322
Triethyl citrate (GRAS)	182.1911	—

in the aqueous matrix of the dough. Amylopectin is a highly branched structure with a molecular weight of 10–40 MDa, that exists in the native starch granule as microcrystalline regions connected by amorphous segments of the molecule. During gelatinization, these microcrystallites "melt" but remain connected. Bread fresh from the oven has an extremely soft crumb. Upon cooling the amylose begins to crystallize (retrograde) and is essentially all crystalline within 24 h. The amylopectin microcrystallites begin to reform, but much more slowly; the half-life for this recrystallization is on the order of 2–3 days at room temperature. The progressive firming of bread crumb during storage is closely connected with this amylopectin retrogradation process (40).

As discussed above, linear alkyl chains such as the fatty acid moiety of monoglycerides can form a complex with helical segments of the solubilized starch molecules. The recrystallization of amylose is apparently little affected by complexation with monoglycerides. Differential scanning calorimetry (DSC) studies of day-old bread crumb show that all the amylose seems to be present in a crystalline state. Complex formation with monoglyceride by the short side chains of amylopectin, however, markedly slow the rate of retrogradation. The half-life for reformation of amylopectin crystals (measured by DSC) is increased twofold to threefold when 1% (relative to flour weight) glycerol monostearate (GMS) is incorporated in the dough. The rate of increase of crumb firming (measured as resistance to compression in an Instron or similar apparatus) is also halved by the addition of 1% GMS.

In bread or roll production monoglyceride is added at a level of 0.5–1% of the flour weight. There is some evidence that formation of the starch–monoglyceride complex occurs best when the monoglyceride is present as the lamellar mesophase. For this reason, many bakers prefer to use hydrated GMS, stabilized in this form. Plastic monoglycerides and diglycerides (roughly half α-monoglyceride) and hydratable distilled monoglycerides are also used, with about equally good antistaling results. It is probable that these latter types hydrate during dough mixing and processing, forming the lamellar mesophase, which can then react with gelatinizing starch during baking. In any event, about 1% α-monoglyceride is usually the upper level used, balancing antistaling properties versus cost.

Dough Strengthening. As discussed, some anionic and polyoxyethylene surfactants increase the ability of a proofed bread dough to withstand mechanical shocks during transfer from the proof box to the oven. These are termed dough strengtheners or dough conditioners. The ones in common use in commercial bakeries are sodium stearoyl lactylate (SSL) and diacetyl tartrate esters of monoglycerides (DATEM). Less frequently calcium stearoyl lactylate (CSL), ethoxylated monoglyceride, and polysorbate 60 are used. DATEM may be used at any level in accordance with good manufacturing practices; the practical upper level is 1% (flour weight basis). The use of the other dough strengtheners is limited to no more than 0.5% of the flour weight. Sometimes bakery suppliers sell a combination ingredient, containing both a softener (monoglyceride) and strengthener, slightly simplifying weighing out ingredients for each batch of dough.

Cake. A plastic shortening intended for making cakes contains 3–4% α-monoglyceride. This may be produced by superglycerination or by addition of monoglyceride to the melted oil before it is plasticized. The primary function is to aid subdivision of air bubbles in the shortening phase during mixing, giving more nucleation of leavening gases during baking and improved grain and volume in the finished cake. A secondary function is to stabilize the dispersion of sugar in the fat phase, allowing the incorporation of more sugar than flour in the batter (so-called high ratio formulas).

A liquid oil is not well suited to entrapping air bubbles, but it was found that the addition of α-tending emulsifiers to the oil before it was incorporated into the cake mix allowed the production of one-stage, high ratio cakes with good volume, fine grain, and excellent keeping qualities. The α-tending emulsifiers solidify in a stable α-crystalline (waxy) form. The main examples in commercial use today are acetyl monoglycerides (AMG), lactyl monoglycerides (LMG), and propylene glycol monoesters (PGME). A number of other emulsifiers have also been used, including polysorbate 60, stearoyl lactylic acid, and sucrose esters, but most commercial cake emulsifiers now offered by suppliers are based on PGME and/or AMG. At concentrations above a certain level, these emulsifiers form a solid film at the oil–water interface (see Figure 12.10). The addition of a second surfactant may enhance film formation; a mixture of PGME and stearic acid (80:20) is a stronger film former than pure PGME at the same weight concentration.

Cake mixes made with oil containing an α-tending emulsifier may be used in a one-stage method, i.e., the dry mix is placed in a bowl, liquid is added, and the batter is mixed at low speed (to blend ingredients) and then at high speed to incorporate air. Air incorporation is primarily as a foam stabilized by protein contributed by flour, milk, and egg whites. The presence of oil inhibits foam stabilization by a protein (e.g., a trace of oil makes it almost impossible to beat egg whites into a meringue), but the solid film at the oil–water interface effectively encapsulates the oil during air incorporation, thus preventing destabilization.

Icings. Many icings used in the bakery are aerated w/o emulsions. The simplest example is a butter cream icing, which contains butter, sugar, and egg whites. This icing has a limited stability, losing air rather readily on standing. Improved versions use an emulsified shortening. A typical emulsified plastic shortening for icing manufacture might contain 1–2% monoglyceride and up to 0.6% polysorbate 60. This emulsifier system improves the air-holding properties of the semisolid water-in-fat suspension; the icing retains its eye appeal and good eating qualities for several days.

10.2 Dairy-type Emulsions

Whipped Toppings. Numerous formulations of whipped toppings exist in which the butterfat of ordinary whipping cream is replaced with a fat having a melting point lower than 38°C, but a rather steep solid fat content (SFC) curve (palm kernel oil and hydrogenated coconut oil are frequently used). The aqueous phase contains milk protein, either sodium caseinate or whey, and sugars. About 5% of an α-tending emulsifier (PGME, LMG, AMG) is also used. Since the fat phase is around 25% of total weight, the concentration of emulsifier in the lipid phase is up to 17%. This is more than sufficient to form the solid interfacial film mentioned earlier. Thus the soluble milk proteins are able to form a stable foam when the emulsion is whipped.

Margarine. This spread, originally formulated as a less-expensive butter substitute, is now made with a variety of partially hydrogenated vegetable oils and often presented as a "healthy" spread, largely because of the presence of some polyunsaturated long-chain fatty acids and the absence of cholesterol. Standardized margarine contains at least 80% fat, and is an w/o emulsion. The fat phase contains 0.1–0.3% monoglyceride, sometimes with the addition of 0.05–0.1% lecithin, β-carotene (for color), and vitamins A and D (if included). The fat is melted, then the aqueous phase (water, flavors, salt, sometimes whey solids) is dispersed in a large agitation tank. The emulsion is then chilled (to solidify the fat) and worked in a scraped-surface heat exchanger. During the working step, the average water drop size is reduced to 2–4 μm. The finished product is solid, and this fat crystal matrix prevents water drop coalescence. The rheological properties (solidity, spreadability) of margarine are governed by the properties of the fat used. With certain fats, the crystals tend to convert from β' crystals (which confer a smooth texture) to β crystals (giving a sandy texture). This may be counteracted by the inclusion of sorbitan monostearate or monoglyceride citrate in the fat phase. Monoglyceride citrate at a level of 0.1–0.2% also counteracts the tendency of margarine to spatter when it is used as a frying fat in the kitchen.

Ice Cream and Mellorine. Ice cream contains no less than 10% milk fat and is basically a mixture of cream, milk, sugar, and flavors that, after emulsification, is frozen (21 *CFR* § 135.110). Emulsification of the milk fat is primarily due to casein proteins present in the milk plus the natural interfacial film present on the milk fat globules. In "economy" ice creams up to 0.1% of emulsifier is sometimes added, primarily to improve stiffness, dryness, and texture in the final product. GMS and polysorbates 65 and 80 are the ones most often used.

Mellorine is similar to ice cream, but contains at least 6% of fat other than milk fat (usually vegetable oil). Milk protein is, again, the major emulsifier, but 0.1% of polysorbates 65 and/or 80 may also be used.

Coffee Creamer. Emulsions of vegetable oil in an aqueous phase, including caseinates, are sold (as pasteurized liquid or as a spray-dried powder) as substitutes for cream for addition to coffee. For the liquid form, an emulsifier system of some combination of polysorbate 60, polysorbate 65, and sorbitan monostearate is used at a concentration of up to 0.4%. The spray-dried whitener uses the same emulsifier system, at a level of 1–3% of the dried powder. In both cases, sodium caseinate is probably the major stabilizer of the o/w emulsion.

10.3 Salad Dressings

Mayonnaise and Starch-based Dressings. Egg yolk is the only emulsifier allowed in the making of mayonnaise. This food is an o/w emulsion, in which the internal phase constitutes a minimum of 70%, and sometimes up to 82%, of the total volume. At this level of ϕ, the emulsion may invert (i.e., become an w/o emulsion) if it is subjected to unfavorable stresses, in particular excess shear in the colloid mill used to subdivide the oil droplets, and if the emulsifier is inadequate for the job. Egg yolk contains about 16% protein (much of it lipoprotein), 35% total lipid, and 10% phospholipids. The protein and phospholipids are both surface active, and readily coat the oil–water interface. The protein–phospholipid combination forms an elastic film that stabilizes the emulsion. Mayonnaise manufacture involves two steps: (*1*) formation of a rather coarse emulsion of oil in the aqueous phase (egg yolk, vinegar, spices) and (*2*) reduction of oil droplet size by passage through a colloid mill. If all the internal phase is present as spheres of the same diameter, the maximum theoretical ϕ is 0.74. In mayonnaise, the spheres have a range of diameters, so the small spheres can fit into the spaces between larger drops, allowing ϕ to be greater than this theoretical limit. The amount of emulsifier must be sufficient to coat all the additional interface formed during drop size reduction. If the clearance in the colloid mill is too small (shear is too large), the oil drops become too small, the interface area is greater than can be covered by emulsifier, and the mayonnaise inverts.

The oil used for mayonnaise manufacture must remain liquid even during storage at refrigerator temperatures. If some portion of the oil solidifies, the fat crystals disrupt the surface film and the emulsion breaks. The oil is usually winterized, i.e., it is held at a low temperature for 1 or 2 days and the solid fraction is removed. Polyglycerol esters may also be added to the oil to inhibit fat crystal formation to prevent this problem.

A less expensive alternative (spoonable salad dressing) contains 40–50% oil, and contains gelatinized modified starch as the stabilizing agent. After treatment in a colloid mill, the emulsion is cooled; the starch gels and the semisolid matrix prevents oil droplet coalescence. Up to 0.75% of emulsifiers (polysorbates, monoglyceride citrate, or DATEM) may be included in the formulation to enhance emulsification of oil during manufacture of the first-stage emulsion.

Pourable Dressings. Numerous pourable dressings (e.g., French, Italian, Thousand Island) contain 30–40% oil in an aqueous phase of vinegar, spices, flavorings, and stabilizers. For the most part, the stabilizers are gums of various sorts that increase the viscosity of the aqueous phase and decrease the rate of oil coalescence when the dressing is shaken before being poured on the salad. Polysorbate 60 may be used at up to 0.3% of the total dressing weight to enhance the ease of oil dispersion on shaking.

REFERENCES

1. A.W. Adamson, *Physical Chemistry of Surfaces,* Wiley-Interscience, New York, 1967.
2. K. Larsson and S.E. Friberg, eds., *Food Emulsions,* Marcel Dekker, Inc., New York, 1990.
3. P. Becher, *Encyclopedia of Emulsion Technology,* Marcel Dekker, Inc., New York, 1985.

4. W.C. Griffin and M.J. Lynch in T.E. Furia, ed., *Handbook of Food Additives,* 2nd ed., CRC Press, Inc., Cleveland, Ohio, 1972.
5. A.R. Handleman, J.F. Conn, and J.W. Lyons, *Cereal Chem.* **38,** 294–305 (1961).
6. J.W. McBain and L.A. Wood, *Proc. R. Soc. London* **A174,** 286–293 (1940).
7. C.E. Stauffer, *J. Colloid Interface Sci.* **27,** 625–633 (1968).
8. E.J.W. Verwey and J.T.G. Overbeek, *Theory of the Stability of Lyophobic Colloids,* Elsevier Science Publishing Co., Inc., New York, 1948.
9. J.D. Dziezak, *Food Technol.* **42**(10), 172–186 (1988).
10. J. Boyd, C. Parkinson, and P. Sherman, *J. Colloid Interface Sci.* **41,** 359–370 (1972).
11. M. El-Nokaly and D. Cornell, eds., *Microemulsions and Emulsions in Foods,* American Chemical Society, Washington, D.C., 1991.
12. M. El-Nokaly, G. Hiler Sr., and J. McGrady in Ref. 11.
13. J.C. Wootton, N.B. Howard, J.B. Martin, D.E. McOsker, and J. Holme, *Cereal Chem.* **44,** 333–343 (1967).
14. E.S. Lutton, C.E. Stauffer, J.B. Martin, and A.J. Fehl, *J. Colloid Interface Sci.* **30,** 283–290 (1969).
15. N. Krog, *Cereal Chem.* **58,** 158–164 (1981).
16. N. Krog and A.P. Borup, *J. Sci. Food Agr.* **24,** 691–701 (1973).
17. N. Krog and B. Nybo Jensen, *J. Food Technol.* **5,** 77–87 (1970).
18. N. Krog in R.B. Duckworth, ed., *Water Relations of Foods,* Academic Press, Inc., New York, 1975.
19. T.L.-G. Carlson, K. Larsson, and Y. Miezis, *Cereal Chem.* **55,** 168–179 (1978).
20. T.L.-G. Carlson, K. Larsson, K. Miezis, and S. Poovarodom, *Cereal Chem.* **56,** 417–419 (1979).
21. H. Birnbaum, *Bakers Dig.* **55**(6), 6–18 (1981).
22. D.T. Rusch, *Cereal Foods World* **26,** 111–115 (1981).
23. W.H. Knightly, *Cereal Foods World* **33,** 405–412 (1988).
24. J. Birk Lauridsen, paper presented at the AOCS Short Course on Physical Chemistry of Fats and Oils, Hawaii, May, 11–14, 1986.
25. C.E. Walker, *Cereal Foods World* **29,** 286–289 (1984).
26. B.F. Szuhaj, *J. Am. Oil Chem. Soc.* **60,** 258A–261A (1983).
27. B.F. Szuhaj, *Lecithins: Sources, Manufacture & Uses,* American Oil Chemists' Society, Champaign, Ill., 1989.
28. W. Van Nieuwenhuyzen, *J. Am. Oil Chem. Soc.* **53,** 425–427 (1976).
29. O.L. Brekke in D.R. Erickson, E.H. Pryde, O.L. Brekke, T.L. Mounts, and R.A. Falb, eds., *Handbook of Soy Oil Processing and Utilization,* American Soybean Association, St. Louis, Mo., 1980.
30. J. Lagendijk and H.J. Pennings, *Cereal Sci. Today* **15,** 354–356, 365 (1970).
31. L.V. Batres and P.J. White, *J. Am. Oil Chem. Soc.* **63,** 1537–1540 (1986).
32. A.-L. Eliasson and G. Ljunger, *J. Sci. Food Agric.* **44,** 353–361 (1988).
33. N. Krog, *Starch/Stärke* **23,** 206–210 (1971).
34. N. Krog, *Starch/Stärke* **25,** 22–27 (1973).
35. K. Larsson, *Starch/Stärke* **32,** 125–133 (1980).
36. K. Ghiasi, R.C. Hoseney, and E. Varriano-Marston, *Cereal Chem.* **59,** 81–85 (1982).
37. C. Tanford, *The Hydrophobic Effect: Formation of Micelles and Biological Membranes,* John Wiley & Sons, Inc., New York, 1973.
38. E. Dickinson in Ref. 11.
39. B. Ericsson in Ref. 2.
40. C.E. Stauffer, *Functional Additives for Bakery Foods,* Van Nostrand Reinhold, New York, 1990.

13

Antioxidants: Technical and Regulatory Considerations

Antioxidants have possibly been used to preserve fats long before recorded history. In prehistoric times, it is conceivable that aromatic plants were used to give fats, to be used for medicinal purposes and foods, pleasant odors and flavors. Many of these plants are now known to contain potent antioxidants.

Retardation of oxidative reactors by certain compounds was first recorded by Berthollet (1) in 1797 and further explained by Davy (2) in 1817. Early workers explain the phenomenon as "catalysts poisoning."

In the mid-nineteenth century Chevreul (3) observed that linseed oil dried much more slowly in oak wood than on glass or natural surfacing; other woods had the same effect but not to the same extent. He stated that oak wood was an antidrying agent for linseed oil.

The course of rancidification of fats remained unknown until Duclaux (4) demonstrated that atmospheric oxygen was the major causative agent of free fatty acid oxidation. Several years later, Tsujimoto (5) found oxidation of highly unsaturated glycerides could bring about rancid odors in fish oil.

Perhaps the first report of antioxidants used in fat was by Deschamps (6) in 1843, who observed that gum benzoin and poplar tree extracts could retard rancidity of ointments made with lard. Wright (7) in 1852 observed that American Indians in the Ohio Valley preserved bear fat using bark of the slippery elm. He found that elm was effective to preserve lard and butterfat. Elm bark was patented as an antioxidant 30 years later.

Present knowledge of the properties of various chemicals to prevent oxidative breakdown of fats and fatty foods began with the classic studies by Moureu and Dufraise (1,8–11). During World War I and shortly thereafter these workers tested more than 500 compounds for antioxidant activity. This basic research, combined with the vast importance of oxidation in practically all manufacturing operations, triggered a search for chemical additives to regulate oxidation. This search is still in progress.

Of the hundreds of compounds that have been proposed to inhibit oxidation deterioration of oxidizable substance, only a few can be used in products for human consumption. Antioxidants are required to be approved for the intended use. In selecting antioxidants, the following properties are desirable:

1. Effectiveness at low concentrations (0.001–0.01%).
2. Absence of undesirable effects on color, odor, taste, and other characteristics of the food.
3. Compatibility with the food and ease of application.
4. Stability under the conditions of processing (carry through) and/or storage of food.
5. The compound and its oxidation products must be nontoxic, even at doses much larger than those that normally would be ingested in food.

CLASSIFICATION

Considerable confusion still exists in classifying food antioxidants. Ingold (12) classified antioxidants into two groups: chain-breaking or primary antioxidants, which react with lipid radicals to produce stable products, and preventive or secondary antioxidants, which retard the rate of chain initiation by various mechanisms. Chipault (13) classified them as primary antioxidant and synergist.

Antioxidants have also been classified into five types (14): primary antioxidants, oxygen scavengers, secondary antioxidants, enzymic antioxidant, and chelating agents or sequistrants. This classification is complete, except it does not completely account for synergistic antioxidants. The following is the classification used in this chapter.

Primary antioxidants are compounds that terminate free radical chains in typical oxidation. The major antioxidants in this group are shown in Figure 13.1.

Synergists are compounds that by themselves have little or no effect on lipid oxidation. They markedly enhance the activity of primary antioxidants. Some primary antioxidants when used in combination may act synergistically.

Oxygen scavengers are compounds that remove oxygen from a closed system. Ascorbic acid, its isomers, and its derivatives are the best examples of this group. Ascorbic acid may also act as a synergist in the regeneration of primary antioxidants

Biological antioxidants include various enzymes, e.g., glucose oxidase, superoxide dismurtase, catalases, and glutathione peroxidase. These may remove oxygen or highly reactive compounds from food systems. Also included in this group would be the tocopherols and naturally occurring compounds from

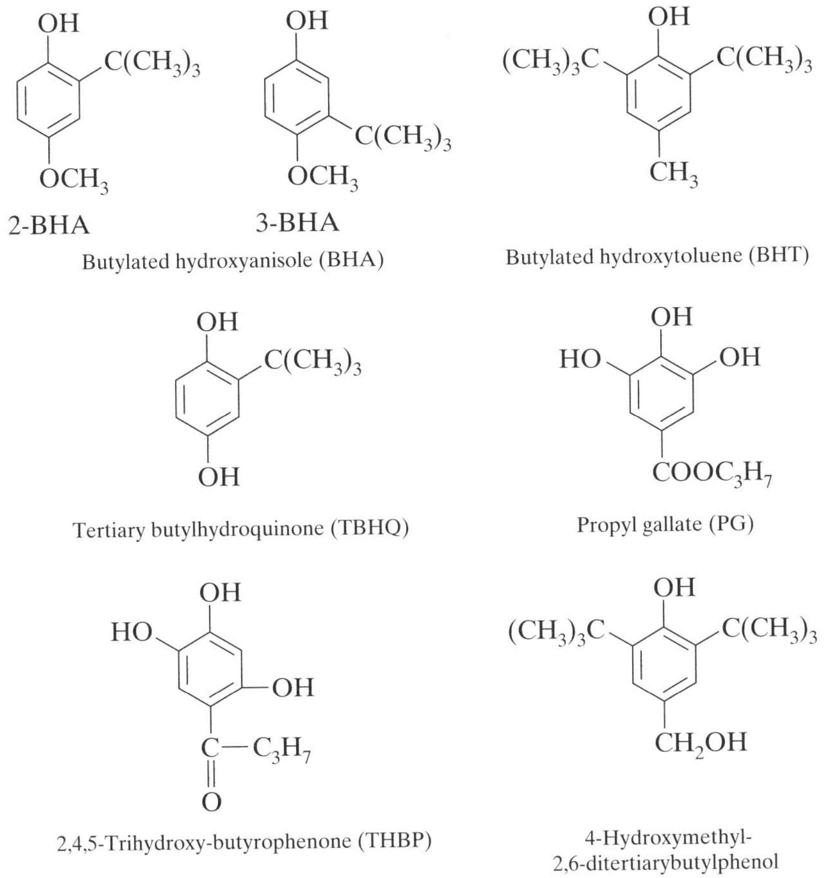

Figure 13.1 Major synthetic antioxidants.

- 2-BHA / 3-BHA: Butylated hydroxyanisole (BHA)
- Butylated hydroxytoluene (BHT)
- Tertiary butylhydroquinone (TBHQ)
- Propyl gallate (PG)
- 2,4,5-Trihydroxy-butyrophenone (THBP)
- 4-Hydroxymethyl-2,6-ditertiarybutylphenol

plant and animal sources. These latter compounds may also be classified as primary antioxidants and/or synergists.

Chelating agents (sequestrants) complex metallic ions, primarily copper and iron, that catalyze lipid oxidation. Ethylenediamenetetra-acetic acid, citric acid, certain plant phenolics, and amino acids are perhaps the major types of compounds in this classification. These are often listed as synergists, because they often show marked synergism.

Miscellaneous antioxidants include plant and animal materials that have been studied extensively as food antioxidants. Many plant oils, i.e., soy (15–18), palm (19,20), kapok (21), rice (22), olive (23), cottonseed (24), chia (25), sesame (26), wheat germ (27), peanut (28), and pimento (29), have been shown to process antioxidant actively. Several amino and protein hydrolysites are potent antioxidants. Several flavonoids and cinnamic acid derivatives (caffeic acid and chlorogenic acid ester) have also been shown to be potential sources of antioxidants (30,31).

MECHANISM OF ANTIOXIDATION

Primary antioxidants interfere with autoxidation by interrupting the chain propagation mechanism. In turn, they are oxidized slowly in the process, and normal autoxidation proceeds when the antioxidant has been destroyed completely. The best known and most effective primary antioxidant substances are polyphenols. Antioxidant activity, however, is not restricted to the phenols; other compounds, usually containing nitrogen or sulfur functional groups with electronic configuration similar to that of the polyphenols may be potent antioxidants.

Phenolic antioxidants commonly used in foods, with one exception, have two hydroxyl groups or one hydroxyl and one substituted hydroxyl group in *ortho* or *para* positions. These compounds are effective at extremely low concentrations; some lose effectiveness as their concentration is increased. At high concentration some may accelerate the rate of autoxidation. Primary antioxidants are most effective in animal fats that contain little natural stabilizer, while they are much less effective in vegetable oils that already may contain optimum amounts of naturally occurring antioxidants.

In general, the most effective antioxidants are highly reactive and are readily destroyed by heat. They protect fats under ordinary storage conditions but are inactivated during the baking or frying of foods and confer little or no protection to these products. The introduction of alkyl groups, particularly of branched groups such as tertiary butyl radicals, in positions next to the hydroxyl groups decreases their activity; however, it retards their destruction by higher temperatures and other conditions employed during the preparation of baked and fried foods. These compounds, known as carry-through, or carry-over, antioxidants, remain effective in foods prepared from fats in which they are present.

Lipid radicals are highly reactive and can readily undergo propagation reactions either by abstraction of hydrogen or by reaction with oxygen in its ground state.

$$RH + R\cdot \rightleftharpoons R\cdot + RH$$
$$R\cdot + {}^3O_2 \rightleftharpoons ROO\cdot$$
$$ROO\cdot + RH \rightleftharpoons ROOH + R\cdot$$

The oxygenation reaction is fast, with almost zero activation energy, therefore, the concentration of $ROO\cdot$ is much higher than that of $R\cdot$ in most food systems in which oxygen is present.

With the exception of edible oil heated at elevated temperatures, the termination reaction is limited. Even though the enthalpy of activation is low,

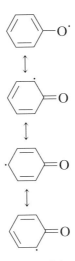

Figure 13.2 Resonance of the phenoxyl radical.

reaction is restricted both by the low concentration of radicals, which makes an encounter between two radicals uncommon, and by steric factors, when radicals are required to collide with the correct orientation.

An abundance of evidence indicates that the reaction between lipids and oxygen, termed autoxidation, is a free radical chain reaction (Figures 13.2 and 13.3). The mechanism of chain reactions can be discussed in terms of initiation, during which free radicals are formed, propagation, during which free radicals are converted into other radicals, and termination, during which two radicals combine with the formation of stable products.

Initiation
$$ROOH \rightleftarrows ROO\cdot + H^+$$
$$ROOH \rightleftarrows RO\cdot + OH^-$$
$$2ROOH \rightleftarrows RO\cdot + H_2O + ROO\cdot$$

Propagation
$$R\cdot + O_2 \rightleftarrows ROO\cdot$$
$$ROO\cdot + RH \rightleftarrows ROOH + R\cdot$$

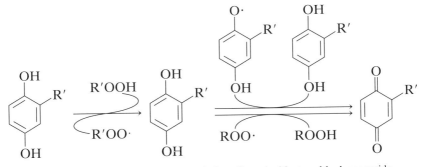

Figure 13.3 Reaction mechanism of phenolic antioxidant and hydroperoxide.

$$\text{Termination} \quad \begin{array}{l} ROO\cdot + ROO \rightleftarrows ROOR + O_2 \\ RO\cdot + R \rightleftarrows ROR \end{array}$$

There are two more probable processes to be considered in the formation of free radicals. The major initiation step involves the decomposition of hydroperoxides. Hydroperoxides are formed during the propagation reactions, but they may also be formed by the reaction of a lipid molecule with an oxygen molecule in its singlet excited state or by an enzyme-catalyzed reaction. The conversion of triplet oxygen to singlet oxygen may occur when a photosensitizer is present. Photosensitizers absorb light in the visible or near uv region, becoming electronically excited. They then transfer excess energy to an oxygen molecule in the following manner:

$$^1Sens, \rightarrow {}^1Sens(excited) \xrightarrow{\text{Intersystem crossing}} {}^3Sens(excited) \xrightarrow{{}^3O_2} {}^1Sens + {}^1O_2$$

Singlet oxygen may then react with a lipid molecule to yield a hydroperoxide $RH + {}^1O_2 \rightarrow ROOH$. Chain initiation may also occur by a direct reaction between a metal catalyst and a fatty acid by the reaction $M^{(n+1)+} + RH \rightarrow M^{n+} + H^+ + R\cdot$.

MAJOR FOOD ANTIOXIDANTS

3.1 Butylated Hydroxyanisole (BHA) and Butylated Hydroxytoluene (BHT)

Properties of major antioxidants are shown in Table 13.1. BHA is an effective antioxidant in fats from animal sources and in food containing animal products. As with most phenolic antioxidants, it is of limited effectiveness in unsaturated

Table 13.1 *Physical properties of major chain-breaking antioxidants*

Characteristic	BHA	BHT	TBHO	Propyl Gallate	Ethoxyquin
Chemical name	3-Tertiary-butyl-4-hydroxy-anisole 2-tertiary-butyl-4-hydroxyanisole	2,6-Ditertiary-butyl-4-menthyl phenol	Tertiary butyl-hydroquinone	Esters of 3,4,5-trihydroxybenzoic acid (octyl and dodecyl esters also used)	6-Ethoxy-1,2-dihydro-2,2,4-trimethyl-quinoline
Molecular weight	180.25	220	166.22	212.20	212.20
Physical appearance	White waxy tablets or flakes	White granular crystal	White to light tan crystals	White crystalline powder	Brownish red liquid
Boiling range, 733 mm, °C	264–270	265	300	Decomposes above 148	123–125
Melting range, °C	48–63	69.7	126.5–128.5	146–150	Liquid
Odor	Slight	Slight	Slight	Slight	Medium strong
Solubility					
Lipids	Soluble	Soluble	Soluble	Poorly soluble	Soluble
Water	Insoluble	Insoluble	Slightly soluble	Slightly soluble	Insoluble
Ethanol	Soluble	Soluble	Soluble	Soluble	Soluble

vegetable or seed oils. BHA maintains its activity during processing and is effective in baked goods. The latter is due to its stability to mild alkaline conditions. It is volatile, which makes it desirable to use in food packaging materials but undersirable in deep-fat frying fats.

BHT is the only common phenolic antioxidant that is not substituted *ortho*- or *para*-polyphenol (Figure 13.4). BHT does not have an optimum concentration. It has similar properties to BHA. Combination of BHA and BHT show marked syncigiso in plain fats, but this effect has not been demonstrated to carry through to baked products. While BHA is a synergist for propyl gallate, BHT is not (32,33). BHA and BHT may impart odor in foods when used at high temperatures (frying temperatures) for extended time periods.

BHA is more effective in suppressing oxidation occurring in animal fats than that occurring in vegetable oils. Among its multiple applications, BHA is particularly useful in protecting the flavor and color of essential oils: in fact, it is the most effective of all food-approved antioxidants for this application (34). BHA is particularly effective in controlling the oxidation of short-chain fatty acids, such as those contained in coconut and palm kernel oils, which are typically used in cereal and confectionery products.

TBHQ is regarded as the best antioxidant for protecting frying oils against oxidation, and like BHA and BHT, it provides carry through protection to the finished fried product (34). Increased oxidative stability and color improvement can also be achieved via hydrogenation, a process that saturates the double bonds of liquid oils, thereby converting them to semisolid plastic fats. The hydrogenated end products are suitable for use in shortenings and in other applications requiring higher melting fats (34). This process is, however, expensive and is limited in its hydrogenation capacity. Eastman Chemical Products, Inc. (34) describes TBHQ as an alternative or supplement to oil hydrogenation for increasing oxidative stability.

3.2 Gallic Acid and the Gallates

Gallic acid (3,4,5-trihydroxybenzoic acid) is widely distributed in the vegetable kingdom, mostly as a component of the vegetable tannins, although free gallic acid has been reported to occur in tea (35). The lower gallates such as ethyl, propyl, and butyl gallate remain slightly soluble in both water and fats, but

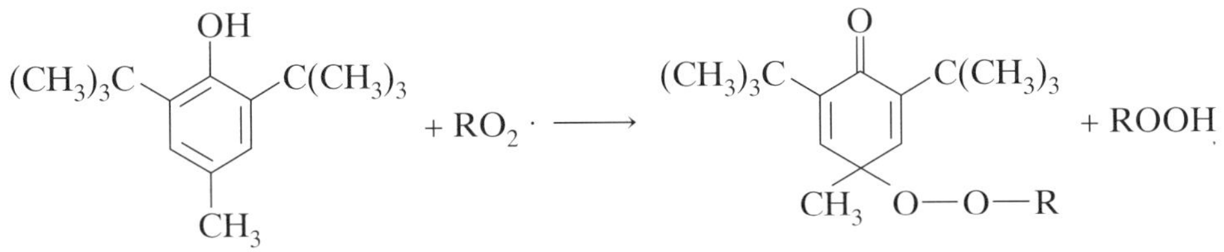

Figure 13.4 Antioxidant mechanism of BHT.

the higher octyl, decyl, and dodecyl esters are practically insoluble in water but easily dissolve in fats and oils.

The antioxidant properties of gallic acid were described by Golumbie and Mattill (35) who found this compound peculiar in that it behaved as a primary phenolic antioxidant and as a synergist. Others have found propyl gallate effective in retarding lipoxidase oxidation of linoleate (36,37), but it had little effect in preventing hematin-catalyzed oxidative fat rancidity (38). Like the tocopherols, the esters of gallic acid show optimum concentrations for antioxidant activity and may act as prooxidants when used at high levels.

Propyl gallate is the only ester of gallic acid permitted in foods in the United States, but several European countries allow the use of higher alkyl gallates. The various gallates have approximately equal antioxidant activities when used in equimolar concentrations, but the increased fat solubility of the higher esters is a definite advantage.

The gallates are not carry-through antioxidants and their power to stabilize fried foods, baked pastries, and crackers prepared from fats containing these compounds is low. The gallates form color complexes with nonchelated metal ions (iron and copper).

3.3 Chelating Agents

Chelating agents are not usually classified as antioxidants, yet they play a valuable role in stabilizing foods. Often classified as synergists, chelating agents complex with prooxidative metal ions such as iron and copper. An unshared pair of electrons in their molecular structures promotes the complexing action. Among the most commonly used chelators are citric acid and its salts, phosphates, and salts of ethylenediaminetetra-acetic acid (EDTA).

3.4 Secondary Antioxidants

Thiodipropionic acid and dilauryl thiodipropionate are unique in their activities relative to the other preservatives discussed here. They decompose the hydrogen peroxide elicited during lipid oxidation into stable end products. Although both components are approved by FDA for use as chemical preservatives (21 *CFR* 182.3109, for thiodipropionic acid, and 21 *CFR* 182.3280, for dilauryl thiodipropionate), Lindsay (38) reported that these preservatives are used more in stabilizing polyolefin resins than foods.

3.5 Synergist

Synergists enhance or prolong the antioxygenic activity of primary antioxidants. Often these compounds by themselves have little effect, if any, on retarding fat oxidation. A significant synergistic effect often occurs between primary and secondary, or preventive, antioxidants. Many low molecular weight hydroxy and amino acids possess marked synergistic activity. However, synergism may also occur between primary antioxidants. Synergism of the primary antioxidants BHA and BHT is important in practical use (39); it results from increased hydrogen donation greater than that expected from the sum of two antioxidants. It has been proposed that, in the presence of BHA, BHT oxidation is enhanced while BHA is regenerated (40). Propyl and other gallates are effective synergists with both BHA and BHT.

The mechanism of synergism is certainly not the same for compounds. Synergists may function as regenerating agents of primary antioxidants, as inactivates of metallic prooxidants, as hydroperoxide decomposers, and as other means of sparing primary antioxidants.

The synergistic effect of tocopherols with ascorbic acid and its esters is of particular interest. This phenomenon was studied using α-tocopherol and ascorbic acid or ascorbyl palmitate (41,42). α-Tocopherol acts as the chain-breaking antioxidant, and the resulting α-tocopheryl radical reacts with ascorbic acid to regenerate α-tocopherol.

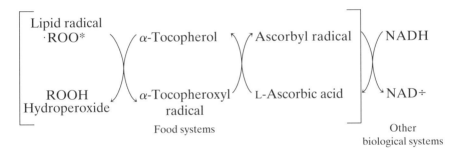

Support for this mechanism was reported by Niki (42), who demonstrated that in an *in vitro* oxidation system using methyl linoleate ascorbic acid was consumed first while α-tocopherol remained unchanged until most of the ascorbic acid was depleted. It was suggested that α-tocopherol scavenges peroxyl radicals much more rapidly than ascorbic acid but that the α-tocopherol radical reacts with ascorbic acid to regenerate α-tocopherol.

Synergism of the primary antioxidants BHA and BHT is important in practical use (39); it results from increased hydrogen greater donation than the expected sum from the two antioxidants. It has been proposed that in the presence of BHA BHT oxidation is enhanced while BHA is regenerated (40).

Many of the flavonoids and related compounds have strong antioxidant characteristics in lipid–aqueous and lipid food systems (43–48) (Figure 13.5). Certain flavones, flavonols (Table 13.2), flavonones (Table 13.3), flavanonals, and cinnamic acid derivatives (Table 13.4) have considerable antioxidant activity. The low solubility of these compounds in lipids is often considered a disadvantage and has been considered a serious disadvantage if an aqueous phase is also present (49) However, flavonoids suspended in the aqueous phase of lipid–aqueous system offer appreciable protection to lipid oxidation. Nearly 40 years ago, it was found that flavonols were effective antioxidants

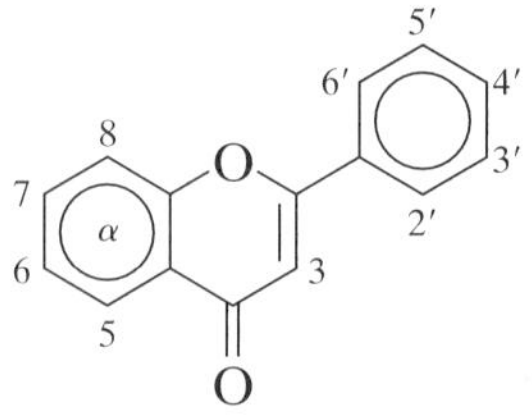

Flavones

Quercetin 3, 5, 7, 3', 4' - penta OH
Fisetin 3, 7, 3', 4' - tetra OH
Luteolin 5, 7, 3, 4 - tetra OH
Quercetin 5, 7, 3, 4 - tetra OH

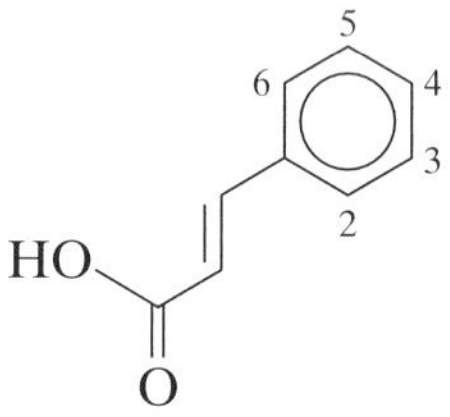

Cinnamic Acids

Caffeic acid 3,4 - di OH
Ferulic acid 4 - OH, 3 - OMe

Figure 13.5 Structures of flavones and cinnamic acid derivatives.

when suspended in lipid systems (50). Sources of natural antioxidants are as follows: algae; alma; browning products; citrus pulp and peel; cocoa powder or shell; heated products; herbs and spices; microbial products; oat flour; olives; osage orange; plant (extracts); protein hydrolysate; resin; and soy products (51).

Table 13.2 Antioxidant activity of flavonols (52)

Compound[a]	Hours to Reach a Peroxide Value of 50
Control (stripped corn oil)	105
Aglycones	
Quercetin (3,5,7,3',4'-pentahydroxy)	475
Fisetin (3,7,3',4'-tetrahydroxy)	450
Myricetin (3,5,7,3',4',5'-hexahydroxy)	552
Robinetin (3,7,3',4',5'-pentahydroxy)	750
Rhammnetin (3,5,3',4'-tetrahydroxy 7 methoxy)	375
Glycosides	
Quercitrin (quercetin 3-rhamnoside)	475
Rutin (quercetin 3-rhamnoglucoside)	195

[a] 5×10^{-4} M in stripped corn oil.

Table 13.3 Antioxidant activity of flavones (52)

Compound[a]	Hours to Reach a Peroxide Value of 50
Control (stripped corn oil)	105
Aglycones	
Naringenin (5,7,3'-trihydroxy)	198
Dihydroquercetin (3,5,7,3',4'-pentahydroxy)	470
Hesperitin (5,7,3'-trihydroxy-4'-methoxy)	125
Glycosides	
Hesperitin (hesperitin 7-rhamnoglucoside)	125
Neohesperidin (hesperitin 7-glucoside)	135

[a] 5×10^{-4} M in stripped corn oil.

Perhaps the major reason for interest in natural antioxidants is that there is a question as to the safety of the presently used synthetic antioxidants (52). However, natural compounds usually are not subjected to the scrutiny and scientific evaluation that the synthetic compounds (BHA, BHT, TBHQ) are. Therefore, caution must be taken in the use of natural compounds. Their potential as mutagens, carcinogens, teratogens, and/or other pathogens must be investigated. For example, nordihydroguariaretic acid (NDGA) was obtained from a natural source (the bush *Larrea divaricata*) yet this antioxidant was found to be a carcinogen and was removed from the GRAS list by FDA in 1967. Certain problems other than toxicity may be associated with the use of natural antioxidants: (*1*) the amount of active ingredient may vary with the source and method of extraction, (*2*) some may impart off colors and off

Table 13.4 Antioxidant activity of some polyphenolics (52)

Compound[a]	Hours to Reach a Peroxide Value of 50
Control (stripped corn oil)	110
Hesperidin methyl chalcone	135
D-Catechin	410
Chlorogenic acid	505
Caffeic acid	495
Quinic acid	105
Propyl gallate	435
p-Comaric acid	120
Ferrulic acid	145

[a] 5×10^{-4} M in stripped corn oil.

odors, (3) the cost of extraction may often be nearly inhibitive, and undesirable reactions may occur with nutrients in the product.

Antioxidants have been isolated from or detected in most edible plants, but they are not always in the edible portions. The same compound, or group of compounds, are not always present throughout the plant. Natural antioxidants occur in all higher plants and in all parts of the plant (wood, bark, stems, pods, leaves, fruit, roots, flowers, pollen, and seeds). The antioxidants are usually phenolic or polyphenolic compounds. Typical compounds that possess antioxidant activity include tocopherols, flavonoids, cinnamic acid derivatives, phosphatides, and polyfunctional organic acids. The flavonoids include flavones, flavonols, isoflavones, catechins, flavonones, and chalcones. The cinnamic acid derivatives include caffeic acid, terralic acid, and chlorgenic acid.

Methods of incorporating antioxidants into food systems is of utmost importance. They must be thoroughly blended in the separation or in the lipid portion. Type of food, processing conditions, and equipment available all influence the method of addition. Antioxidants may be added directly, as a concentrate in oil or fat, as a food-grade solvent, or in emulsified form. Antioxidants and concentrate solutions may be proportioned in a continuous-flow system, sprayed, or added by various other methods such as through the packaging system or by pasting on the surface of fresh meat and fish products (39,53). Meat and fish products present different problems, because antioxidants can only be applied uniformly to the minced or ground product.

ANTIOXIDANT FORMULATIONS

Antioxidants used in fats and oils are usually in the form of one of many solution (liquid) formulations that have been devised to satisfy specific needs and take advantage of the different properties of the antioxidant compounds. For example, combinations of BHA and BHT may be used to achieve maximum carry-through effect in baked goods. Formulations with TBHQ or propyl gallate combined with BHA and/or BHT and including citric acid may provide improved shelf life to a fat or oil (due to the TBHQ or propyl gallate) and increase the oxidative stability of baked goods prepared with the fat or oil (due to the BHA or BHT), while at the same time taking advantage of the metal chelating and synergistic properties of the citric acid. Combinations of antioxidants, such as BHA and TBHQ, may be used to obtain maximum allowable treatment levels in meat and poultry fats within certain restraints imposed by regulations. These are just a few of many reasons to use formulations, with major considerations being the following:

1. They combine antioxidants to take advantage of their different types of potency.
2. They enable better control and accuracy in applying antioxidants.
3. They provide the synergistic effects of antioxidant and acid combinations.

4. They enable more complete distribution or solution of antioxidants in fats and oils.
5. They minimize discoloration problems sometimes associated with specific antioxidants.
6. They are more convenient to handle than the pure antioxidant.

REGULATORY STATUS AND SAFETY ISSUES

Synthetic antioxidants have been tested for safety, and approval for food on the basis of complex toxicity studies varies from country to country. Their toxicity is generally comparable with, or lower than that of, analogous natural antioxidants and is much lower than that of lipid oxidation products formed when antioxidants have not been added before storage. Some doubts have recently been reported about the possible carcinogenicity of some antioxidants, but the experiments should be confirmed under the conditions of their application, as many substances (including natural food components) could be found to be toxic when tested under nonphysiological conditions.

In the United States, use of antioxidants is regulated by the Federal Drug Administration (FDA) and the U.S. Department of Agriculture (USDA). A Coast Guard regulation applies to ethoxyquin use in fishmeal. While many countries have adopted regulations similar to those used in the United States, significant differences exist among countries, not only in the antioxidants approved but also in their applications and use levels. Regulatory approvals for BHA, BHT, propyl gallate, TBHQ, and tocopherols are shown in Table 13.5. In Canada, approved uses of antioxidants are listed in the Food and Drug Regulations (National Health and Welfare), and TBHQ has not been approved. In the European Economic Community (EEC), directives regulate the use of antioxidants (54). Individual member countries still regulate use levels (Table 13.6), a practice expected to be discontinued. In the EEC, TBHQ has not been approved. In Japan, the food sanitation law specifies the use of antioxidants, and TBHQ has not been approved.

In the United States, questions have arisen concerning the safety of synthetic antioxidants. Recently, the FDA was considering a petition for banning use of BHA in foods, based primarily on reports of forestomach carcinomas in rodents and slight profilerative effect on the esophagus of pigs and monkeys when high levels of BHA were fed. The FDA invited public comment on this petition. California's Department of Health Services (DHS) considered requiring warning labels under Proposition 65. The DHS, however, published a proposed risk assessment of 4 mg BHA per product per day level before the deadline for the warning labels. This essentially eliminated the need for warning labels because the intake of BHA is less than 4 mg per product per day.

The FDA's Cancer Assessment Committee is reportedly reviewing the carcinogenicity of BHT. This review was prompted, at least in part, by a Danish study showing increased liver tumors in rats fed high levels of BHT

Table 13.5 *Regulatory responsibility for major antioxidants*

	U.S. FDA[a]	USDA[b]	Canada (NHW)[c]	Europe (EEC)[d]
BHA	21 *CFR* 182.3169	9 *CFR* 318.7	Table XI, Class IV, B.2, 67–48	E320
BHT	21 *CFR* 182.3173	9 *CFR* 318.7	Table XI, Class IV, B.2, 67–48A	E321
Gallates	21 *CFR* 184.1660	9 *CFR* 381.147	Table XI, Class IV, P.1, 67–50A	E310–312
TBHQ	21 *CFR* 172.185	9 *CFR* 361.147	Not approved	Not approved
Tocopherols	21 *CFR* 182.3890	9 *CFR* 318.147	Table XI, Class IV, T.2, 67–50B	E306–309

[a] Food and Drug Administration.
[b] Department of Agriculture.
[c] National Health and Welfare.
[d] European Economic Community.

Table 13.6 *Regulatory approval status of major antioxidants*

Country	Antioxidant				
	BHA	BHT	Gallates	TBHQ	Tocopherols
Afghanistan	+	+	−	−	+
Argentina	+	+	−	+	+
Australia	+	+	+	+	+
Austria	+	−	−	−	+
Bahrain	+	+	+	+	+
Barbados	−	−	−	+	−
Belgium	+	+	−	−	+
Brazil	+	+	+	+	+
Chile	+	+	+	+	+
China, People's Republic	+	+	+	−	−
China, Taiwan	+	+	+	+	+
Columbia	+	+	+	+	+
Cyprus	+	+	−	−	+
Denmark	+	+	+	−	+
Ecuador	+	+	−	−	+
Finland	+	+	+	−	+
France	+	+	+	−	+

Table 13.6 (Continued)

Country	Antioxidant				
	BHA	BHT	Gallates	TBHQ	Tocopherols
Germany	+	+	+	−	+
Gibraltar	+	+	−	−	+
Greece	−	−	+	−	+
Hong Kong	+	+	−	−	+
Hungary	+	+	−	−	−
Indonesia	+	−	−	−	−
Ireland	+	+	+	−	+
Israel	+	+	−	+	+
Italy	+	+	+	−	+
Jamaica	−	−	−	−	+
Japan	+	+	+	−	+
Kenya	+	−	+	−	−
Korea, South	+	+	+	+	+
Luxembourg	+	+	+	−	+
Malaysia	+	+	+	+	+
Malta	+	+	−	+	+
Mauritius	+	+	+	−	+
Mexico	+	+	+	+	+
Morocco	−	−	−	+	−
Netherlands	+	+	+	−	+
New Zealand	+	+	+	+	+
Nigeria	+	+	−	−	−
Norway	+	+	+	−	+
Pakistan, West	+	+	−	−	+
Panama	+	+	+	+	+
Papua New Guinea	+	−	−	−	+
Peru	+	+	−	−	+
Philippines	+	+	+	+	+
Portugal	+	+	−	−	−
Saudi Arabia	+	+	+	+	+
Singapore	+	+	+	−	+
South Africa	+	+	+	+	+
Spain	+	−	−	−	+
Sweden	+	+	+	−	+
Switzerland	+	+	+	−	+
Thailand	+	+	+	+	+
Trinidad/Tobago	+	+	+	−	−
Turkey	−	−	+	+	−
United Kingdom	+	+	+	−	+
Uruguay	+	+	−	−	+
Venezuela	+	+	+	+	+
Zimbabwe	+	−	−	−	−

(55). The study was found inconclusive by the International Agency for Research on Cancer working group because survival among the treated group was higher than for the controls (56).

Gallates are not mutagenic and do not appear to induce forestomach tumors (52). There is no regulatory review of their safety at this time. TBHQ has no marked effects on the liver at high doses (57) and has little or no effect on rodent stomach hyperplasia. It is unlikely to cause tumors in these organs (58).

Tocopherol toxicity is low. At extremely high levels, α-tocopherol impairs blood clotting by interacting with vitamin K and may cause toxic effects in the liver. Intake of tocopherols from food antioxidant use is low compared either with total intake from the diet (59) or with levels that may cause adverse effects over long periods.

Toxicity of L-ascorbic acid and its derivatives is low. High levels of 1 g/day may cause adverse effects, including oxalate formation and excretion. Because intake from its use as an antioxidant is low compared with daily dietary intake, no maximum acceptable intake has been established. It should be noted, however, that stereoisomers of ascorbic acid such as D-isoascorbic may have little or no vitamin C activity and their antioxidant and safety characteristics are different. In many countries, approval is limited to L-ascorbic acid and its derivatives.

EVALUATION OF ANTIOXIDANT POTENCY

Many techniques have been developed and used to detect and measure the extent of oxidation in fats and fat-containing foods and to ascertain the resistance of fats to oxidative deterioration. Misleading results may be obtained because the mechanism of oxidation changes under the test conditions used to accelerate oxidation, e.g., under high temperature. Some of the more important methods include the following.

6.1 Shelf Storage Tests

The most realistic estimate of the effectiveness of antioxidants in prolonging the life may be obtained by storing the test materials under conditions identical or similar to those in commercial practice. Periodic evaluation, by sensory and chemical tests, is used to determine the onset of oxidation. Because this method, known as the shelf storage test, is time-consuming, several other methods have been developed to enable rapid evaluation of antioxidants (60–62).

6.2 Oxygen Absorption Tests

Oxidation involves measurable uptake of oxygen. Oxygen absorption tests have been developed based on the quantitation of oxygen, using manometric and bomb calorimetric techniques. Some tests simply measure the change in the weight of the sample.

6.3 Oven Storage Test

Heat is used to accelerate formation of peroxides. Testing is usually conducted in electrically heated convection ovens at 60–63°C and is often referred to as the Schaal oven test. While sensory criteria such as odor and flavor are commonly used, chemical determination of peroxides and other factors may also be used to determine rancidity.

6.4 Active Oxygen Method

The active oxygen method (AOM) is widely used to measure oxidative stability of fats and to estimate the potency of antioxicants in them. It has been referred to as the swift stability test and involves bubling air through the heated test sample to accelerate oxidation. Periodic analyses of peroxide content are conducted to determine the time required for the fat to oxidize under the AOM conditions.

6.5 Other Methods

Other methods include measuring persistence and intensity of carotene in an agar plate diffusion system, oxygen uptake by electrode or gas chromatographic reactor, and detecting and measuring antioxidant components in foods by thin-layer chromatography. Several methods have been developed that measure the coupled oxidative potency of β-carotene and linoleic acid (or other PUFA) when emulsified with the test antioxidant (16,17,63–65).

THERMAL DEGRADATION OF PHENOLIC ANTIOXIDANTS

Antioxidants escape from the frying medium by volatilization and steam distillation due to the high temperatures and the large amount of water that is boiled out. Loss of antioxidant during frying has often been reported as a factor contributing to the loss of antioxidant potency. It has been reported that BHT was not effective in frying operations (66). The ineffectiveness of BHA and BHT has been observed under frying conditions; TBHQ was seen to be relatively more effective under the same conditions (67). It was postulated that some phenolic antioxidants are ineffective during frying due to thermal destruction and loss through steam distillation (33).

Peled and co-workers (68) found that the low effectiveness of BHT in retarding thermal deterioration of the oil in frying could be attributed to BHT losses due to volatilization and/or destruction by oxidation. They also found that when the oil was heated in the presence of air, the antioxidant destruction was lessened by the presence of water in the system, but a large portion of BHT was lost through steam distillation.

In another study, Warner and co-workers (69,70) determined the extent of volatilization and decomposition of antioxidants and the extent to which unknown antioxidant decomposition products are retained by the food by introducing ring-labeled [^{14}C]BHA, ring-labeled [^{14}C]TBHQ, and [7−^{14}C]BHT in deep-fat frying under actual cooking conditions. BHA or its decomposition products were largely (>80%) retained by the lard. For [7−^{14}C]BHT, the radioactivity decreased markedly with frying. After the equivalent of 4 batches of French-fried potatoes, 50% of the label was retained, while after 12 batches, less than 20% remained in the lard. BHT, which was found to be the most volatile of the phenolic antioxidants, volatilized to the extent of 28%, while the corresponding losses for BHA and TBHQ were 11% and 6.1%, respectively.

Unlike BHA and BHT, which were stable in unheated lard throughout the experiment, TBHQ appeared to be unstable in lard even at room temperature. Furthermore, TBHQ was undetectable in the heated lard and was present only in a minute amount (2%) in the volatiles. Overall, the heat treatment resulted in virtually total decomposition (99%) of the added TBHQ in lard, and the resultant products appeared to be less polar than TBHQ itself.

Perhaps the most comprehensive study on the thermal decomposition of phenolic antioxidant was done recently by Hamama and Nawar (71). They not only identified products of thermal decomposition but also established the order of stability for BHT, BHA, TBHQ, and propyl gallate.

It has been reported that other factors (e.g., changes in reaction pathways and interactions of food components) may play a major role in influencing antioxidant potency under frying conditions (72). It has also been noted that antioxidants absorbed in food are lost during storage (61,73). These decreasing levels of antioxidants may be attributed to the reaction of antioxidants with reactive species in the food or the oil.

TBHQ and BHA are oxidized to tert-butylbenzoquinone (2TBBQ) at 190°C (69,70). In addition, quinones are subject to a wide variety of reactions of the 1,4-Michael addition type (Figure 13.6). The reactions of TBBQ with various nucleophilic substances (REH), such as water, carboxylic acids, thiols, amines, and alcohols, result in the substituted tert-butylhydroquinone (SBHQ). Apparently, the SBHQ must be an excellent antioxidant, since it is a sterically hindered hydroquinone with a lower oxidation potential than TBHQ, because of the ring substitution of the atom E with one or more unshared electron pairs. This may explain why TBHQ yields a fried product with a longer shelf life than the identical product fried in antioxidant-free oil, even though TBHQ is unstable in heated fat. Warner and co-workers (69,70) proposed that TBBQ as an initial product has only a transitory existence and would undergo extensive decomposition in a frying medium because hydroxyquinone is an unstable compound.

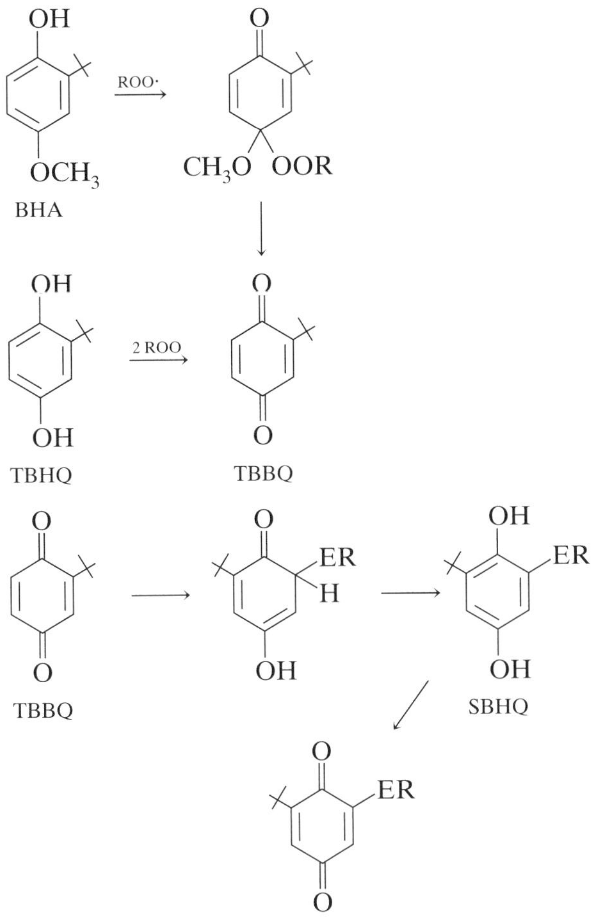

Figure 13.6 Reactions of BHA and TBHQ with alkylperoxy radicals and reaction of quinone with nucleophilic substances to form substituted hydroquinone (SBHQ).

Warner and co-workers (69,70) also determined two thermolysis pathways of 2, 6-di-tert-butyl-4-hydroperoxy-4-methylcyclohexa-2, 5-dienone (HBHT), the primary oxidation product of BHT. Their first route leads to the predominant product, which is BHT itself. This observation suggests the intriguing possibility that BHT can capture two chain-propagating alkylperoxy radicals and in subsequent thermolysis reappear as the parent antioxidant. The second route of decomposition, involving the intermediate alkoxy radical, would yield the products given in Figure 13.7. Only a trace of DBBQ (mol. wt. 220) was detected. Of the molecular weight 236 compounds, 2, 6-di-tert-butyl-r-hydroxy-4-methylcyclohexa-2, 5-dione (MHCD) was the predominant product. For BHA, under the well-characterized reaction conditions typical of organic reaction, BHA is readily oxidized to a coupling product, 3,3'-bis-(1,1-dimethylethyl)-5,5'-dimethoxy-1,1'-biphenyl-1,1'-diol.

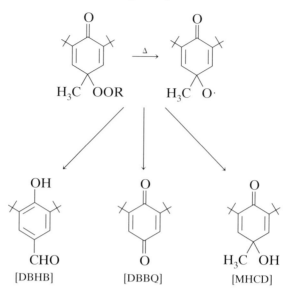

Figure 13.7 Reaction of the alkoxy radical formed by cleavage of the oxygen bond in HBHT.

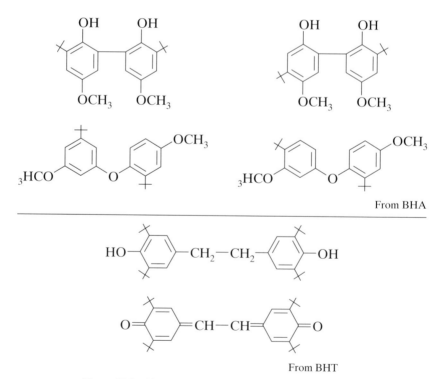

Figure 13.8 Dimers formed from BHA and BHT oxidation.

Costcove and Waters (74) reported that BHT could be oxidized to dimers 1, 2-bis-(3, 5-di-tert-butyl-4-hydroxyphenyl) ethane and 3, 3'-5, 5-sp-tetra-tert butylstilbene-4, 4' luinone. These substances were isolated and identified from photooxidized polyethylene film (75).

Furthermore, Maga and Monte (76) succeeded in isolating and identifying (2', 3-di-tert-butyl-2-hydroxy-4', 5-dimethoxy-biphenyl ether (DHDP) and its structural isomers in fresh corn oil systems containing 0.02% BHA at 218°C (Figure 13.5). These results are consistent with findings in the model systems of Kurechi and Kato (77) (Figure 13.8).

REFERENCES

1. C. Moureu and C. Dufraisse, *Chem. Rev.* **3,** 113 (1926).
2. H. Davy, *Ann. Chim. Phys.* **4**(2), 347 (1817).
3. M.E. Chevreul, *Ann. Chim. Phys.* **47**(3), 209 (1858).
4. M.E. Duclaux, *CR* **102,** 1077 (1886).
5. M. Tsujimoto, *J. Coll. Engl. Imp. Univ. Tokyo* **4,** 181 (1908).
6. K.C. Bailey, *The Retardation of Chemical Reactions,* Longman Green, New York, 1937, p. 192.
7. C.W. Wright, *Am. J. Pharm.* **24,** 180 (1852).
8. C. Moureu, C. Dufraisse, P. Robin, and J. Pougnet, *CR* **170,** 26 (1920).
9. C. Moureu and C. Dufraisse, *CR* **174,** 258 (1922).
10. C. Moureu and C. Dufraisse, *CR Soc. Biol.* **86,** 321 (1922).
11. C. Moureu and C. Dufraisse, *J. Chem. Soc.* **127,** 1 (1925).
12. K.U. Ingold, *Chem. Rev.* **61,** 563 (1961).
13. J.P. Chipault in W.O. Lundberg, ed., *Autoxidation and Antioxidants,* Vol. 2, Wiley-Interscience, New York, 1962, pp. 477–540.
14. S.P. Kochran and J.B. Rossell in B.J.F. Hudson, ed., *Food Antioxidants,* Elsevier Applied Science Publishers, Ltd., Borking, U.K., 1990.
15. M. Naim, B. Gestetner, S. Zilkah, Y. Birk, and A. Bondi, *J. Agr. Food Chem.* **22,** 806 (1974).
16. P.A. Hammerschmidt and D.E. Pratt, *J. Food Sci.* **43,** 556 (1978).
17. D.E. Pratt and P.M. Birac, *J. Food Sci.* **44,** 1720 (1979).
18. P. Gyorgy, K. Murata, and H. Ikehata, *Nature* **203,** 870 (1964).
19. R.C. Newton and W.D. Richardson, Can. Pat. 323, 764 (June 28, 1932).
20. R.C. Newton and W.D. Richardson, U.S. Pat. 1,890,585 (Dec. 13, 1933).
21. V.C. Mehlenbacher, U.S. Pat. 2,108,922 (Feb. 22, 1938).
22. H. Nobori, *J. Soc. Chem. Ind. Jpn.* **46,** 646 (1943).
23. Swift and Co., Ger. Pat. 618,836 (Sept. 17, 1935).
24. C.C. Whiltern, EE. Miller, and D.E. Pratt, *J. Am. Oil Chem. Soc.* **61,** 1075 (1984).
25. M.S. Taga, E.E. Miller, and D.E. Pratt, *J. Am. Oil Chem. Soc.* **65,** 928 (1984).
26. G.W. Fiero, *Pharm. Arch.* **11,** 1 (1940).
27. E. Sandell, *Sv. Farm. Tidskr.* **54,** 473, 501, 525 (1950).
28. R.M. Sikora, Master's thesis, Purdue University, West Lafayette, Ind., 1977.
29. J.R. Chipault, G.R. Mizuno, J.M. Hawkins, and W.O. Lundberg, *Food Res.* **17,** 46 (1952).
30. D.E. Pratt and B.M. Watts, *J. Food Sci.* **29,** 27 (1964).

31. L.R. Dugan Jr., L. Marx, C.E. Weir, and H.R. Kraybill, *Am. Meat Inst. Found. Bull.*, (18) (1954).
32. F.D. Tollenaar and H.J. Voss, *Fette Seifen Anstrichmit.* **58**, 112 (1956).
33. B.H. Stuckey in T.E. Furia, ed., *CRC Handbook of Food Additives,* Chemical Rubber Co., Cleveland, Ohio, 1972.
34. D.F. Buck, *Cereal Food World* **29**, 301 (1984).
35. C. Golumbie and H.A. Mattill, *Oil Soap* **19**, 144 (1942).
36. A.L. Tappel, W.O. Lundberg, and P.D. Boyer, *Arch. Biochem. Biophys.* **42**, 293 (1953).
37. Y.T. Lew and A.L. Tappel, *Food Technol.* **10**, 285 (1956).
38. R.C. Lindsay in O.R. Fennenna, ed., *Principles of Food Science, Part I, Food Chemistry,* Marcel Dekkar, Inc., New York, 1976, p. 465.
39. K. Kikugawa, A. Kunugi, and T. Kurechi in Ref. 14.
40. W.M. Cort, *J. Am. Oil Chem. Soc.* **51**, 321–325 (1974).
41. J.E. Packer, T.F. Slater, and R.L. Wilson, *Nature* **278**, 737–738 (1979).
42. E. Niki, *Chem. Phys. Lipids* **44**, 227–253 (1987).
43. D.E. Pratt, *J. Food Sci.* **30**, 737 (1965).
44. B.J.F. Hudson and J.I. Lewis, *Food Chem.* **10**, 47 (1965).
45. S.Z.F. Dziedzic and B.J.F. Hudson, *Food Chem.* **14**, 45 (1984).
46. S.Z.F. Dziedzic and B.J.F. Hudson, *Food Chem.* **12**, 205 (1983).
47. S.Z. Dziedzic, B.J.F. Hudson, and G.J. Barnes *J. Agr. Food Chem.* **33**, 244 (1985).
48. D.E. Pratt and B.J.F. Hudson in Ref. 14, Chapt. 7.
49. W. Heimann, A. Heimann, and H. Holland, *Fette Seifen* **55**, 394 (1953).
50. C.H. Lea and P.A.T. Swoboda, *Chem. Ind.* (1289) (1958).
51. D.E. Pratt, *Toxicol. Ind. Health* **9**, 63–75 (1993).
52. S.M. Barlow in Ref. 14.
53. P. Schuler in Ref. 14.
54. Anonymous, *Food Additives and the Consumer,* Catalogue No. CB-25-78-744-EN-C, Commission of the European Communities, Office for Official Publications of the European Communities, Luxembourg, 1980.
55. P. Olsen, D. Meyer, N. Billie, and G. Wurtzen, *Food Chem. Toxicol.* **24**, 1–12, (1986).
56. IARC, *IARC Monograph* **40**, 161 (1986).
57. B.D. Astill and co-workers, *J. Am. Oil Chem. Soc.* **52**, 53–58 (1975).
58. G.J. Van Esch, *Food Chem. Toxicol.* **24**, 1063–1065 (1986).
59. J.G. Bieri and R.P. Evarts, *J. Am. Diet. Assoc.* **62**, 147–151 (1973).
60. B.J.F. Hudson in Allen and R.I. Hamilton, eds., *Rancidity in Foods,* Elsevier Applied Science, Publishers, Ltd., Barking, U.K., 1989.
61. E.R. Sherwin, *J. Amer. Oil. Chem. Soc.* **55**, 809–814 (1978).
62. E.R. Sherwin in D.B. Min and T.H. Smouse, eds., *Flavor Chemistry of Fats and Oils, American Oil Chemists' Society,* 1985.
63. G.J. Marco, *J. Am. Oil Chem. Soc.* **45**, 594–598 (1968).
64. W.R. Nawar in Ref. 62.
65. D.D.M. Wayner, and G.W. Burton in J. Miquel, ed., *Handbook of Free Radicals and Antioxidants in Biomedicine,* CRC Press, Boca Raton, Fla., 1988.
66. I.P. Freeman, F.H. Padley, and W.L. Sheppart *J. Am. Oil Chem. Soc.* **50**, 101 (1973).
67. M.A. Augustin and S.K. Berry, *J. Am. Oil Chem. Soc.* **61**, 873 (1984).
68. M. Peled, T. Gutfinger, and A. Letar, *J. Am. Oil Chem. Soc.* **26**, 1655 (1975).

69. C.R. Warner, W.C. Brumley, D.H. Daniel, F.L. Joe, and T. Fazio, *Food. Chem. Toxicol.* **24,** 1015 (1986).
70. C.R. Warner, D.H. Daniel, F.S.D. Lin, F.L. Joe, and T. Fazio, *J. Agr. Food Chem.* **34,** 1 (1986).
71. A.A. Hamama and W.W. Nawar *J. Agr. Food Chem.* **39,** 1063 (1991).
72. T.P. Labuza, *CRC Crit. Rev. Food Technol.* **2,** 355–405 (1971).
73. E.R. Sherwin, *J. Am. Oil Chem. Soc.* **53,** 430 (1976).
74. S.L. Costcove and A.W. Water *J. Chem. Soc. London* 388 (1951).
75. H. Daun, S.G. Gilbert, and J. Giacin *J. Am. Oil Chem. Soc.* **51,** 404 (1974).
76. J.A. Maga and W.C. Monte *Lebensm Wiss. Technol.* **10,** 102 (1977).
77. T. Kurechi and T. Kato *Chem. Pharm. Bull.*, 81 (1983).

Index

Absorption:
  cholesterol reduction, milk fat, 19–20
  frying technology, filtration techniques, 445
Accelerated testing, cooking and salad oils, 199–200
Active oxygen method (AOM):
  antioxidant potency and, 539
  cooking and salad oils, quality control, 200
  feedstuffs and pet foods, quality control, 307
  oil stability data, 169–170
Additives:
  cooking and salad oils, 201–202
  shortening, 168–171
    antifoamers, 169
    antioxidants, 169
    colorants, 170–171
    flavors, 171
    liquid shortening, 179–180
    metal inactivators, 169–170
Adhesion, emulsion formation, 493–494
Aflatoxins, feedstuffs and pet foods, 261
Alkalinization, cocoa, 356
All-purpose shortenings, solid fat index (SFI) profile, 119
Alnarp treatment, butter manufacturing, 43–45
American Feed Industry Association (AFIA), feedstuffs and pet foods, definitions and standards, 271–272
American Oil Chemists' Society, shortening manufacturing quality control standards, 149
Ames test, frying technology research, 470–471
AMF system, mayonnaise manufacturing, 209

Amphiphiles:
  emulsifiers as, 483–484
  See also Surfactant
Amylose, emulsifier interaction with starch, 512–513
Amylose-complexing index (ACI), 513
Anco Cooling Roll, shortening manufacturing with, 127–129
Anhydrous milk fat (AMF):
  butter manufacturing, 45–47
  butter quality control, 26
  butter texture and spreadability, 44–45
  confectionery—liquors and liqueur, 52–53
Animal fat:
  AAFCO definition, 266
  AFIA feed grade definition, 271
Anionic emulsifiers, structure and characteristics, 507–508
Anisidine value (AV):
  cooking and salad oils, 198–199
  frying technology, degradation marker, 462
Annatto, as margarine coloring agent, 96
Antifoamers, shortening, 169
Antioxidants:
  AAFCO feed grade definitions, 270
  biological antioxidants, 524–525
  butylated hydroxyanisole and hydroxytoluene (BHA and BHT), 528–529
  chelating agents, 530
  classification, 524–526
  cooking and salad oils, 202
  feedstuffs and pet foods, storage conditions, 308–309
  formulations, 534–535
  gallic acid and gallates, 529–530
  margarine, preservatives, 94

547

Antioxidants (*Continued*)
  mechanism of, 526–528
  miscellaneous antioxidants, 525–526
  oil breakdown during frying, snack foods, 413
  overview, 523–524
  oxygen scavengers, 524
  potency evaluation, 538–539
  primary antioxidants, defined, 524
  regulatory issues, 535–538
  secondary antioxidants, 530
  shortening, 169
  synergists, 524, 530–534
  thermal degradation of phenolic antioxidants, 539–543
Apocarotenal, shortening colorant, 171
Apparent digestible energy (DE), feedstuffs and pet foods, NRC definition, 292
Apparent metabolizable energy (ME), feedstuffs and pet foods, NRC definition, 292
APV Pasilac technique, reduced fat butter production, 37–38
Aquatic animals, fish oil in diets of, 239–241
Aqueous-phase gelling agents, as emulsifier, 93
Association of American Feed Control Officials (AAFCO), feedstuffs and pet foods, definitions and standards:
  extracted and whole seed products, 272–273
  source oils, 266–271
  regulations, 259–260
Automation, shortening processing, 154–157
Autoxidation:
  antioxidant interference with, 526–528
  baked products, 349–351
  milk fat quality, 13–14
Avians, fat digestion in, 289–290

Babcock test, butterfat percentage determined by, 1–2
Bakery products:
  breads and rolls, fats in, 332–334
    frozen doughs, 333–334
    oil *vs.* plastic fat, 333
    texture, effect on, 332
    volume, effect on, 332–333
  cake donuts, 339–340
  cakes, 336–339
    creme icings, 339
    emulsified plastic fats, 336–337
    muffins and cupcakes, 337–338
    oil shortening with α-tending emulsifiers, 337–338

  cookies, 340–343
    coating fat, 343
    dough processing, 341–342
    filling fat, 342
    texture, spread, fats' effect on, 340–341
  crackers, 343
  emulsions in, 516–521
    antistaling properties, 516–518
    cake products, 518–519
    dough strengthening, 518
  extruded snacks, 425
  icing, filler and coating fats in, 332
  layered doughs, 334–336
    croissants, 335
    Danish pastry, 334–335
    puff pastry, 335–336
  pie crust, biscuits, 344
  shortening used in, 331
    liquid shortening, 180
    oxidative stability, 349–351
    plasticity, 346–349
    solid fat index/content (SFI/SFC), 344–346
  specifications for, 344–352
Base stock system, shortening manufacture, 171–174
  advantages, 171–172
  soybean oil base stocks, 172–174
  uniformity control, 172
Batch processing:
  butter manufacturing, 30–31
  margarine, emulsification, 96–97
  pourable dressings, 214
Beef cattle, feedstuffs and pet foods for, 296
Benzidine value, cooking and salad oils, 198–199
Benzoic acid, as margarine preservative, 95
Bicontinuous products, low-calorie spreads, 104
Biscuits, oils and fats in, 344
Blanching, deep-fat frying, 431
Bleaching, fish oils, 246
Body, butter manufacturing, 41
Bread and rolls, fats used in, 332–334
  frozen doughs, 333–334
  oil *vs.* fats, 333
  texture, 332
  volume, 332–333
Break-in oil, frying technology, 452
Brookfield Helipath viscometer, mayonnaise quality measurements, 208
Bulking agents, fat-free dressings, 218
Butter:
  chemical composition, 2–13

Index **549**

in confectionery products, 401–402
consumption patterns, 58–60
  comparisons with margarine, 66
defined, 24
history of, 1–2
manufacturing processes, 28–49
  anhydrous milk fat manufacture, 45–47
  batch butter manufacture, 30–31
  continuous butter manufacture, 31–34
  crystallization, 29–30
  cultured butter manufacture, 34
  heat treatment, 30
  milk and cream separation, 28–29
  neutralization, 30
  packaging, 47–48
  physical and organolleptic characteristics, 40–41
  reduced fat butter, 34–40
  storage and transport, 48–49
  texturization and spreadability, 41–45
as margarine ingredient, 91–92, 95
marketing, 57–60
milk fat modification, 13–22
  cholesterol reduction, 16–22
  hydrogenation, 13–14
  interesterification, 15–16
  triglyceride melting and crystallization, 13
quality control, 22–28
  composition control, 24
  grading, standards and definition, 24–26
  hazard analysis and critical control points (HACCP), 24
  lipase activity, 27
  oxidation, 27–28
  specialized analytical methods, 26–27
shortening compared with, 116
Butter flavors, manufacturing, 54–55
Butter oil, defined, 26
Butterfat products:
  butter flavoring agents, 54–55
  butterfat–vegetable oil blends, 49–51
  comparison with other fats, 7, 11
  confectionery—liquors and liqueur, 52–53
  decholesterolized milk fat, 55
  desaturated milk fat, 55–56
  extended-life creams, 55
  ghee, 51–52
  hypoallergenic butter, 54
  lactone content, 7, 11
  nonfood applications, 56
  nutraceauticals and healthy fats, 56
  pastry, cake, and biscuit products, ingredient in, 52

  powders, 53–54
  recombined products, ingredient in, 52
  short-life creams, 55
  ultrahigh temperature creams, 55
Butterfat–vegetable oil blends, 49–51
Buttermilk, in pourable dressings, 212
Butylated hydroxyanisole (BHA):
  physical properties of, 528–529
  regulatory status, 535–538
  in shortening, 169
  synergism of, 531
  thermal degradation, 539–543
Butylated hydroxytoluene (BHT):
  physical properties of, 528–529
  regulatory status, 535–538
  shortening, 169
  synergism of, 531
  thermal degradation, 539–543

C18 fatty acid chain, defined, 256–257
Cakes:
  emulsifiers used in, 518–519
  fats and oils in, 336–339
    creme icings, 339
    emulsified plastic fats, 336–337
    muffins and cupcakes, 338–339
    oil shortening with α-tending emulsifiers, 337–338
Calcium salts, AAFCO feed grade definition, 267
Canola oil:
  crystal habits (sandiness) in, 81–82
  as margarine source oil, 77–78
Capacitance, frying technology, degradation marker, 464
Caprenin, in fat-reduced confections, 405
Carbon dioxide extraction, fish oils, 246
Carotene:
  antioxidant potency and, 539
  in butter, 8–9
β-Carotene:
  antioxidant potency and, 539
  in margarine, 95–96
  shortening colorant, 170–171
Carotenoids:
  in margarine, 95–96
  shortening colorant, 170–171
Cats, feedstuffs and pet foods for, 295–297
"Chair" configuration, emulsifier interaction with starch, 511–512
Charlotte colloid mill, mayonnaise manufacturing, 209
Chelating agents:
  defined, 525

550 Index

Chelating agents (*Continued*)
  physical properties of, 530
Chemetator scraped-surface heat exchanger, shortening manufacturing, 146
Cherry-Burrell Gold'n Flow buttermaking process, 32, 34
Chick edema, feedstuffs and pet foods contamination, 261
Chill rolls, shortening chips and flakes, 190
Chilling:
  margarine processing, 97–98
  shortening consistency, 184–185
Chocolate:
  analogues, 378–380
  conching of, 359–361
  cooling, 374–375
  emulsifiers, 394–398
  fat bloom on, 403–404
  liquor, production of, 355–358
  milk fat as ingredient, 53
  milk products and, 362–363
  overview, 354–355
  refining of, 358–359
  rheology, 363–366
  standards of identity, 375–378
  sweet dark chocolate, components, 358
  tempering, 370–374
  tropicalization, 398–399
  See also Cocoa butter; Cocoa powder
Cholesterol:
  butter, standards for reduction, 25
  in fish oils, 228–229
  frying technology research, 469–471
  margarine, reduced cholesterol labeling, 73
  milk fat, reduction techniques, 16–22
    absorption, 19–21
    enzymatic reduction, 22
    regulatory and nutritional guidelines, 22
    short-path molecular distillation, 17, 19
    solvents, 21
    supercritical fluid extraction, 21
    vacuum steam distillation, 16, 18
Choline, AAFCO feed grade definitions, 269
Citric acid, cooking and salad oils, 202
Coalescence kinetics, ionic repulsion, 494–495
  oil–water *vs.* water–oil emulsion formation, 492
  steric hindrance, 494
Cocoa bean:
  composition of, 354
  products of, 355–358
Cocoa butter:
  as chocolate analogues, 378–380

  in confectionery products, 401–402
  physical and chemical properties, 366–371
  production of, 355–358
  tempering, 372–374
Cocoa butter equivalents and extenders (CBEs):
  heat stability, 399
  physical and chemical properties, 380–384
Cocoa butter replacers (CBRs):
  heat stability, 399
  physical and chemical properties, 384–389
Cocoa butter substitutes (CBSs):
  heat stability, 399
  physical and chemical characteristics, 389–393
Cocoa powder:
  production of, 355–356
  properties and composition, 357
Coconut oils:
  interesterification, 127
  in shortenings, 165
Codex Alimentarius:
  chocolate identity standard, 376–378
  margarine identity standard, 69–70
Coffee creamers, emulsifiers in, 519–520
Cold-pressed oils, production and consumption patterns, 195
Color:
  butter manufacturing, 41
  cooking and salad oils, quality control, 201
  deep-fat frying, 431
  frying technology, oil color, 463
  in margarine, 95–96
  margarine, deterioration and shelf life, 105–106
  in pourable dressings, 213
  shortening, 170–171
Colostrum, in bovine milk, 7, 11
Composition control, butter manufacture, 24
Compound coatings (chocolate coating analogues), 378–380
Computerized control, shortening processing, 154–157
Concentrated butter, packaging systems, 47–48
Conching, chocolate, 359–361
Cone penetrometer, shortening, quality control analysis, 151
Confections, oils and fats in:
  butterfat as ingredient in, 52–53
  caramel, toffee and butterscotch, 401–402
  chocolate, 354–378
    conching, 359–361
    cooling, 374–375

Index **551**

emulsifiers, 394–398
liquor, cocoa butter and powder, 355–358
milk products from, 362–363
refining, 358–359
rheology, 363–366
standards of identity, 375–378
tempering, 370–374
tropicalization of, 398–399
coatings for processing, 399–403
cocoa butter, alternative products, 378–394
equivalents (CBEs), 380–384
physical and chemical properties, 366–370
replacers (CBRs), 384–387
substitutes (CBSs), 389–394
tempering, 370–374
fat reduction in, 404–405
ice cream coatings, 402–403
oil migration, 403–404
overview, 353
Consistency:
butter manufacturing, 40–41
mayonnaise, 208
shortening, 184–188
chilling, 184–185
creaming gas, 185–186
quality control, 187–188
tempering, 186–187
working and filler pressure, 185
Consumption patterns:
butter, 58–60
cooking and salad oils, 193–196
margarine, 65–69
See also Production statistics
Contamination standards, feedstuffs and pet foods, 261
Continuous processing:
butter manufacturing, 31–34
margarine, emulsification, 97
mayonnaise manufacturing, 209–210
Conveyance systems, deep-fat frying, 435–436
Cookies, oils and fats in, 340–343
Cooking and salad oils:
additives, 201–202
composition and functionality, 202–203
consumption patterns, 193–194
natural and processed oils, 194–196
new product development, 202
nutrition-oriented products, 202
overview, 193–194
quality control, 198–201
accelerated test methods, 199–200

benzidine or anisidine value, 198–199
color, 201
free fatty acid content, 200–201
oxidized fatty acids, 201
pentane and hexanal values, 199
peroxide value (PV), 198
polymers, 201
thiobarbituric acid test, 199
volatile profile analysis, 199
stability, 196–197
Coproduct:
defined, 256
feedstuffs and pet foods, stability of, 262
Corn chips, processing technology, 422–423
Corn oil, endosperm, AAFCO feed grade definition, 267
Cosmetics, milk fat in, 56
Cost controls, feedstuffs and pet foods, 260–261
Cottonseed oil:
in cooking and salad oils, 196
feedstuffs and pet foods, whole cottonseed as feed, 317
in shortenings, 164
soapstock, AFIA feed grade definition, 271–272
Cottonseeds, AAFCO definitions, feed-grade products, 274–276
Crackers, oils and fats in, 343
Cream:
defined, 24
extended-life creams, 55
separation, 28–29
short-life creams, 55
tempering process for, 43–45
ultrahigh temperature (UHT) creams, 55
Creaming, emulsion formation, 493–494
Creaming gas, shortening consistency, 185–186
Creaming techniques, buttermaking, 1–2
Critical micelle concentration (CMC), surface tensions of emulsion, 486–487
Croissants, fats and oils in, 335
α-Crystal polymorphs:
margarine fat crystals, 80–81
shortening, crystal habit, 183–184
β-Crystal polymorphs:
baked products, shortening plasticity, 347–349
cocoa butter triglycerides, 369–370
margarine fat crystals, 80
shortening, 122–123
crystal habit, 183–184
liquid opaque shortening, 188–189

Crystallization (of fat):
  butter manufacturing, 29–30
  cocoa butter triglycerides, 369–371
  cooking and salad oils, 195
  liquid crystals, phase diagrams, 499–500
  margarine, crystallization rate, 82–83
    organoleptic properties and, 100
    polymorphism, 79–82
    solid fat, 77, 79
    working unit for processing, 98
  milk fat, 13
  shortening, 181–184
    chips and flakes, 190–191
    crystal habit, 182–184
    crystal structure, 167
    fat plasticity, 182
    liquid opaque shortening, 188–189
    manufacturing process, 136
    nature of, 122–123
    plasticity and crystal size, 121–122
    quality control linked to, 150
Cultured butter manufacturing, 34
Cupcakes, oils and fats in, 338–339
Cyclodextrins, cholesterol reduction, milk fat, 19–21

Dairy cattle feedstuffs and pet foods:
  fat feeding considerations, 296–299
  roasted soybeans, effects on milk yield, 321–323
Dairy products. *See* Milk products and protein
Danish pastry, fats and oils in, 334–335
DATEM emulsifier:
  chocolate, 396–398
  dough strengthening with, 518
  starch interaction with, 513–514
De novo fatty acids, fat digestion by animals, 282–285
Decholesterolized milk fat, manufacturing, 55
Deep-fat frying industry:
  benefits of, 430–431
  production statistics for, 429–430
  See also Frying technology
Deep-setting system of buttermaking, 1–2
Degradation:
  cooking and salad oils, 200
  frying technology, 453
    chemical degradation, 459–465
    chemical markers for, 441–442
    index of, 460
    markers, 461–463
    oil model, 458
  See also Thermal degradation

Degumming, fish oils, 244–245
Desaturated milk fat, 55–56
Designer foods, feedstuffs and pet foods:
  evolution of, 258
  fatty acid profiles, 324–326
Deterioration, margarines, 104–106
Dewaxing, cooking and salad oils, 195
Diacetyl:
  margarine, flavor compound, 95
  shortening, flavor compound, 171
Dietary oils, margarine with, 88–89
Diglycerides:
  margarine emulsifier, 93
  margarine fat crystals, 81–82
Dimethylpolysiloxane:
  cooking and salad oils, 202
  shortening antifoaming agent, 169
Direct expansion refrigeration:
  gravity refrigeration system, 133–135
  shortening manufacture, 133–136
Dixie mixer, mayonnaise manufacturing, 209
DLVO theory, surface interactions, 488–490
Docosahexaenoic acid (DHA), nutritional value of, 228
Dogs, feedstuffs and pet foods for, 294–295
Donuts, oils and fats in, 339–340
Drainage mechanism, foam formation, 496–497
Dry fractionation, shortenings, 125
Dry shortenings, solid fat index (SFI) profile, 120
Dust control, feedstuffs and pet foods, 321–322
Dwell time, frying technology, snack foods, 416–417

EDTA (ethylenediamine-tetraacetic acid):
  margarine preservative, 94
  in mayonnaise, 205
Eggs and egg products:
  in mayonnaise, 206
  poultry feedstuffs and pet foods, designer eggs and, 325–326
Eicosapentaenoic acid (EPA), nutritional value of, 228
Electrical double layer, surfactant, 489–490
Elein, margarine development and, 66
Emulsification process:
  margarine, 96–97
  shortening, 167–169
Emulsifiers:
  α-tending emulsifiers:
    baking applications, 337, 494–496
    structure, 494–496

as amphiphiles, 483–484
anionic emulsifiers, 507–508
in cakes:
  oil shortening with α-tending emulsifiers, 337
  plastic fats, 336–337
chocolate, 394–398
coffee creamers, 519–520
emulsifier plus water, physical state, 499–502
  mesophase structures, 499–501
  phase diagrams, 499
emulsion formation, 490–493
  division of internal phase, 490–491
  hydrophilic–lipophilic balance (HLB) system, 492–493
  microemulsions, 493
  oil–water vs. water–oil emulsions, 491–492
fat-continuous, in confections, 404–405
flocculation, 493–494
  creaming and adhesion, 493–494
foams, 495–498
  formation, film drainage, 495–497
  stabilization, 497–498
food applications, 516–521
  baked goods, 516–518
  food surfaces and interfaces, 484–485
  pourable dressings, 518
ice cream, 519–520
lecithin, 509–511
margarine, 92–93, 519–520
in mayonnaise, 519–521
mellorine, 519–520
monoglycerides, 502–505
polyhydric emulsifiers, 508–509
in pourable dressings, 212
protein interactions, 514–516
regulatory issues, 517
shortening, 168–169
sorbitan emulsifiers, 505–506
stability:
  coalescence and, 494–495
  margarine, milk products and protein in, 92
starch interaction, 511–514
surface activity, 485–490
  energetics, 485–486
  interface concentration, 486–488
  surface interactions, 488–490
tightness, margarine, melting characteristics, 76
wetting, 498–499
See also Foams

Energetics, surface tension and surface free energy, 485–486
Energy requirements, feedstuffs and pet foods, 291–294
Enzymatic cholesterol reduction, milk fat, 22
Enzymatic modification, cocoa butter equivalents and extenders (CBEs), 382–383
Essential fatty acids (EFAs), feedstuffs and pet foods:
  defined, 256
  in fish meals, 239
Esters, AAFCO feed grade definition, 266–267
Extended-life creams, manufacturing, 55
Extrusion:
  feedstuffs and pet foods:
    free-flowing fats, 314–315
    full-fat oilseeds, 317–320
    high-fat content feeds, 312–314
  snack foods, 423–425
  frying technology, 424

Fat coatings:
  cookies, 343
  feedstuffs and pet foods, 310–312
Fat composition:
  butter texturization and spreadability, 42–45
  margarine, labeling standards, 72
Fat corrected meal (FCM), feedstuffs and pet foods, 321
Fat digestion systems, feedstuffs and pet foods:
  avian systems, 289–290
  comparisons of, 279–291
  fatty acid synthesis elongation and desaturation, 282–285
  fish systems, 290–291
  nonruminant mammalian systems, 285–287
  ruminant mammalian systems, 287–289
Fat mimetics, fat-free dressings, 217
Fat products:
  AAFCO feed grade definition, 267
  crystal habits of, 81, 122
Fat selection, feedstuffs and pet foods, 305–308
Fat substitutes, fat-free dressings, 217
Fat-free products, labeling standards, 72–74
  salad dressings, 216–219
    composition, 217
    fat replacement, 217
    flavor quality, 217–218
    stability, 218
Fats and oils, AAFCO definition, 266

554 Index

Fatty acids:
  AAFCO feed grade definitions:
    long-chain calcium salts, 267
    volatile fatty acid salts, 267
  butter manufacturing, texturization and spreadability, 43–44
  in butterfat, 2–6
    distribution of specific acids, 6, 8–9
    seasonal changes in, 6–7
  cocoa butter equivalents and extenders (CBEs), 382–384
  cocoa butter substitutes (CBSs), 392–393
  cooking and salad oils:
    alteration of, 196
    free fatty acids, 200–201
    oxidized fatty acids, 202
  fat digestion by animals:
    nonruminant animals, 285–287
    ruminant animals, 287–289
    synthesis, elongation and desaturation, 282–285
  in feedstuffs and pet foods:
    fish food, 305
    lecithins, 278–279
    modifications of, designer foods, 324–326
    profiles of fat sources, 321–322
    tallows, fish and oilseeds oils, 277–278
  in fish oils, 227, 239–241
  margarine, oil blending and, 85–86
  in menhaden oil, 247–248
  in milk, 4–5
    lipids, 10
    positional and geometric isomers, 5–6
  milk fat, effects of feeding corn and peanut oils on, 7, 10
    hydrogenation and, 14
    interesterification and, 15–16
  modification of, with biotechnology, 165
  in monoglycerides, 503–504
  reactivity during frying, 412–413
  shortening, distribution in, 123–125
  See also Highly unsaturated fatty acids (HUFA); Polyunsaturated fatty acids (PUFA); Trans fatty acids
Federal Standard of Identity, margarine manufacture, 67, 69–71
Feed Composition Data Bank (FCDB), feedstuffs and pet foods, 260
Feedstuffs and pet foods, fats and oils in:
  comparative digestion systems, 279–291
    avians, 289–290
    fatty acids synthesis, elongation and desaturation, 282–285
    fish, 290–291
    nonruminant mammalian systems, 285–287
    overview, 280–282
    ruminant mammalian systems, 287–289
  composition of, 277–279
  energy requirements and availability, 291–294
    analytical techniques, 293–294
    apparent digestible energy (DE), 292
    apparent metabolizable energy (ME), 292
    gross energy (GE), 291
    net energy (NE), 292
    net energy of lactation (NEL), 293
    NRC definitions, 291
    true metabolizable energy (TME), 292
  fat storage and application, 308–324
    dry roasting, 320–321
    dust control, 321–324
    extrusion of full-fat oilseeds, 317–320
    extrusion of high-fat content feeds, 312–314
    fat coating, 310–312
    free-flowing fats, 314–315
    liquid fat application to mixed feeds, 309–310
    rumen-protected fats, 315–316
    storage tank design, 308–309
    whole cottonseed feeding, 317
  fatty acid profile modification, designer foods, 324–326
  historical background, 257–258
  information sources, 258–261
  NRC species considerations, 294–305
    beef cattle, 296
    cats, 295–297
    dairy cattle, 296–299
    dogs, 294–295
    fish, 304–305
    horses, 299–300
    pet foods, 294
    poultry, 300–303
    swine, 300
  overview, 255–257
  product definitions, extracted and whole seed products, 272–277
    fats and related products, 263–272
  supply and uses of, 262–263
  utilization practices, fat selection principles, 305–308
Filled print packaging, stick margarine, 99–100
Filler pressure, shortening consistency, 185
Filtration, frying technology:
  absolute filtration, 444

Index   **555**

benefits, 445–446
depth filtration, 445
disadvantages, 447–448
passive and active treatments, 443
regular filtration, 441
selection of active treatment systems, 448–449
snack foods, 414–415
Fish:
fat digestion in, 290–291
feedstuffs and pet foods for, fat feeding considerations, 304–305
oil content of, by species, 247–248
Fish meal:
conventional processing, 230–233
as marine oil source, 229–230
Fish oil:
AAFCO feed grade definition, 268
composition of, 226–227
in animal diets, 238–242
aquatic animals, 239–241
land animals, 238
refined oil as additive, 241–242
feedstuffs and pet foods, fatty acid composition, 277–278
in human diet, 242–252
stabilization techniques, 243–249
as margarine source oils, 77–78
nutritional values, 227–228
Flash frying, defined, 439
Flavonoids, antioxidant properties of, 531–533
Flavor:
butter manufacturing, techniques, 41
flavored butter, 54
conching process for chocolate and, 360–361
fat-free dressings, 217–218
margarines, 95
deterioration and shelf life, 104–106
specifications, 89–90
marine oils, 249–250
in pourable dressings, 213
shortening, 166–167
colorant, additives for, 171
Flocculation, emulsion formation, 493–496
Foams:
formation, film drainage, 495–497
frying technology, foam height as degradation marker, 464
stabilization, 497–498
Fondant, fats and oils in, 402
Food and Drug Administration (FDA):
antioxidant regulations, 535–538

chocolate identity standard, 375–378
labeling requirements, margarine, 71–74
margarine identity standard, 69–71
Food supplements, marine oils and, 249–252
Foods:
emulsions in, 516–521
frying technology, oil interaction with, 451–452
Forced circulation refrigeration system, shortening manufacture, 135
Fractionation:
cocoa butter replacers (CBRs), 386–387
cocoa butter substitutes (CBSs), 392–394
shortenings, 125
See also Dry fractionation
Free fatty acids (FFAs):
cooking and salad oils, 200–201
frying technology:
degradation markers, 461
snack foods, 415
Free radical chain reaction, antioxidant mechanisms and, 527–528
Fresh oil, frying technology, 452
Fritz processing, butter manufacturing, 30–34
Frozen doughs, fats used in, 333–334
Frying technology:
active treatment systems:
disadvantages of, 447–448
selection, 448–449
chemical degradation of fats and oils, 459–465
degradation products, 459–460
index of degradation, 460
markers of degradation, 461–464
comparison with other cooking techniques, 432
conveyance systems, 435–436
corn chips, 422–423
deep-fat frying, benefits of, 430–432
filtration techniques, 443–445
benefits of, 445–446
food interaction with oils, 451–452
fryer design, 432–433, 439–440
heat transfer during cooking, 451
heating systems, 436–439
liquid shortening used for, 180
mass transfer during cooking, 450
oil absorption:
high absorption, in snack foods, 416
low absorption, in snack foods, 416–417
oil chemistry during, 442–443, 452–454
optimization techniques, 456–458
overview, 429–430
passive and active filters/treatments, 443

Frying technology (*Continued*)
  quality control, food and oil, 455–456
  regulatory issues, 465–478
    first generation heated oils research, 465–467
    future issues, 472–473, 476
    guidelines and limits, 473–474
    industry and food service practices, 471–472
    inspection and monitoring, 474–475
    legal regulations, 472
    second generation heated oils research, 467–471
    U.S. regulations, 475–476
  snack foods:
    fryer design, 415–416, 419–420
    oil breakdown during, 411–415
  styles of frying, 439
  surfactant control, 458–459
  system care and maintenance, 439–442
  system components, 433–449
  theories of, 454–455
  tortilla chips, 421–422
Fudge, fats and oils in, 402
Fuels, fish oil for, 238
Functionality:
  cooking and salad oils, 202–203
  shortening, 116–117

Gallic acid and gallates:
  physical properties of, 529–530
  regulatory status, 538
Gelatin:
  low-calorie spreads, 103–104
  margarine, as emulsifier, 93
Gelling agents, low-calorie spreads, 103–104
Gerstenberg & Agger margarine/shortening systems, shortening manufacturing, 146–149
Ghee:
  manufacturing, 51–52
  shortening compared with, 116
Gibbs convention, surface solvent concentration, 488
Girdler CR mixer, mayonnaise manufacturing, 209–210
Glucopyranoside, emulsifier interaction with starch, 511–512
Glycerides, milk fat, interesterification and, 15
Glycerol monostearate (GMS):
  antistaling properties, 518
  structure and characteristics, 502–505
Gossypol levels, cottonseed products, 275–276

Grading processes, butter manufacture, quality control, 24–26
Graininess, margarine fat crystals, polymorphism, 82
GRAS (generally recognized as safe) status:
  fish oils, 244
  margarine source oils, 77–78
Gravity refrigeration system, shortening manufacture, 133–135
Gross energy (GE), feedstuffs and pet foods, NRC definition, 291
Growth-stimulating hormones, feedstuffs and pet foods, 257–258
Gum slurry, pourable dressings, 214–215
Gums:
  food processing with, 400
  in pourable dressings, 212–213

Hard fat effect, shortening:
  powdered shortenings, 181
  wide plastic range shortenings, 176–178
Hardness:
  quality control linked to, 151
  shortening, 167
Hazard analysis and critical control points (HACCP), butter quality control, 24
Health claims, margarine, labeling standards, 73–74
Heat stability, chocolate, 398–399
Heat transfer:
  deep-fat frying, 431
  frying technology, 451
  optimization of, 457–458
Heat treatment, butter manufacturing, 30–31
Heated oils research:
  first generation research, 465–467
  second generation research, 467–471
Heating systems:
  deep-fat frying, 436–439
  external heat exchanger, 437–438
  indirect heating systems, 437
Heat-stable dressings, 219
Heavy metals, margarine, preservatives, 94
Hexanal value, cooking and salad oils, 199
High-density lipoprotein (HDL), in fish oils, 228–229
High-liquid-oil blends, margarine, 86
Highly unsaturated fatty acids (HUFAs):
  fat digestion by animals, 283–285
  in fish oils, 227
  nutritional value of, 228–229
    in aquatic animals, 239–242
    forms of, for human consumption, 251–252
    in human diet, 242–252

in land animals, 238
  variation in oil content by species and, 247–249
High-ratio shortenings, history of, 163
High-stability shortenings, solid fat index (SFI) profile, 118–119
Homogenization, pourable dressings, 215
Horses, feedstuffs and pet foods for, 299–300
Hydrogenation:
  cocoa butter replacers (CBRs), 384–389
  margarine, low-trans oil blends for, 87–88
  margarine development and, 66–67
  milk fat, 13–14
  shortening, 125–126
    flakes and chips, 180–181
    physical characteristics, 167–168
    soybean oil base stock system, 173–174
    wide plastic range shortenings, 175–178
Hydrolyzed fat or oil, AAFCO feed grade definition, 266
Hydrolyzed sucrose polyesters, AAFCO feed grade definition, 267–268
Hydroperoxide:
  antioxidant mechanisms and, 527–528
  cooking and salad oils, 197
Hydrophilic–lipophilic balance (HLB), emulsion formation, 492–493
Hypoallergenic butter, manufacturing, 54

Ice cream:
  coatings for, 402–403
  emulsifiers in, 519–520
Icings, oils and fats in:
  creme icings, 339
  stabilizers, shortening, flakes and chips, 181
Industrial applications:
  frying technology research, 471–472
  marine oils, 233–238
    cutting oils for metal surfaces, 237
    insecticides, 236–237
    leather tanning, 234
    lubricants, 235–236
    paint, varnish and protective coatings, 234–235
    pharmaceuticals, 235
    soap manufacture, 236
Information sources, feedstuffs and pet foods, 258–261
Insecticides, fish oil for, 236–237
Interesterification:
  cocoa butter equivalents and extenders (CBEs), 383–384
  cocoa butter substitutes (CBSs), 389–393
  margarine:
    fat crystals, polymorphism, 81–82

    low-trans oil blends for, 87–88
    oil blending and, 85–86
  milk fat, 15–16
  shortenings, 126–127
Internal phase division, emulsion formation, 490–491
International feed numbers (IFNs), feedstuffs and pet foods, 260
Iodine value (IV):
  margarine, emulsifier, 93
  selected oils, 124
  shortening:
    base stock system of uniformity control, 172
    narrow plastic range shortening, 178
Ionic repulsion, emulsion formation, 494–495

Labeling requirements:
  butter, nutritional labeling standards, 24
  margarine, 71–74
Lactic acid:
  anionic emulsifiers, 507–508
  buttermaking and, 2
  as margarine preservative, 95
Lactobacillus acidophilus, butterfat manufacturing, 56
Lactobacillus bifidus, butterfat manufacturing, 56
Lamellar mesophase, structure, 499–502
LAN cylinder, cholesterol reduction, milk fat, 16, 18
Land animals, fish oil in diets of, 238
Land O'Lakes, reduced fat butter production, 38–39
Lard:
  history of, 162
  shortening compared with, 116
  in shortenings, 164–165
Leather tanning, fish oil for, 234
Lecithin:
  cocoa butter replacement requirements, 364–366
  as cocoa powder additive, 356–358
  as emulsifier, 509–511
  feedstuffs and pet foods, 278–279
  as margarine emulsifier, 92–93
  shortening, flakes and chips, 181
Light Butter, history of, 35–36
"Light-struck" flavor, margarine deterioration and shelf life, 105–106
Linoleic acid:
  antioxidant potency and, 539
  feedstuffs and pet foods, poultry food, 303
  margarine, high-liquid-oil blends, 86
Linolenic acid, feedstuffs and pet foods, 257

γ-Linolenic acids, margarine with, 89
Lipase activity:
  butter, quality control, 27
  cocoa powder and, 400
Lipids:
  antioxidant interaction with, 526–528
  in bovine milk, 2–3
    fatty acid composition, 10
    glycerol ethers in, 7, 11
  in cooking and salad oils, 193
  in marine oils, 225–226
Liquid fats, feedstuffs and pet foods, 309–310
Liquid margarine, processing, 101–102
Liquid overfeed (LOF) refrigeration systems, 135–136
Liquid shortenings:
  formulations, 178–180
  opaque shortenings, 188–190
  solid fat index (SFI) profile, 118–119
Low-calorie spreads, processing, 102–140
Low-density lipoprotein (LDL), in fish oils, 228–229
Low-erucic-acid-rapeseed (LEAR), shortenings with, 125
Low fat products, margarine, labeling standards, 73
  See also Reduced fat products
Low-trans oil blends, margarine, 86–88
Lubricants:
  fish oil for, 235–236
  food production with, 400–401

Margarine:
  consumption patterns, 65–69
  deterioration and shelf life, 104–106
  emulsifiers, 92–93, 519–520
  fat crystallization, 77, 79–83
    polymorphism, 79–82
    rate of, 82–83
    solid fat, 77, 79
  flavors, 95
  historical development, 65–67
  labeling requirements, 71–74
  low-calorie spreads, 102–104
  milk products and protein in, 91–92
  oil blending, 83–89
    high-liquid-oil blends, 86
    low-trans oil blends, 86–88
    special dietary oils, 88–89
  oil specifications, 89–90
  preservatives, 93–95
  processing operations, 96–99
    chilling, 97–98
    emulsification, 96–97
    Gerstenberg & Agger margarine/shortening systems, 146–149
    liquid margarine, 101–102
    packaging, 98–99
    resting, 98
    soft margarine, 101
    stick margarine, 99–100
    Votator stainless steel margarine/shortening system, 142–145
    whipped margarine, 100–101
    working, 98
  product characteristics, 74–76
    melting qualities, 75–76
    oil separation, 75
    spreadability, 74–75
  shortening compared with, 116
  solid fat profiles, 118
  source oils, 77–78
  typical formulations for, 90–91
  vitamins and colors, 95–96
  worldwide production of, 116–117
Marine oils:
  extraction and refining, 229–233
  feedstuffs and pet foods, fatty acid composition, 277–278
  fish meal processing, 230–233
    cooking as received raw material, 231–232
    miscella removal, 232
    press cake drying, 232–233
  as food supplement, 249–252
    flavors and odors, 249–251
    HUFA for human consumption, 251–252
  industrial applications, 233–238
    cutting oils for metal surfaces, 237
    insecticides, 236–237
    leather tanning, 234
    lubricants, 235–236
    paint, varnish and protective coatings, 234–235
    pharmaceuticals, 235
    soap manufacture, 236
  nutritional value of, 229
  overview, 225–226
  refining, 233
  sources of, 229–230
  world supply, 225–226
  See also Fish oils
Market share:
  butterfat products, 57–58
  butterfat–vegetable oil blends, 51
  margarine/spread products, 68–69
Mass transfer:
  frying technology, 450
  optimization of, 457–458

Mayonnaise:
  composition, 204–206
  defined, 204
  emulsifiers in, 519–520
  manufacturing process, 208–210
  quality control, 207–208
Meleshin process, buttermaking, 34
Mellorine emulsifier, 519–520
Melting point:
  margarine, oil specifications, 89–90
  selected oils, 124
  shortening, 167
    base stock system of uniformity control, 172
Melting profile:
  cocoa butter, 367–368
  cocoa butter replacers (CBRs), 385–387
  cocoa butter substitutes (CBSs), 389–393
  margarine, 75–76
  milk and cream separation, 29
  milk fat, 13
Menhaden oil:
  fatty acid content, 247–248
  as margarine source oils, 77–78
Mesophase structures, emulsifiers plus water, 499–502
Metabolism. *See* Fat digestion systems
Metal inactivators, shortening, 169–170
Metal protection, fish oil for, 237
Metals, frying technology, elimination from oil sources, 441
Mettler dropping point:
  margarine, oil specifications, 89–90
  shortening, base stock system of uniformity control, 172
Microemulsions, formation, 493
Microencapsulation techniques, fish oil, 249
Milk fat:
  chemical composition, 2–3
  cocoa butter replacers (CBRs), 387–388
  decholesterolized, 55
  desaturated milk fat, 55–56
  globules, chemical composition, 2–3
  modification of, 13–22
    cholesterol reduction, 16–22
    hydrogenation, 13–14
    interesterification, 15–16
    triglyceride melting and crystallization, 13
  seasonal changes in composition of, 12
  shelf life, 48–49
Milk fat globule membrane (MFGM), anhydrous milk fat (AMF) manufacturing, 46–47

Milk products and protein:
  chocolate containing, 362–363
  in margarine, 91–92
  pourable dressings, 212
Milk separation, background, 28–29
Minarine (reduced fat butter), 35
Miscella, fish meal processing for marine oil, 231–233
Model foods, frying technology, 456–457
Model oils, frying technology, 457
Moisture content:
  margarine, deterioration and shelf life, 106
  oil breakdown during frying, snack foods, 413–414
  snack foods, extruded snacks, 424–425
Molded print packaging, stick margarine, 99–100
Monogastric digestive systems, fat digestion in, 280–281
Monoglycerides:
  baking applications for, 502–505
  margarine, crystallization rate, 83
    dietary oils for, 88–89
    as emulsifier, 93
  mesophase structures, 499–502
  staling delayed by, 502–503
Monounsaturated fatty acids (MUFA), in seal oil, 228
Muffins, oils and fats in, 338–339
Mustard, in mayonnaise, 206

National Renderers Association (NRA), feedstuffs and pet foods, definitions and standards, 264–265
National Research Council (NRC), feedstuffs and pet food standards:
  beef cattle, 296
  cats, 295–297
  dairy cattle, 296–299
  dogs, 294–295
  energy requirements and definitions, 291
  fish, 304–305
  horses, 299–300
  nutrient requirements, 259
  poultry, 300–303
  species fat feeding considerations, 294–305
  swine, 300
Net energy (NE), feedstuffs and pet foods, NRC definition, 292–293
Net energy of lactation (NEL), feedstuffs and pet foods, NRC definition, 293
Net heat load, deep-fat frying, 438–439
Neutralization, butter manufacturing, 30
Newtonian flow behavior, chocolate, 363–366

560 Index

Nitrogen:
  protection, snack foods, fats in, 410–411
  in whipped margarine, 100–101
NIZO method of buttermaking, 32
Nondairy products, liquid shortening used for, 180
Nonfood products, milk fat in, 56
Nonruminant animals, fat digestion in, 282, 285–287
Nontriglycerides, fish oils, 243–244
Nordihydroguariaretic acid (NDGA), antioxidant activity of, 533–534
Nuclear magnetic resonance (NMR) analysis:
  margarine, melting characteristics, 76
  solid fat values, 79
Nutraceauticals, butterfat manufacturing, 56
Nutrition:
  butter:
    labeling standards, 24
    vitamin A content, 8–9
  cholesterol reduction, milk fat, 22
  cooking and salad oils, 202
  fish oils and, 227–229
    in aquatic animals, 239–242
    in human diet, 242–252
    in land animals, 238
  frying technology research, 465–471
  margarine, health claims, 73–74
    labeling standards, 72–73
Nuts, roasting technology, 426

Octadecadienoic acids, in butterfat, 5–6
Odor, marine oils, 249–251
Oil blending, margarine, 83–84
Oil migration, confectionery products, 403–404
Oil separation:
  margarine, 75
  shortening quality control, 188
Oil specifications, margarine, 89–90
Oil stability index (OSI):
  cooking and salad oils, 200
  feedstuffs and pet foods, quality control, 307–308
Oil-based dressings, 202–203
  emulsifiers in, 520–521
  fat-free dressings, 216–219
    composition, 217
    fat replacement, 217
    flavor quality, 217–218
    processing, 219
    stability, 218
  heat-stable dressings, 219
  pourable dressings, 210–215
    composition, 211–213
    manufacture, 214–215
    quality control, 213–214
  reduced-calorie dressings, 215–216
  refrigerated dressings, 219
  viscous or spoonable dressings, 203–210
    manufacturing, 208–210
    mayonnaise, 204–206
    quality control, 207–208
    salad dressings, 206–207
Oil-continuous products, low-calorie spreads, 102–103
Oil-in-water emulsion:
  formation compared with water-in-oil emulsion, 491–492
  low-calorie spreads, 104
  surfaces and interfaces, 484–485
Oils and fats, frying technology, food interaction during, 452–453
  See also Cooking oils; salad oils
Oilseeds, AAFCO definitions:
  cottonseed and cottonseed products, 274–276
  extracted products, 272–273
  full-fat soybeans, 273
  feedstuffs and pet foods:
    definitions, 272–277
    extrusion for full-fat seeds, 317–320
    overview, 258
  mechanically extracted meals, 272
Oil-soluble polymers, margarine, low-trans oil blends, 88
Oleomargarine. See Margarine
Olestra, reduced fat butter production with, 39
Olive oil, production and consumption patterns, 195
Omega-3 acids, margarine with, 89
One-phase pourable salad dressings, 210–211
Optimization, frying technology, 456–458
Optimum oil, frying technology, 453
Organoleptic properties:
  butter manufacturing, 40–41
  low-calorie spreads, 102–103
  margarine:
    packaging procedures and, 100
    specifications, 89–90
Oriented wedge theory, oil–water vs. water–oil emulsion formation, 491–492
Oxidation:
  antioxidant potency and, 538–539
  fatty acids, cooking and salad oils, 201
  oil breakdown during frying, snack foods, 412–415

Oxidative stability:
  baked products, 349–351
  butter, quality control, 27–28
    storage and transport, 48–49
  cooking and salad oils, 197
  marine oils, flavor and odor qualities, 250–251
Oxygenation, frying technology, 455

Packaging:
  butter manufacture, 47–48
  margarine, 98–99
    deterioration and shelf life, 105–106
  shortening, 151–154
Paints, varnish and protective coatings, fish oil for, 234–235
Palm kernel oil:
  cocoa butter substitutes (CBSs), 391–393
  interesterification, 127
Palm oil:
  interesterification, 127
  in shortenings, 164
Palmitic acid, fatty acid synthesis, 282–283
Pan and grill liquid shortening, 180
Particle addition, pourable dressings, 215
Pasteurization, buttermaking and, 2
Pastry, cake and biscuit products, butterfat as ingredient in, 52
Pentane value, cooking and salad oils, 199
Peroxide value (PV):
  baked products:
    oxidative stability, 350–351
    shortening specifications, 350–352
  cooking and salad oils, 198
  feedstuffs and pet foods, 306–307
  frying technology, degradation marker, 462
Pet foods, fats and oils in. *See* Feedstuffs and pet foods, fats and oils in
Pharmaceuticals, fish oil for, 235
Phase diagrams, emulsifier plus water, 499–500
Phase instability temperature, margarine melting characteristics, 76
Phospholipids, in bovine milk, glycerol ethers in, 7, 12
Pickering stabilization, margarine processing, 98
Pie crusts, oils and fats in, 344
Plastic solids, defined, 120–121
Plasticity:
  baked products, shortenings, 346–349
  conditions for, 121
  crystal size and, 121–122
  plastic solids, 120

process definition of shortening, 121
shortening, consistency, 184–188
  crystallization, 182
  liquid shortenings, 178–180
  narrow plastic range shortenings, 178
  powdered shortenings, 181
  quality control linked to, 151
  shortening flakes and chips, 180–181
  wide plastic range shortenings, 174–178
Polar materials, frying technology:
  degradation index, 460
  future trends, 477–478
Polybrominated biphenyls (PBB), feedstuffs and pet foods, contamination with, 261
Polyglycerol esters:
  margarine, as emulsifier, 93
  polyhydric emulsifiers, 508–509
Polyglycerol polyricinoleate (PGPR), chocolate and, 365–366, 394–396
Polyhydric emulsifiers, 508–509
Polymers:
  cooking and salad oils, quality control, 201
  frying technology, degradation marker, 462–463
Polymorphism, margarine fat crystals, 79–82
Polyphenolics, antioxidant activity of, 533–534
Polysorbate 60, baking applications, 506
Polyunsaturated fatty acids (PUFAs):
  fat digestion by animals, 283–284
  feedstuffs and pet foods, defined, 256
  in fish oils, 227
Pool boiling refrigeration systems, shortening manufacture, 135
Pooling system for cream production, 1–2
Popcorn, processing technology, 425–426
Pork rinds, frying technology, 426
Potato chips, processing systems, 417–421
  destoner, peeler–washer and inspection table, 419
  fryer design, 421–422
  seasoning-salter, 420–421
  slicer and washer–tumbler, 421
Poultry, feedstuffs and pet foods:
  fatty acid profiles, 325–326
  fat feeding considerations, 300–303
  fish oils in, 238
Pourable salad dressings, 210–215
  composition, 211–213
    colorants, 213
    dairy products, 212
    emulsifiers, 212
    flavors, 213
    gums, 212–213

Pourable salad dressings (*Continued*)
   preservatives, 213
   salt, 212
   source oils, 211
   spices, 212
   stabilizers, 213
   vinegar, 211
   water content, 211
  emulsions in, 521
  manufacturing processes, 214–215
  quality control and stability, 213–214
  See also Oil-based salad dressings
Pourable shortenings, solid fat index (SFI) profile, 119–120
Powdered fats, manufacturing, 53–54
Powdered shortenings, formulation, 181
Prechilling, margarine, emulsification, 97
Preservatives:
  margarine, 93–95
  in pourable dressings, 213
Press cake, fish meal processing for marine oil, 232–233
Press liquor. *See* Miscella
Protein, emulsifier interaction with, 514–516
Puff pastry, fats and oils in, 335–336

Quality control:
  butter manufacture, 22–28
  cooking and salad oils, 198–201
    accelerated testing, 199–200
    benzidine/anisidine value, 198–199
    color, 201
    free fatty acid content, 200–201
    oxidized fatty acids, 201
    pentane/hexanal value, 199
    peroxide value, 198
    polymers, 201
    thiobarbituric acid test, 199
    volatile profile method, 199
  feedstuffs and pet foods, 306–308
  frying technology, food and oil quality, 455–456
  mayonnaise, 207–208
  potato chip processing, potatoes, physical properties of, 417–418
  pourable dressings, 213–214
  shortening:
    consistency evaluation, 187–188
    manufacturing, 148–151
  snack foods, prior to usage, 409–411
"Quick titer" shortening, base stock system of uniformity control, 172

Recombined products, butterfat as ingredient in, 52

Reduced fat products:
  butter, 34–40
    production schematic, 37
  butterfat–vegetable oil blends, 50–51
  confections, 404–405
  margarine:
    development of, 68–69
    labeling standards, 72–73
  reduced-calorie dressings:
    composition, 215–216
    production statistics, 210
Refining methods:
  chocolate, 358–359
  fish oils, 245–246
Refractometer readings, shortening, base stock system of uniformity control, 172
Refrigerated dressings, 219
Refrigeration systems. *See* specific systems, e.g., Direct expansion refrigeration
Regulatory issues:
  antioxidants, 535–538
  cholesterol reduction, milk fat, 22
  emulsifiers, 517
  frying technology, 465–478
    future trends, 476
    guidelines and limits, 473–474
    heated oils research, 472
    inspection and compliance, 474
    U.S. regulations, 475–476
  margarine production, 67, 69–76
    labeling requirements, 71–74
    standards of identity, 67, 69–71
Resting stage, margarine processing, 98
Reversion process, cooking and salad oils, 197
Rheological properties:
  chocolate, 363–366
  emulsifiers, 394–398
Roasting processes, feedstuffs and pet foods, 320–321
Roll-in shortenings, solid fat index (SFI) profile, 120
Rumen-protected fats, feedstuffs and pet foods, 315–316
Ruminant animals:
  fat digestion in, 287–289
  overview, 281–282
Runaway oil, frying technology, 453

Salad oil, defined, 194
  See also Cooking and salad oils; Oil-based dressings; Pourable salad dressings
Salt, frying technology:
  elimination from oil sources, 441
  in pourable dressings, 212

volatile fatty acids, AAFCO feed grade
  definition, 268
Sandiness:
  margarine fat crystals, canola oil margarines, 81–82
  shortening quality control, 188
Saturated fat free standard, margarine, labeling standards, 73
Schaal oven test, cooking and salad oils, 200
Shelf life, margarines, 104–106
Shortening:
  in bakery products:
    icing, filler and coating fats in, 332
    liquid shortening, 180
    oxidative stability, 349–351
    physical properties, 340
    plasticity, 346–349
    solid fat index/content (SFI/SFC), 344–346
    specifications for, 344–352
  base stock system, 171–174
    advantages of, 171–172
    soybean oil base stocks, 172–174
    uniformity control, 172
  characteristics, 115, 165–171
    chemical adjuncts, 168–171
    crystal structure, 167
    emulsification, 167–168
    flavor, 166–167
    physical characteristics, 167
    similarity with other products, 116
  chips and flakes, 190–191
    chill rolls, 190
    conditions for, 191
    crystallization, 190–191
    formulations, 174, 180–181
  crystallization properties, 181–184
    crystal habit, 182–184
    plasticity, 182
  definition, 115
  formulation, 122–127
    crystallinity, 122–123
    fatty acid distribution, 123–125
    fractionation, 125
    hydrogenation, 125–126
    interesterification, 126–127
    standards, 174–181
    flakes and chips, 180–181
    liquid shortening, 178–180
    narrow plastic range shortenings, 178
    powdered shortenings, 181
    wide plastic range shortenings, 174–178
  functionality, 116–117
  future trends in manufacturing, 157–159
  historical overview, 161–163

  liquid opaque shortenings, 188–190
    crystallization, 188–189
    tempering and storage, 189–190
  manufacturing processes and equipment, 127–139
    Anco cooling roll, 127–129
    crystallization, 136
    general overview, 127
    supercooling and direct expansion refrigeration, 133–136
    tempering, 138–139
    Votator agitated working unit, 136–138
    Votator process, 129–130
    Votator scraped-surface heat exchanger, 129, 131–133
  packaging and storage, 151–154
  plastic theory regarding, 120–122
    conditions fostering plasticity, 121
    crystal size and, 121–122
    plastic solids, 120–121
    process definition, 121
  plasticized consistency, 184–188
    chilling, 184–185
    creaming gas, 185–186
    quality control, 187–188
    tempering, 186–187
    working and filler pressure, 185
  production systems, 140–148
    Chemetator scraped-surface heat exchanger, 146
    Gerstenberg & Agger margarine/shortening systems, 146–150
    Votator carbon steel lard and shortening systems, 140–142
    Votator stainless steel margarine/shortening system, 142–145
  quality control, 148–151
  recent processing innovations, 154–159
    automation and computer control, 154–156
  solid fat profiles, 118–120
    all-purpose shortenings, 119
    high-stability shortenings, 119
    pourable shortenings, 119–120
    specialty shortenings, 120
  source oils for, 163–165
  "superglycerated" shortening, 503
  worldwide production of, 116–117
Short-life creams, manufacturing, 55
Short-path molecular distillation, cholesterol reduction, milk fat, 17, 19
Simplesse, reduced fat butter production with, 39
Smoke point, frying technology, degradation marker, 464

564 Index

Snack foods, oils and fats in:
  corn chips, 422–423
  extruded snacks, 423–425
  fryer design, 415
  high oil absorption, 416
  low oil absorption, 416–417
  nut roasting, 426
  popcorn, 425–426
  pork rinds, 426
  potato chips, processing, 417–421
  quality control:
    degradation, during frying, 411–415
    prior to usage, 409–411
  source oils for, 417
  tortilla chips, 421–422
Soapmaking, fish oil for, 236
Sodium dodecyl sulfate (SDS), structure, 508
Sodium stearoyl lactylate:
  dough strengthening with, 518
  emulsifier interaction with protein, 515–516
  structure, 507
Sodium stearyl fumarate, structure, 507–508
Soft margarine products:
  development of, 67–68
  processing, 101
Solid fat content (SFC):
  baked products, shortenings, 344–346
  cocoa butter replacers (CBRs), 386–387
  margarine, crystallization properties, 79
  shortening, quality control linked to, 150
Solid fat index (SFI):
  baked products, cookies, 341–343
    layered doughs, 334–336
    plasticity properties, 347–349
    shortenings, 344–347
  margarine:
    crystallization properties, 79
    oil blending linked with, 83–86
    oil specifications, 89–90
    profiles, 118
    spreadability characteristics and, 74–75
  shortening, 118–120
    base stock system of uniformity control, 172
    flakes and chips, 180
    liquid shortening, 179–180
    narrow plastic range shortenings, 178
    plasticity formulation, 174
    quality control linked to, 150
    wide plastic range shortenings, 175–178
Solvents:
  cholesterol reduction, milk fat, 21
  extraction, fish oils, 246
  fraction, shortenings, 125

Sorbic acid, as margarine preservative, 95
Sorbitan emulsifiers:
  as chocolate emulsifiers, 395–398
  structure and characteristics, 505–506
Source oils:
  feedstuffs and pet foods, 262–264
    AAFCO definitions, 266–271
    AFIA definitions, 271–272
    compositions, 277–279
    NRA recommended standards and definitions, 265–266
  margarine, 77–78
  in mayonnaise, 205, 207–208
  pourable dressings, 211
  shortenings, 163–165
Soy phosphate/lecithin, AAFCO feed grade definition, 268
Soy protein, as margarine ingredient, 91–92
Soybean oil:
  in cooking and salad oils, 196
  as margarine source oil, 77–78
    blending processes, 83–86
  shortening, 164
    base stock system with, 172–174
Soybeans:
  AAFCO definitions, full-fat soybeans, 273–274
  cocoa butter replacers (CBRs) with, 387–388
  feedstuffs and pet foods, dry roasting, 320–321
  soapstock, AFIA feed grade definition, 272
Special-purpose products, AAFCO feed grade definitions, 270–271
Spices:
  in mayonnaise, 206
  in pourable dressings, 212
Spray chilled or beaded hard fat blends, formulation, 181
Spray dried fat emulsions, formulation, 181
Spreadability:
  of butter:
    manufacturing techniques, 41–45
    milk fat composition as factor in, 9–10
  margarine, 74–75
Stability:
  cooking and salad oils, 196–197
  emulsion formation, 494–496
  fat-free dressings, 218
  feedstuffs and pet foods, coproducts supplies, 262
  foam formation, 497–498
  mayonnaise, 207–208
  pourable dressings, 213–214

Index **565**

Standards:
  butter manufacture, quality control, 24–26
  chocolate, standards of identity, 375–378
  feedstuffs and pet foods:
    AAFCO standards, 266–271
    AFIA standards, 271–272
    National Renderers Association (NRA) standards and definitions, 264–265
    margarine, standards of identity (FDA and USDA), 69–71
  See also GRAS (generally recognized as safe) status
Starch, emulsifier interaction with, 511–514
Steric hindrance, emulsion formation, 494–496
Stick margarine, processing, 99–100
Storage conditions:
  antioxidant potency and, 538–539
  butter, 48–49
  feedstuffs and pet foods, 308–309
  margarine, deterioration and shelf life, 106
  shortening, 151–154
    liquid opaque shortening, 189–190
  snack foods, fats in, prior to usage, 410–411
Streaking, shortening quality control, 187–188
Substituted hydroquinone (SBHQ), thermal degradation, 541–543
Sucrose esters, polyhydric emulsifiers, 509–510
Sugar, in pourable dressings, 212
Supercooling, shortening manufacture, 133–136
Supercritical fluid extraction (SFE), cholesterol reduction, milk fat, 21
Surface excess of emulsifier, 486–488
Surface tension, emulsions, 485–487
Surfactant:
  defined, 483
    chocolate flow properties, 365–366
    electrical double layer, 489–490
    emulsifier interaction with protein, 515–516
    excess concentration at water–oil interface, 487–488
    frying technology:
      food–oil interactions, 453–454
      quality control through, 458–459
    lipophilic/hydrophilic characteristics, 483–484
    mesophase structure, 499–502
    polyglycerol esters, 508–509
    sucrose esters, 509–510

Van der Waal's forces, 489–490
  as wetting agents, 498–499
Suspended solids, frying technology, 464
Swine, feedstuffs and pet foods for, fat feeding considerations, 300
Synergists, antioxidant properties of, 530–534

Table spread products:
  development of, 67–69
  shortening compared with, 116
  See also Margarine
Tallow:
  feedstuffs and pet foods:
    defined, 256
    fatty acid composition, 277–278
    liquid fat application, 309–310
    sources of, 262–263
  shortening compared with, 116
  in shortenings, 165
Temperature, oil breakdown during frying:
  high absorption, 416
  low absorption, 416–417
  snack foods, 411–415
Tempering:
  chocolate, 370–374
    "mush" technique, 372
    "partial melt" technique, 372
    "seeding" technique, 372
  cocoa butter extenders (CBEs), 380
  shortening:
    consistency, 186–187
    liquid opaque shortening, 189–190
    manufacturing technique, 138–139
Tertiary butylhydroquinone (TBHQ):
  cooking and salad oils, 202
  physical properties of, 529, 533–534
  regulatory status, 535–538
  shortening, 169
  thermal degradation, 539–543
Texture:
  baked products, breads and rolls, fats used in, 332
  butter, 12–13, 41–45
Thermal degradation:
  cooking and salad oils, 197
  phenolic antioxidants, 539–543
Thermosyphon effect, shortening manufacture, 133–135
Thiobarbituric acid test, cooking and salad oils, 199
Thiodipropionic acid, physical properties of, 530
Tirtiaux dry fractionation, butter manufacturing, texturization and spreadability, 44–45

Tocopherols:
  regulatory status, 538
  synergism of, 531–534
Tortilla chips:
  processing technology, 421–422
  cooking-steeping process, 421–422
Total oxidation (TOTOX) value, cooking and salad oils, 199
Total polar materials (TPM), frying technology, degradation index, 460
Total quality management (TQM), feedstuffs and pet foods, 260–261
Total volatile carbonyl (TVC), oil breakdown during frying, snack foods, 413–414
Trans fatty acids, margarine, low-trans oil blends for, 86–88
Triacylglycerols, fat digestion by animals, 285–287
Triglycerides:
  butterfat, 6
  in cocoa butter, 367–370
  cocoa butter alternatives, 380
  cocoa butter equivalents and extenders (CBEs), 381–384
  fat migration, 403–404
  in fish oils, 226–227, 249
  margarine:
    caprylic and capric acids with, 89
    fat crystals, polymorphism, 80–81
  milk fat, melting and crystallization, 13
  shortening, distribution in, 123–125
True metabolizable energy (TME), feedstuffs and pet foods, NRC definition, 292
Turmeric, in margarine, as coloring agent, 96
Turnover rate, oil breakdown during frying, snack foods, 413–414
Two-phase pourable salad dressings, 210–211

Ultrahigh temperature (UHT):
  butter manufacture, 23–24
  cream manufacturing, 55
Ultraviolet light, frying technology, 441
U.S. Department of Agriculture (USDA):
  antioxidant regulations, 535–538
  margarine identity standard, 69

Vacreation, butter manufacture, quality control, 23–24
Vacuum steam distillation, cholesterol reduction in milk fat, 16, 18
Van der Waal's forces, surfactant interaction, 489–490
Vanaspati, shortening compared with, 116

Vegetable fats and oils:
  AAFCO feed grade definition, 266–267
  margarine containing:
    food production applications, 400
    labeling requirements, 71–72
Very-low-density lipoproteins (VLDL), in fish oils, 228–229
Vinegar:
  in mayonnaise, 205–206
  pourable dressings, 211
Viscosity:
  chocolate, 365–366
  chocolate emulsifiers, 397–398
  frying technology, degradation marker, 464
Viscous/spoonable dressings, 203–210
  alternatives to mayonnaise, 206–207
  mayonnaise, 204–206
  quality control and stability, 207–208
Vitamin A:
  AAFCO feed grade definitions, 268–269
  in butter, 8–9
  in margarine, 95–96
Vitamin D:
  AAFCO feed grade definitions, 268–269
  in butter, 9–10
  in margarine, 95–96
Vitamin E:
  AAFCO feed grade definitions, 269
  feedstuffs and pet foods, cat food, 295–297
  fish oil, 250–251
  in margarine, 95–96
Vitamins:
  AAFCO feed grade definition, 268–269
  in margarine, 95–96
Volatile fatty acids (VFAs):
  chocolate conching, release of, 360–361
  digestive systems:
    overview, 281–282
    ruminant animals, 287–289
Volatile profile techniques, cooking and salad oils, 199
Volume, in baked products, breads and rolls, fats used in, 332–333
Vometaria Negra disease, 233
Votator processing, shortening manufacture, 129–130
  agitated working unit, 136–138
  automation and computer control, 155–157
  carbon steel lard and shortening systems, 140–142
  stainless steel margarine/shortening system, 142–145
Votator C unit, tempering with, 139

Votator scraped-surface heat exchanger, 129, 131–133

Water, oil and, frying technology, 453
Water content, pourable dressings, 211
Water emulsion titratables (WET), frying technology, 461–462
Water-in-oil emulsion:
   formation compared with oil-in-water emulsion, 491–492
   margarine preservative, 94
   surfaces and interfaces, 485
Wetting mechanisms, emulsions, 498–499
Whey protein concentrate (WPC), reduced fat butter production with, 39–40
Whipped products, margarine, processing, 100–101
Wide plastic range, shortening formulations, 174–178
Working unit:
   margarine processing, 98
   shortening consistency, 185
World production statistics:
   margarine, 116–117
   oils and fats:
      future trends, 157–159
      in metric tons, 226
   shortening, 116–117

YN lecithin, chocolate and, 365